UNITY COLLEGE LIBRARY
WITHDRAWN

The Development of Mathematics

THE DEVELOPMENT *of* MATHEMATICS

BY

E. T. Bell

Professor of Mathematics, California Institute of Technology

SECOND EDITION

McGRAW-HILL BOOK COMPANY

New York · London

1945

510.9
B413d

UNITY COLLEGE LIBRARY
WITHDRAWN

THE DEVELOPMENT OF MATHEMATICS

Copyright, 1940, 1945, by
E. T. Bell

PRINTED IN THE UNITED STATES OF AMERICA

All rights reserved. This book, or parts thereof, may not be reproduced in any form without permission of the author.

16 17 18 19 20 - MAMM - 7 5 4 3

04330

To Any Prospective Reader

Nearly fifty years ago an American critic, reviewing the first volume (1888) of Lie's *Theorie der Transformationsgruppen*, set his own pace (and ours) in the following remarks.

> There is probably no other science which presents such different appearances to one who cultivates it and one who does not, as mathematics. To [the noncultivator] it is ancient, venerable, and complete; a body of dry, irrefutable, unambiguous reasoning. To the mathematician, on the other hand, his science is yet in the purple bloom of vigorous youth, everywhere stretching out after the "attainable but unattained," and full of the excitement of nascent thoughts; its logic is beset with ambiguities, and its analytic processes, like Bunyan's road, have a quagmire on one side and a deep ditch on the other, and branch off into innumerable by-paths that end in a wilderness.*

Once we venture beyond the rudiments, we may agree that those who cultivate mathematics have more interesting things to say than those who merely venerate. Accordingly, we shall follow the cultivators in their explorations of a Bunyan's road through the development of mathematics. If occasionally we have no eyes for the purple bloom, it will be because we shall need all our faculties to avoid falling into the ditch or wandering off into a wilderness of trivialities that might be mistaken for mathematics or for its history. And we shall leave to antiquarians the difficult and delicate task of restoring the roses to the cheeks of mathematical mummies.

The course chosen in the following chapters was determined by two factors. The first was the request from numerous correspondents, principally students and instructors, for a broad account of the general development of mathematics, with par-

* C. H. Chapman, *Bulletin of the New York Mathematical Society*, 2, 1892, 61.

ticular reference to the main concepts and methods that have, in some measure, survived. The second was personal association for several years with creative mathematicians in both the pure and the applied divisions.

Not a history of the traditional kind, but a narrative of the decisive epochs in the development of mathematics was wanted. A large majority asked for technical hints, where possible without too great detail, why certain things continue to interest mathematicians, technologists, and scientists, while others are ignored or dismissed as being no longer vital. Many who planned to end their mathematical education with the calculus, or even in some instances earlier, wished to be shown something of the general development of mathematics beyond that outstanding landmark of seventeenth-century thought, as part of a civilized education. Those intending to continue in mathematics or science or technology also asked for a broad general treatment with technical hints. They gave two additional reasons, the second of singular interest to any professed teacher. They believed that a survey of the main directions along which living mathematics has developed would enable them to decide more intelligently in what particular field of mathematics, if any, they might find a lasting satisfaction.

The second reason for their request was characteristic of a generation that has grown rather tired of being told what to think and whom to respect. These candid young critics of their would-be educators hoped that a cursory personal inspection of the land promised them, even from afar off, would enable them to resist the blandishments of persuasive 'subdividers' bent on selling their own tracts to the inexperienced. We seem to have come a long way since 1873, when that erudite English historian of mathematics and indefatigable manufacturer of drier-than-dust college textbooks, Isaac Todhunter (1820–1884), counseled a meek docility, sustained by an avid credulity, as the path of intellectual rectitude:

> If he [a student of mathematics] does not believe the statements of his tutor, probably [in Todhunter's day at Cambridge] a clergyman of mature knowledge, recognized ability and blameless character—his suspicion is irrational, and manifests a want of the power of appreciating evidence, a want fatal to his success in that branch of science which he is supposed to be cultivating.

Be the wisdom of Todhunter's admonition what it may, it is astonishing how few students entering serious work in mathematics or its applications have even the vaguest idea of the highways, the pitfalls, and the blind alleys ahead of them. Consequently, it is the easiest thing in the world for an enthusiastic teacher, "of mature knowledge, recognized ability and blameless character," to sell his misguided pupils a subject that has been dead for forty or a hundred years, under the sincere delusion that he is disciplining their minds. With only the briefest glimpse of what mathematics in this twentieth century —not in 2100 B.C.—is about, any student of normal intelligence should be able to distinguish between live teaching and dead mathematics. He will then be less likely than his confiding companion to drown in the ditch or perish in the wilderness.

Many asked for some reference to the social implications of mathematics. A classic strategy in mathematics is the reduction of an unsolved problem to one already solved. It seems plausible that more than half the problem of mathematics and society is reducible to that of the physical sciences and society. There being as yet no widely accepted solution of the latter problem, we shall leave the former with the reduction indicated. Anyone will thus be able to reach his own conclusions from that solution of the scientific problem which he accepts. Proposed solutions range from Platonic realism at one extreme to Marxian determinism at the other. Occasional remarks may suggest an inquiry into the equally difficult question of what part civilization, with its neuroses, its wars, and its national jealousies, has played in mathematics. These asides may be of interest to those intending to make mathematics their lifework. Incidentally, in this connection, I was told that I might write for adults. Chronological age is not necessarily a measure of adulthood; a first-year student in a university may be less infantile, in everything but mathematics, than the distinguished savant lecturing at him.

The topics selected for description were chosen after consultation with numerous professionals who know from hard personal experience what mathematical invention means. On their advice, *only main trends* of the past six thousand years are considered, and these are presented *only through typical major episodes* in each. As might be anticipated by any worker in mathematics, the conclusions reached by following such advice differ occasionally from those hallowed by the purely historical tradition. Wherever this is so, references to other accounts will enable any reader to form his own opinion. There are no absolutes (except possibly this) in mathematics or in its history.

Most of the differences reflect two possible and sometimes divergent readings of mathematical evolution. Whoever has himself attempted to advance mathematics is inclined to be more skeptical than the average spectator toward any alleged anticipation of notable progress. From his own experience and that of others still living, the professional mathematician suspects that often what looks like an anticipation after the advance was made was not even aimed in the right direction. From many a current instance, he knows further that when at length progress started, it proceeded along lines totally different from those which, in retrospect, it 'should' have followed.

Nothing is easier, on the other hand, than to fit a deceptively smooth curve to the discontinuities of mathematical invention. Everything then appears as an orderly progression from the Egypt of 4000 B.C. and the Babylon of 2000 B.C. to the Göttingen of 1934 and the U.S.A. of 1945, with Cavalieri, for instance, indistinguishable from Newton in the neighborhood of the calculus, or Lagrange from Fourier in that of trigonometric series, or Bhaskara from Lagrange in the region of Fermat's equation. Professional historians may sometimes be inclined to overemphasize the smoothness of the curve; professional mathematicians, mindful of the dominant part played in geometry by the singularities of curves, attend to the discontinuities. This is the origin of most differences of opinion between the majority of those who cultivate mathematics and the majority of those who do not. That such differences should exist is no disaster. Dissent is good for the souls of all concerned.

No apology need be tendered the thousands of dead and living mathematicians whose names are not mentioned. Only a meaningless catalogue could have cited a tenth of those who have created mathematics. Nor, when between 4,000 and 5,000 papers and books devoted to mathematical research—the creation of new mathematics—are being published every 365 days, is there any point in attempting to minimize the omission of certain topics that have interested, and may still interest, hundreds of these unnamed thousands. However, what a sufficient number of competent men consider the vital things are at least mentioned. Anyone desirous of following the detailed history of certain major developments will find the technical histories of special topics, written by mathematicians for mathematicians, ample for a beginning. Some of these severely technical histories extend to hundreds of pages, a few to thousands; they refer to the labors of thousands of men, most of whom are all but completely forgotten. Yet, like the tiny creatures whose empty frames survive in massive coral reefs that can wreck a battleship, these hordes of all but anonymous mathematicians have left something in the structure of mathematics more durable than their own brief and commonplace lives.

As to the mechanical features of the book, the inevitable footnotes have been kept to a minimum by the simple expedient of throwing hundreds away. Some direct those seeking further information on the relevant mathematics to works by creative mathematicians. Other things being equal, preference is given works containing extensive bibliographies compiled by experts having firsthand knowledge of the subjects treated. *A superscript number indicates a footnote; all are collected for easy reference just before the index. All should be ignored till a possible return to some point.*

The index will be found helpful. Men's initials and dates (except a very few, unobtainable without undue labor), seldom repeated in the text, are given in the index; cross references to definitions, etc., are avoided by the same means.

Nationalities are stated; if more than one country has a claim to some man, the place where he did most of his work is given. On a previous occasion (*Men of mathematics*, Simon & Schuster, New York, 1937), I almost precipitated an international incident by calling a Pole a Russian. I trust that few such disastrous blunders will be found here. The book mentioned contains full-length biographies of about thirty-five leading mathematicians of the past.

Dates in the text appended to mathematical events serve two purposes, the first of which is obvious. The second is to avoid elaborate references. The date, if later than 1636 and earlier than 1868, will usually enable anyone seriously interested to locate the matter concerned in the collected works of the author cited; if later than 1867, and whether or not collected works are available, the exact reference, with a concise abstract of the work, is given in the annual *Jahrbuch über die Fortschritte der Mathematik*. For the period beginning in 1931, the *Zentralblatt für Mathematik und ihre Grenzgebiete* serves the same purpose. The American *Mathematical Reviews*, 1940–, is of the same general character as the German abstract journals. Comparatively scarce early periodicals, likely to be found only in specialized libraries, are not cited, although they were frequently consulted. This omission may be partly compensated by referring to the German and French mathematical encyclopedias listed in the notes. Other references to sources before 1637 are given in the proper places.

For the period before 1637, the works of professional historians of mathematics have been used for some matters on which the historians are in approximate agreement among themselves. Theirs is a difficult and exacting pursuit; and if controversies over the trivia of mathematics, of but slight interest to either students or professionals, absorb a considerable part of their energies, the residue of apparently sound facts no doubt justifies the inordinate expense of obtaining it. Without the devoted labors of these scholars, mathematicians would know next to nothing, and perhaps care less, about the first faltering steps of their science. Indeed, an eminent French analyst of the twentieth century declared that neither he nor any but one or two of his fellow professionals had the slightest interest in the history of mathematics as conceived by historians. He amplified his statement by observing that the only history of mathematics that means anything to a mathematician is the thousands of technical papers cramming the journals devoted exclusively to mathematical research. These, he averred, are the true history of mathematics, and the only one either possible or profitable to write. Fortunately, I am not attempting to write a history of mathematics; I hope only to encourage some to go on, and decide for themselves whether the French analyst was right.

Preference has been given in citing purely historical references to works in English, French, or German, as these are the three languages of which those interested in mathematics are most likely to have an adequate reading knowledge. For those especially interested in geometry, Italian also is necessary. Italian historical works are included in the bibliographical material of the histories listed.

To the many professional friends who have advised me on their respective specialties and whose generous help I have attempted to pass on to others, I am very grateful. A special word of thanks is due Professor W. H. Gage, of the University of British Columbia, who removed many obscurities and greatly improved several of the presentations

This has been an opportunity to do something a little off the beaten track to show prospective readers how the mathematics familiar to them got where it is, and where it is going from there. I trust that students will tolerate the departure from the traditional textbook. For one thing, at any rate, the more sensible should be grateful: only the most ingenious instructor could set an examination on the book.

It has, unhappily, been necessary in writing the book to consider many things besides the masterpieces of mathematics. Rising from a protracted and not always pleasant session with the works of bickering historians, scholarly pedants, and contentious mathematicians, often savagely contradicting or meanly disparaging one another, I pass on, for what it may be worth, the principal thing I have learned to appreciate as never before. It is contained in Buddha's last injunction to his followers:

Believe nothing on hearsay. Do not believe in traditions because they are old, or in anything on the mere authority of myself or any other teacher.

E. T. Bell.

Note to the Second Edition

About fifty pages of new material have been added in this edition. The additions include numerous short amplifications of miscellaneous topics from Greek mathematics to mathematical logic, with longer notes on symbolism, algebraic and differential geometry, lattices, and other subjects in which there have been striking recent advances.

E. T. Bell.

California Institute of Technology,
Pasadena, California,
July, 1945.

Contents

The Development of Mathematics

CHAPTER 1

General Prospectus

In all historic times all civilized peoples have striven toward mathematics. The prehistoric origins are as irrecoverable as those of language and art, and even the civilized beginnings can only be conjectured from the behavior of primitive peoples today. Whatever its source, mathematics has come down to the present by the two main streams of number and form. The first carried along arithmetic and algebra, the second, geometry. In the seventeenth century these two united, forming the ever-broadening river of mathematical analysis. We shall look back in the following chapters on this great river of intellectual progress and, in the diminishing perspective of time, endeavor to see the more outstanding of those elements in the general advance from the past to the present which have endured.

'Form,' it may be noted here to prevent a possible misapprehension at the outset, has long been understood mathematically in a sense more general than that associated with the shapes of plane figures and solid bodies. The older, geometrical meaning is still pertinent. The newer refers to the structure of mathematical relations and theories. It developed, not from a study of spacial form as such, but from an analysis of the proofs occurring in geometry, algebra, and other divisions of mathematics.

Awareness of number and spacial form is not an exclusively human privilege. Several of the higher animals exhibit a rudimentary sense of number, while others approach genius in their mastery of form. Thus a certain cat made no objection when she was relieved of two of her six kittens, but was plainly distressed when she was deprived of three. She was relatively as advanced arithmetically as the savages of an Amazon tribe who can count up to two, but who confuse all greater numbers in a nebulous 'many.'

Again, the intellectual rats that find their way through the mazes devised by psychologists are passing difficult examinations in topology. At the human level, a classic puzzle which usually suffices to show the highly intelligent the limitations of their spacial intuition is that of constructing a surface with only one side and one boundary.

Although human beings and the other animals thus meet on a common ground of mathematical sense, mathematics as it has been understood for at least twenty-five centuries is on a far higher plane of intelligence.

Necessity for proof; emergence of mathematics

Between the workable empiricism of the early land measurers who parceled out the fields of ancient Egypt and the geometry of the Greeks in the sixth century before Christ there is a great chasm. On the remoter side lies what preceded mathematics, on the nearer, mathematics; and the chasm is bridged by deductive reasoning applied consciously and deliberately to the practical inductions of daily life. Without the strictest deductive proof from admitted assumptions, explicitly stated as such, mathematics does not exist. This does not deny that intuition, experiment, induction, and plain guessing are important elements in mathematical invention. It merely states the criterion by which the final product of all the guessing, by whatever name it be dignified, is judged to be or not to be mathematics. Thus, for example, the useful rule, known to the ancient Babylonians, that the area of a rectangular field can be computed by 'length times breadth,' may agree with experience to the utmost refinement of physical measurement; but the rule is not a part of mathematics until it has been deduced from explicit assumptions.

It may be significant to record that this sharp distinction between mathematics and other sciences began to blur slightly under the sudden impact of a greatly accelerated applied mathematics, so called, in the second world war. Semiempirical procedures of calculation, certified by their pragmatic utility in war, were accorded full mathematical prestige. This relaxation of traditional demands brought the resulting techniques closer in both method and spirit to engineering and the physical sciences. It was acclaimed by some of its practitioners as a long-overdue democratization of the most aristocratic of the sciences. Others, of a more conservative persuasion, deplored the passing of the ideal of strict deduction, as a profitless confusion of a simple issue which had at last been clarified after several centuries of futile disputation. One fact, however, emerged from the difference of opinion: It is difficult, in modern warfare, to wreck, to maim, or to kill efficiently without a considerable expenditure of mathematics, much of which was designed originally for the development of those sciences and arts which create and conserve rather than destroy and waste.

It is not known where or when the distinction between inductive inference—the summation of raw experience—and deductive proof from a set of postulates was first made, but it was sharply recognized by the Greek mathematicians as early as 550 B.C. As will appear later, there may be some grounds for believing that the Egyptians and the Babylonians of about 2000 B.C. had recognized the necessity for deductive proof. For proof in even the rough and unready calculations of daily life is indeed a necessity, as may be seen from the mensuration of rectangles.

If a rectangle is 2 feet broad and 3 long, an easy proof sustains the verdict of experience, founded on direct measurement, that the area is 6 square feet. But if the breadth is $\sqrt{2}$ and the length $\sqrt{3}$ feet, the area cannot be determined as before by cutting the rectangle into unit squares; and it is a profoundly difficult problem to prove that the area is $\sqrt{6}$ feet, or even to give intelligible, usable meanings to $\sqrt{2}$, $\sqrt{3}$, $\sqrt{6}$, and 'area.' By taking smaller and smaller squares as unit areas, closer and closer approximations to the area are obtained, but a barrier is soon reached beyond which direct measurement cannot proceed. This raises a question of cardinal importance for a just understanding of the development of all mathematics, both pure and applied.

Continuing with the $\sqrt{2} \times \sqrt{3}$ rectangle, we shall suppose that refined measurement has given 2.4494897 as the area. This is correct to the seventh decimal, but it is not right, because $\sqrt{6}$, the exact area, is not expressible as a terminated decimal fraction. If seven-place accuracy is the utmost demanded, the area has been found. This degree of precision suffices for many practical applications, including precise surveying. But it is inadequate for others, such as some in the physical sciences and modern statistics. And before the seven-place approximation can be used intelligently, its order of error must be ascertained. Direct measurement cannot enlighten us; for after a certain limit, quickly passed, all measurements blur in a common uncertainty. Some universal agreement on what is meant by the exact area must be reached before progress is possible. Experience, both practical and theoretical, has shown that a consistent and useful mensuration of rectangles is obtained when the rule 'length times breadth' is deduced from postulates abstracted from a lower level of experience and accepted as valid. The last is the methodology of all mathematics.

Mathematicians insist on deductive proof for practically workable rules obtained inductively because they know that analogies between phenomena at different levels of experience are not to be accepted at their face value. Deductive reasoning is the only means yet devised for isolating and examining hidden assumptions, and for following the subtle implications of hypotheses which may be less factual than they seem. In its modern technical uses of the deductive method, mathematics employs much sharper tools than those of the traditional logic inherited from ancient and medieval times.

Proof is insisted upon for another eminently practical reason. The difficult technology of today is likely to become the easy routine of tomorrow; and a vague guess about the order of magnitude of an unavoidable error in measurement is worthless in the technological precision demanded by modern civilization. Working technologists cannot be skilled mathematicians. But unless the rules these men apply in their technologies have been certified mathematically and scientifically by competent experts, they are too dangerous for use.

There is still another important social reason for insistence on mathematical demonstration, as may be seen again from the early history of surveying. In ancient Egypt, the primitive theory of land measurement, without which the practice would have been more crudely wasteful than it actually was, sufficed for the economy of the time. Crude both practically and theoretically though this surveying was, it taxed the intelligence of the Egyptian mathematicians. Today the routine of precise surveying can be mastered by a boy of seventeen; and those applications of the trigonometry that evolved from primitive surveying and astronomy which are of greatest significance in our own civilization have no connection with surveying. Some concern mechanics and electrical technology, others, the most advanced parts of the physical sciences from which the industries of twenty or a hundred years hence may evolve.

Now, contrary to what might be supposed, modern trigonometry did not develop in response to any practical need. Modern trigonometry is impossible without the calculus and the mathematics of $\sqrt{-1}$. To cite but one of the commoner applications, over a century and a half elapsed before this trigonometry became indispensable in the theory and practice of alternating currents. Long before anyone had dreamed of an electric dynamo, the necessary mathematics of dynamo design was available. It had developed largely because the analysts of the eighteenth century sought to understand mathematically the somewhat meager legacy of trigonometry bequeathed them by the astronomers of ancient Greece, the Hindus, and the mathematicians of Islam. Neither astronomy nor any other science of the eighteenth century suggested the introduction of $\sqrt{-1}$, which completed trigonometry, as no such science ever made any use of the finished product.

The importance of mathematics, from Babylon and Egypt to the present, as the primary source of workable approximations to the complexites of daily life is generally appreciated. In fact, a mathematician might believe it is almost too generally appreciated. It has been preached at the public, in school and out, by socially conscious educators until almost anyone may be pardoned for believing that the rule of life is rule of thumb. Because routine surveying, say, requires only mediocre intelligence, and because surveying is a minor department of applied mathematics, therefore only that mathematics which can be manipulated by rather ordinary people is of any social value. But no growing economy can be sustained by rule of thumb. If new applications of a furiously expanding science are to be possible, difficult and abstruse mathematical theories far beyond the college level must continue to be developed by those having the requisite talents. In this living mathematics it is imagination and rigorous proof which count, not the numerical accuracy of the machine shop or the computing laboratory.

A familiar example from common things will show the necessity for mathematics as distinguished from calculation. A nautical almanac is one of the indispensables of modern navigation and hence of commerce. Machines are now commonly used for the heavy labor of computing. Ultimately the computations depend upon the motions of the planets, and these are calculated from the infinite (non-terminating) series of numbers given by the Newtonian theory of gravitation. For the actual work of computation a machine is superior to any human brain; but no machine yet invented has had brains enough to reject nonsense fed into it. From a grotesquely absurd set of data the best of machines will return a final computation that looks as reasonable as any other. Unless the series used in dynamical astronomy converge to definite limiting numbers (asymptotic series also are used, but not properly divergent), it is futile to calculate by means of them. A table computed by properly divergent series would be indistinguishable to the untrained eye from any other; but the aviator trusting it for a flight from Boston to New York might arrive at the North Pole. Despite its inerrant accuracy and attractive appearance, even the most highly polished mechanism is no substitute for brains. The research mathematician and the scientific engineer supply the brains; the machine does the rest.

Nobody with a grain of common sense would demand a strict proof for every tentative application of complicated mathematics to new situations. Occasionally in problems of excessive difficulty, like some of those in nuclear physics, calculations are performed blindly without reference to mathematical validity; but even the boldest calculator trusts that his temerity will some day be certified rationally. This is a task for the mathematicians, not for the scientists. And if science is to be more than a midden of uncorrelated facts, the task must be carried through.

Necessity for abstractness

With the recognition that strict deductive reasoning has both practical and aesthetic values, mathematics began to emerge some six centuries before the Christian era. The emergence was complete when human beings realized that common experience is too complex for accurate description.

Again it is not known when or where this conclusion was first reached, but the Greek geometers of the fourth century B.C. at latest had accepted it, as is shown by their work. Thus Euclid in that century stated the familiar definition: "A circle is a plane figure contained by one line, called the circumference, and is such that all straight lines drawn from a certain point, called the center, within the figure to the circumference are equal."

There is no record of any such figure as Euclid's circle ever having been observed by any human being. Yet Euclid's ideal circle is not only that of school geometry, but is also the circle of the handbooks used by engineers in calculating the performance of machines. Euclid's mathematical circle is the outcome of a deliberate simplification and abstraction of observed disks, like the full moon's, which appear 'circular' to unaided vision.

This abstracting of common experience is one of the principal sources of the utility of mathematics and the secret of its scientific power. The world that impinges on the senses of all but introverted solipsists is too intricate for any exact description yet imagined by human beings. By abstracting and simplifying the evidence of the senses, mathematics brings the worlds of science and daily life into focus with our myopic comprehension, and makes possible a rational description of our experiences which accords remarkably well with observation.

Abstractness, sometimes hurled as a reproach at mathematics, is its chief glory and its surest title to practical usefulness. It is also the source of such beauty as may spring from mathematics.

History and proof

In any account of the development of mathematics there is a peculiar difficulty, exemplified in the two following assertions, about many statements concerning proof.

(*A*) It is proved in Proposition 17, Book 1, of Euclid's *Elements*, that the square on the longest side of a right-angled triangle is equal to the sum of the squares on the other two sides (the so-called Pythagorean theorem).

(*B*) Euclid proved the Pythagorean theorem in Proposition 47 of Book I of his *Elements*.

In ordinary discourse, (*A*), (*B*) would usually be considered equivalent—both true or both false. Here (*A*) is false and (*B*) true. For a clear understanding of the development of mathematics it is important to see that this distinction is not a quibble. It is also essential to recognize that comprehension here is more important than knowing the date (c. 330–320 B.C.) at which the *Elements* were written, or any other detail of equal antiquarian interest. In short, the crux of the matter is mathematics, which is at least as important as history, even in histories of mathematics.

The statement (*A*) is false because the attempted proof in the *Elements* is invalid. The attempt is vitiated by tacit assumptions that Euclid ignored in laying down the postulates from which he undertook to deduce the theorems in his geometry. From those same postulates it is easy to deduce, by irrefragable logic, spectacularly paradoxical consequences, such as "all triangles are equilateral." Thus when an eminent scholar of Greek mathematics asserts that owing to the "inerring logic" of the Greeks, "there has been no need to reconstruct, still less to reject as unsound, any essential part of their doctrine," mathematicians must qualify assent by referring to the evidence. The "essential part of their doctrine" has indeed come down to us unchanged, that part being insistence on deductive proof. But in the specific instance of Euclid's proofs, many have been demolished in detail, and it would be easy to destroy more were it worth the trouble.

The statement (*B*) is true because the validity of a proof is a function of time. The standard of mathematical proof has risen steadily since 1821, and finality is no longer sought or desired. In Euclid's day, and for centuries thereafter, the attempted proof of the Pythagorean proposition satisfied all the current requirements of logical and mathematical rigor. A sound proof today does not differ greatly in outward appearance from Euclid's; but if we inspect the postulates required to validate the proof, we notice several which Euclid overlooked. A carefully taught child of fourteen today can easily detect fatal omissions in many of the demonstrations in elementary geometry accepted as sound less than fifty years ago.

It is clear that we must have some convention regarding 'proof.' Otherwise, few historical statements about mathematics will have any meaning. Whenever in the sequel it is stated that a certain result was proved, this is to be understood for the sense as in (*B*), namely, that the proof was accepted as valid by professional mathematicians at the time it was given. If, for example, it is asserted that a work of Newton or of Euler contains a proof of the binomial theorem for exponents other than positive integers, the assertion is false for the (*A*) meaning, true for the (*B*). The proofs which these great mathematicians gave in the seventeenth and eighteenth centuries were valid *at that time*, although they would not be accepted today by a competent teacher from a student in the first college course.

It need scarcely be remarked that few modest mathematicians today expect all of their own proofs to survive the criticisms of their successors unscathed. Mathematics thrives on intelligent criticism, and it is no disparagement of the great work of the past to point out that its very defects have inspired work as great.

Failure to observe that mathematical validity depends upon its epoch may generate scholarly but vacuous disputes over historical minutiae. Thus a meticulous historian who asserts that the Greeks of Euclid's time failed to solve quadratic equations by their geometric method because they 'overlooked' possible negative roots, to say nothing of imaginaries, himself overlooks one of the most interesting phenomena in the entire history of mathematics.

Until positive rational fractions and negative numbers were invented by mathematicians (or 'discovered,' if the inventors happened to be Platonic realists), a quadratic equation with rational integer coefficients had precisely one root, or precisely two, or precisely none. A Babylonian of a sufficiently remote century who gave 4 as *the* root of $x^2 = x + 12$ had solved his equation completely, because -3, which we now say is the other root, did not exist for him. Negative numbers were not in his number system. The successive enlargements of the number system necessary to provide all algebraic equations with roots equal in number to the respective degrees of the equations was one of the outstanding landmarks in mathematical progress, and it took about four thousand years of civilized mathematics to establish it. The final necessary extension was delayed till the nineteenth century.

An educated algebraist today, wishing to surpass the meticulous critic in pedantry, would point out that "how many roots has $x^2 = x$?" is a meaningless question until the domain in which the roots may lie has been specified. If the domain is that of complex numbers, this equation has precisely two roots, 0, 1. But if the domain is that of Boolean algebra, this same quadratic (since 1847) has had n roots where n is any integer equal to, or greater than, 2. Boolean algebra, it may be remarked, is as legitimately a province of algebra today as is the theory of quadratic equations in elementary schoolbooks. In short, criticizing our predecessors because they completely solved their problems within the limitations which they themselves imposed is as pointless as deploring our own inability to imagine the mathematics of seven thousand years hence.

Some of the most significant episodes in the entire history of mathematics will be missed unless this dependence of validity upon time is kept in mind as we proceed. In ancient Greece, for example, the entire development of by far the greater part of such Greek mathematics as is still of vital interest stems from this fact. The discontinuities in the time curve of acceptable proof, where standards of rigor changed abruptly, are perhaps the points of greatest interest in the development of mathematics. The four most abrupt appear to have been in Greece in the fifth century B.C., in Europe in the 1820's and in the 1870's, and again in Europe in the twentieth century.

None of this implies that mathematics is a shifting quicksand. Mathematics is as stable and as firmly grounded as anything in human experience, and far more so than most things. Euclid's Proposition I, 47 stands, as it has stood for over 2,200 years. Under the proper assumptions it has been rigorously proved. Our successors may detect flaws in our reasoning and create new mathematics in their efforts to construct a proof satisfying to themselves. But unless the whole process of mathematical development suffers a violent mutation, there will remain some proposition recognizably like that which Euclid proved in his generation.

Not all of the mathematics of the past has survived, even in suitably modernized form. Much has been discarded as trivial, inadequate, or cumbersome, and some has been buried as definitely fallacious. There could be no falser picture of mathematics than that of "the science which has never had to retrace a step." If that were true, mathematics would be the one perfect achievement of a race admittedly incapable of perfection. Instead of this absurdity, we shall endeavor to portray mathematics as the constantly growing, human thing that it is, advancing in spite of its errors and partly because of them.

Five streams

The picture will be clearer if its main outlines are first roughly blocked in and retained while details are being inspected.

Into the two main streams of number and form flowed many tributaries. At first mere trickles, some quickly swelled to the dignity of independent rivers. Two in particular influenced the whole course of mathematics from almost the earliest recorded history to the twentieth century. Counting by the natural numbers 1, 2, 3, . . . introduced mathematicians to the concept of *discreteness*. The invention of irrational numbers, such as $\sqrt{2}$, $\sqrt{3}$, $\sqrt{6}$; attempts to compute plane areas bounded by curves or by incommensurable straight lines; the like for surfaces and volumes; also a long struggle to give a coherent account of motion, growth, and other sensually continuous change, forced mathematicians to invent the concept of *continuity*.

The whole of mathematical history may be interpreted as a battle for supremacy between these two concepts. This conflict may be but an echo of the older strife so prominent in early Greek philosophy, the struggle of the One to subdue the Many. But the image of a battle is not wholly appropriate, in mathematics at least, as the continuous and the discrete have frequently helped one another to progress.

One type of mathematical mind prefers the problems associated with continuity. Geometers, analysts, and appliers of mathematics to science and technology are of this type. The complementary type, preferring discreteness, takes naturally to the theory of numbers in all its ramifications, to algebra, and to mathematical logic. No sharp line divides the two, and the master mathematicians have worked with equal ease in both the continuous and the discrete.

In addition to number, form, discreteness, and continuity, a fifth stream has been of capital importance in mathematical history, especially since the seventeenth century. As the sciences, beginning with astronomy and engineering in ancient times and ending with biology, psychology, and sociology in our own, became more and more exact, they made constantly increasing demands on mathematical inventiveness, and were mainly responsible for a large part of the enormous expansion of all mathematics since 1637. Again, as industry and invention became increasingly scientific after the industrial revolution of the late eighteenth and early nineteenth centuries, they too stimulated mathematical creation, often posing problems beyond the existing resources of mathematics. A current instance is the problem of turbulent flow, of the first importance in aerodynamics. Here, as in many similar situations, attempts to solve an essentially new technological problem have led to further expansions of pure mathematics.

The time-scale

It will be well to have some idea of the distribution of mathematics in time before looking at individual advances.

The time curve of mathematical productivity is roughly similar to the exponential curve of biologic growth, starting to rise almost imperceptibly in the remote past and shooting up with ever greater rapidity as the present is approached. The curve is by no means smooth; for, like art, mathematics has had its depressions. There was a deep one in the Middle Ages, owing to the mathematical barbarism of Europe being only partly balanced by the Moslem civilization, itself (mathematically) a sharp recession from the great epoch (third century B.C.) of Archimedes. But in spite of depressions, the general trend from the past to the present has been in the upward direction of a steady increase of valid mathematics.

We should not expect the curve for mathematics to follow those of other civilized activities, say art and music, too closely. Masterpieces of sculpture once shattered are difficult to restore or even to remember. The greater ideas of mathematics survive and are carried along in the continual flow, permanent additions immune to the accidents of fashion. Being expressed in the one universally intelligible language as yet devised by human beings, the creations of mathematics are independent of national taste, as those of literature are not. Who today except a few scholars is interested or amused by the ancient Egyptian novelette of the two thieves? And how many can understand hieroglyphics sufficiently to elicit from the story whatever significance it may once have had for a people dead all of three thousand years? But tell any engineer, or any schoolboy who has had some mensuration, the Egyptian rule for the volume of a truncated square pyramid, and he will recognize it instantly. Not only are the valid creations of mathematics preserved; their mere presence in the stream of progress induces new currents of mathematical thought.

The majority of working mathematicians acquainted in some measure with the mathematics created since 1800 agree that the time curve rises more sharply thereafter than before. An open mind on this question is necessary for anyone wishing to see mathematical history as the majority of mathematicians see it. Many who have no firsthand knowledge of living mathematics beyond the calculus believe on grossly inadequate evidence that mathematics experienced its golden age in some more or less remote past. Mathematicians think not. The recent era, beginning in the nineteenth century, is usually regarded as the golden age by those personally conversant with mathematics and at least some of its history.

An unorthodox but reasonable apportionment of the time-scale of mathematical development cuts all history into three periods of unequal lengths. These may be called the remote, the middle, and the recent. The remote extends from the earliest times of which we have reliable knowledge to A.D. 1637, the middle from 1638 to 1800. The recent period, that of modern mathematics as professionals today understand mathematics, extends from 1801 to the present. Some might prefer 1821 instead of 1801.

There are definite reasons for the precise dates. Geometry became analytic in 1637 with the publication of Descartes' masterpiece. About half a century later the calculus of Newton and Leibniz, also the dynamics of Galileo and Newton, began to become the common property of all creative mathematicians. Leibniz certainly was competent to estimate the magnitude of this advance. He is reported to have said that, of all mathematics from the beginning of the world to the time of Newton, what Newton had done was much the better half.

The eighteenth century exploited the methods of Descartes, Newton, and Leibniz in all departments of mathematics as they then existed. Perhaps the most significant feature of this century was the beginning of the abstract, completely general attack. Although adequate realization of the power of the abstract method was delayed till the twentieth century, there are notable anticipations in Lagrange's work on algebraic equations and, above all, in his analytic mechanics. In the latter, a direct, universal method unified mechanics as it then was, and has remained to this day one of the most powerful tools in the physical sciences. There was nothing like this before Lagrange.

The last date, 1801, marks the beginning of a new era of unprecedented inventiveness, opening with the publication of Gauss' masterpiece. The alternative, 1821, is the year in which Cauchy began the first satisfactory treatment of the differential and integral calculus.

As one instance of the greatly accelerated productivity in the nineteenth century, consequent to a thorough mastery and amplification of the methods devised in the middle period, an episode in the development of geometry is typical. Each of five men—Lobachewsky, Bolyai, Plücker, Riemann, Lie—invented as part of his lifework as much (or more) new geometry as was created by all the Greek mathematicians in the two or three centuries of their greatest activity. There are good grounds for the frequent assertion that the nineteenth century alone contributed about five times as much to mathematics as had all preceding history. This applies not only to quantity but, what is of incomparably greater importance, to power.

Granting that the mathematicians before the middle period may have encountered the difficulties attendant on all pioneering, we need not magnify their great achievements to universe-filling proportions. It must be remembered that the advances of the recent period have swept up and included nearly all the valid mathematics that preceded 1800 as very special instances of general theories and methods. Of course nobody who works in mathematics believes that our age has reached the end, as Lagrange thought his had just before the great outburst of the recent period. But this does not alter the fact that most of our predecessors did reach very definite ends, as we too no doubt shall. Their limited methods precluded further significant progress, and it is possible, let us hope probable, that a century hence our own more powerful methods will have given place to others yet more powerful.

Seven periods

A more conventional division of the time-scale separates all mathematical history into seven periods:

(1) From the earliest times to ancient Babylonia and Egypt, inclusive.

(2) The Greek contribution, about 600 B.C. to about A.D. 300, the best being in the fourth and third centuries B.C.

(3) The oriental and Semitic peoples—Hindus, Chinese, Persians, Moslems, Jews, etc., partly before, partly after (2), and extending to (4).

(4) Europe during the Renaissance and the Reformation, roughly the fifteenth and sixteenth centuries.

(5) The seventeenth and eighteenth centuries.

(6) The nineteenth century.

(7) The twentieth century.

This division follows loosely the general development of Western civilization and its indebtedness to the Near East. Possibly (6), (7) are only one, although profoundly significant new trends became evident shortly after 1900. In the sequel, we shall observe what appears to have been the main contribution in each of the seven periods. A few anticipatory remarks here may clarify the picture for those seeing it for the first time.

Although the peoples of the Near East were more active than the Europeans during the third of the seven periods, mathematics as it exists today is predominantly a product of Western civilization. Ancient advances in China, for example, either did not enter the general stream or did so by commerce not yet traced. Even such definite techniques as were devised either belong to the trivia of mathematics or were withheld from European mathematicians until long after their demonstrably independent invention in Europe. For example, Horner's method for the numerical solution of equations may have been known to the Chinese, but Horner did not know that it was. And, as a matter of fact, mathematics would not be much the poorer if neither the Chinese nor Horner had ever hit on the method.

European mathematics followed a course approximately parallel to that of the general culture in the several countries. Thus the narrowly practical civilization of ancient Rome contributed nothing to mathematics; when Italy was great in art, it excelled in algebra; when the last surge of the Elizabethan age in England had spent itself, supremacy in mathematics passed to Switzerland and France. Frequently, however, there were sporadic outbursts of isolated genius in politically minor countries, as in the independent creation of non-Euclidean geometry in Hungary in the early nineteenth century. Sudden upsurges of national vitality were occasionally accompanied by increased mathematical activity, as in the Napoleonic wars following the French Revolution, also in Germany after the disturbances of 1848. But the world war of 1914–18 appears to have been a brake on mathematical progress in Europe and to a lesser degree elsewhere, as also were the subsequent manifestations of nationalism in Russia, Germany, and Italy. These events hastened the rapid progress which mathematics had been making since about 1890 in the United States of America, thrusting that country into a leading position.

The correlation between mathematical excellence and brilliance in other aspects of general culture was sometimes negative. Several instances might be given; the most important for the development of mathematics falls in the Middle Ages. When Gothic architecture and Christian civilization were at their zenith in the twelfth century (some would say in the thirteenth), European mathematics was just beginning the ascent from its nadir. It will be extremely interesting to historians eight centuries hence if it shall appear that the official disrepute into which mathematics and impartial science had fallen in certain European countries some years before the triumph of medieval ideals in September, 1939, was the dawn of a new faith about to enshrine itself in the unmathematical simplicities of a scienceless architecture. Our shaggy ancestors got along for hundreds of thousands of years without science or mathematics in their filthy caves, and there is no obvious reason why our brutalized descendants—if they are to be such—should not do the same.

Attending here only to acquisitions of the very first magnitude in all seven of the periods, we may signalize three. All will be noted in some detail later.

The most enduringly influential contribution to mathematics of all the periods prior to the Renaissance was the Greek invention of strict deductive reasoning. Next in mathematical importance is the Italian and French development of symbolic algebra during the Renaissance. The Hindus of the seventh to the twelfth centuries A.D. had almost invented algebraic symbolism; the Moslems reverted in their classic age to an almost completely rhetorical algebra. The third major advance has already been indicated, but may be emphasized here: in the earlier part of the fifth period—seventeenth century—the three main streams of number, form, and continuity united. This generated the calculus and mathematical analysis in general; it also transformed geometry and made possible the later creation of the higher spaces necessary for modern applied mathematics. The leaders here were French, English, and German.

The fifth period is usually considered as the fountainhead of modern pure mathematics. It brackets the beginning of modern science; and another major advance was the extensive application of the newly created pure mathematics to dynamical astronomy, following the work of Newton, and, a little later, to the physical sciences, following the methodology of Galileo and Newton. Finally, in the nineteenth century, the great river burst its banks, deluging wildernesses where no mathematics had flourished and making them fruitful.

If the mathematics of the twentieth century differs significantly from that of nineteenth, possibly the most important distinctions are a marked increase in abstractness with a consequent gain in generality, and a growing preoccupation with the morphology and comparative anatomy of mathematical structures; a sharpening of critical insight; and a dawning recognition of the limitations of classical deductive reasoning. If 'limitations' suggests frustration after about seven thousand years of human strivings to think clearly, the suggestion is misleading. But it is true that the critical evaluations of accepted mathematical reasoning which distinguished the first four decades of the twentieth century necessitated extensive revisions of earlier mathematics, and inspired much new work of profound interest for both mathematics and epistemology. They also led to what appeared to be the final abandonment of the theory that mathematics is an image of the Eternal Truth.

The division of mathematical history into about seven periods is more or less traditional and undoubtedly is illuminating, especially in relation to the fluctuating light which we call civilization. But the unorthodox remote, middle, and recent periods, described earlier, seem to give a truer presentation of the development of mathematics itself and a more vivid suggestion of its innate vitality.

Some general characteristics

In each of the seven periods there was a well-defined rise to maturity and a subsequent decline in each of several limited modes of mathematical thought. Without fertilization by creative new ideas, each was doomed to sterility. In the Greek period, for example, synthetic metric geometry, as a method, got as far as seems humanly possible with our present mental equipment. It was revivified into something new by the ideas of analytic geometry in the seventeenth century, by those of projective geometry in the seventeenth and nineteenth centuries, and finally, in the eighteenth and nineteenth centuries, by those of differential geometry.

Such revitalizations were necessary not only for the continued growth of mathematics but also for the development of science. Thus it would be impossible for mathematicians to apprehend the subtle complexities of the geometries applied to modern science by the methods of Euclid and Apollonius. And in pure mathematics, much of the geometry of the nineteenth century was thrust aside by the more vigorous geometries of abstract spaces and the non-Riemannian geometries developed in the twentieth. Considerably less than forty years after the close of the nineteenth century, some of the geometrical masterpieces of that heroic age of geometry were already beginning to seem otiose and antiquated. This appears to be the case for much of classical differential geometry and synthetic projective geometry. If mathematics continues to advance, the new geometries of the twentieth century will likely be displaced in their turn, or be subsumed under still rarer abstractions. In mathematics, of all places, finality is a chimera. Its rare appearances are witnessed only by the mathematically dead.

As a period closes, there is a tendency to overelaboration of merely difficult things which the succeeding period either ignores as unlikely to be of lasting value, or includes as exercises in more powerful methods. Thus a host of special curves investigated with astonishing vigor and enthusiasm by the early masters of analytic geometry live, if at all, only as problems in elementary textbooks. Perhaps the most extensive of all mathematical cemeteries are the treatises which perpetuate artificially difficult problems in mechanics to be worked as if Lagrange, Hamilton, and Jacobi had never lived.

Again, as we approach the present, new provinces of mathematics are more and more rapidly stripped of their superficial riches, leaving only a hypothetical mother lode to be sought by the better-equipped prospectors of a later generation. The law of diminishing returns operates here in mathematics as in economics: without the introduction of radically new improvements in method, the income does not balance the outgo. A conspicuous example is the highly developed theory of algebraic invariants, one of the major acquisitions of the nineteenth century; another, the classical theory of multiply periodic functions, of the same century. The first of these contributed indirectly to the emergence of general relativity; the second inspired much work in analysis and algebraic geometry.

A last phenomenon of the entire development may be noted. At first the mathematical disciplines were not sharply defined. As knowledge increased, individual subjects split off from the parent mass and became autonomous. Later, some were overtaken and reabsorbed in vaster generalizations of the mass from which they had sprung. Thus trigonometry issued from surveying, astronomy, and geometry only to be absorbed, centuries later, in the analysis which had generalized geometry.

This recurrent escape and recapture has inspired some to dream of a final, unified mathematics which shall embrace all. Early in the twentieth century it was believed by some for a time that the desired unification had been achieved in mathematical logic. But mathematics, too irrepressibly creative to be restrained by any formalism, escaped.

Motivation in mathematics

Several items in the foregoing prospectus suggest that much of the impulse behind mathematics has been economic. In the third and fourth decades of the twentieth century, for obvious political reasons, attempts were made to show that all vital mathematics, particularly in applications, is of economic origin.

To overemphasize the immediately practical in the development of mathematics at the expense of sheer intellectual curiosity is to miss at least half the fact. As any moderately competent mathematician whose education has not stopped short with the calculus and its commoner applications may verify for himself, it simply is not true that the economic motive has been more frequent than the purely intellectual in the creation of mathematics. This holds for practical mathematics as applied in commerce, including all insurance, science, and the technologies, as well as for those divisions of mathematics which at present are economically valueless. Instances might be multiplied indefinitely; four must suffice here, one from the theory of numbers, two from geometry, and one from algebra.

About twenty centuries before the polygonal numbers were generalized, and considerably later applied to insurance and to statistics, in both instances through combinatorial analysis, the former by way of the mathematical theory of probability, their amusing peculiarities were extensively investigated by arithmeticians without the least suspicion that far in the future these numbers were to prove useful in practical affairs. The polygonal numbers appealed to the Pythagoreans of the sixth century B.C. and to their bemused successors on account of the supposedly mystical virtues of such numbers. The impulse here might be called religious. Anyone familiar with the readily available history of these numbers and acquainted with Plato's dialogues can trace for himself the thread of number mysticism from the crude numerology of the Pythagoreans to the Platonic doctrine of Ideas. None of this greatly resembles insurance or statistics.

Later mathematicians, including one of the greatest, regarded these numbers as legitimate objects of intellectual curiosity. Fermat, cofounder with Pascal in the seventeenth century of the mathematical theory of probability, and therefore one of the grandfathers of insurance, amused himself with the polygonal and figurate numbers for years before either he or Pascal ever dreamed of defining probability mathematically.

As a second and somewhat hackneyed instance, the conic sections were substantially exhausted by the Greeks about seventeen centuries before their applications to ballistics and astronomy, and through the latter to navigation, were suspected. These applications might have been made without the Greek geometry, had Descartes' analytics and Newton's dynamics been available. But the fact is that by heavy borrowings from Greek conics the right way was first found. Again the initial motive was intellectual curiosity.

The third instance is that of polydimensional space. In analytic geometry, a plane curve is represented by an equation containing two variables, a surface by an equation containing three. Cayley in 1843 transferred the language of geometry to systems of equations in more than three variables, thus inventing a geometry of any finite number of dimensions. This generalization was suggested directly by the formal algebra of common analytic geometry, and was elaborated for its intrinsic interest before uses for it were found in thermodynamics, statistical mechanics, and other departments of science, including statistics, both theoretical and industrial, as in applied physical chemistry. In passing, it may be noted that one method in statistical mechanics makes incidental use of the arithmetical theory of partitions, which treats of such problems as determining in how many ways a given positive integer is a sum of positive integers. This theory was initiated by Euler in the eighteenth century, and for over 150 years was nothing but a plaything for experts in the perfectly useless theory of numbers.

The fourth instance concerns abstract algebra as it has developed since 1910. Any modern algebraist may easily verify that much of his work has a main root in one of the most fantastically useless problems ever imagined by curious man, namely, in Fermat's famous assertion of the seventeenth century that $x^n + y^n = z^n$ is impossible in integers x, y, z all different from zero if n is an integer greater than two. Some of this recent algebra quickly found use in the physical sciences, particularly in modern quantum mechanics. It was developed without any suspicion that it might be scientifically useful. Indeed, not one of the algebraists concerned was competent to make any significant application of his work to science, much less to foresee that such applications would some day be possible. As late as the autumn of 1925, only two or three physicists in the entire world had any inkling of the new channel much of physics was to follow in 1926 and the succeeding decade.

Residues of epochs

In following the development of mathematics, or of any science, it is essential to remember that although some particular work may now be buried it is not necessarily dead. Each epoch has left a mass of detailed results, most of which are now of only antiquarian interest. For the remoter periods, these survive as curiosities in specialized histories of mathematics. For the middle and recent periods—since the early decades of the seventeenth century—innumerable theorems and even highly developed theories are entombed in the technical journals and transactions of learned societies, and are seldom if ever mentioned even by professionals. The mere existence of many is all but forgotten. The lives of thousands of workers have gone into this moribund literature. In what sense do these half-forgotten things live? And how can it be truthfully said that the labor of all those toilers was not wasted?

The answers to these somewhat discouraging questions are obvious to anyone who works in mathematics. Out of all the uncoordinated details at last emerges a general method or a new concept. The method or the concept is what survives. By means of the general method the laborious details from which it evolved are obtained uniformly and with comparative ease. The new concept is seen to be more significant for the whole of mathematics than are the obscure phenomena from which it was abstracted. But such is the nature of the human mind that it almost invariably takes the longest way round, shunning the straight road to its goal. There is no principle of least action in scientific discovery. Indeed, the goal in mathematics frequently is unperceived until some explorer more fortunate than his rivals blunders onto it in spite of his human inclination to follow the crookedest path. Simplicity and directness are usually the last things to be attained.

In illustration of these facts we may cite once more the theory of algebraic invariants. When this theory was first developed in the nineteenth century, scores of devoted workers slaved at the detailed calculation of particular invariants and covariants. Their work is buried. But its very complexity drove their successors in algebra to simplicity: masses of apparently isolated phenomena were recognized as instances of simple underlying general principles. Whether these principles would ever have been sought, much less discovered, without the urge imparted by the massed calculations, is at least debatable. The historical fact is that they were so sought and discovered.

In saying that the formidable lists of covariants and invariants of the early period are buried, we do not mean to imply that they are permanently useless; for the future of mathematics is as unpredictable as is that of any other social activity. But the methods and principles of the later period make it possible to obtain all such results with much greater ease should they ever be required, and it is a waste of time and effort today to add to them.

One residue of all this vast effort is the concept of invariance. So far as can be seen at present, invariance is likely to be illuminating in both pure and applied mathematics for many decades to come. In our survey we shall endeavor to observe the methods and the concepts which have been sublimated from other masses of details, and which offer similar prospects of endurance. It is not epochs that matter, but their residues. Nor, as epochs recede into the past, do the men who made them obscure the permanence and impersonality of their work with their hopes, their fears, their jealousies, and their petty quarrels. Some of the greatest things that were ever done in mathematics are wholly anonymous. We shall never know who first imagined the numbers 1, 2, 3, . . . , or who first perceived that a single 'three' isolates what is common to three goads, three oxen, three gods, three altars, and three men.

Two recent opinions on the general history of science are apposite for that of mathematics, and may stand here as an introduction to what is to follow. In his *Autobiographia* (1923), the Spanish histologist Santiago Ramón y Cajal had this to say of scientific history:

> In spite of all the allegations of self-love, the facts at first associated with the name of a particular man end by being anonymous, lost forever in the ocean of Universal Science. Thus the monograph imbued with individual human quality becomes incorporated, stripped of sentimentalisms, in the abstract doctrine of the general treatise. To the hot sun of actuality will succeed —if they do succeed—the cold beams of the history of learning.

The next is singularly pertinent, coming as it does from the man who advanced beyond Newton in the mathematical theory of gravitation. Speaking of Newton's work in optics, Einstein says:

> Newton's age has long since passed through the sieve of oblivion, the doubtful striving and suffering of his generation have vanished from our ken; the works of some few great thinkers and artists have remained, to delight and ennoble those who come after us. Newton's discoveries have passed into the stock of accepted knowledge.

Finally, we shall try to observe the caution suggested in the observation of an M.D. and writer who is not a mathematician, Halladay Sutherland: "There is always the danger of seeing the past in the light of a golden sunset."

CHAPTER 2

The Age of Empiricism

It is not known where, when, or by whom it was first perceived that a mastery of number and form is as useful as language for civilized living. The historical record begins, in Egypt and in Mesopotamia (Babylonia, including Sumer and Akkad), with both number and form far advanced beyond the primitive stage of culture, and even here the cardinal dates have been disputed. Those dates are 4241 ± 200 B.C. at the earliest and 2781 B.C. at the latest for Egypt,[1] and about 5700 B.C. for Mesopotamia. Both refer to the earliest calendric reckoning, and each is more or less substantiated by astronomical evidence.

The basis of both the Egyptian and the Mesopotamian civilizations was agriculture. In an agricultural economy a reliable calendar is a necessity. A calendar implies both astronomical and arithmetical accuracy far beyond the facilities of mythology and haphazard observation, and it is not come at in a year. Some primitive peoples who have never been driven to farming have only the vaguest notions of the connection between the periodicity of seasons and the aspect of the heavens. By 5700 B.C. the Sumerian predecessors of the Semitic Babylonians were dating the beginning of their year from the vernal equinox. A thousand years later the first month of the year was named after the Bull, the sun being in the constellation Taurus at the vernal equinox of about 4700 B.C. Thus the inhabitants of Mesopotamia must have had a workable elementary arithmetic.

These same pioneers toward mathematics also invented or helped to transmit two major curses which continue to blight the unscientific mind, numerology (number mysticism) and astrology. It is an open question which of astrology or astronomy preceded the other. Arithmetic of some sort necessarily came before numerology.

For Egypt, the early historical record is somewhat more detailed. The more liberal of rival Egyptian chronologies assigns 4241 B.C. as the earliest precise date in history, this coinciding with the adoption of the Egyptian calendar of twelve thirty-day months with five days of feasting to complete the 365. This date also is supported by inconclusive astronomical evidence, correlating the heliacal rising of the Dog Star Sothis, our Sirius, with the date at which the annual inundation by the Nile could be expected. Here again the impulse to develop astronomy, and hence also arithmetic, was agricultural necessity unless, of course, it was astrology.

The geographical location of Sumer was more propitious than that of Egypt for a rapid development of the mathematics conceived in agriculture and born in astronomy. Egypt lay far off the main trade route between East and West. Sumer, the non-Semitic predecessor of Semitic Babylonia, lay directly across the path of the merchants at the north end of the Persian Gulf. Commerce stimulated mathematical invention in Sumer and ancient Mesopotamia as it probably never has since. Europe of the late Middle Ages also profited mathematically through trade; but the gain was in a diffusion of knowledge rather than in the creation of new mathematics necessitated by commerce.

Possibly of greater importance than trade for the development of mathematics were the demands of primitive engineering. Both the Babylonians and the Egyptians were indefatigable builders and skilled irrigation engineers, and their extensive labors in these fields may have stimulated empirical calculation. But it would be gratuitous generosity to infer that because the Egyptians, say, succeeded in raising huge obelisks, they were therefore engineers in any sense that would now be recognized as scientific. Ten thousand slaves can muddle through the work of one head; and the apparent marvels of ancient engineering that impress us today may be only monuments to a lavish expenditure of brawn and a strict conservation of brains. The Israelites and others whom the Egyptians persuaded to take up practical engineering do not seem to have been greatly impressed by the technical skill of their overseers.

Reliable evidence shows that arithmetic and mensuration in Babylonia developed from the early work of the non-Semitic Sumerians. This gifted people also invented a pictorial script which evolved into the efficient cuneiform characters that were to prove adequate for the expression of their arithmetic and mensuration. The political absorption of the Sumerians by physically but not intellectually more vigorous peoples occurred about 2000 B.C. Astronomy and arithmetic continued to flourish and, what is of singular significance, a sort of algebra evolved with incredible speed. This early appearance of algebra is one of the most remarkable phenomena in the history of mathematics.

For all that is known to the contrary, other early civilizations may have made progress toward mathematics comparable to that of Mesopotamia and Egypt, records of these two having survived largely by physical accident. The semiarid climate of Egypt and the inordinate reverence of the Egyptians for all their dead, including bulls, crocodiles, cats, and human beings, united to preserve the papyri that must have perished in a harsher atmosphere, and kept the memories of common things colorful for thousands of years on the walls of tombs and temples. Some of the most interesting historical documents as yet recovered from the past survived only because the Egyptian morticians discovered that useless papyri made excellent stuffing to plump out the mummies of sacred crocodiles to lifelike obesity.

The Babylonians impressed their records on a yet more durable medium, clay tablets, cylinders, and prisms, baked in the sun or in kilns. Sharpened sticks, one like an implement still used by children in modeling, indented the wedge-shaped characters in the soft clay, and the baking fixed a record more durably than any printers' ink on the toughest paper. Wars and the long decadence of a great civilization for once conspired to preserve some of the best in that civilization. The baked tablets, resistant to damp, rust, and pressure, and immune to the attacks of worms and insects, were buried beneath the mud ruins of dissolving temples and libraries. It would be easier for some science-hating zealot to obliterate modern mathematics than it would be for us to destroy the mathematical records of Babylonia. There is no reason to suppose that all the mathematical bricks have been exhumed.

If the records themselves are solid and tangible beyond dispute, the like cannot be claimed for their interpretation. The reading of the most suggestive parts of the Sumerian and Babylonian records is a matter of great difficulty, demanding an unusual combination of linguistic, historical, and mathematical talents. Several points of interest are still in dispute among the scholars who since 1929 have finally broken the seals on ancient Babylonian mathematics. We shall not find it necessary to use any of this disputed material in order to give an idea, sufficient for our purposes, of what the Babylonians accomplished. What remains after the few doubtful items are discarded is impressive enough.

Those far-off centuries in Babylon and Egypt are the first and last great age of the empiricism that led to mathematics. Above a multitude of details, five epochal landmarks survived for the guidance of later centuries. Number was subdued to the service of astronomy and commerce; the perception of form was clarified in an empirical mensuration and applied to astronomy, surveying, and engineering; the vast extensions of the natural number system which mathematics uses today were initiated; a method more powerful than arithmetic was begun in an algebra more than well begun; and last, also perhaps most significantly, practical difficulties in mensuration compelled some of those early empiricists to grapple at least subconsciously with the concept of the mathematical infinite. From that day to this, a stretch of nearly four thousand years, the struggle to compass the infinite has continued, and the record of the struggle is mathematical analysis.

Possibly of greater significance for the future of the race than all the technical advance toward mathematics was another for which that advance was to be largely responsible. It dawned on the human mind that man might dispense with the thousands of capricious deities created by human beings in the childhood of their race, and give a rational account of the physical universe. Although an explicit statement of this possibility was to be reserved for one of the earliest and greatest of the Greek mathematicians, it was anticipated by the astronomers and scientists of Egypt and Babylon, and it was there that our race began to grow up.

Arithmetic[2] *to* 600 B.C.

Since 1929 our knowledge of mathematics in ancient Babylonia has been increased many times over all that was previously known, largely through the pioneering work of O. Neugebauer (1899–). Apart from their great intrinsic interest, these new accessions are extremely suggestive as possible clues to the origins of Greek mathematics. If, as now seems probable, the Greek mathematicians of the sixth and fifth centuries B.C. were directly indebted to the Babylonian tradition, the development of living mathematics can be traced with scarcely a break for about four thousand years.

To anticipate, and at the same time indicate what is to be noticed in later chapters, the clue is through the first recorded empirical observation concerning similar figures, to the arithmetization of geometrical 'magnitudes' by the Pythagoreans in the fifth century B.C.; thence, inversely, to the geometrization of number by Eudoxus in the fourth century B.C., following the revolt of the Greek sophists; and finally, during and after the seventeenth century A.D., back to the arithmetization of form in analytic geometry and mathematical analysis. From the wealth of detailed material now available we shall select only enough to substantiate this suggested development. The general character of the early evidence must first be noted.

The mathematical records of Babylonia cover about two thousand years, from the first Babylonian dynasty, 2186–1961 B.C., almost to the beginning of the Christian era, the greatest number belonging to the period of about 2000–1200 B.C. Records of eclipses compiled in the reign of Sargon about 2700 B.C. might also be included, but the most significant contributions fall in the interval stated. As already remarked, the Sumerians of approximately 3500–2500 B.C. were responsible for the initial ideas. The richness of the period of the great legislator Hammurapi (c. 2100 B.C.) suggests that arithmetic and algebra were already old when the Sumerians were absorbed by more aggressive peoples. But without extrapolating from the tangible evidence of baked mathematical texts to an undiscovered antiquity, we have an incontrovertible mass of facts on which expert scholars are in substantial agreement. These, fortunately, are precisely those items of cardinal importance for any attempt to trace the development of vital mathematics. We shall consider first the arithmetic.

By 2500 B.C. the merchants of Sumeria were familiar with weights and measures, the arithmetic of a ruthless usury, both simple and compound, and equivalents for much of what we call commercial paper. In their commerce and trade their arithmetic reveals them as a tight-fisted, grasping lot well aware of the might of property long before coins were invented. Their scale of numeration, which they passed on to the Babylonians, was the sexagesimal (60 as base), with a slight admixture of the decimal system with 10 as base.

It has been conjectured that the 10 commemorates finger counting, while the 60 is 6×10, the 6 being adopted so that the useful fractions $1/2^a3^b5^c$ (a, b, c non-negative integers) would be expressible in finite terms. Traces of the sexagesimal system survive in our reckoning of time and in the corresponding division of the circumference of a circle into 6×60 degrees. But it is no longer universally supposed that such considerations induced the Sumerians to choose 60 as a base, and still less that the zodiac influenced their choice.

The place-value system was used for both positive and negative powers of the base. Thus, in the appropriate cuneiform symbols, 17, 35; 6, 1, 43, the semicolon indicating the beginning of the fractional part, denotes $17 \times 60 + 35 + 6/60 + 1/60^2 + 43/60^3$. The method of writing sometimes introduces ambiguities; but the pertinacity of modern scholarship has removed these and made sense of the residue. According to the mood of the scribe, a blank might or might not indicate absence of the corresponding power of 60. This particular difficulty was overcome by the introduction of a special character for zero but not, probably, until the time of the Greeks.

The great practical invention of zero has usually been attributed to the Hindus, and it may still be debatable whether they or the Babylonians were first. If, as seems highly probable, the Babylonians were original in their invention, zero is an interesting example of the independent origins of mathematical ideas in different cultures. Zero also appeared in the arithmetic of another gifted people, the Mayas of Central America, who used 20 as a base and had a system of place-value. The Maya numeration has been assigned to the period A.D. 200–600. Their calendric cycles go back to 3373 B.C., but this does not imply that the Mayas were civilized or even in existence that early.

The Babylonians were among the most indefatigable compilers of arithmetical tables in history. Since it was easier to multiply than to divide, they tabulated $1/n$ for integers n adapted to the base 60. Other reciprocals, 'irregulars,' like $\frac{1}{7}$, $\frac{1}{11}$, naturally caused more trouble, but were competently avoided by manufacturing problems in which such awkward divisors would automatically drop out in the course of the work. This is not the only instance in Babylonian mathematics where the teacher or the pupil appears to have applied a technique once classical in mathematical physics: given the solution, find the problem. There were also multiplication tables for such multipliers as 7, 10, $12\frac{1}{2}$, 16, 24, etc. Tables of squares appropriately read served as tables of square roots, and likewise for cubes. Another table listed values of $n^3 + n^2$ for $n = 1, 2, \ldots, 30$. The peculiar significance of this strange tabulation will appear when we come to Babylonian algebra.

From all this and a great deal more of a similar character, it is evident that the Babylonians of about 2000 B.C. were highly skilled calculators. It may not be too generous to credit them with an instinct for functionality; for a function has been succinctly defined as a table or a correspondence.

Historically, the most remarkable thing about this rapid progress in the subjugation of number is that it appears to have been ignored by the Greeks of the sixth century B.C. For what now seems to us the simplest, most natural development of mathematics this was a calamity. The fact that it happened casts a slight shadow of doubt on the vaunted intelligence of the early Greek mind. But since to press this point would be tantamount to historical blasphemy, we merely suggest that the mathematically informed observer examine the evidence and reach his own conclusions, even at the risk of upsetting sacrosanct tradition.

Egyptian arithmetic shows even more starkly its laboriously empirical origin. As early as 3500 B.C., the Egyptians freely handled numbers in the hundreds of thousands. Their hieroglyphics of this early date actually record the capture of 120,000 human prisoners, 400,000 oxen, and 1,422,000 goats. The last is probably a poetic flight of the conqueror's imagination—the catalogue occurs on a royal mace—for even today the experts of the U.S. Census Bureau would be exercised to enumerate that many goats in the brief interval between victory and celebration. But the enthusiastic exaggeration, like that of the ancient Hindus in multiplying their gods practically to infinity, shows at least that the Egyptians of 3500 B.C. had completely overcome the inability of primitive peoples to think boldly in terms of numbers. The significance of this advance can be appreciated only by comparison with the arithmetical backwardness of peoples well beyond barbarism today and also, as will appear, once more with the Greeks.

Egyptian numeration followed the decimal system, but without place-values. The arithmetic of about 1650 B.C. was capable of addition, subtraction, multiplication, and division, and was applied to numerous extremely simple problems involving all these operations. In fractions, $\frac{2}{3}$ was denoted by a special symbol; other fractions were reduced to sums of fractions of the form $1/n$, n an integer. In the Rhind papyrus of about 1650 B.C., copied by the scribe Ahmes (A'h-mose) from an older work, divisions are performed by means of these 'unit fractions,' the technique being the expression of m/n, $m > 1$, as a sum of unit fractions; for example, $\frac{2}{97} = \frac{1}{56} + \frac{1}{679} + \frac{1}{776}$. How such curious resolutions were first obtained seems not to be known. They may represent the fossil experience of centuries carefully preserved in tables for future use, as we today store up logarithms. Ahmes transmitted resolutions of all fractions $2/n$, where n is any odd number from 5 to 101. These could have been derived from successive applications of the solution in positive integers of the so-called optical equation, $1/x + 1/y = 1/n$, but it is most improbable that they were. Of the many other conjectures, none is acceptable to a majority of competent scholars.

All of the problems solved are childishly simple. Some are quite delightful for their unintentional revelations of Egyptian manners and customs, as when Ahmes goes into the arithmetic of bartering beer for bread and vice versa. Either the Egyptians were less puritanical in their schools than we, or Ahmes intended his treatise only for expert mathematicians. Less inflammatory problems are concerned with the rationing of oxen and various kinds of birds from geese to cranes and quail. A more fanciful kind, obviously of no earthly use to anyone, recalls the older type of English examination question and the more antique efforts of our College Entrance Examination Board. Loaves of bread are partitioned among several imaginary beings who are to receive amounts in arithmetical progression. There is nothing new, provided it be silly enough, under the sun.

The most significant detail for the development of mathematical thought in all the Egyptian arithmetic is the occasional checking of a calculation. This seems to show that the Egyptians of at latest the seventeenth century B.C. understood the value of proof in arithmetic. If this is a justifiable conclusion, those ancients were well on their way to mathematics when, unaccountably, they stopped.

It is said that the Egyptians' arithmetic was sufficiently advanced for the simple demands of their daily affairs. Of greater interest, perhaps, for the evolution of mathematics are those very problems of Ahmes and other ancients which can be called practical only by courtesy. These hint at an awakening of curiosity about numbers for their own interest, one of the strongest historical impulses toward the creation of mathematics, both pure and applied. Ahmes, for example, asked for a number which, added to its fifth, makes 21, and gave a singularly cumbersome verbal solution. Conceivably this might be of some practical value, but the next, dated between 2200 and 1700 B.C., is above suspicion of utility. In current symbolism the problem is to solve $x^2 + y^2 = 100$, $y = \frac{3}{4}x$ for x, y, one of the earliest instances of a simultaneous system. The Egyptian who solved these equations used a method, later called false position, which survived well into the fifteenth century of our era. 'Algebra,' to the beginner today, connotes a genetic, algoristic symbolism, although he may not realize that the notations he manipulates so glibly do most of his thinking for him; false position would severely strain his ingenuity, and his progress in the art of solving easy equations might be inconsiderable. For reasons such as these, a modern student of algebra is sometimes rather puzzled to learn that the rhetorical solutions offered by all but a few of the predecessors of Vieta (sixteenth century A.D.) were indeed algebra. Without symbolism as it exists today, much of even elementary algebra and arithmetic would be beyond the capacities of all but the most highly gifted in reasoning ability. The more honor, then, to the ancients who persevered through jungles of words to attain what the moderns reach with a few almost mechanical strokes of the pen.

We have reserved for the last items in this sketch of pre-Greek arithmetic two outstanding achievements of the Babylonian mathematicians. All of twenty-five centuries ahead of its time, one was the first recorded hint of our modern generalized conception of number. Negative numbers appear in their own right as numbers on a parity with positive numbers in at least three instances where the problem is that of solving a pair of simultaneous equations in two unknowns. If this is not controverted by a more critical reading of the texts concerned, and there seems to be only a remote possibility of this, it shows that the Babylonians had some conception of negative numbers *as numbers*.

Less significant perhaps, but still far ahead of its time, was the explicit use by the Babylonian astronomers of the fourth century B.C. of the correct rules of signs in multiplication. How the Greeks overlooked all this is a mystery.

Algebra without symbolism[a]

Passing on to Babylonian algebra of about 2000 B.C., we come to what historians consider the most remarkable anticipation in the development of mathematics.

First is the question of proof, without which mathematics in the accepted sense does not exist. Did the Babylonian successors of the Sumerians have any conception of deductive reasoning? No categorical answer can be given. So far (1945) there has not been discovered any Babylonian record of a mathematical demonstration. But this is not necessarily conclusive against at least a mute intuition for proof, evidence for which is overwhelming.

To put the case as favorably as seems justified by the undisputed evidence, we may picture a teacher of elementary algebra today grading an examination on quadratic equations. The pupils have been asked to solve $12x^2 - 7x = -1$. Some have substituted in the standard formula $x = (-b \pm \sqrt{b^2 - 4ac})/2a$ for the solution of $ax^2 + bx + c = 0$ and have been satisfied with one root; others have 'completed the square'; and one original genius has 'normalized' the equation by multiplying throughout by the coefficient of x^2, getting

$$(12x)^2 - 7(12x) = -12,$$

before solving for $12x$, whence he easily finds x by division. Falling short of the better Egyptians 3,500 years before him, not one of the pupils has sought to verify his solution by substituting in the equation, as the harried teacher had forgotten to demand a check. Nor has anyone offered a word of proof in support of his formal calculations. All have gone through the steps to a solution as if the teacher, with an open book in her hand, were dictating from first to last what to do next.

Equivalents of all this, including the absence of verification, occur in the Babylonian tablets of about 2000 B.C. Verbal instructions direct the solver to follow a path leading to the solution by our standard formula, or to normalize the equation, or to complete the square. It is algebra by rule and without algebraic symbolism. The scribes who indented these paradigms on the soft clay, or who directed others, certainly had general procedures in mind. But it must be admitted that correct general rules, even when successfully applied to hundreds of special examples, do not constitute mathematical proof. All of the Babylonian algebra is of this character: detailed solution of one numerical problem after another by verbal instructions following definite patterns. No pattern is ever isolated as a general procedure.

The empirical character of the algebra of the Babylonians, and perhaps also their social outlook, is even more strikingly evident in their astonishing solutions of cubic equations with numerical coefficients. Expressed in our terminology, equations of the type $x^3 + px^2 + q = 0$ are reduced to the normal form $y^3 + y^2 = r$, with $y \equiv x/p$, $r \equiv -q/p^3$, by multiplying the original equation throughout by $1/p^3$. If the resulting r is positive, the value of y, and hence that of x, is obtainable from tabulated values of $n^3 + n^2$, provided r is in the table.

From the equations actually solved in this manner, it is conceivable that the scribe proceeded from certain tabulated r's to construct his equations $x^3 + px^2 = q$ so that they would be solvable. Then he triumphantly produced the solution. If so, his pupils must have been as thoroughly mystified as is any student with a grain of mathematical intelligence today when his mysterious instructor pulls mathematical rabbits out of invisible hats. Brilliant trickery is no longer considered reputable mathematics. But if it is true, as has been asserted, that the mathematics of Babylonia and Egypt was the jealously guarded secret of a priestly sect, the mystery vanishes. One of the greatest services the Greek mathematicians rendered civilization was their shattering of the tradition of secrecy fostered by self-perpetuating priesthoods. The attempt of Pythagoras to carry on the secretive tradition of Babylon and Egypt was quickly dissipated, and enlightenment was put within the grasp of any unsanctified vulgarian with the will and the intelligence to reach for it.

It has been conjectured that the reduction of the general cubic to the above normal form was within the powers of the Babylonian algebraists. But we need not assume this much to grant that the Babylonians had taken a long stride toward mathematics in what they actually did. For the spirit animating mathematical discovery is the recognition of uniformity in a host of apparently diverse phenomena. The project of reducing a multitude of particular equations to a standard form, or even the easier inverse problem of constructing special equations indefinitely to fit a prescribed solution, would occur only to an intelligence that was essentially mathematical.

This methodology of transformation and reduction, generalized and many times refined with advancing knowledge, runs like a scarlet thread through all the greater epochs of mathematics. A relatively difficult problem is reduced by reversible transformations to a more easily approachable one; the solution of the latter then drags along with it the solution of the former and of all problems of which it is the type. The Babylonian reduction of cubics appears to be the first recorded instance of this methodology. We shall observe it again in the Italian algebra of the early sixteenth century, and in Vieta's signal advance half a century later. In geometry, to cite but one instance, the method first appears in the device of central projection, whereby geometers in the seventeenth century derived the properties of conics from those of the circle.

Taking the view, as we shall in general, that uniform methods are of more lasting significance than the sum total of the individual results, however brilliant or useful, obtained up to any given epoch by their use, we might rest the case for the Babylonians as mathematicians on their reduction of cubics to a normal form. But the spectacular ingenuity of their algebra—when we consider that nothing surpassing it was known in Europe till the sixteenth century A.D.—demands a summary indication of certain particulars.

Always relying on their extensive numerical tables, the Babylonian algebraists solved simultaneous linear equations in two unknowns, also simultaneous quadratics of the type $xy = 600$, $(ax + by)^2 + cx + dy = e$ for 55 sets of special numerical values of a, b, c, d, e, each of the sets leading to a quadratic in x. They also proposed a problem leading to the general quartic which, it need hardly be said, they did not solve; and likewise for a general cubic arising from a problem on frustums of pyramids. A cubic in x^2 also appears. In their solutions of quadratics the Babylonians were usually content with one root, although in one example both roots (positive) are given. A multiplicity of unknowns does not seem to have dismayed them, one problem leading to ten linear equations in ten unknowns.

Even more remarkable, perhaps, is the successful solution by initial trial and subsequent interpolation of an exponential equation to determine the time required for a sum of money to double itself at a stated rate of compound interest. Such equations are solved today by logarithms. But to infer that the Babylonians understood logarithms even to base 2 would be as fantastic as that classic fable of archaeology which declares that the ancient Egyptians were familiar with wireless telegraphy because not a scrap of wire has been found in their tombs.

In another direction, the Babylonians partly anticipated the summation of a geometric progression by Archimedes in the third century B.C., giving the correct result for ten terms by a special case of the general rule.

Of greater significance for the future of mathematics was the highly intelligent reaction of the Babylonian algebraists to irrationals. Their tables and their equations taught them that not every rational number in their tables had a tabulated square root. Faced with this fundamental fact, they proceeded to approximate by means of the rules $(a^2 + b^2)^{\frac{1}{2}} = a + b^2/2a$, or $= a^2 + 2ab^2$. The first is reasonable and reappears about two thousand years later with Heron of Alexandria; the second is hopelessly wrong, being dimensionally impossible. We note in passing that the reasonable approximation is obtainable from Newton's binomial series; but again this does not imply anticipation. In further approximations to quadratic surds they used what may be interpreted as the first steps toward conversion into periodic continued fractions. For $\sqrt{2}$ they gave the approximation $1\frac{5}{12}$, correct to two decimals. As will be seen when we consider the Pythagoreans, $\sqrt{2}$ marks one of the cardinal turning points in the history of mathematics.

Contemplating work of this caliber, done for the most part about two thousand years before the Christian era, we can only marvel how it was done, for we do not know. The detailed numerical solution of specific examples gives no hint of the thought inspiring the uniform procedures. Neugebauer emphasizes that the technique is based on elaborate numerical tables. At the lowest estimate, the high skill in using such tables indicates an extraordinary capacity for detecting uniformities in masses of empirical data. The Babylonians were the world's first exact astronomers; and so accurate were their first observations and their calculations that Kidinnu, about 340 B.C., made the capital discovery of the precession of the equinoxes, anticipating Hipparchus by about two centuries. It seems reasonable to assume that unrecorded centuries of observing the planets led to the accumulation of numerical data from which a purely rhetorical algebra evolved. For Babylonian algebra was entirely unsymbolic. More remarkable still, the processes which now would be summarized in formulas were never, so far as is known, reduced to written rules. If these elaborate procedures were transmitted wholly by word of mouth, the strain on even a strong memory must have been considerable.

Another unsolved problem is even more puzzling. Up to 1900 it was customary to ascribe the beginnings of algebra to the Greek Diophantus in the third century A.D., over two thousand years after the Babylonians had bettered some of his best. Where was algebra buried in the meantime? It has been conjectured that the Greeks of the sixth and fifth centuries B.C. must have been acquainted with what the Babylonians had done in algebra because, as will appear, a considerable amount of Babylonian empirical geometry almost certainly was known to those same Greeks. Direct evidence is lacking that the early Greeks were *not* acquainted with Babylonian algebra, but the indirect evidence is at least worth noting. For if the early Greeks *were* cognizant of Babylonian algebra, they made no attempt to develop or even to use it, and thereby they stand convicted of the supreme stupidity in the history of mathematics. But it is commonly agreed that the early Greek mathematicians and philosophers were among the most intelligent human beings that ever lived.

This awkward historical dilemma can be circumvented by a slight anticipation. The ancient Babylonians had a rare capacity for numerical calculation; the majority of the Greeks were either mystical or obtuse in their first approach to number. What the Greeks lacked in number, the Babylonians lacked in logic and geometry, and where the Babylonians fell short, the Greeks excelled. Only in the modern mathematical mind of the seventeenth and succeeding centuries were number and form first clearly perceived as different aspects of one mathematics.

Nothing has been said about Egyptian algebra because it was far less advanced than the (probably) earlier work of the Babylonians. Between 1850 and 1650 B.C. the Egyptians solved easy numerical equations of the first degree by trial, or by what was called the rule of false position in the Middle Ages. The last makes it plausible that the Egyptians understood proportion. If they did, and historical experts do not doubt it, they share with the Babylonians the honor of having uncovered a main root of mathematical analysis.

Toward geometry and analysis[3]

Babylonian mensuration[2] of about 2200–2200 B.C. is almost as astonishing as the contemporary algebra. Mathematically, it is of the same character as the algebra in its disregard of proof. Correct rules are applied for finding the area of any rectangle, right triangle, isosceles triangle, trapezoid with one side perpendicular to the base, and "if π be taken equal to 3," any circle. This approximation to π is famous also for its occurrence in the Old Testament. A little later, between 1850 and 1650 B.C., the Egyptians had the closer approximation $\frac{256}{81}$, ~ 3.16. It would be interesting to know what suggested the curious $(\frac{4}{3})^4$.

In their mensuration of solids the Babylonians of about 2000 B.C. gave correct solutions of numerical problems involving rectangular parallelepipeds, right circular cylinders, and right prisms with trapezoidal bases. Some of this had obvious applications to earthwork problems in the excavation of canals for drainage or irrigation. Their rule for the volume of a truncated square pyramid was incorrect. The correct rule of the Egyptians, one of the most remarkable achievements of pre-Greek mathematics, will be considered later by itself.

Passing to theorems of pure geometry known to the Babylonians of the same period, we select three for their outstanding historical suggestiveness. The first two are: the angle in a semicircle is a right angle; the Pythagorean theorem $c^2 = a^2 + b^2$, where c,a,b are the sides of a right triangle, for certain numerical values of c,a,b, as 20, 16, 12 and 17, 15, 8. The first of these, often considered one of the most beautiful theorems in elementary geometry, is said to have been proved by the Greek Thales about 600 B.C. It would be guessed immediately on inscribing rectangles in circles. The Babylonians offered no justification. From the second, and certain numerical calculations for the sides of right triangles, it has been argued that the Babylonians knew the Pythagorean theorem in the general case, but the evidence seems inconclusive. Until 1923 it was supposed that the Egyptians knew this theorem at least in the case $5^2 = 4^2 + 3^2$, because the Egyptian 'rope stretchers' were formerly said to have used this property of 5, 4, 3 in laying out right angles for the orientation of buildings. But it is now claimed that even though 5, 4, 3 *may* have been used thus, the Egyptians knew not a single instance of the Pythagorean $c^2 = a^2 + b^2$, because there is no documentary evidence that they did. Since c, a, b are the sides of a right triangle if and only if $c^2 = a^2 + b^2$, we have here an interesting historical puzzle as to how the Egyptians guessed what they needed.

Regarding the Pythagorean theorem itself, whoever first guessed it, we recall that it is the cornerstone of Euclidean metric geometry and one of the bases of all metrics. It too, like similar triangles, threads all mathematical history, not only in geometry, but also in algebra, the theory of numbers, and mathematical physics.

The third significant empirical theorem of the Babylonians in pure geometry is the earliest recorded trace of the origins of mathematical analysis: the sides about corresponding angles of similar triangles are proportional. This theorem implies equality of ratios. It has been said to follow that the Babylonians had some conception of ratio. But, if we wish to be as precise here as we were a moment ago in the case of the Pythagorean theorem versus the nonsuited Egyptians, we have no right to assert that the Babylonians actually had even the remotest conception of ratio. For 'equal ratios' and 'ratio' are distinct concepts in mathematics, and an extensive theory of 'equal ratios' is easily possible without any definition of 'ratio.' 'Ratio' is on a higher level of abstraction than 'equal ratios.' Euclid attempted, not too successfully, to define ratio. His definition has been translated by De Morgan thus: "ratio is a certain mutual habitude of two magnitudes of the same kind depending upon their quantuplicity." Fortunately Euclid never had to appeal to this abstruse definition, his 'theory of ratio' being wholly a theory of proportion, that is, of equal ratios. That the Babylonians, or anyone else before the nineteenth century, had a workable conception of ratio seems extremely improbable. The ratio of m to n, written usually as m/n, is understandable only—so far as we know at present—as a number-couple (m, n) with certain postulated properties for the four rational operations on such couples. So far as documentary evidence goes, there is none, apparently, to show that the Babylonians ever got within hailing distance of Euclid, who, if he did not succeed in giving an unmystical definition of ratio, at least acted as if he were aware that 'ratio' and 'proportion' are different concepts. But in this whole matter we have no desire to be as precise as we were in the case of the Egyptians; we have tried merely to indicate where the mathematical crux of the history lies. With their numerical examples of four numbers in proportion, the Babylonians took the first step toward the Greek theory of proportion which has lasted, practically unmodified, to this day.

Another possible source of much modern mathematics will be noted presently in connection with the pyramid. But the subsequent history of what evolved from similar triangles is so clear, and of such outstanding importance for all mathematics, that we shall leave Babylonian metrics here with this as its crown.

With one exception to be discussed presently, the Egyptian empirical mensuration is less impressive than the Babylonian. From their prodigious architecture it might be reasonable to infer that the ancient Egyptians were skilled construction engineers and hence at least respectable geometers. They were neither. Brute force in the form of unlimited slave labor made brains all but superfluous. Until it was discovered how they raised the huge stone blocks to build their pyramids, it was supposed that the Egyptian overseers were acquainted with at least the rudiments of scientific engineering. What they actually did[4] puts them on the intellectual level of the ants. As the successive tiers of a pyramid rose, the slaves laboriously buried under thousands of tons of sand the face of the work already done. The swarms of slaves lugged the blocks up the long ramp. When the task of building was finished, the slaves removed the mountain of sand burying the pyramid, to put it all back where they had got it in the first place. The dazzling result of their labors shone out in all its splendor, another time-outlasting monument to the unconquerable spirit of man's temporal rulers and the unbreakable backs of those who do the work. The sagacity of the slave-driving Egyptian overseers has been rated highly by competent enthusiasts.

In accuracy of practical measurements both the stonemasons and the irrigation engineers of Egypt of the third millennium B.C. reached great heights. It is asserted, for instance, that the maximum error in a side and in a corner angle of the Great Pyramid are only small fractions of one per cent. Again, the surveyors responsible for observing the Nile succeeded in placing their water gauges in one plane for a distance of about 700 miles round all the bends of the river. With a sufficient number of centuries for observation, this could be done by trial and error, and it does not necessarily imply any great knowledge of scientific surveying. The Egyptians had plenty of time.

In the direction of geometry they seem to have known that the area of any triangle is obtainable by the rule $\frac{1}{2}$ base $\times$ altitude. They also computed the volume of a cylindrical granary correctly. These results are as advanced as anything the Egyptians are known definitely to have obtained, with the exception of their work on the pyramid to be noted next. For a people who achieved magnificent art, it must be admitted that the Egyptians' efforts toward geometry are mostly trivial and disappointing. This, probably, is only to be expected, as acceptable art is created by peoples but little above savagery.

The greatest Egyptian pyramid

Every list of the seven wonders of the ancient world includes the Great Pyramid. But since the translation of the Moscow papyrus[5] in A.D. 1930, this pyramid has been overtopped by a greater than any the slaves in Egypt could ever have reared. This greatest of Egypt's pyramids existed only in the mind of a nameless mathematician who discovered or guessed the most remarkable result in pre-Greek geometry. He gave *a numerical example* of the correct formula, $\frac{1}{3}h(a^2 + ab + b^2)$, for the volume of the frustum of a truncated square pyramid, h being the altitude and a,b the sides of the top and bottom bases. This numerical application of a special case of the prismoidal formula dates about 1850 B.C. It is not known how this formula was obtained. Of several plausible conjectures, none is accepted by a majority of reconstructive scholars.

Had the forgotten Egyptian responsible for this result proved his procedure, he would rank high among the greater creators of mathematics. Even the empirical discovery of such a process or its verbal equivalent is evidence of extraordinary mathematical insight. In some guise the essential method underlying the formula has reappeared in all the great ages of mathematics. The Greeks called it exhaustion;[6] Cavalieri in the seventeenth century called it the method of indivisibles and, as will appear in the proper place, got no closer to proof than the ancient Egyptians of at latest 1850 B.C. To us it is the theory of limits and, later, the integral calculus. The reasons for believing that no Egyptian could ever have even distantly approached a proof recur many times in mathematical history; the final and conclusive one was stated only in A.D. 1900.

The complete method of exhaustion is sufficiently described through the simpler problem of determining the area of a circle. Regular polygons of n sides are inscribed and circumscribed to the circle; the required area is less than that of the circumscribed polygon and greater than that of the inscribed; as n is increased, the difference between the areas of the polygons diminishes until, in the limit, as n tends to infinity, the difference vanishes, or is 'exhausted,' and the common area of the limiting polygons is equal to that of the circle. In many partial applications of the method, only inscribed polygons were considered. In either variant, it is necessary to know the area of a regular polygon of n sides. This is immediate once the area of an isosceles triangle is known. If the limits described exist, and if they can be calculated, the problem is solved.

At any stage, say $n = 96$, where Archimedes stopped in the third century B.C., an approximation to the area of the circle is obtained from the calculable polygons. Moreover, this approximation is comprised between determinate bounds given by the areas of the inscribed and circumscribed polygons of 96 sides. But the crucial step in obtaining the exact formula for the area, *or even defining what is meant by the area*, is taken only by passage to the limit as n becomes indefinitely great.

For the truncated square pyramid we might proceed similarly by inscribing and circumscribing stairways whose steps are rectangular prisms with square bases; and it is conceivable that the Egyptian inferred his rule from the easily calculated approximations given by stairways with a few steps. Indeed, the earlier pyramids were of this type, and the Great Pyramid itself presented just such an appearance before the final smooth sheathing of dressed stone was applied. But however the Egyptian reached his rule, his intuition gave him the correct result that is provable *only* by the integral calculus in some guise. For all proofs of the prismoidal formula and its special cases ultimately appeal to the formula for the volume of a triangular pyramid. The trivial generalization of this result for a pyramid on any polygonal base was attributed by Archimedes to Democritus, the founder of atomism, in the fifth century B.C. Criticism by the Greek sophists of the limiting processes used by Democritus and others was partly responsible for the particular course which mathematics followed in ancient Greece; and this was one of the major turning points in the evolution of mathematics. It would be interesting to know whether Democritus was influenced by the Egyptian result. He was one of the most widely traveled and more boastful of the early Greeks, bragging that although the Egyptian rope stretchers taught him all they knew, he himself knew far more.

Might it not be possible, it may be asked, that the Egyptian obtained his rule by some device obviating the theorem of Democritus? This would be so only if it were possible to prove by elementary means that triangular pyramids of equal altitudes are to one another as their bases. Euclid's proof in his *Elements* is not elementary in that it is by the method of exhaustion, implying the concept of continuity. It was this concept to which the sophists objected so pertinently that they succeeded in deflecting mathematics into a narrow channel which, to a modern mathematician, seems forced and unnatural. The consideration of this will be our principal concern when we follow mathematical thought through ancient Greece.

Conceivably, a strictly finite proof of Euclid's basic theorem for triangular pyramids might be possible. The lack of such a proof might be due only to mathematical incapacity and not to the nature of mathematics. Were such a proof possible, the Egyptian might well have proved his rule, or at least have perceived, however dimly, some mathematical grounds for it.

Recognizing the fundamental mathematical significance of the possibility of a strictly finite proof for Euclid's theorem, C. F. Gauss (A.D. 1777–1855, German), usually rated with Archimedes (287?–212 B.C., Greek) and Isaac Newton (A.D. 1642–1727, English) as one of the three greatest mathematicians in history, in 1844 urged that a proof not depending upon continuity be sought. Thus it was by no means obvious to Gauss that such a proof might not be found. In 1900, M. W. Dehn proved that no such proof is possible.

It seems unlikely, then, that the Egyptian had anything resembling a proof for his rule. If it was just a lucky guess, he was so good at guessing that he needed no mathematics. Several of the greater mathematicians have emphasized intuition in mathematics as the necessary spark without which there is no discovery. Some have even discounted proof almost to zero, claiming that any competent hack can grind out a proof once the result has been guessed. Measured by this demotic standard, the nameless Egyptian was a very great mathematician indeed.

The contribution of Babylon and Egypt

It has been said that no subject loses more when divorced from its history than mathematics. This may be true, but there is a sort of converse which is equally true. The history of no subject loses more when divorced from its subject than does the history of mathematics. With this in mind, we recall that we are primarily interested in the development of mathematical thought, rather than in the exhibits in a museum of antiquities. It is now time to apply this primary interest to our first collection of specimens. With the memorable achievements of Babylon and Egypt behind us, let us glance back for a moment, forget all the human struggling that made these long-dead things once live, and estimate them solely in the light of mathematics. One item of all the treasure will suffice as typical of all. The Babylonian mensuration of the circle throws into sharp relief the distinction between what is mathematics and what merely resembles mathematics.

In the familiar formulas $2\pi r$, πr^2 for the circumference and area of a circle of radius r, π denotes a *constant* number which, to seven decimals, is 3.1415926. The last of course is of great practical importance. But the long chronicle of π signifies vastly more for the history of mathematics than a rather dreary record of successive approximations from the crude 3 of the Babylonians of about 2000 B.C. to the 707 decimal places, all but a few of them quite useless, of W. Shanks in A.D. 1853.

Any tyro in geometry understands what is intended by such an elliptical statement as "the ancient Babylonians took π equal to 3." But accepted literally, this statement implicitly denies the existence of mathematics and makes nonsense of its history. So far as is known, nobody before the ancient Greek mathematicians ever "took π equal to" anything. Until it had been proved that the ratio of the circumference of any circle to its radius is independent of the radius, or that (Euclid, XII, 2) the areas of any two circles are to one another as the squares on their diameters, there was no "π" to be taken.

Induction from physically measured circles may have suggested to some empiricist that the circumference of any circle is greater than $3\frac{10}{71}$ diameters and less than $3\frac{1}{7}$, the bounds proved by Archimedes to exist. But only a mind very immature scientifically would trust these bounds for circles either so small or so large that they could not be measured by the means used for the others. Certainly no mind with the faintest stirrings of a mathematical instinct would trust them. Induction from practical experience is not enough here; mathematics is demanded.

If, in this particular matter of π, it be argued that the ancients before Greece had no need for mathematical—not merely numerical—precision, and that close induction from experience sufficed, several replies may be given, all pertinent and applicable to the entire history of mathematics. First, on the severely practical side, any civilized people using a calendar would need sooner or later to know, or at least to believe, that there *is* a constant, say c, such that the circumference of a circle is c times its radius. Otherwise, as their astronomy became more exact, they would live in constant dread that their calendar might begin fluctuating disastrously, and with it, their commerce and agriculture.

Second, mathematical precision and numerical precision are very different things, in spite of what some practical souls may imagine to the contrary. A fair degree of numerical precision was demanded in ancient times. Had civilization crystallized in the second millennium before our era, no greater precision in numerical calculations than that which sufficed in Babylon would have been required. But, to cite only three instances of the need for greater numerical precision as civilization evolved, the calendar, geography, and navigation demanded an increasingly precise astronomy, and this was forthcoming only when arithmetic and geometry had progressed far beyond the sharpest exactness possible to mathematical empiricism. The validation of a formula is its proof, without which precision even in the narrowest sense of practical utility is impossible once the earlier stages of civilization are passed.

A third distinction which sharply separates the Archimedean mensuration of the circle from the Babylonian is exactly the distinction between scientific and prescientific thinking. A mind which rests content with a collection of facts is no scientific mind. The formulas in a mathematical handbook are no more mathematics than are the words in a dictionary a literary masterpiece. Until some unifying principle is conceived by which an amorphous mass of details can be given structure, neither science nor mathematics has begun.

The first and most extensive of all the structures unifying number and form is deductive reasoning. There is no conclusive evidence that such reasoning was used in mathematics before the Greeks. They also advanced far beyond mythology in their attempts to unify their observations of nature. Their cosmic speculations may have been too naive to be of much scientific value; nevertheless, they were deliberate steps away from mythology and superstition and toward science. With the conscious recognition that unity and generality are desirable values both practically and aesthetically, mathematics and science became possible.

All this may sound rather dogmatic, but it is not so intended. It is merely one of two possible points of view; and the reader is recommended to take the opposite side, follow it out consistently, and observe to what conclusions he is led in his estimates of mathematics and its history. The like applies to our entire future course, and in particular to the following paragraph, with which no doubt many will disagree. It is the estimate[7] of pre-Greek mathematics to which our argument has led us.

Until, if ever, evidence is uncovered proving that the Greeks were anticipated in their conception of mathematics as a deductive science, the greatest contribution of the Babylonians and Egyptians must remain their unconscious part in helping to make possible the golden ages of Eudoxus and Archimedes. It was enough, and should preserve their memory as long as mathematics lasts.

CHAPTER 3

Firmly Established

Greece, 600 B.C.–A.D. 300

At this point we depart from the traditional historiography of mathematics. Those seeking detailed accounts of the lives and works of mathematicians mentioned will find them in the places cited.[1] As indicated in the Prospectus, our main concern is with those creations of the past which have survived, if not unaltered at least in recognizable form, in the body of still-living mathematics. Consequently, the items to be selected presently from the rich treasure of Greece are not necessarily those which would be of greatest interest on other accounts. A bald obituary of the very dead Greek computation may serve first as a specific instance typical of all inclusions or omissions.

Mathematics and computation

The ancient Greeks separated their work on the rational numbers into logistic and arithmetica. Logistic embraced the techniques of numerical calculation as practiced in trade, commerce, and the sciences, particularly astronomy. Arithmetica, our higher arithmetic or theory of numbers, was concerned with the properties of numbers as such. To dispose here of arithmetica before Diophantus (date uncertain, c. A.D. 250), the Greeks made two major contributions. Euclid proved the basic theorems on arithmetical divisibility from which Gauss in 1801 deduced the fundamental theorem of arithmetic: a positive integer is a product of primes in one way only apart from permutations of the factors. Euclid also proved that there is no greatest prime. If Euclid himself was not first to discover these theorems, he at least transmitted proofs of them in his *Elements*.[2]

For their percussions on the higher arithmetic of Fermat and others in the eighteenth to the twentieth centuries, the figurate numbers of the Pythagoreans (sixth to fifth centuries B.C.) may be remembered as one of the most suggestive contributions of arithmetica to the modern higher arithmetic. These numbers also achieved a certain prestige in Plato's science, as for example in his *Timaeus*. The triangular numbers in particular, when insinuated into the Empedoclean chemistry of the four 'elements,' earth, air, fire, and water, were partly responsible for the remarkable metaphysical conclusion that "all matter is essentially triangles." The figurate numbers are supposed to have originated from the representation of the regular polygons by placing a pebble on each vertex and then bordering the polygons in such a manner that regularity and number of sides were preserved. This possible origin has been instanced as an early occurrence of the connections between number and space. Whether this superficial connection is more than a mathematical pun, the square numbers pebbled out as described may be responsible for the persistence of 'squares' in our algebra, where geometric imagery is not only obsolescent but irrelevant.

Another numerical item that might be credited to arithmetica as practiced by the Pythagoreans is the law of musical intervals, traditionally attributed to Pythagoras himself. The law relates the pitches of notes emitted by plucked strings of the same kind, under equal tensions, to the lengths of the strings. This discovery, the first in mathematical physics, revealed an unexpected interdependence of number, space, and harmony. It is scarcely surprising, then, that it precipitated a deluge of number mysticism. Human credulity being what it is, the resulting crop of esoteric philosophies and bizarre creeds which sprang up in ancient times, and which continue to flourish in our own, might have been anticipated. The Pythagorean law was also responsible for the retention of 'music' in the standard medieval curriculum. In fact, nearly every conceivable use, except a sensible one from a modern point of view, was made of the epochal discovery that musical sounds and numbers are related. The fact that had been discovered by experiment became the occasion for abandoning experiment in favor of the unaided human reason. Consequently, the experiment that might have started a scientific age, in the modern sense, aided most effectively in retarding that age for about 2,000 years.

In logistic—computation—the Greeks did nothing that is

not best forgotten as quickly as possible by a mathematician. Their best attempt to symbolize numbers was a childish scheme little better than juxtaposition of the initial letters of number names. Yet the development of Greek numeration, such as it was, might legitimately merit a great expenditure of erudition, time, and space in the antiquarian history of mathematics. Its interest here is negligible because, fortunately for mathematics, Greek numeration quickly perished. Only one of its many disabilities was its incapacity to represent even moderately large numbers concisely. Archimedes in the third century B.C. overcame this in a scheme of counting by eighth powers of ten. But as he just missed the place-system of numeration, his ingenious idea also perished.

It is supposed that the Greeks themselves, except a few experts, made little use, if any, of their alphabetic numbers in computation, but resorted to the abacus. Sporadic attempts to rehabilitate the battered reputation of Greek logistic as a workable system appear to be based on misapprehensions of what the Greeks actually did, and the majority opinion remains that of the conservative and sympathetic historian of Greek mathematics who characterized Greek numeration as vile. Some good thing undoubtedly came out of Nazareth, but it seems unlikely that any decent arithmetic could have issued from an inherently vile way of symbolizing numbers, and those who have gone most deeply into the matter assert that none did.

The problem of numeration was finally solved by the Hindus[3] at some controversial date before A.D. 800. The introduction of zero as a symbol denoting the absence of units or of certain powers of ten in a number represented by the Hindu numerals has been rated as one of the greatest practical inventions of all time. Certainly the merchants and traders of medieval Europe thought so when they learned the Hindu system from the Moslems. Not only did the numerals stimulate commerce; they later shortened astronomical computations tremendously, and hence were partly responsible for the computation in a reasonable time of the tables demanded during and after the Renaissance by an increasingly more extensive navigation, which in turn accelerated trade, which in its turn hastened practical refinements in computation. In brief, it is difficult to exaggerate the practical importance of a simple and universally applicable system of numeration. But it is easy to overestimate its importance in the development of mathematics, applied as well as pure.

Computation begins only after the mathematics has been done. Neither the Hindu numerals nor any others are of any importance whatever in vast tracts of modern mathematics. No numerical computations are performed.

Gauss is reported to have lamented that his ancient compeer Archimedes just failed to anticipate the Hindu system of numeration. With his own prodigious astronomical calculations behind him, Gauss speculated how much farther advanced the science of the nineteenth century might have been had Archimedes succeeded. If this report is accurate, it admirably points the parting of the ways. For Gauss had computational astronomy in mind; and it was Gauss the expert calculator, not Gauss the creative mathematician, who was lamenting the failure of Archimedes to take that last simple but essential step.

Ex Oriente lux

In an older day the sudden rise to maturity of Greek mathematics was classed with the miracles. Before the twentieth-century research on the records of Babylon and Egypt, it appeared that mathematics in Greece had grown from conception to vigorous manhood in a mere flash of about three centuries. Today we know that the respect which Greek writers themselves expressed for the wisdom of the East, even while extolling their own, was justified. The sudden maturity is no longer incredible. Modern science, beginning with Galileo and Newton, has developed with equal rapidity from origins relatively no more promising than those from which Greek mathematics evolved.

The route by which the learning of the East reached Greece has yet to be uncovered in detail. But the battles of Marathon (490 B.C.), Thermopylae, and Salamis (480 B.C.), where the Greeks broke the Persians on land and sea, may have been a turning point in mathematical history as they were in that of all Western civilization. Those battles prove at least that young Greece was in close contact with ancient Persia, the imperial successor of Egypt, Babylonia, Phoenicia, Syria, and all Asia Minor.

Marathon and Salamis are universally acclaimed by partisan humanists as unmixed benefits for the development of civilized culture. The career of mathematics hints that they may have been the beginning of a long detour round the origins of much that today is more vital than some of the Greek masterpieces.

Until the route from East to West is traced, we shall not

know definitely how much Greek mathematics owed to its predecessors. Without in any way belittling the Greek contribution, we may safely believe that the extreme miracle of spontaneous generation did not happen in Greece. In this connection, two opposing theories of the anthropologists may be mentioned. According to the first, closely similar cultures will evolve spontaneously in similar environments, no matter how widely separated. According to the second, all civilization is propagated from foci of culture, which in turn were civilized from more remote foci, and so on, until all culture is traced to one initial focus, usually Egypt.

A third theory combines the patent advantages of both. The spontaneous theory is echoed in the frequent remark that mathematical and scientific discoveries are often made independently and almost simultaneously by two or three men. On the diffusion theory this can be explained by observing—what is the fact—that in such instances the discoveries usually germinate in a body of knowledge accessible to all. Something roughly like this is now believed to have been responsible for the sudden efflorescence of Greek mathematics.

The lore of the East was available to any curious Greek who could afford a journey to Egypt and Babylonia. Trusting Greek tradition, we may assert that many early Greeks, urged by their notorious and childlike curiosity, traveled extensively in the East and profited enormously by their travels. Greek mathematics is sufficient evidence of the insatiable hunger of the awakening Greek mind for exact knowledge, and the most adequate measure of its intellectual capacity.

Two supreme achievements

All the immaterial riches of the generous East were anyone's for the asking and the taking. The early Greeks appear to have asked for everything and to have taken nearly all, fools' gold along with the rest. In their youthful eagerness to acquire, they overlooked two obvious opportunities, each of the first importance for the futures of science, mathematics, and philosophy then possible.

The sixth century before Christ was the time, and Greece the place, for human beings to reject once for all the pernicious number mysticism of the East. Instead, Pythagoras and his followers eagerly accepted it all as the celestial revelation of a higher mathematical harmony. Adding vast masses of sheer numerological nonsense of their own to an already enormous bulk, they transmitted this ancient superstition to the golden age of Greek thought, which passed it on in the first century A.D. to the decadent arithmologist Nicomachus. He, enriching his already opulent legacy with a wealth of original rubbish, left it to be sifted by the Roman Boethius, the dim mathematical light of the Middle Ages, thereby darkening the mind of Christian Europe with the venerated nonsense, and encouraging the gematria of the Talmudists to flourish like a weed.[4]

It is customary in the histories of science and mathematics to ignore these vagaries of the human mind. However, it would seem to impartial onlookers that only a distorted image of the not too flattering facts can result from any historical account which reports only what are now considered successes and ignores the failures. Frequently the sense of one epoch has become the nonsense of a later, and what no longer is meaningful may once have been of prime scientific or social importance. A case in point is the phlogiston theory of heat; another, the tripartite infinity of theology, a lineal descendant of the mystical arithmetic of the Pythagoreans; and yet another, the Platonic theory of mathematical truths, long since abandoned by unmystical mathematicians. Without some attention to such misadventures in ideas, the development of mathematics, no less than that of science, appears as an uninterrupted parade of triumphs with never a recession to relieve the glorious monotony. Gratifying as such a presentation may be, it is not therefore necessarily adequate.

Had the Pythagoreans rejected the number mysticism of the East when they had the opportunity, Plato's notorious number,[5] Aristotle's rare excursions into number magic, the puerilities of medieval and modern numerology, and other equally futile divagations of pseudo mathematics would probably not have survived to this day to plague speculative scientists and bewildered philosophers. Nor would a mathematical astronomer[6] of the early twentieth century have beheld the astounding spectacle of God masquerading as a mathematician. Among the gains accruing from the ancient numerology is the inspiration for much of Plato's theory of Eternal Ideas.

If on the other hand the early Greeks had accepted and understood Babylonian algebra, the time-scale of mathematical development might well have been compressed by more than a thousand years. But to a people just starting to grow up mathematically, the attractions of a mystical, all-embracing philosophy were naturally more seductive than those of an austere algebra.

A greater disaster at the height of the Greek golden age held mathematics and science back immeasurably. Instead of following the bold lead of Archimedes and developing a fluent, dynamic mathematics applicable to the ceaseless flux of nature, the lesser Greek mathematicians of the third century B.C., and after, lingered behind with the Platonists and cast their thought in geometric shapes as perfect and as rigidly static as the Parthenon. In the entire history of Greek mathematics, all but the incomparable Archimedes and a few of the more heterodox sophists appear to have hated or feared the mathematical infinite. Analysis was thwarted when it might have prospered.

Such are the debits in the account of Greek mathematics with time. They are heavy enough; but beside the credits, they are of little moment. It has fallen to the lot of but one people, the ancient Greeks, to endow human thought with two outlooks on the universe neither of which has blurred appreciably in more than two thousand years. From all the mass of their great achievement, these two, each of superlative excellence, may be exhibited here by themselves, not to diminish their magnitude by a crowd of lesser masterpieces, all great but not the greatest.

The first was the explicit recognition that proof by deductive reasoning offers a foundation for the structures of number and form.

The second was the daring conjecture that nature can be understood by human beings through mathematics, and that mathematics is the language most adequate for idealizing the complexity of nature into apprehensible simplicity.

Both are attributed by persistent Greek tradition to Pythagoras in the sixth century before Christ. No contemporary record of these epochal advances survives; and there is an equally persistent tradition that it was Thales in the sixth century B.C. who first proved a theorem in geometry. But there seems to be no claim that Thales, earliest of the 'seven wise men of Greece,' proposed the inerrant tactic of definitions, postulates, deductive proof, theorem as a universal method in mathematics. Again, in attributing any specific advance to Pythagoras himself, it must be remembered that the Pythagorean brotherhood was one of the world's earliest unpriestly cooperative scientific societies, if not the first, and that its members assigned the common work of all by mutual consent to their master. It is sufficient to remember that these advances were made as early as 400 B.C. at the latest, and that both were Greek.

Chronology of Greek mathematics

Before considering in some detail a few items of more than antiquarian appeal, we shall give a short prospectus of the leading schools of Greek mathematics with their dates, a few key names, and brief mention of the principal advances made by each. Some of these will not be noted again. All dates except those attached to men's names are only approximate; those without A.D. are B.C.

The birth, maturity, and senescence of Greek mathematics cover about ten centuries, roughly from 600 B.C. to A.D. 400. The earliest period, 640–550, was that of Thales (624?–550?), of the Ionian school, and Pythagoras (569?–500?). Its outstanding achievements are the founding of mathematics as a deductive system and the program of mathematicizing natural phenomena.

In the fifth century, the Greek sophists of Elea in Italy hardly constituted a mathematical school, yet were of fundamental importance for the development of all mathematical thought. By his ingenious paradoxes on infinite divisibility, Zeno (495?–435?) cast doubt on some of the reasoning of his predecessors, and was partly responsible for the characteristically Greek course which mathematics entered with the succeeding school and followed thereafter. The sophist revolt against plausible reasoning thus marks one of the cardinal turning points in the history of mathematics.

The third and fourth schools, Athens and Cyzicus, 420–300, are one except geographically. Of the very first importance for all the future of mathematics was the disposal of some of the sophists' objections by Eudoxus (408–355), a pupil and at one time a friend of Plato, in his theory of proportion. Essentially a theory of the real number system, this Greek work of the fourth century B.C. was not substantially modified till the latter half of the nineteenth century A.D., when critical difficulties in analysis necessitated a thorough reexamination of the concept of real number.

In this period, Plato (429–348) was to assume a mathematical importance greatly in excess of any warranted by his own slight contributions. The general professional opinion is that Plato's too rigid ideal of mathematics[7] as a high philosophic art was to cramp and trammel mathematicians abler than himself. However, discounting the excessive purity of the Platonic ideal, Menaechmus (375?–325?), a pupil of Eudoxus and reputedly a tutor of Alexander the Great, inaugurated the geometry of conic sections. There is a tradition that Plato encouraged Menaechmus. If this is true, Plato made a fundamental contribution to mathematics.

A basic technique was added to mathematical reasoning in this period by Hippocrates[8] (of Chios, 470?–?; not to be confused with the great physician of the same name, of Cos). By exploiting it in his own geometry, Hippocrates demonstrated the power of the indirect method (*reductio ad absurdum*, reduction to an absurdity or deduction of a contradiction from an assumed hypothesis which it is desired to disprove). The universal validity of this method remained unchallenged till the twentieth century, when objections were raised to its indiscriminate use in reasoning about infinite classes.

The fifth school was the First Alexandrian, 300–30, in the city founded by Alexander the Great in 332. This was the culmination of Greek mathematics. With the exception of Diophantus (possibly not a Greek, date conjectured from second to fourth century), the rest is anticlimax. In this great age, Euclid (365?–275?) wove elementary plane and solid synthetic geometry into the close deductive system that was to remain the school standard for over 2,200 years. He also systematized the Greek arithmetica as it existed in his time, and wrote on geometrical optics.

In this age lived Archimedes (287–212), the greatest scientific and mathematical intellect of the ancient world, also, by virtue of the uncramped freedom of his methods, the first modern mathematician. Not till England produced Newton in the seventeenth century, and Germany Gauss in the nineteenth, did the exact sciences show this ancient Greek his peers. With magnificent indifference to the mathematical proprieties of his age, Archimedes used whatever came to mind or hand to advance mathematics. Unlike many of his fellow Greeks, he did not disdain experiment. He founded the mathematical sciences of statics and hydrostatics. He anticipated the integral calculus, also, in the one problem of drawing a tangent to his equiangular spiral, the differential calculus.

In this age also lived the supreme master of the synthetic method in geometry. Apollonius (260?–200?) left but little for his successors in that method to do in the metric geometry of conics.

Astronomy became a mathematical science during this period, in the work of Hipparchus (of Rhodes, second half of second century B.C.). From Hipparchus on through Ptolemy (second century after Christ), Copernicus (fifteenth century), Tycho Brahe (sixteenth century), and Kepler (sixteenth century), astronomy did not deviate from the Hipparchian program of a geometry to describe the motions of the planets. With Newton, this geometry evolved in the seventeenth century into dynamics. Hipparchus also was the first to use a sort of trigonometry systematically, and is said to have produced the equivalent of a rudimentary table of sines.

Geodesy advanced in this period with the measurement, as accurate as the available data and instruments permitted, of a degree of the earth's surface by Eratosthenes. This man is remembered also for his reform of the calendar and for his method of sifting out the primes from the sequence of all the integers.

Finally, this rich period produced one of history's most ingenious scientific engineers, Heron (of Alexandria, second century, perhaps not Greek). The formula

$$[s(s - a)(s - b)(s - c)]^{\frac{1}{2}}$$

for the area of a triangle with sides a,b,c and $2s \equiv a + b + c$, often attributed to Heron, is of course important in trigonometry. But its peculiar historical significance is elsewhere. It marks a carefree departure from the too rigid niceties of orthodox Greek mathematics, a departure which was halted all too soon. No academic Greek geometer would have presumed to 'multiply together four lines,' as in the formula; for the product has no geometrical meaning in Euclid's space of three dimensions. The engineer Heron was not deterred by such obstacles. He discovered or transmitted the right result and, like the Egyptian he may have been, left it for future generations of mathematicians to show that he had not erred in his own proof. If, however, as is now claimed, the formula is due to Archimedes, the mystery of it all vanishes.

The sixth and last school was the Second Alexandrian, 30 B.C.–A.D. 640. The first date marks the absorption of Egypt by Rome, the second[9] the destruction by the Moslems of what little Roman virility, Greek neglect, and early Christian intolerance had left—some say nothing—of the great library at Alexandria. But Greek mathematics had lost most of its creative power long before the library disappeared, and only three men in the six centuries of the Second Alexandrian school would have been noticed as mathematicians by the giants of the First.

Ancient astronomy culminated in the second century A.D. in the eccentrics and epicycles of Ptolemy. For about fourteen centuries, Ptolemy's geocentric description of the solar system was to be accepted as ultimate. Geometry and arithmetic had long since become independent provinces of mathematics when Ptolemy, compelled by exigencies of astronomical computation, all but split off trigonometry as a distinct mathematical science in his geometrical theorems equivalent to the addition formulas for the sine and cosine, and in his computation of a table of chords. The failure of trigonometry to attain its freedom was due to Ptolemy's lack of algebra and the disabilities of logistic.

Almost at the end of the creative period, a belated geometer, Pappus (second half of the third century), either transmitted or himself discovered three prophetic theorems. He proved the focus-directrix property for the ellipse, parabola, and hyperbola, thereby foreshadowing the general equation of the second degree for all conics in analytic geometry. He also proved in effect that the cross ratio (or anharmonic ratio) of four collinear points is a projective invariant, thus isolating a cardinal theorem in the projective geometry of the seventeenth and nineteenth centuries. Finally, he used one of the numerous disguises of the integral calculus to obtain the theorem often ascribed to P. Guldin (A.D. 1577–1643, Swiss) that the volume generated by a plane figure F rotated about a fixed axis is AL, $A \equiv$ the area of F, $L \equiv$ the length of the path traced by the centroid of F. Obviously it is impossible to give an acceptable proof without a full use of the calculus.

At last, in the rudimentary algebra of Diophantus, also in his higher arithmetic, mathematics all but entered a renaissance. But it was getting late, and the Greek spirit was too tired to return to its point of departure and resume the march begun by others some twenty-four centuries before in Babylon. If there is such a thing as the Zeitgeist, it must have permitted itself a sardonic smile as it prepared the sourest jest in the history of mathematics. Not Diophantus, but his historical predecessor of the first century after Christ, the numerologist Nicomachus, was to transmit arithmetica to Christian Europe.

Number from Pythagoras to Diophantus

A few items in the foregoing prospectus overtop the rest in importance for what was to be the future of mathematics. These will now be examined in closer detail.

The Pythagorean brotherhood's conception of mathematics was broad and human. All of their philosophy, of which mathematics was only a subordinate if important part, was directed to but the one end of sane, civilized living. Arithmetic, geometry, astronomy, and music were the four divisions of their mathematics. This tetrad was to survive for centuries, passing through the Middle Ages in the attenuated quadrivium which formed four-sevenths of a liberal education, the rest being the trivium of grammar, rhetoric, and logic. By then, however, the Pythagorean liberality of spirit had been stifled, and living often was neither sane nor civilized in any sense that Pythagoras would have recognized.

Of Pythagoras himself only legends remain. In middle life he migrated from his native Samos to Crotona in southern Italy, where the best work of his brotherhood was done. For the rest, fable makes him an enthusiastic if somewhat pompous mystagogue who had traveled extensively in the East, and who used his mystical lore to impress everybody from blacksmiths to young women. In liberality of mind he was centuries ahead of his time. As but one indication of many that the Pythagoreans sought to enlighten their contemporaries, women were admitted to the master's lectures; and Pythagoras himself seems to have had no use for the very peculiar masculinity of the Athens of Socrates and Plato. It is said that Pythagoras and his immediate disciples perished in the flames kindled by those whom they had striven to deliver from brute ignorance, prejudice, and bigotry. In any event, the Pythagoreans were driven out to seed their wisdom elsewhere.

The evil that accrued from the Pythagorean numerology has been sufficiently noted. But some good also issued from it at a very long last. From nonsensical hypotheses the Pythagoreans deduced that both the sun and the moon shine by light reflected from a Central Fire. Some twenty centuries later, Copernicus (or his officious editor) in his dedicatory epistle to the reigning Pope stated that this wild deduction gave him a hint for his own heliocentric theory of the solar system.

Again, readily discovered relations between the positive

integers of certain categories, such as the odd or even, or the polygonal numbers, might easily delude an imaginative mind into ascribing human and superhuman powers to number. When Pythagoras discovered the ratios $\frac{2}{1}$, $\frac{3}{2}$, $\frac{4}{3}$ for the lengths of plucked strings under the same tension to give the octave, the fifth, and the fourth of a note, the first recorded fact in mathematical physics, it was an understandable extrapolation that "Number rules the Universe," and that the 'essence' of all things is number. In the pardonable enthusiasm of that too-inclusive generalization, the modern theory of the continuum of real numbers originated.

To follow the clue from Pythagoras to the present, we must return to Thales. He too had learned much from the wise men of the East. The story that he predicted a solar eclipse in 585 B.C. appears to be apocryphal.[10] The like may be true for the equally famous and mathematically more important legend[10] that, while in Egypt, Thales estimated the height of the Great Pyramid by an obvious application of similar triangles to the shadow of the pyramid and that cast by his staff when held perpendicular to the ground.

Whether or not the legend records a fact, the Pythagoreans by the fifth century B.C. had reached a critical stage in the development of the number concept. For they proceeded to prove that if a,b,c and a',b',c' are corresponding sides of similar triangles, then $a/b = a'/b'$, $b/c = b'/c'$. (The remarks on history and proof in the Prospectus are particularly relevant here.)

By the fourth century B.C. it was perceived that the Pythagorean proof concealed the subtle assumption that the numbers measuring the sides a,b,c,a',b',c' are rational, that is, each is expressible as the ratio (quotient) of two integers. But these sides had been assumed to be of any finite lengths whatever. Hence it had been assumed that there is a one-one correspondence between the lengths of straight-line segments and the rational numbers. In particular, it had been assumed that the length of the diagonal of a square whose side is a rational number is itself a rational number. If the side is 1 unit, the diagonal is $\sqrt{2}$ units in length. But the Pythagoreans easily proved that $\sqrt{2}$ is not expressible in the form m/n, where m, n are integers.

It would be of great interest to know who[11] first proved the irrationality of $\sqrt{2}$, but probably we never shall. In reply to a question by Socrates, Theaetetus says, "Theodorus was writing out for us something about roots, such as the roots of three or five feet, showing that in linear measurement (that is, comparing the sides of the squares) they are incommensurable by the unit; he selected the numbers which are roots, up to seventeen, but he went no farther; and as there are innumerable roots, the notion occurred to us of attempting to include them all under one name or class." But this does not settle the vexed question of who first proved the irrationality of $\sqrt{2}$; and anyone who wishes may still believe without danger of contradiction by incontrovertible evidence that Pythagoras himself did. All we need keep in mind is that the Pythagoreans by the end of the fifth century B.C. knew that $\sqrt{2}$ is irrational. With great ingenuity they approximated to $\sqrt{2}$ by successive solutions of the equations $2x^2 - y^2 = \pm 1$.

Two ways of proceeding lay open. The choice was between some lengths corresponding to no number, or $\sqrt{2}$ and other positive irrationals being numbers. Choosing the second, the geometers of the fourth century B.C. passed one of the epochal milestones in the history of all thought. "The grand continuum" of analysis, the real number system, was already in view. So also were the paradoxes of the infinite. Unless the new numbers, the positive irrationals, could be incorporated with the positive rationals in a unified domain of 'numbers' or 'magnitudes' so that all should form a self-consistent system under the operations addition, subtraction, multiplication, and division as then understood for the rationals, the newly imagined irrationals would be illusory. Further, operations in the enlarged number system must yield the same results for the rational numbers as before the adjunction of irrationals to the rational number system. The demand for internal consistency in the enlarged domain was automatically imposed, for it had already been agreed that mathematics should not defy strict deduction.

No further extension was made until the seventeenth century, when the negative numbers were fully incorporated (but without mathematical understanding) into the real number system. About 1800 the final step was taken when the imaginaries were adjoined to the completed real number system, and the domain of complex numbers ($a + b\sqrt{-1}$, a, b real) was created. In both of these later extensions, the underlying methodology of generalization and internal consistency had not changed since the fourth century B.C. The Greeks appear to have been guided by subconscious mathematical tact. Explicit formulation and perhaps clear understanding of the methodology of extending the number system came only in the late nineteenth century. We shall return to this presently.

Thus far only half the project of adjoining the irrationals, as the Greeks saw the problem, had been imagined. They thought in spacial imagery and had generalized their conception of geometrical 'magnitude' to include both rational and irrational magnitudes. It did not occur to them immediately that the more difficult half of the project remained undone. They had still to prove that their enlarged system of magnitudes was self-consistent; and they appear either to have overlooked this necessity entirely at first, or to have considered it obviously satisfied. Looking critically at what appeared obvious—one of the almost infallible ways of making a fundamental addition to mathematics—they discovered that it was not all obvious. On close inspection they perceived difficulties that have not been completely resolved even today. As already emphasized, the manner in which Eudoxus surmounted these difficulties marks a major turning point in the long history of mathematics.

It was impossible for the Greeks or anyone else to understand either geometry or the real number system without some theory of continuity in the mathematical sense. This incidentally necessitated a clarification of the limiting processes, such as exhaustion, already described in connection with the Egyptian mensuration of the pyramid. Until such processes were strictly validated it was nonsense to speak of the area of a circle, or of the volume of any solid, or of the length of any line, straight or curved, except only when the numerical measures of such areas, volumes, and lengths were rational numbers. As irrational measures are infinitely more numerous (to the power of the continuum) than the rational, mensuration and the geometrical theory of proportion scarcely existed before Eudoxus.

The necessity for drastic revision was strikingly emphasized by Zeno in four ingenious paradoxes or, as some might say, sophistries. A sophistry in mathematics is a logical argument that some dislike but cannot refute. Zeno's classic four have probably occasioned more inconclusive disputation than any equal amount of disguised mathematics in history.

Zeno's service to mathematics is so outstanding that it would be interesting to know something of the man himself. Very little has survived. By tradition, he was a pugnacious dialectician with a passion for being different from everyone else. In middle age he was "of a noble figure and fair aspect." His paradoxes are evidence enough of an independent mind, and it is told that his uncompromising intellectual honesty finally cost him his life. He had conspired with the political faction which lost, and met his death by torture with heroic fortitude. The first of his paradoxes will suffice here.

Zeno argued that you cannot get to the end of a racecourse, because you must traverse half of any given distance before you traverse the whole, and half of that again before you can traverse it, and so on, *ad infinitum*. Hence there are an infinite number of points in any given line, and "you cannot touch an infinite number one by one in a finite time." Therefore you will never get to the end this side of eternity.

In this and another of the same type (Achilles and the tortoise), Zeno argued against the infinite divisibility of space and time. To show his philosophic impartiality, he devised two equally exasperating paradoxes on the other side: if finite spaces and times contain only a finite number of points and instants, we again deduce consequences contradicted by experience.

Mathematics had freely used the concept of infinite divisibility. Thus Zeno's paradoxes, in addition to affording grounds "for almost all theories of space and time and infinity which have been constructed from his day to our own," showed that geometry and mensuration in the fifth century B.C. needed a new foundation. Eudoxus provided this in his theory of proportion, applicable to any real 'magnitudes.'

Before recalling how Eudoxus met this ancient crisis, we note the nineteenth-century way out of Zeno's difficulties. By a simple application of infinite series it is easily shown that the runner will reach his goal and that Achilles will pass the tortoise. But—a reservation of the first importance—the logic of continuity supporting the modern theory of convergence descended in the twentieth century to a deeper level than any that had been explored when the nineteenth-century analysts imagined they had disposed of Zeno's paradoxes.

Eudoxus based his theory on his definition of 'same ratio:' The ratio P/Q is said to be the same as the ratio X/Y, when, *m and n being any* (positive) *integers whatever*, mX is greater than, equal to, or less than nY according as mP is greater than, equal to, or less than, nQ. If the ratios P/Q, X/Y are the same, P, Q, X, Y are called proportionals. The theory is expounded in

Euclid's *Elements*, Book V. Book VI contains the application to similar figures.

Some modern critics, particularly among the French, have been unable to appreciate the radical distinction between this Greek theory of the real number system and that now current, due to J. W. R. Dedekind (1831–1916, German). Beginning in the 1870's, the criticisms continued well into the twentieth century, especially during the nationalistic fervors generated by the first world war. The later critics appeared to be unaware that Dedekind had disposed of their contentions in 1876. The point of historical interest is that no current theory of real numbers is that which sufficed for the fourth century B.C.

The Eudoxian theory of proportion indirectly validated the empirical rule of the Egyptians for the volume of a truncated pyramid, and completed the work of the Pythagoreans on similar figures. It also certified the method of exhaustion and, after Dedekind (1872), the use of the integral calculus in the determination of lengths, areas, and volumes. In short, it provided a foundation for the real number system of mathematical analysis.

It may be significant that Eudoxus was another of the great Greek mathematicians who is said to have visited the East. At one time a protégé and friend of Plato, he left Plato's Athenian Academy to found his own school at Cyzicus when Plato—so it is said—began to show signs of most unphilosophical envy and jealousy. But this is hardly credible of the man who composed the *Lysis* and the *Symposium*.

The italicized phrase in the definition of 'same ratio' illustrates the fact that finality is as hard to reach in mathematics as it is in philosophy. For not all schools of mathematical thought in the twentieth century have admitted 'any integers whatever' as a legitimate concept in deductive reasoning. The phrase conceals an infinity of trials on *all* the integers m, n to test the inequalities $mX \gtreqless nY$, $mP \gtreqless nQ$. Thus a consisent finitist, if there is one, might say that Eudoxus produced a milder paradox of Zeno's racecourse; for "you cannot test for an infinite number of pairs of integers in a finite time." However, the influential mathematical schools ignore such sophistries, and continue to create new mathematics of great interest and indubitable scientific utility.

Having passed this outstanding landmark in the development of mathematical thought, we shall look forward from it for a moment before proceeding to the next stage in the Greek elaboration of number. Its specific importance in relation to geometry, mensuration, and the real number system has been sufficiently indicated; and anyone with the slightest feeling for great mathematics will admit its greatness without any reservation. Contemplating this masterpiece, mathematicians may be pardoned a little pride that it was their guild which fashioned it. But to leave the Greek masterpiece in sculptured isolation as a monument for all time to the perspicacity of the mathematical intellect would be to give a totally false impression of the manner in which mathematics has developed. Its history is not the record of one brilliant victory after another. Rather is it a somewhat sobering chronicle of intelligence fighting desperately against tremendous odds to overcome the all but ineluctable stupidity of the human mind. That such progress as has been made should have been possible at all is the miracle of the ages.

With the detailed example of the Greek attack on incommensurables (irrationals) set out before them with all the elaborate precision of a copybook for children, mathematicians stumbled about for twenty centuries before imitating the Greek methodology in their struggle to incorporate negative and complex numbers with the positive real numbers in a single self-consistent system. The irrationals first appeared in geometry, the negatives in arithmetic and algebra, and the imaginaries in algebra. Both negatives and imaginaries entered when it was gratuitously assumed that certain rules of operation, known to produce consistent results in special circumstances, would retain their validity in all circumstances of a superficially similar kind. In Greek geometry, the rules in question were those used in proofs concerning similar triangles with rational sides; the tacit assumption was that these same rules would give consistent results for all triangles. In algebra, the negatives and imaginaries entered in an analogous way with the solution of equations; and just as the Pythagoreans were reluctant to grant the status of number to the irrationals, so were the earlier algebraists unwilling to admit negatives and imaginaries as legitimate roots of algebraic equations.

The Greeks recognized that they were confronted by a fundamental problem, isolated it, and solved it. Perhaps the decisive step was their bold hypothesis that a 'magnitude' (number represented geometrically) need not be rational in order to be a 'magnitude.' They generalized the concept of magnitude as it first presented itself to experience and intuition. The algebraists up to the seventeenth century failed to recognize that the negatives and imaginaries presented a problem at all. They either blindly manipulated such things when the rules for solving equations turned them out, or rejected them without attempting to justify the rejection. A contemporary of Eudoxus would have wanted to know why some equations produced only intelligible roots, others only some intelligible, and others none. In their lack of common mathematical curiosity, the algebraists of Islam and the European Renaissance were contemporaries of the ancient Egyptians. They wondered and were perplexed, of course; but there they stopped, because they lacked the Greek instinct for logical completeness and generality.

It was only in the nineteenth and twentieth centuries that these difficulties were satisfactorily met, and then it was by a methodology abstractly identical with that of the Greeks. First, it was ascertained what algebraists were subconsciously striving to do. They were attempting, with no mathematical justification, to include all the reals and all the imaginaries in one system closed under the four rational operations (addition, subtraction, multiplication, division) of common algebra. They actually proceeded on the tacit assumption that such closure was a mathematical fact, namely, that it was self-consistent. This is what they wished it to be, in order to certify their empirical calculations. No progress was made until they followed the Greek lead in geometry and stated explicit postulates for the real and complex numbers, thereby *defining* the 'numbers' presented in algebraic experience. Thus the concept of 'number' was extended to cover all sets closed under the four rational operations. Finally, as will be seen, it was proved in the late nineteenth century that the most general set of this kind, in which $xy = 0$ only if at least one of $x, y = 0$, is that of all complex numbers $a + b\sqrt{-1}$, a, b real; and that the only such sets are this itself and certain of its subsets, for example, the set of all rational numbers.

When we reflect that it took the Greeks less than two centuries to recognize and reach their goal, we may well wonder whether mathematics today is not starting on another of its two-thousand-year quests in search of simplicity. With all its prolific inventiveness, mathematics seems to have lost some of its youthful directness. Nearly always it is the recondite and complicated which is elaborated first; and it is only when some relatively unsophisticated mind attacks a problem that its deep simplicity is revealed.

In their further encounters with number, the Greeks found much that underlies some living mathematics, but nothing, perhaps, of such abiding significance as the work of Eudoxus. Like the similar triangles of Thales which were partly responsible for the Eudoxian theory, one origin of some of the most interesting Greek higher arithmetic was in Egypt.

The Egyptian 'rope stretchers' laid out right angles for the orientation of buildings by means of a triangle of sides 3,4,5. A string of length $3 + 4 + 5$ was marked or knotted at the points 3,4. With this and three pegs a right-angled triangle was obtained in an obvious way. Instead of the particular positive integer solution 3,4,5 of $a^2 = b^2 + c^2$, they might have used any other, provided they knew any. The general positive integer solution a, b, c was given by Euclid in his *Elements* (X, 28, Lemmas). This appears to be the first *proved* complete integer solution of an indeterminate equation. Whether first or not, it is the germ of vast theories in the modern higher arithmetic and, less directly, of the like in algebra. To complete the record, it must be noted that since 1923 it has been customary to deny that the Egyptians ever used 3, 4, 5 to lay out right angles. The argument on which this denial is based appears to run as follows. Because the rope stretchers stretched their ropes for purposes other than laying out right angles, therefore they did not lay out right angles by rope stretching. Further, because right angles were laid out by other means, therefore they were not laid out by— etc. It can be asserted only that the history here may be sounder than its supporting logic.

The solution in integers, or in rational numbers, of indeterminate equations belongs to diophantine analysis. The name honors Diophantus, whose treatise of thirteen books, of which only six survive, was the first on the subject. The Latin translation (A.D. 1621) of this suggestive fragment directly inspired Fermat to his creation of the modern higher arithmetic. It also inspired something much less desirable. Diophantus contented himself with special solutions of his problems; the majority of his numerous successors have done likewise, until diophantine analysis today is choked by a jungle of trivialities bearing no resemblance to cultivated mathematics. It is long past time that the standards of Diophantus be forgotten though he himself be remembered with becoming reverence. For the opinion on this matter of an expert in both the history and the practice of diophantine analysis, those interested may consult L. E. Dickson's *History of the theory of numbers*, vol. 2, 1920.

On another account also this work of Diophantus is memorable. It was the first Greek mathematics, if indeed it was Greek, to show a genuine talent for algebra. Following the Pythagoreans, Euclid had given geometrical equivalents for simple identities of the second degree, such as $a(a + b) = a^2 + ab$, $(a + b)^2 = a^2 + b^2 + 2ab$, and had solved $x^2 + ax = a^2$, a positive, geometrically. Diophantus gave essentially algebraic solutions of special linear equations in two and three unknowns, such as $x + y = 100$, $x - y = 40$. More important, he had begun to use symbols operationally. This long stride forward is all the more remarkable because his algebraic notation, compared to that of today or of the seventeenth century when Descartes practically perfected it, was almost as awkward as Greek logistic. That he accomplished what he did with the available technique places him beyond question among the great algebraists.

His operational advance was profoundly significant. In algebra a formula, say $a + b - c$, directs us to perform certain operations on given numbers (or, in modern algebra, abstract marks), here an addition and a subtraction on a, b, c, in a prescribed order. That is, algebra escapes from verbal instructions to symbolic directions and ceases to be purely rhetorical. Diophantus had even invented a species of minus sign, and permitted a negative number to function in an equation on a parity with positive numbers. He also used symbols for the unknowns and for powers. All this was a long step toward symbolic algebra. It seems probable that some of Diophantus' algebra was of Babylonian origin, although the connection has yet to be traced. Unfortunately for the development of algebra and of mathematics generally, Diophantus was at least four centuries later than Archimedes.

To conclude this account of Greek arithmetica, we may return to its origin in geometry and instance the timelessness of great mathematics by an episode in arithmetic and geometry from the twentieth century.

The Pythagorean theorem that $x^2 + y^2 = z^2$, where x,y,z are the sides of a right triangle, is the basis of metric geometry in Euclidean space. In the spaces defined by Riemann (1854), the quadratic algebraic form $x^2 + y^2$ in two variables, x, y is replaced by a quadratic differential form in n variables; $n = 4$ is the case of interest in relativity. The significance of $x^2 + y^2 = z^2$ in diophantine analysis has been remarked. There, this equation is also generalized, and it is required to solve the general quadratic equation in n unknowns, with integer coefficients, in integers. This arithmetical problem, together with that of reducing the general equations of the second degree in the analytic geometry of conics and quadrics to canonical form, suggested (nineteenth century) the purely algebraic problem of reducing a quadratic form in n variables to a sum of squares each multiplied by an appropriate coefficient. Incidentally, this problem is important in dynamics.

Toward the close of the nineteenth century (A.D. 1882), a notable advance was made in the diophantine problem by Minkowski, then a youth of eighteen. In treating this problem, Minkowski acquired a mastery of the algebraic theory of the reduction of quadratic forms. Becoming interested about the turn of the century in mathematical electromagnetism, he applied his algebraic skill to special differential forms. By the peculiar accident of his interests, he was the ideal candidate to recast the mathematics of Einstein's special relativity of A.D. 1905 into a shape which still retains its attractiveness.

Between the orientation of the Egyptian temples and the welding of space and time into space-time stretch some four or five thousand years of troubled history. In mathematics the two events appear almost contemporaneous.

The postulational method

Had the Greeks done nothing more than put a foundation under the real number system, they would have been assured of perpetual remembrance in mathematics. But they did a great deal more. Indeed, 'Greek mathematics' inevitably suggests synthetic geometry, and it was in the Greeks' elucidation of spacial form that many see their greatest contribution.

The development of geometry from a practically workable empiricism to a strict deductive science was extraordinarily rapid. The earliest *proof* in geometry is traditionally ascribed to Thales, about 600 B.C. He is said to have proved, as one of some half-dozen theorems, that a circle is bisected by any of its diameters. A century and a half later the Pythagoreans had gone about as far in plane geometry as students today in the first half of an American school course. Among other details, they knew the Pythagorean theorem, the properties of parallels, the angle-sum for any triangle and possibly for any convex rectilinear polygon, the principal facts about similar figures; and they had adequate geometrical equivalents for addition, subtraction, multiplication, division, the extraction of square roots, and the Euclidean solution of $x^2 + ax = a^2$. Some of this, naturally, was within the limitations implicitly imposed by their conception of the number system. Scarcely modified, devices of the Pythagorean graphical arithmetic and algebra survive in the techniques of our drafting rooms.

In solid geometry, the Pythagoreans knew at least three of the five regular solids, and possibly all. If they did know all, their faith in Number as the ruler of the Cosmos may have suffered a setback; for the first three solids occur naturally in common minerals that would attract the eye of any geometer, while the dodecahedron and the icosahedron, having fivefold axes, do not occur in nature. (Copper antimony sulfide, or tetrahedrite, and zinc blende crystallize in tetrahedra; galena, rock salt, and fluorite in cubes; magnetite in octahedra. None are rarities.)

But a proof that precisely five regular solids are possible requires a well developed theory of Euclidean space. The proof is ascribed to Theaetetus, about the middle of the fourth century B.C. Euclid in the same century completed the elementary theory of these solids in his Book XIII, as the superb climax of his geometry. To call these solids 'the Platonic bodies,' as some of the Greeks themselves did, not only violates history but also insults mathematics. It is true that Plato describes a familiar construction of the five regular solids from the appropriate regular polygons. But it is also true that he used these solids as pulpits from which to preach Pythagorean numerology

With the completion of Euclid's *Elements*, Greek elementary geometry, exclusive of the conics, attained its rigid perfection. It was wholly synthetic and metric. Its lasting contribution—and Euclid's—to mathematics was not so much the rich store of 465 propositions which it offered as the epoch-making methodology of it all.

For the first time in history masses of isolated discoveries were unified and correlated by a single guiding principle, that of rigorous deduction from explicitly stated assumptions. Some of the Pythagoreans and Eudoxus before Euclid had executed important details of the grand design, but it remained for Euclid to see it all and see it whole. He is therefore the great perfector, if not the sole creator, of what is today called the postulational method, the central nervous system of living mathematics.

It seems strange that Euclid's method should have had to wait till the nineteenth century for the only kind of appreciation that counts for anything in mathematics, application. Synthetic metric geometry of course continued in the postulational tradition. But this, apparently, was mere inertia; for it was decades after the explosive outburst of projective geometry in the nineteenth century before that subject received a sound basis. And it was only in the A.D. 1830's that any serious attempt was made to provide a postulational foundation for elementary algebra. Not until A.D. 1899, in the work of another great geometer, D. Hilbert (1862–1943, German), was the full impact of Euclid's methodology felt in all mathematics.

Concurrently with the pragmatic demonstration of the creative power of the postulational method in arithmetic, geometry, algebra, topology, the theory of point sets, and analysis which distinguished the first four decades of the twentieth century, the method became almost popular in theoretical physics in the A.D. 1930's through the work of P. A. M. Dirac (A.D. 1902–, English). Earlier scientific essays in the method, notably by E. Mach (A.D. 1838–1916, Austrian) in mechanics and A. Einstein (A.D. 1878–) in relativity, had shown that the postulational approach is not only clarifying but creative. Mathematicians and scientists of the conservative persuasion may feel that a science constrained by an explicitly formulated set of assumptions has lost some of its freedom and is almost dead. Experience shows that the only loss is denial of the privilege of making avoidable mistakes in reasoning. As is perhaps but humanly natural, each new encroachment of the postulational method is vigorously resisted by some as an invasion of hallowed tradition. Objection to the method is neither more nor less than objection to mathematics. It may be true that the life-sciences, for example, are still too lush a wilderness for the sowing of a few intelligible postulates here and there, but the attempt has begun, as in the work (1937) of J. H. Woodger. If the Pythagorean dream of a mathematicized science is to be realized, all of the sciences must eventually submit to the discipline that geometry accepted from Euclid.

Flight from intellectual prudery

As Plato (430–349 B.C.) preceded Euclid (365?–275? B.C.) by about twenty years, it is possible but improbable that the geometer was influenced by the philosopher. It may be regrettable, but it appears to be true, that creative mathematicians pay little attention to philosophers whose mathematical education has not gone much beyond the elementary vocabulary.

Of all changes that mathematical thought has suffered in the past 2,300 years, the profoundest is the twentieth-century conviction, apparently final, that Plato's conception of mathematics was and is fantastic nonsense of no possible value to anyone, philosopher, mathematician, or mere human being.

Not all, however, are iconoclasts of Platonic realism. Some, whose mathematical achievements entitle them to an opinion on the matter, have expressed themselves quite forcibly on the enduring validity of realistic mathematics. G. H. Hardy (1877–, English), for example, stated (1940) his belief that "mathematical reality lies outside us, that our function is to discover or *observe* it, and that the theorems which we prove, and which we describe grandiloquently as our 'creations,' are simply the notes of our observations." It will be recalled that similar beliefs regarding other intangibles caused some rather unpleasant mischief in the Middle Ages and the Renaissance.

Plato himself may not be responsible for the more outrageous absurdities concerning mathematics in his dialogues; there is always the half-mythical figure of Pythagoras in the background. But it was the high poetic quality of the dialogues that preserved the ancient nonsense for later generations of mathematicians and philosophers to admire and imitate. This worked great mischief in geometry. In Platonic realism, the straight lines and circles of mundane geometry are unimportant; it is the Eternal Idea of a straight line or of a circle that alone is worthy of philosophic contemplation. Thus in this particular philosophy the useful abstractness of mathematics is vaporized into a nothing of ethereal beauty that has yet to make its first contribution to geometry.

To a Platonic geometer it is self-evident that the Archetypal Circle is more rotundly round than any other curve in the Eternal Mind, also that no Idea is straighter than the Ideal Straight Line in the same everlasting locus. Hence it follows that terrestrial geometry should be restricted in all its constructions to a straightedge and a pair of compasses. If, for example, an angle is to be trisected, it must be done with these implements. It follows also that in comparison with the geometry of straight lines and circles, that of ellipses, parabolas, and hyperbolas is slightly disreputable, or at least not ideally immaculate. A geometry using any mechanical contrivances other than the sacrosanct two was severely reprimanded for "thus turning its back on the ideal objects of pure intelligence."

It is not surprising that Plato disdained applied mathematics. In his philosophy of mathematics, Plato was the finished intellectual aristocrat, purer than the purest of pure mathematicians. Fortunately for both pure and applied mathematics, such an excess of purity, not to say prudery, did not appeal to the real aristocrat Archimedes.

Archimedes by himself was an epoch in the development of mathematics. There is immortality enough for a dozen in his great discoveries, and these are overshadowed by the methods which he invented or perfected and which, unfortunately, perished with him. Centuries were to pass before science and mathematics overtook him.

The legend of his life is familiar from Plutarch's incidental account. He was a close friend and perhaps a kinsman of Hiero, tyrant of Syracuse, where he was born and where he died. During the siege of Syracuse by Marcellus in the second Punic war, the mechanical armaments of Archimedes delayed and all but defeated the Romans. When the city fell (212 B.C.), the defenseless old mathematician was killed by a Roman soldier.

Rome won the war, finally destroyed Carthage (*delenda est Carthago!*), and marched on to almost unimaginable heights of splendor, but not in science or mathematics. As bluntly practical as the soldier who dispatched Archimedes, the Romans were the first wholehearted exponents of virile living and bucolic thinking, and the first important people to realize that a modicum of brains can be purchased by those who have only money or power. When they needed any science or mathematics not already reduced to easy rule of thumb, the Romans enslaved a Greek. But they blundered when they killed Archimedes. He was only seventy-five and still in full possession of his powers. In the five years or more of which the soldier robbed him, his truly practical mind might have taught the Romans something to ward off the fatty degeneration of the intellect which finally rendered them innocuous.

All the work of Archimedes is characterized by rigor, imagination, and power. He may rightfully be called the second mathematical physicist in history, and one of the greatest. Pythagoras was the first. In this capacity Archimedes is almost unique, in that he used his physics to advance mathematics. The usual procedure, in which he also excelled, is the reverse. A sample of his great work may suffice to suggest the magnitude of the whole.

In applying the method of exhaustion to the mensuration (of both surfaces and volumes) of the sphere, cylinder, cone, spherical segments, spheroids, and hyperboloids and paraboloids of revolution, Archimedes proved himself the complete master of mathematical rigor and the perfect artist. Some of this involved (in modern notation) the evaluation of the definite integrals $\int_0^\pi \sin x\, dx$, $\int_0^c (ax + x^2)dx$. His problem of cutting a sphere by a plane so that the segments shall be in a given ratio presented him with a cubic equation of the type $x^3 + ab^2 = bx^2$, which he may have solved geometrically by the intersection of conics. The catholicity of his interests is shown by his famous 'cattle problem,' which demands incidentally the solution in integers x, y of $x^2 - 4{,}729{,}494y^2 = 1$. Finally, in pure mathematics, Archimedes anticipated the method of the differential calculus in his construction of a tangent to the spiral ($\rho = a\theta$) known by his name.

His most original work perhaps was in his applied mathematics. Here, so far as is known, he was a pioneer. Menaechmus and others had successfully applied the method of exhaustion to difficult problems (Archimedes himself mentions Eudoxus and attributes to Democritus the statement of the result for the volume of a pyramid); but none had applied mechanics to mathematics. Before Archimedes, no scientific mechanics existed. There may have been empirical rules, but such are in a different universe. His discovery of the law of buoyancy practically created the science of hydrostatics, and his formulation of the theory of the lever did the same for statics. So powerful were his methods that he determined the positions of equilibrium and stability of a floating paraboloid of revolution in various positions. True to the Greek tradition, Archimedes based his mechanics on postulates. His determinations of centroids were about as difficult as those in a course in the calculus today. For example, he found the centroid of a semicircle, a hemisphere, a segment of a sphere, and a right segment of a paraboloid of revolution. It is small wonder that the Moslems held Archimedes in almost superstitious veneration. There was not his like for two thousand years.

Archimedes' sublime disregard of convention is seen in what is his most curious work. It is the problem, which he solved, of finding the area of a parabolic segment. The proof, of course, is rigorous. It amounts to an integration, somewhat disguised as exhaustion in the official proof. It is the unofficial proof which is of greater interest. This came to light in 1906, when a work by Archimedes describing his heuristic method was found in Constantinople. To *discover* what the required area was, Archimedes translated the problem in *geometry* into an equivalent in *mechanics*. Having solved the latter, he states that the result has not been "actually proved." He then proceeds to give a geometrical proof in which, incidentally, he performs the first summation of an infinite series in history. The series is $\sum_{0}^{\infty} 4^{-n}$, and he uses the fact that 4^{-n} tends to zero as n tends to infinity. He had already summed a finite series, $\sum_{s=1}^{n} as^2$.

An isolated gem may show that Archimedes was as perspicacious as he was inventive. To the untutored mind it is obvious that by laying off a given segment, no matter how small, a finite number of times, any point on a line may be reached or passed. It was obvious to Archimedes only that this is an *assumption* which should be stated explicitly as one of the postulates of geometry. He did so; and non-Archimedean geometries, in which the postulate is rejected, were constructed in the nineteenth and twentieth centuries. Like Euclid in his explicit statement of the parallel postulate, Archimedes had the true mathematician's caution in the presence of the obvious.

Modern mathematics was born with Archimedes and died with him for all of two thousand years. It came to life again in Descartes and Newton.

Through geometry to metaphysics

The negative obligation of mathematics to ancient philosophy has been indicated. As will appear when we discuss medieval Europe, it is possible that the obligation was reversed in that mathematical desert. For the present, it will be of interest to note how Greek mathematics was indirectly responsible for some profoundly interesting work in epistemology since about A.D. 1920.

The Greek geometers left undecided four elementary problems that were to defy mathematical ingenuity for over two thousand years. None of the four is of mathematical importance today. Historically, no more prolific problems were ever proposed, with the possible exception of Zeno's. Repeated failures to settle the first three disclosed fundamental difficulties unsuspected by the ancients, and necessitated a sharpening of the number concept. Unsuccessful attempts for about 2,300 years to dispose of the fourth at last suggested a great advance in mathematical methodology which now seems trivially obvious, but which eluded some of the keenest minds in history.

The problems are as follows. In each of the first three, with due deference to Plato, the desired construction is to be performed wholly by means of a finite number of straight lines and circles.

PROBLEM 1. To trisect any angle.

PROBLEM 2. To construct the side of a cube whose volume shall be twice that of a given cube.

PROBLEM 3. To construct a square equal in area to any given circle.

PROBLEM 4. To deduce Euclid's fifth postulate from the others.

The fifth postulate is formally equivalent to the following. Through any point P not on the straight line L there can be drawn, in the plane determined by P and L, precisely one straight line which does not meet L.

Problem 2 is equivalent to demanding a geometrical construction, by the means prescribed, for the real root of $x^3 - 2 = 0$; Problem 1 is similar. These two were not settled till P. L. Wantzel (A.D. 1814–1848, French) in 1837 obtained necessary and sufficient conditions for the solution of an algebraic equation with rational coefficients to be geometrically constructible in the manner specified. Neither of the cubics concerned satisfies the conditions. Thus the problems were proved to be impossible.

If the restriction that the only permitted means are a finite number of straight lines and circles be removed, solutions of Problems 1, 2 are readily obtained, for example by conics, as done by the Greeks, or by linkages. The historical importance of these two is the impetus they gave, long after Greece, to the investigation of the arithmetical nature of the roots of algebraic equations with integer coefficients. Such roots are called algebraic numbers; a number which is not algebraic is said to be transcendental.

The third problem tapped a deeper spring. By Wantzel's theorem, if Problem 3 is solvable, its algebraic equivalent must be a finite number of equations satisfying his conditions. The problem will be impossible if $\pi(= 3.14 \ldots)$ is transcendental. In A.D. 1882, C. L. F. Lindemann (1852–1939, German) proved that π is transcendental. His proof, with its curious dependence on rational arithmetica, would have delighted Pythagoras.

Problem 3, as 1, 2, is solvable when modified to permit the use of curves other than circles. The quadratrix (ρ, θ polar equation $\pi\rho = 2r\theta \csc \theta$) invented by Hippias in the fourth century B.C. for the trisection problem suffices. This, however, is of but trivial interest; the significance of circle squaring is its connection with transcendental numbers.

Squaring the circle implies an irrationality of a kind radically distinct from that which taught the Pythagoreans that not all numbers are rational; $\sqrt{2}$ is algebraic, π is not. It would seem a reasonable guess to one exploring the number system for the first time that all real numbers are algebraic, or at least that the transcendentals are extremely rare. Cantor proved in A.D. 1872 that the algebraic numbers are the rare exceptions; the transcendentals are infinitely (to the power of the continuum) more numerous. It is an interesting exercise to trace the implications of the restriction to a *finite* number of straight lines and circles in the conditions of the three problems.

Problem 4—to prove Euclid's parallel postulate—will reappear when we follow geometry through the nineteenth century. It is one of Euclid's greater achievements to have perceived that this postulate demands explicit statement as an assumption.

The new quirk in methodology which finally disposed of the problem in A.D. 1826 may be described here, as it is one of those profoundly simple, powerful devices which are so obvious after they have once been pointed out that they are first imagined only by minds of the highest originality. A problem which has resisted the best efforts of genius for centuries *may* be impossible, or meaningless, or improperly posed. The quirk is simply to admit that one of these three may be the fact. This admitted, what seems the likeliest of the three is developed mathematically.

The parallel postulate was circumvented by the third possibility: a self-consistent geometry was constructed without it. Problems 1, 2 evaporated when the suspected impossibility was pursued, and a contradiction was deduced from assumed possibility. The problem of squaring the circle suffered the same fate, but was much harder to dissipate.

A brilliant application of the quirk to a modern problem was Abel's proof in A.D. 1824 that the general algebraic equation of degree higher than the fourth is unsolvable by radicals. He appears to have been the first to state the methodology explicitly as a general procedure. As an item of historical interest, the Persian poet, mathematician, and connoisseur of wine, women, and song, Omar Khayyam, is said to have conjectured in the twelfth century that the algebraic solution of the general cubic is impossible. He was mistaken. We shall recur to this in a later chapter.

Beginning about A.D. 1930, the Viennese school of mathematical logicians attacked some of the classical problems of philosophy, particularly metaphysics, by this methodology, attempting to show that the problems were either meaningless or improperly posed. Of course history may prove them as mistaken as Omar was. Needless to say, the attack was vigorously resisted, especially by those who refused to master enough elementary symbolism to enable them to read a proof in symbolic logic.

Thus four elementary problems of Greek geometry were partly responsible for a subversive movement in philosophy that would have shocked the ancient Greek philosophers as profoundly as it shocked some of the moderns. There was at least a prospect in A.D. 1945 that the incipient revolution—if it may be called that with propriety—might necessitate some revision of accepted epistemology. The non-Euclidean geometry of the nineteenth century that issued from Problem 4 abolished Kant's theory of mathematical 'truths.'

Plane, solid, and linear loci

Toward the end of Greek mathematics, a hesitant step toward unity and generality was taken by Pappus (probably third century) in his Μαθηματικῶν συναγωγῶν βιβλία. This collection of eight books, of which only the last six and a mere fragment of the second are extant, was a compendium of much of the mathematical knowledge of its time. The missing parts may have dealt with arithmetic; the six known books include proportion, parts of solid geometry, selected higher plane curves, isoperimetric problems, spherics, centroids, special curves of double curvature and their orthogonal projections, and finally mechanics, which seems to have signified yet more geometry to the ingenious compiler. Those items of the collection that may have been due to Pappus himself have been called brilliant by competent critics; certainly they display a boldness of conception and an uncramped freedom of method reminiscent of Archimedes rather than of Euclid. If a kinematic generation of a curve seemed natural to Pappus, he did not hesitate to avail himself of it. Probably much earlier than his day it had been suspected that the three classic problems of Greek geometry are unsolvable by Euclidean methods, although no Greek mathematician is known to have stated the impossibility of Euclidean solutions as a working hypothesis. Accepting the suspected fact, Pappus proceeded to a masterly investigation of the higher plane curves which about five centuries of experience had shown to be sufficient for solving the problems. The spiral of Archimedes, the conchoid of Nicomedes (second century B.C.), the cissoid of Diocles (same century), and the quadratrix of Hippias (fifth century B.C.) were accorded full geometric status. These outlaws of the rigid classical geometry were shown to be as worthy of serious attention as the hackneyed conics. The conchoid had been invented to solve the trisection problem, the quadratrix for the rectification and quadrature of the circle, and the cissoid for the classic Greek problem of inserting two geometric means between two given 'magnitudes' represented as straight-line segments. The conchoid and cissoid are algebraic curves; the quadratrix is transcendental. Nevertheless, all three are thrown into the vaguely inclusive class of 'linear' loci. This ill-defined receptacle held all loci other than the plane and solid.

Circles and straight lines were 'plane' loci; the conics, 'solid' loci, doubtless so named on account of their origin as sections of cones (of the second degree). In passing, one of the decisive achievements of Apollonius was his replacement of the three species of cone, used by his predecessors to obtain the various conics, by the right circular cone of which all are sections. If derivation from conical surfaces was the ground for calling conics solid loci, it seems rather peculiar that Pappus should have cast the quadratrix into the nebulous limbo of linear loci. For two of his most striking personal contributions were his definitions of the quadratrix as orthogonal projections of certain skew curves. In one, the curve is the intersection of a cone of revolution and a right cylinder whose base is a spiral of Archimedes. Here was synthetic geometry in the grand manner almost of Archimedes himself.

Though the classification of loci as plane, solid, and linear may not seem very significant to a modern geometer, nevertheless it was a conscious attempt to put some system and order into the chaos of imaginable plane curves. Without algebraic symbolism, little either reasonable or useful was possible, and almost nothing general. If Apollonius be granted the use of coordinates claimed for him by some of his admirers, this in itself but emphasizes the inadequacies of the clumsy substitutes for a genuine symbolism used by the Greek geometers. That they accomplished so much that has retained its interest after twenty centuries or more is a tribute to their genius rather than a recommendation of their technique. But lest we overvalue our own acquisitions at the expense of theirs, we may remember that there is as yet no satisfactory classification of the uncountable infinity of transcendental plane curves. If such a classification is not a dead problem, it is a project for the analysis rather than the geometry of the future.

A wrong turning?

Greek mathematics stands as one of the half dozen or so supreme intellectual achievements of our race. Its best is now well over two thousand years behind us. Looking back on it in the light of the mathematics that has developed since the early decades of the seventeenth century, we shall try to see it dispassionately in "the cold beams of the history of learning." Its two greatest achievements—those traditionally attributed to the Pythagoreans—shine out as clearly as ever, and with them, Euclid's. Apart from Archimedes, two thousand years ahead of his age, what of the rest?

For better or worse, our technical deveopment of science and mathematics differs radically from that of the Greeks. Their mathematics is intelligible to us, and any modern can appreciate it at what they themselves considered its true value. Ours, beyond the mere rudiments, would appear to them—with the exception of Archimedes—as conclusive evidence of insanity. A straight line to them, for example, meant a finite segment capable of prolongation; to us a straight line is defined once for all from minus infinity to plus infinity.

The limited modes of Greek thought are not ours. With the maturing of elementary algebra in the sixteenth and seventeenth centuries of our era, and the introduction of analytic methods in the seventeenth, mathematics in its return to number drew closer to Babylon, Egypt, and India than it ever did to Greece after the fall of Alexandria. Except for insistence on proof, our preferences in mathematics as in religion are more oriental than Greek.

It may be a hard saying, but it appears to be none the less true, that on the long view Greek geometry was in part a tactical blunder. No necessity compelled Thales, Pythagoras, Euclid, Apollonius, and all their disciples to develop the synthetic method exclusively. At the beginning of their arduous mathematical journey, two possible roads had been plainly indicated to the Greeks by their Eastern predecessors. Either by conscious predilection or ironic mischance they all took the same turning and hewed their way through tremendous obstacles to the end of a blind alley. The synthetic geometry of conics marks the end of the journey.

Further significant progress—that which mathematics follows today—became possible only when the harsh way was retraced or temporarily forgotten, and the road which the Greeks had passed by in the sixth century B.C. was entered in the seventeenth of our era. Returning to the thought of the East, from which Thales and Pythagoras had started, European mathematics detoured almost completely round the territory consolidated by the Greek geometers. Resuming a march interrupted twenty-three centuries earlier, mathematics during and after the seventeenth century proceeded with incredible speed to the conquest of world after new world beyond the farthest reach of Greek thought.

The geometry of Ptolemy's *Almagest* appears to us as an all but superhuman effort of mathematical genius. It and the 387 propositions in the conics of Apollonius are the masterpieces of the synthetic method. But the new science inaugurated by Galileo and Newton in the seventeenth century needed more than one or two masters of mathematics every four or five hundred years if it was to exploit its opportunities with reasonable speed.

Few mathematicians who have followed the Grecian proofs in Newton's *Principia* believe that all the propositions demonstrated could ever have been discovered in one lifetime by the methods of Greek geometry. There are limits even to the mind of a Newton; and we have his own word for it that he used analytic methods—his calculus—for discovery. The rigid synthetic proofs were devised partly to reassure himself but principally so that he might be understood by others. Still, it *might* be possible to claim the *Principia* as a monument to synthetic geometry. But not the most generous imagination would concede the dynamics of Lagrange, Hamilton, Jacobi, and Lie to a hypothetical application of Greek geometry, although these creations of the eighteenth and nineteenth centuries evolved with apparent inevitability from the dynamics of the *Principia*. And last, in reference to his method which Newton saw fit to translate into Greek, discovery after all is more important in science than strict deductive proof. Without discovery there is nothing for deduction to attack and reduce to order.

Returning for a moment to the sixth century B.C., we may try to imagine what mathematical history might have been if the Greeks had taken the Babylonian highway. Like other 'might have beens,' this one is futile, except possibly as it may indicate which of several roads we ourselves might the more profitably explore.

The orientals had a more catholic taste than the Greeks for number. At least some of the orientals were not terrified by mere magnitude; Indian mythology with its millions of deities, its "tangled trinities," and its aeons of aeons is an adumbration of the mathematical infinite. For that matter, the Egyptian trinity exhibits some of the seeming contradictions of the modern concept of the infinite, with its one-one correspondence between part and whole; and the like is evident in Christian theology, the heir, not of dead Greek mythology, but of oriental religions. A minor but significant indication of the ineptitude of the Greek mind for mathematical analysis is the fact that for centuries it remained content with a system of numeration which, compared with the best of the oriental work, was puerile.

It is not definitely known that the Early Greeks were acquainted with the advances and speculations in number of other peoples; but from internal evidence it is highly probable that they must have heard of them. There are too many oriental inclusions in Greek mathematics to make the miracle of a curiously partial transmission of knowledge credible. For the sake of our hypothesis we shall assume that the Greeks were not entirely ignorant of what their neighbors to the East had done.

Had the early Greek mind been sympathetic to the algebra and arithmetic of the Babylonians, it would have found plenty to exercise its logical acumen, and might easily have produced a masterpiece of the deductive reasoning it worshipped logically sounder than Euclid's greatly overrated *Elements*. The hypotheses of elementary algebra are fewer and simpler than those of synthetic geometry. The algebraic-analytic method in mensuration and geometry was well within the capacity of the Greek mathematicians, and they could have developed it with any degree of logical rigor they desired. Had they done so, Apollonius would have been Descartes, and Archimedes Newton.

As it was, the very perfection—for its age, and for long after

—of Greek geometry retarded progress for centuries. It was admired, as it merited, by the Moslems who finally restored it in the Middle Ages to a forgetful Europe; and much of the genius that might have gone into expanding their own arithmetic, algebra, and trigonometry was lavished on translation and commentary. If there is any truth in Bergson's élan vital, or in Hegel's philosophy of history, Greek geometry was a splendid disaster for both.

This, needless to say, is not the traditional conclusion. The superiority of purely synthetic methods over the algebraic and analytic, as being more intuitive, has been urged by numerous distinguished mathematicians, particularly of the British school, since the time of Newton. And we find the same contention being put forward in the seventeenth and nineteenth centuries for the superiority of the synthetic method in projective geometry over the analytic. No working mathematician would deny the utility and suggestiveness of diagrams; but that is not the point at issue. It has been said that in geometry the synthetic and the analytic methods are like a pair of hands; and this undoubtedly is true, provided the geometry handled is simple enough. But inspection of a treatise on modern physics, or of one on partial correlation in statistics, even when the analysis may be described in the language of geometry, reveals few synthetic proofs, if any. The like is true of the greater part of living geometry itself. And Lagrange, the great master of dynamics after Newton, prided himself that his analytic mechanics contained not a single diagram.

One of the most vigorous defenses of the geometrical method of Euclid, Apollonius, and Ptolemy is that of Thomas Young (A.D. 1773–1829, English), the universal genius who is remembered for his contributions to medicine, Egyptology, elasticity, and the wave theory of light. His arguments are frequently repeated even today, especially in the intermediate instruction of science students. By a singular historical irony, Young's defense was first printed in A.D. 1800, the year which marked the end of the middle period of mathematics and the beginning of the recent. It was republished, together with a slashing attack on the analytic mechanics of Lagrange, in A.D. 1855, the year in which Gauss, the inaugurator of the recent period, died. But if Young after all was right, his was a voice crying in the wilderness that few appear to have heard. For better or worse, mathematics in the seventeenth century committed itself to analysis, and the Greek methods became of only historical interest.

CHAPTER 4

The European Depression

It is customary in mathematical history to date the beginning of the sterile period from the onset of the Dark Ages in Christian Europe. But mathematical decadence had begun much earlier, in one of the greatest material civilizations the world has known, in the Roman Empire at the height of its splendor. Mathematically, the Roman mind was crass.

Beyond the cumbersome Roman numerals, which can be called a mathematical creation only by undiscriminating charity, the Romans created nothing even faintly resembling mathematics. They took what little they needed for war, surveying, and brute-force engineering from the Greeks they had crushed by weight of arms, and were content. When Julius Caesar reformed the calendar in 46 B.C., it was no Roman who proposed leap year with its extra day in February, but the Alexandrian Sosigenes. The Roman contribution to civilization was in law, government, and peace at the sword's point.

The military *Pax Romana* began collapsing in earnest in A.D. 410, when the invaders penetrated the city of the Caesars, and the last garrison was recalled from Britain to stand with the defenders against an onrushing flood of barbarians. The debacle of Roman grandeur came about sixty years later, and five centuries of darkness descended on Christian Europe.

Five years after the recall of the Roman garrisons, a riot in the last capital of Greek learning foreshadowed the centuries of confusion, and marked the end of the first great epoch in creative mathematics.

One of the last of the Greek mathematicians was a woman, Hypatia. Like her male colleagues at Alexandria, Hypatia was a critic and commentator rather than a creator. Her death symbolizes the end of pagan science and mathematics, and the beginning of an age of faith. In 415, when Hypatia died, there were good works more urgent than geometry and arithmetica to be done. The hordes from the north were in need of civilizing and conversion to a gentler religion.

To the zealous tillers of this all but virgin field it seemed obvious that the decaying remnants of an effete Greek culture must first be cleared out of the way. Had not Greek intellectualism and immorality sapped the virility of Rome? Therefore Greek thought must be swept back into the past. As a representative of the older enlightenment, Hypatia was a conspicuous obstacle in the path of the new. Encouraged by their uncompromising bishop, the willing Christians of Alexandria effectively removed the obstacle by inducing her to enter a church, where they murdered her in a needlessly barbarous manner.[1]

Mathematics lived on, just breathing, in Christian Europe. The next significant epoch was inaugurated in the eighth century by the infidel followers of the prophet Mahomet.

European mathematics from Boethius to Aquinas

Before passing on to the one thing of any suggestiveness for the development of mathematical thought that may have had a root in the sterile centuries, we must propitiate tradition by doing honor to the learned Europeans of that period whose names adorn the classical histories of mathematics. From a long list of historical celebrities we select the following as a fair sample, with their names, their dates, *all* A.D. *henceforth*, and the places where they flourished: Boethius (c. 475–524, Rome, Italy); Isodorus (c. 570–636, Seville); the Venerable Bede (c. 673–735, England); Alcuin (735–804, born at York, labored in France); Gerbert (950–1003, Rome); Psellus (1020–1100, Greece, Constantinople); Adelard (early eleventh century, England); Robert of Chester (early twelfth century, England, Spain). This list may be very considerably lengthened without adding any undue burden of mathematics.

No census of the leading European mathematicians of the Middle Ages would be complete without the memorable name of Thomas Aquinas (1226–1274, Naples, Paris, Rome, Pisa, Bologna). Although this Newton of scholastic theology is not usually counted among the elite of medieval mathematics, we shall see that he might be.

In contemplating the barren record from Bede to Aquinas, it is well to remember that while European civilization rotted, another culture, the Moslem,[2] was conserving the Greek classics and developing the algebra and arithmetic of India in preparation for the European Renaissance. We are immediately concerned only with the contributions of Christian scholars.

Of the European background, it is sufficient to recall the persevering struggle of the church to dominate the people and mildly educate a few of them, and the dawning enlightenment that accompanied the crusades of the eleventh to the thirteenth centuries. The crusades no doubt accelerated throughout a wakening Europe the diffusion of knowledge that began with the Moslem conquest of Spain in 711. These influences are reflected in a gradual change, beginning in the twelfth century, in the character of European mathematics.

Medievalists disagree on which century, the twelfth or the thirteenth, was of greater significance in the awakening of Europe. The distinction, if any, is unimportant for mathematics. There was one item of any moment, and only one. Latin versions of Greek mathematical classics, made for the most part from translations by the Moslems into Arabic or Persian, became available to European scholars. While remembering with gratitude the devoted labors of the translators, we need not forget that translation is not creation. The best of the translations added nothing new to mathematics; the worst, by men who might be erudite scholars but who were wretched mathematicians, added only misunderstanding.[3]

To see how low mathematics sank, and to guess how low it may sink again if the enthusiasts for all things medieval prevail, we resume our sample from Boethius to Aquinas, and note what some of these giants did. In their own semicivilized times, several of the men cited were conspicuous conservers of such civilization as there was. Some helped to found elementary schools, others taught, while the more thoughtful wrote shabby textbooks and zealously cultivated theological numerology. Before the great depression got well under way, Boethius described the consolations of philosophy in a homily which was to solace many in dire need of solace in the Middle Ages. Gerbert, one of the more enlightened popes—unjustly accused at one time of collaboration with the Devil[4]—donned the tiara in 999 and steered the church safely through that ominous year 1000 whose widely heralded Satanic disasters unaccountably failed to materialize. In deference to scholarship, it must be recorded in passing that one school of medievalists proves conclusively that no disasters were ever prophesied, while an equally positive school proves conclusively that they were. Whatever the facts, Gerbert wrote on division and on computation by the abacus, collected trifles on polygonal numbers, compiled an alleged geometry from Boethius and another still less enlightened source, and is said to have had a part in popularizing the Hindu numerals. He is also reputed to have been a man of vast learning and acute intellect. Some of his letters reveal him as singularly dense in the most elementary arithmetic. If Gerbert's contributions to mathematics are passed over in silence, it is for the sufficient reason that he made none, despite the fact that no history of mathematics is complete without his illustrious name. The like applies to Bede and Alcuin,[5] justly reckoned among the pedagogical heroes of the Middle Ages. Climaxing this phase, Psellus seems to be remembered chiefly because it is doubtful whether he ever did anything at all in mathematics. His introductions to Nicomachus and Euclid, perhaps fortunately for his reputation as a mathematician, are of uncertain authenticity. But his version of the quadrivium persisted through the fifteenth century.

With Adelard and Robert of Chester, we advance to the next stage. An indefatigable traveler and painstaking scholar, Adelard was an intelligent collector and translator of mathematical classics. The path of the bibliophile in the eleventh century was less smooth than it is now, and Adelard frequently risked his skin to secure his coveted manuscripts. He is credited with one of the first European translations of Euclid into Latin and with a translation of Al-Khowarizmi's astronomical tables. Adelard's one putative original contribution to mathematics was an utterly trivial problem in elementary geometry. Robert translated Al-Khowarizmi's algebra.

All of Adelard's predecessors together managed to keep some semblance of life in the rudimentary mathematics of Christian Europe. Beyond that, the best of these worthy men made only clumsy calculations in the simplest arithmetic, or attempted to approach elementary geometry in a spirit that would have disgraced a Greek schoolboy of fourteen. The mathematical awakening of Europe was due to no effort of theirs, and their illustrious names might be dropped from the history of mathematics without loss. But tradition, rightly or wrongly, forbids. We therefore continue our descent to the nadir of mathematics, and follow the learned Boethius into the abyss.

It was the elementary schoolbooks of Boethius that set the mathematical pace of the Middle Ages in Europe. Returning to the Pythagorean synthesis, Boethius expounded a denatured quadrivium of arithmetic, music, geometry, and astronomy. Considered apart from their medieval content, the names of these four divisions of the Pythagorean tetrad are impressive. But could Pythagoras have looked behind the names, he might have been somewhat disappointed. The geometry, for example, made a brave show by starting from Euclid. But it did not get very far. Only the enunciations of the propositions in Book I and a few in III, IV were offered to the eager students. At the lowest ebb of mathematical intelligence, the liberally educated graduated from geometry when they had learned by rote the enunciations of the first five propositions in Book I of the *Elements*. Later, when more of Euclid became available, ambitious would-be clerics were encouraged to memorize the proofs of these propositions. The fifth, appropriately enough, was nicknamed the Asses' Bridge (*pons asinorum*). Few attempted the hazardous passage over the equal angles at the base of an isosceles triangle.

In his arithmetic, Boethius followed the Alexandrian Nicomachus. As we have seen, Nicomachus in his turn had followed Pythagoras at his most mystical, producing a shoddy treatise on the elementary properties of numbers that might have been composed by an amiable philosopher with a passion for numerology. In passing, mathematicians rather deprecate such effusions being called "the theory of numbers" in some of the traditional accounts.[7] To confuse astrology with astronomy would be less wide of the mark. However, Boethius reproduced the sieve of Eratosthenes and offered some amusing trifles on figurate numbers. Proof seems to have had no greater attraction for him than for his master Nicomachus. Boethius also is credited by some controversialists with a problematical introduction of the Hindu numerals to supplement the abacus and the counting board of trade. The practical outcome of all this was a cumbersome reckoning sufficient for simple transactions involving money, and for keeping the calendar in order so that the date of Easter might not elude annual recapture. To call any of this computation—or of the debased geometry—mathematics is a gross exaggeration. The significance of mathematics as a deductive system had been forgotten. Science having sunk to the level of superstition, the other half of the Pythagorean vision survived only in the fantastic absurdities of sacred and profane numerology. Number indeed ruled the darkened universe of the European Middle Ages.

The erudite Boethius however made another contribution to learning that may have had more influence on the development of mathematics than all the editions of his sorry arithmetic and geometry ever had. He made a part of Aristotle's system of logic available to European scholars in a Latin translation.

What follows is only a speculation. But even so, it is less depressing and possibly less futile than the dreary record of the homunculi mathematici whose saintly lives and lack of works constitute the official history of mathematics in Christian Europe of the Middle Ages.

Submathematical analysis

To the scientific or mathematical mind up to the twentieth century, the logical disputes which absorbed a major part of the mental energy of the Middle Ages had long seemed the acme of futility. But in the two decades following the world war of 1914–18, all things medieval became more popular than at any time since the rise of modern science. For sufficient reasons, many had lost their illusion of progress. Blaming science, with an occasional diatribe against mathematics, disillusioned idealists groped blindly back to the twelfth century, or even the ninth, seeking an authoritative assurance of a security more satisfying than any science. Should followers of these hopeless travelers to the past equip themselves properly, they may restore to the future a treasure-trove for the history of mathematical thought comparable to that which has been recovered from Babylon. They may discover where European mathematics went underground after the death of Hypatia, and what shape it assumed during its long burial.

Two things in particular may be sought with some prospect of reward: the struggle of the Greek philosophical concept of the infinite, attributed to Anaximander in the sixth century B.C., to get itself transmuted into the modern mathematical infinite; the closely cognate struggle of mathematical analysis to get itself born.

The mathematical mind was not dead in the Middle Ages. It was merely sleeping. In its uneasy rest it imagined something curiously like mathematics; but it was powerless to throw off its dreams and wake. The theological subtleties and the scholastic quibblings which absorbed the intellect of generation after generation of potential geometers and analysts were the troubled dreams of a torpid mathematics.

Possibly the suddenness with which the mathematical mind awoke after its long sleep will seem a less abrupt discontinuity when scholars shall have had the patience to explore medieval thought in the light of modern mathematics. Somewhere and somehow in the Middle Ages mathematics suffered a profound mutation. When it went to sleep, mathematics was Greek; when it awoke, it rapidly developed into something that was not Greek. The Moslems were not responsible for the change. Their mathematics, with its shunning of the infinite, is no closer in spirit to analysis than is that of the Babylonians.

The transition from ancient thought to modern appears to have been more difficult in mathematics than it was in science. There seems to be no clear-cut instance of a mathematical mind in the late Middle Ages two or three centuries ahead of its time, as Roger Bacon's (1214–1294?) was in science. Even Bacon, all but fully awake in science, was still as fast asleep in the mathematics which he eulogized as were any of his European opponents of the thirteenth century. His mathematical reasoning[8] is still that of the Aristotelian scholastics whom he believed he was confounding with his logic: "I say therefore that if matter can be the same in two substances, it can be the same in an infinite number . . . Therefore matter is of infinite power. Wherefore also of infinite essence, as will be proved, and therefore it must be God"—which, of course, Bacon is refuting. He seeks to accomplish his purpose by an argument based on a Euclidean postulate which the sharper scholastics had rejected. They did not assume, as did he, that "the whole exceeds any of its parts" is valid for infinite assemblages.

The similarity between Greek art, from sculpture to architecture, and Greek mathematics has often been remarked. There is no need to pursue this resemblance here; either it is felt as more than a vague metaphor, or it is dismissed as having no possible meaning. A like comparison between Gothic architecture and modern mathematics had impressed many not primarily interested in mathematics before O. Spengler exploited his Faustian theory of mathematics since the Greeks. Thus, writing in 1905, Henry Adams said of the cathedral at Chartres that "Chartres expressed . . . an emotion, the deepest man ever felt—the struggle of his own littleness to grasp the infinite."[9]

Adams also gives a sympathetic parody of the medieval perversion of elementary mathematical reasoning to the uses of scholasticism, in an imaginary but convincing debate between those two formidable champions of submathematical analysis, Abélard (1079–1142) and William of Champeaux (1070–1122). But as a specimen of Gothic mathematics at its best, his quotation from Archbishop Hildebert (eleventh century) is more suggestive: "God is over all things, under all things, outside all; inside all; within but not enclosed; without but not extended; above but not raised up; below but not depressed; wholly above, presiding; wholly beneath, sustaining; wholly without, embracing; wholly within, filling." This goes far beyond any logic of Aristotle; it all but makes its subject the Excluded Middle.

Lest, after this, the project of disinterring mathematics from the dialectic of the Middle Ages seem fantastic, two most significant facts may be recalled to encourage those who would proceed. Georg Cantor (1845–1918, Germany), founder of the modern theory of the mathematical infinite, was a close student of medieval theology. In this connection, it appears that the traditionally religious type of mind is the most strongly attracted by Cantorian mathematics. The other significant fact is the discovery[10] by K. Michalski (Polish) in 1936 that William of Occam (1270–1349, English) proposed a three-valued logic, thus anticipating to a slight extent the work of non-Aristotelian mathematical logicians in many-valued logics since 1920. Aristotle's is a two-valued logic, the 'values' being 'truth,' 'falsity' assigned to propositions. In Occam's logic Aristotle's excluded middle is admitted.

To the gentle and scholarly Boethius with his translation of Aristotle belongs a large share of whatever credit there may be for having put mathematics to sleep in medieval Europe. The rest may be awarded to the hordes of tireless logicians who strove for centuries to weld theology and philosophy into a self-consistent whole. When at last Thomas Aquinas (1227–1274)—"the dumb ox of Sicily," as he was called by jealous and envious rivals, but master of them all—succeeded, interest in the stupendous project had already waned. Roused by the waspish Moslems, Christian Europe woke, and turned, possibly with a sigh of relief, to some science and mathematics.

Glancing ahead, we recall that in the first week of September, 1939, the medieval mind at last came into its own once more in Christian Europe. It would be interesting to know what our regenerated descendants will remember of our science and mathematics in 2039, and what people, if any, are to be the Moslems of the future.

CHAPTER 5

Detour through India, Arabia, and Spain

A.D. 400–1300

The sudden rise and the almost equally sudden decline of the Moslem culture in the seventh to the twelfth centuries is one of the most dramatic episodes in history.[1] Here we are interested only in seeing what enduring influence the culture of this period had on mathematics; and we must not let the sudden brilliance of Mahometan civilization, contrasted against darkened Europe, dazzle us into seeing more in Moslem[2] mathematics than was actually there.

By 622 the followers of Mahomet were well started on their travels. Their swarming under the green banner was the greatest religious revival on record, its only close competitor being the counter-revival of the crusades in the twelfth and thirteenth centuries with their avowed purpose of supplanting the banner by the cross. From the capture of Damascus in 635, the victorious Moslems proceeded to the siege of Jerusalem, taking that holy city in 637. Four years later they had subdued Egypt, incidentally putting the final touches to the destruction of the Alexandrian library. This, however, was only a youthful indiscretion, as the Moslems were shortly to settle down and become the most assiduous patrons of Greek learning in history. Having subdued Egypt, they next (642) took Persia and all its civilized erudition.

Seventy years later (711), the conquerors entered Spain, where they furthered civilization for about eight centuries before being expelled by the Europeans they had at last stung awake. In addition to sowing the fertile seeds for centuries of war, they had brought the arithmetic and algebra of India and Greece, and Greek geometry, to Europe. Bagdad on the Tigris under the Abbasid caliphs from 750 to 1258 became the capital of culture in the East, Cordova in Spain the intellectual queen of the West. After the Moors' defeat (1212), Jewish scholars—many of whom had acquired their learning from the tolerant Moslems—vied with Christian teachers in spreading the science and mathematics that were to relegate scholasticism to the limbo of forgettable but unforgotten misadventures in intellection.

The final act was delayed till 1936, when the degenerated followers of the Prophet returned triumphantly to Spain under a red and gold banner, to harry the descendants of the people who had driven out their ancestors some three and a half centuries before. During the Moslems' long absence, Spain had contributed nothing to mathematics. With the involuntary departure of the Jews in the late fifteenth century, savage intolerance for all free thought, whether of Jew or gentile, succeeded sane liberality, leaving four sterile centuries as its monument to science.

Partial emergence of algebra

Perhaps the most significant advance of the period was the gradual emergence of algebra as a mathematical discipline in its own right, all but independent of arithmetic and geometry but closely affiliated with both.

Trigonometry also became clearly recognizable as a separate division of mathematics; and some see in the trigonometry of the Moslems their greatest and most original work. Part of its originality may be granted. But for reasons that will appear as we proceed, the trigonometry is not comparable in importance for living mathematics with the Hindu-Moslem algebra, with its frustrated struggle toward operational symbolism. Before trigonometry could function vitally in modern mathematics it, like geometry, had to become analytic. There is no hint of such a transformation before the seventeenth century, and actually it was fully accomplished only in the eighteenth. The Moslem trigonometry is still essentially Ptolemy's, amplified and refined by some algebraic reasoning and an extensive application of Hindu-Moslem arithmetic to the computation of tables. From its very nature as mathematics of the discrete, algebra could not become a province of mathematical analysis; and hence it was beyond disturbance by the analytic upheavals of the seventeenth century.

Moslem algebra appears to have evolved from the late Greek, as in Diophantus, and the much sharper technique of the Hindus. Estimates of Indian algebra differ widely, but on two points there is substantial agreement. Proof was as distasteful to the Indian temperament as it was congenial to the Greek; the Hindus were as apt in calculation as the Greeks were inept. Only by an aggressively sympathetic scrutiny of Hindu algebra can anything resembling proof be detected. Rules were clearly stated, but the statement of rules is not proof. A third feature of early Hindu algebra strikes a modern observer as extremely curious: the first skillful algebraists seemed to find indeterminate (diophantine) equations much easier than the determinate equations of elementary algebra. The reverse is the situation today.

A small sample from Hindu algebra will suffice here to indicate the quality of what the Moslems inherited, conserved, and partly spoiled. In the sixth century, Aryabhatta summed arithmetical progressions, solved determinate quadratics in one unknown and indeterminate linear equations in two unknowns, and used continued fractions. Shortly after, Hindu algebra experienced what some consider its golden age, with the work of Brahmagupta in the early seventh century, just as the Moslems were about to start on their travels. Brahmagupta stated the usual algebraic rules for negatives, obtained one root of quadratics, and, most remarkable, gave the complete integer solution of $ax \pm by = c$, where a, b, c are constant integers. He discussed also the indeterminate equation $ax^2 + 1 = y^2$. The last is misnamed the Pellian equation; it inspired Lagrange in 1766–9 to some of his greatest work in pure mathematics. It is fundamental in the arithmetical theories of binary quadratic forms and quadratic fields. Its place in the history of mathematics will be noted presently.

Again it seems strange that algebraists who did not hesitate to attack problems of real difficulty failed to see completely through simple quadratics. As remarked in connection with Eudoxus, the early algebraists were halted by a deficiency in the Greek logical faculty. Without an extended number system, it was impossible for the Hindus to create much that even resembled a scientific algebra. Thus Mahavira in the ninth century unhesitatingly discarded as inexistent the imaginaries he encountered, without attempting to account for their appearance. Three centuries later, Bhaskara recognized that formalism produces two roots for quadratics, but rejected the negatives.

Hindu algebra, however, took a hesitant step toward operational symbolism. To what has already been noted regarding symbolism, the following summary of the principal advances of the Hindus toward symbolic algebra may be added. Critics disagree on how far the Hindus got in this direction, but what follows seems to be established fact. Aryabhatta (sixth century) suggested the use of letters to represent unknowns. Brahmagupta (seventh century) used abbreviations for each of several unknowns occurring in special problems, also for squares and square roots. A negative number was distinguished by a dot; and fractions were written in our way, but without the bar, thus, $\begin{matrix}3\\4\end{matrix}$. A manuscript assigned to the period 700–1100 displays a cross, like our plus sign, written after the number affected, to indicate minus. Bhaskara (twelfth century) imitated Brahmagupta in the notation for fractions, also in the custom of putting one member of an equation under the other, and in a systematic, syncopated script for successive powers. There was no sign for equality. Brahmagupta also effected the reduction of Diophantus' three types of quadratic equations in one unknown to the standard form now current.

Differences of opinion concern the weight to be given these devices. By the most liberal estimate, the Hindus had the gist of algebraic symbolism as an operational technique proceeding according to fixed rules and in standardized patterns: the technique for solving problems of certain types was indicated in the mere writing of the problems. All the elaborate verbal directions for taking the successive steps toward a solution are explained away as insurance against stupidity. For even at its best, Hindu algebra, in spite of its free use of abbreviations, was still largely rhetorical in that operational directions were not fully symbolized. The least generous appraisal admits no advance in methodology beyond Diophantus. The Hindus themselves appear to have left no record supporting the first estimate. Possibly they imagined the meaning of what they did so obvious as to render comment on the methodology superfluous. Introspection in mathematics is a modern neurosis.

There is also the vexed question of how much of the Hindus' algebra was their own and how much Greek. Until competent scholars reach some shadow of agreement, there is little point in others reproducing their divergent conjectures. With the discovery of Babylonian algebra, the dispute seems less likely than ever to be settled in a finite time. There are also—a fascinating possibility—the ancient Chinese to be considered. Did they, or did they not, influence the Sumerians? Possibly the Sumerians taught the Chinese? Perhaps the Indians taught them all? Or did they all teach the Indians? And what part did Syria play? One type of argument supporting a favored conjecture may be noted in passing. If civilization A is assumed to be older than civilization B, and if in B a certain type of problem was discussed at a later date than in A, it follows that B got the problem from A. This is according to the diffusion theory of culture.[4] On the alternative spontaneous theory, no conclusion can be drawn; and even on the diffusion theory it remains to be shown that the data are uncontaminated by intrusions of spontaneity. The scarcity of documentary evidence complicates the problem.

Fortunately for our immediate concern, these profound questions need not be settled before the influence of Hindu arithmetic and algebra on Moslem mathematics can be substantiated. The Moslems themselves admit having translated Hindu works. It is therefore reasonable to infer that the Moslems were influenced by the Hindus. Assuming this, we note once more the human propensity to take the longest way home. Like the Greeks in their indifference to Babylonian algebra, the Moslems finally turned their backs on the rudimentary hints of an operational symbolism in the Hindu algebra and in their own, and wrote out everything, even the names of numbers, in full. The Moslem retrogression in this respect was as long a backward step as any in the history of mathematics. Absorbed in the intelligent collection and painstaking examination of numerous interesting specimens, they missed the main thing completely. Only in 1489, in Germany, with J. W. Widmann's invention of $+$, $-$, did algebra begin to become more operationally symbolic than it had been for Diophantus and the Hindus.

Before leaving Hindu algebra, we note what is usually considered its high tide. We shall inspect this rather closely, first in the haze of a golden sunset, then in the unsentimentalized light of mathematics. The two appearances are strangely dissimilar; it may be left to individual taste which is preferred.

Bhaskara, about 1150, "gave a method of deducing new sets of solutions of $Cx^2 + 1 = y^2$ from one set found by trial."[5] The problem is that of the so-called Pellian equation: to solve $Cx^2 + 1 = y^2$ in integers x, y, where C is a given nonsquare integer and $xy \neq 0$. Bhaskara also discussed $Cx^2 + B = y^2$, C, B being non-square integers. His very elementary devices have excited the liveliest admiration. Contemplating them in the golden haze, we observe[6] that "the first incisive work [on the Pellian equation] is due to Brahmin scholarship," and note that this equation "has exercised the highest faculties of some of our greatest modern analysts." We also see that Bhaskara's attack on the equation "is above all praise; it is certainly the finest thing which was achieved in the theory of numbers before Lagrange."[7]

The first of these three quotations is a verifiable statement of fact. The second seems to imply that Bhaskara's "incisive work" is qualitatively comparable to that of "some of our greatest modern analysts." The third makes Bhaskara's tentative, partial solution a finer thing in the theory of numbers than Euclid's direct, complete solution of $x^2 + y^2 = z^2$.

In the unflattering light of mathematics, it appears that Bhaskara could find any number of solutions provided he was lucky enough to guess one. He possessed no means of determining whether a given Pellian equation was solvable. Nor, even when he had derived further solutions from a lucky guess, could he tell whether he had all solutions. His process for generating solutions from an initial one was ingenious. But he ignored the only points of any difficulty or mathematical interest: the existence of a solution and the completeness of those solutions exhibited.

In contrast with this supreme achievement of Brahmin scholarship, we observe what happened when one of "our greatest modern analysts" exercised his "highest faculties" on the Pellian equation. Lagrange admitted that he had to stretch himself to accomplish what he did. In 1766–9 he settled the problem of existence and gave a direct, nontentative method for finding all solutions. Bhaskara was an empiricist; Lagrange, a mathematician.

The unromantic conclusion is that Bhaskara fell far below the standard set by Euclid, a standard which was not reached again till Lagrange attained it in the eighteenth century. It seems not unjust to draw the same conclusion regarding the rest of Hindu mathematics. But it is generally conceded that the better Hindu algebraists were far ahead of Diophantus in manipulative skill. This, and their frustrated attempt to create an operational symbolism, appear to be the chief contribution of the Hindus to the development of mathematics.

Moslem algebra seems to have hesitated between the tastes of Greece and India, choosing the latter in its most creative period, only to lapse into an impossible rhetoric as it became classic in the ninth century with the masterpiece of its most famous exponent, Al-Khowarizmi.

Indian algebra was translated into Arabic and Persian by the Moslems; and as Arabic was an important language not only in scholarship but also in commerce and war, Greek and Indian algebra, simplified and somewhat systematized by the Moslems, at last penetrated Europe. If the crusades of the twelfth and thirteenth centuries did nothing more, they helped indirectly to spread algebra, trigonometry, the classics of antiquity, and contagious diseases.

Of an impressive list of Moslem translators, commentators, and minor contributors, only two need be mentioned here. Each showed some originality and both, particularly the first, profoundly influenced early European algebra. To give him almost his full name, Mohammed ibn Musa Al-Khowarizmi (died c. 850) of Bagdad and Damascus produced the first treatise (c. 825) in which occurs an equivalent for our 'algebra'—*al-jebr w'almuquabala*, meaning 'restoration and reduction.' The reference is to what now would be called transposition of negatives to yield equations with all terms positive, and to subsequent reduction by collecting like powers of the unknown. This appears to have been Al-Khowarizmi's own idea; the work as a whole is a compost of Greek and Hindu results. His principal advance in the positive direction was an application of the Hindu number-names to the numerical solution of equations.

Al-Khowarizmi's signal progress in the negative direction has been noted.[8] Why he returned to a purely rhetorical algebra unenlivened by any trace of symbolism, seems not to be known. A psychiatrist might say it was the death instinct having its way. All but strangled, algebra survived, thereby demonstrating that more than a resolute attempt at suicide is necessary to deprive mathematics of its life. In what is reputed to be an algebraic masterpiece, Al-Karkhi (c. 1010) continued in the rhetorical tradition. If the fact were not well established, it would be difficult to believe that medieval European algebraists had the persistence to find out what the rhetorical algebraists of Islam were attempting to communicate.

Whether justly or not, algebra without symbolism is rather disappointing to the average layman who has been assured that "the Arabs invented algebra." Unfortunately, an expert knowledge of Arabic was never one of the more graceful accomplishments of a gentleman, nor even of a scholar or a country squire, as was a slight acquaintance with Latin or Greek in the eighteenth century. Consequently the mathematician or the historian of mathematics competent to form a personal estimate of Moslem algebra has always been a rarity; and of the few who have deigned to share their findings with those innocent of Arabic, some have presented the outcome of their researches in the familiar symbolism of algebra as taught to beginners today. In certain respects those sophisticated versions of the original verbiage resemble beggars masquerading in robes of satin. To appreciate the difference between the original and its modern disguise, the curious should prevail upon a professional scholar of Arabic to read them a verbatim translation of an original document in Moslem algebra. In lieu of this, the excerpt presently transcribed from the English translation of Al-Khowarizmi's *Algebra* by F. Rosen (1831) may be exhibited. Rosen's translation reproduces the original Arabic, so that cognoscenti may savor its quality. The passage quoted is from *A history of mathematical notations* (1928) by the American historian of mathematics, Florian Cajori. Rosen remarks that "numerals are in the text of the work always expressed by words: Hindu-Arabic figures [numerals] are only used in some of the diagrams, and in a few marginal notes." The excerpt follows.

> What must be the amount of a square, which, when twenty-one dirhems are added to it, becomes equal to the equivalent of ten roots of that square? Solution: Halve the number of the roots; the moiety is five. Multiply this by itself; the product is twenty-five. Subtract from this the twenty-one which are connected with the square; the remainder is four. Extract its root; it is two. Subtract this from the moiety of the roots, which is five; the remainder is three. This is the root of the square which you required and the square is nine. Or you may add the root to the moiety of the roots; the sum is seven; this is the root of the square which you sought for, and the square itself is forty-nine.

Of course symbolism of itself is not mathematics, and no amount of beautifully appropriate notation can make shoddy or trivial reasoning look like mathematics. Extensive tracts of mathematics contain almost no symbolism, while equally extensive tracts of symbolism contain almost no mathematics.

However, as in the above specimen, the total avoidance of symbolism is not always a virtue to be imitated, especially by neophytes. For laymen who may have difficulty in recognizing algebra when it is spread before them, Rosen transposes Al-Khowarizmi's rhetorical exercise into its symbolic equivalent:

$$x^2 + 21 = 10x;$$
$$x = \tfrac{10}{2} \mp \sqrt{[(\tfrac{10}{2})^2 - 21]} = 5 \mp \sqrt{(25 - 21)},$$
$$= 5 \mp \sqrt{4} = 5 \mp 2 = 3, 7.$$

This particular equation recurs many times in the early history of algebra.

Al-Khowarizmi's treatise is credited with a large share of the mathematical awakening of Christian Europe; and a twelfth-century Latin translation of a lost tract of Al-Khowarizmi's on the Hindu numerals is said to have done much to acquaint Europeans with that great invention. Giving this labor of transmission its full historical weight, and balancing it against Moslem algebra, we may leave the reader to find his own point of equilibrium somewhere among the three following estimates of Moslem mathematics.[9]

"The greatest mathematician of the time [early ninth century], and, if one takes all circumstances into account, one of the greatest of all times was Al-Khowarizmi." In the next two, full weight is given to Moslem trigonometry, to be described presently. "Their [the Moslems'] work was chiefly that of transmission, although they developed considerable originality in algebra and showed some genius in their work on trigonometry." "If the work produced [by the Moslems] be compared with that of Greek or modern European writers it is, as a whole, second-rate both in quantity and quality."

Three centuries after the great Al-Khowarizmi had finished his labors, and therefore toward the close of the cultural period, the Persian poet-mathematician, Omar Khayyam (died c. 1123), reached a considerably higher mathematical level than any of his predecessors. This devil-may-care, somewhat cynical philosopher had imagination. Not content with collections of rules, Omar classified cubic equations and devised a method of geometrical solution for numerical cubics, general within the limitations of the existing number system. Others, said to have taken the hint from Archimedes, had solved cubics by the use of conics long before Omar; indeed, the method was familiar to the Moslems of the ninth century. It was not Omar's technical labors, however, but his erroneous conjectures that cubics are algebraically, and quartics geometrically, unsolvable, that mark him out as more than a faithful transmitter and a skilled tactician and algebraic taxonomist.

But bold and original as he was, Omar steadfastly refused to accept negative roots. His hyperbolas, too, were deficient in negative branches. Once more it was the failure to come to grips with the number concept that thwarted both algebra and geometry.

The emergence of trigonometry

In its literal meaning of 'triangle measurement,' trigonometry is as old as Egypt, of course in an extremely rudimentary form. Greek astronomy demanded spherical geometry, and this, combined with the reduction of observations, necessitated what we should call the computation of trigonometric functions.

Ptolemy in the second century after Christ summarized in his (μεγάλη σύνταξις =) *Almagest* the main features of spherical trigonometry, and indicated a method for the approximate calculation of what amounts to a crude table of sines, or 'half-chords.' Ptolemy used chords; the crudity was unavoidable by the geometrical method necessitating interpolations over too wide an interval. Thus, traditionally, plane trigonometry was merely a computational adjunct to spherical trigonometry, and hence the mathematically more important elements of trigonometry emerged with unnecessary slowness. Perhaps the ultimate source of the Hindu and Moslem development of trigonometry was not applications to surveying but the astronomical necessity for sharper interpolation.

A Hindu work of about the fourth century advanced considerably beyond Greek trigonometry in both method and accuracy, giving a table of sines calculated for every 3.75° of arc up to 90°. The rule used to compute the table was erroneous, but possibly it gave results of sufficient accuracy for the inexact observations of the age. In any event, its reversion to empiricism affords an interesting illustration of the radical distinction between the Greek mathematical attack and the oriental or, for that matter, between the oriental and the modern even at its most crudely practical.

The Moslems adopted and developed the Indian trigonometry. Their first notable advance was due to the astronomer

Al-Battani (died 929) in the ninth century. If not actually the first to apply algebra instead of geometry exclusively to trigonometry, this astronomer-mathematician was the earliest to take a long stride in that direction. In addition to the Hindu sine, he used also the tangent and cotangent. Tables for the last two were computed in the tenth century, when also the secant and cosecant made their appearance as named trigonometric ratios. As the concept of a function was still about six hundred years in the future, none of his work bears much resemblance to elementary trigonometry as it is today.

Three more names may be cited as marking definite stages in the emergence of trigonometry as a distinct mathematical discipline. Abul-Wefa in the latter half of the tenth century began the systematization of all the trigonometry known at the time, and reduced it to a decidedly loose deductive system. The first Moslem text on trigonometry as an independent science was that of the Persian astronomer Nasir-Eddin (1201–1274). The book was more than a mere compendium, giving abundant evidence of a sure mathematical talent. Like the algebra of Diophantus, this work fell too close to the end of its cultural epoch to exert its full weight on the future of mathematics, and Europeans duplicated much of it without, apparently, being aware of its existence.

The last name we shall cite suggests a curious bit of history which might be worth exploring. Leonardo of Pisa (Fibonacci) will reappear in the sequel; here we note that he published his masterpiece, the *Liber abaci*, in 1202 (revised, 1228). Leonardo was largely responsible for acquainting an awakening Europe with the Hindu-Moslem algebra and the Hindu numerals. Among other significant trifles in Leonardo's book is the well-known algebraic identity

$$(a^2 + b^2)(c^2 + d^2) = (ac \pm bd)^2 + (ad \mp bc)^2.$$

It would be interesting to know where Leonardo picked this up on his travels in the East, as it is easily shown to include the addition theorems for the sine and cosine. It also became a germ of the Gaussian theory of arithmetical quadratic forms, and later of interesting developments in modern algebra. With the appropriate restrictions as to uniformity, continuity, and initial values when a,b,c,d are functions of one variable, the identity contains the whole of trigonometry.

Mathematics at the crossroads

While Europe slept and all but forgot Greek mathematics, the Moslem scholars were industriously translating all they could recover of the works of the classic Greek mathematicians. Several of these translations became the first sources from which Christian Europe revived the mathematics it had all but let die. For this timely service to civilization, the Moslems no doubt deserve all the gratitude they have received. But even at the risk of appearing ungracious, any mathematician must temper gratitude with the hard fact that scholarship and creation are in different universes.

Had the Moslems done nothing but preserve and transmit, they would scarcely have merited a passing mention in even the briefest account of the development of mathematics. This may seem too brutually direct; yet, by the only standard whose maintenance insures progress rather than stagnation, it is just. The one criterion by which mathematicians are judged is that of creation. Unless a man adds something new to mathematics he is not a mathematician. By this standard, the Moslems were not mathematicians in their extremely useful work of translation and commentary. Remembering that we are interested chiefly in things that have lasted, we shall consider briefly the Moslems' work of translation and commentary, and with it some of their trigonometry, in the light of living mathematics.

Only a few specialized historians of mathematics ever really digest any of the Greek masterpieces. Not only is life too short for those who would acquire some usable mathematics for them to master Apollonius or Archimedes; it would also be as wasteful an effort as could be imagined. Nothing could possibly come out of it but erudition. The works of the Greek masters were washed up centuries ago on the banks of the living stream; the spirit of their essential thought, and some few results that beginners learn today more easily by modern methods, alone survive. Thus the Mosem contribution of translation and commentary has lasted, not mathematically, but only as a monument to scholarship. It can be argued that without this moribund mass of Greek mathematics, there would have been no inspiration for the new mathematics of the seventeenth century: without Apollonius there would have been no Descartes, without Diophantus no Fermat, and so on, and on. Against this it can be maintained that originality was smothered under a blanket of erudition, and that, barring sentimental reservations, the greatest service the dead past can render mathematics is to bury its dead. The arguments are equally incapable of objective decision; so we shall leave them with the fact that mathematicians no longer study the Greek classics preserved by the Moslems, nor have they for the past two and a half centuries.

The fate of spherical trigonometry, on which the Moslems lavished so much of their skill, illustrates the inevitable recession of things first developed for immediately practical ends. Their utility may remain; but any living scientific interest they may once have had has long since died. Spherical trigonometry is no longer in the living stream of mathematics, either for its content or for any method it may have. Unless a student today requires the subject for some definite routine, such as the old-fashioned positional astronomy, he need not even know that spherical trigonometry exists. Of all the subjects in elementary mathematics, spherical trigonometry is probably the deadest and the most repulsive to anyone with the faintest stirrings of a feeling for vital mathematics. At rare intervals some optimistic enthusiast attempts to breathe a little life into the dry bones; but after a few perfunctory rattles, silence descends once more, and spherical trigonometry is deader than ever. Even the profound revision (1893) of the entire subject in terms of nineteenth-century algebra and analysis by E. Study (1862–1922, German) attracted only passing attention from mathematicians.

Plane trigonometry with the Greeks and the Moslems was encouraged principally because it was a useful servant to its elderly spherical sister, and she in her turn was honored mainly for her services to astronomy. While plane trigonometry was growing up, astronomy was chiefest of the sciences and the only one demanding any considerable application of mathematics. Astronomy then needed only the solution of triangles. When positional astronomy receded to a subordinate routine in modern astronomy, the trigonometric functions—for reasons in no way connected with the solution of triangles—became the indispensable mathematical aid from celestial mechanics to spectroscopy.

As modern science evolved after Galileo, astronomy became but one science of many, some being of perhaps even greater practical importance than astronomy in a scientific civilization. Here again it was the trigonometric (or circular) functions that were to prove indispensable, and again for no reason even remotely concerned with the solution of triangles. The sine and cosine derive their scientific importance from two properties: they are the simplest periodic functions; they furnish the first instances of a set of orthogonal functions. Both properties were centuries in the future when the Moslems had finished their work; the second had to wait for the integral calculus. Orthogonality underlies modern applications of the sine and cosine, making possible, as it does, the solution of important boundary-value problems arising from the differential equations of mathematical physics.

It would be difficult to imagine a physical science without orthogonal functions; but our successors may, and they may then look back on us as we look back on our predecessors. May they be as mindful of us as we are of the Hindus and the Moslems for what they invented, developed, and passed on to us to be further developed.

As we take leave of the Moslems, we see them hesitating rather forlornly at the crossroads of ancient and modern mathematics. Progress passed them before they could make up their minds to turn their backs on the past they had rescued from oblivion.

CHAPTER 6

Four Centuries of Transition

1202–1603

The thirteenth to the sixteenth centuries in Europe is one of the most eventful periods in world history. These four centuries also include the sharply marked transition from ancient mathematics to modern, the break being clearly discernible in the half-century following 1550. As will appear in the Italian solutions of the cubic and quartic equations of about 1545, algebra then was still in the Greek-Hindu-Moslem tradition. The French work (Vieta) of the latter half of the sixteenth century was in a totally different spirit, and one which mathematicians today can recognize as akin to their own. In less than fifty years, the Greek and middle oriental traditions became extinct in creative mathematics.[1]

The precise dates 1202, 1603 in the heading are intended merely to recall two of the most significant landmarks in the four centuries of transition. The first marks the publication of Leonardo's *Liber abaci;* the second, the death of Vieta, the first mathematician of his age to think occasionally as mathematicians habitually think today.

The somewhat narrow scope of Vieta's technical achievements is irrelevant to his importance in the development of mathematics. It was not what he actually accomplished in mathematics, although that was considerable, that counted; it was the quality of his thought. Whether or not the mathematicians of the early seventeenth century consciously looked back on Vieta as their herald, he was. They quickly surpassed what he had done; but their superiority was one of degree, not of kind.

Mathematics was ripe for the transition a full two centuries before it actually happened. The sharp change was delayed by a social chaos in which civilization had all it could do to keep alive. Concurrently, deeper movements were sweeping the superficial barbarism of the times back into the past, and clearing the way for a more humane economy in which mathematics shared. A brief recapitulation of the main events responsible for the delay and the subsequent advance will make the transition seem less miraculous than it otherwise might. We shall point out in passing what some of the major events implied for the future of mathematics.

Opposing currents

All the learning of the ancient world could not continue to flood into Europe for much longer after 1200 without setting up opposing currents. Broadly, the conflict became a struggle between established authority to maintain its vested interests unimpaired, and a quickening impatience with mere authority as the final arbiter between free inquiry and dictated belief, whether in knowledge of the natural universe or in government and religion. With both the rapid assimilation of the ancient learning and "the process of the suns" the thoughts of men were being broadened, and nothing short of complete destruction of the race could halt progress. As it happened, disaster was averted by a rather narrow margin.

On the side of liberality, the universities of Paris (1200), Oxford, Cambridge, Padua, and Naples were founded between 1200 and 1225. Although the early universities bore but little physical rememblance to what they later became, they were extremely significant steps toward intellectual freedom. The thirteenth century also saw the founding of the great orders of the Franciscans and the Dominicans, at least part of whose activities were educational.

A too strict devotion to scholasticism in the early universities precluded any serious study of mathematics; but the phenomenon of thousands of eager students at Paris squatting in mildewed straw and avidly absorbing Abélard's (1079–1142) hair-splitting dialectics and his humanistic contempt for mathematics shows at least that the capacity for abstract thought was not extinct. Some of the universities being direct outgrowths of the cathedral schools, it was but natural that they should favor the curricula they did. As late as the fifteenth century, only a smattering of arithmetic and a few propositions of Euclid satisfied the mathematical demands of a liberal education as certified by a bachelor's degree from Oxford.

Throughout all this period, war was a capital industry. The sack of Constantinople by the Crusaders in 1204, barbarous though it was in itself, might be reckoned with the cultural gains, as a convincing demonstration that greed and religion form a highly explosive compound. There appears to be no doubt, however, concerning what finally issued from the Holy Inquisition, established in one of its milder forms in 1232, shortly after the Moors were disciplined in Spain. Nor is there great difference of opinion in democratic countries about the protracted dissolution of feudalism, and the faint hints of democracy in the consequent rise of the middle and merchant classes in the two centuries following 1250. Although civilization almost dissolved in the process, the decay of feudalism and the gradual concretion of national monarchies accelerated the growth of knowledge after the critical period was safely passed.

The confusion and intolerance already evident in the thirteenth century became worse confounded in the fourteenth. But the picture was not painted wholly in one color. The names of Dante (1265–1321) and Petrarch (1304–1374) suggest that an occasional ray of light penetrated the gloom; while that of Boccaccio (1313–1375) recalls that some could still appreciate a bawdy story even in the presence of the Black Death (1347–1349), which carried off between a third and a half of the population of Europe. The science which less than three centuries in the future was to be fought with all the weapons of intolerance has alone wiped out such plagues.

In this fateful century the Hundred Years' War also got well under way, lasting from 1338 to 1453 by one count, and from 1328 to 1491 by another. Whichever estimate is correct, there seems to be general agreement that war flourished somewhat rankly in Christian Europe for nearly two centuries. For ruthless brutality, cynical disregard of the pledged word, and unblushing degeneracy, the famous Hundred Years of the fourteenth and fifteenth centuries had no superior till the twentieth. Not the will, but only the lack of adequate means of destruction, prevented a complete return to barbarism. It seems incredible that anything faintly resembling civilization could survive such a reversion to brutehood. But it did. Had some poet of the blackest years sung "the world's great age begins anew," as Shelley did shortly before the deepest squalor of the Industrial Revolution, they would have called him mad.

One great scientific invention in the first half of this same fourteenth century of our era passed almost unnoticed except as a curiosity for a few careless years, when Europeans quite suddenly envisioned the limitless horizons of destruction revealed by gunpowder. Radical improvements in the art of war consequent on this warmly appreciated gift of alchemy were to necessitate much refined pure mathematics and higher dynamics in the accurate calculation of trajectories. Without the mathematics of exterior ballistics, old-fashioned gunpowder, or even modern high explosive shells and rockets, would be less effective than the bows and arrows of the English archers at Agincourt. It therefore seems unlikely that war could be abolished by suppressing mathematics.

Although it could not have been foreseen at the time, the fifteenth century was to prove a landmark in mathematics as it was in all knowledge. In 1453 Constantinople fell to the Turks, and Eastern culture found its most hospitable welcome in Italy. The powerful family of the Medici in this period rendered distinguished service to civilization by their patronage of scholars and collectors of manuscripts. So far as mathematics is concerned, the net gain of this liberality was a further increase in erudition. But something of infinitely greater importance than the accumulation of libraries happened at this time, and at last made mathematics accessible to anybody with the capacities to take it. About 1450 the printing of books from movable types started in Europe.

In the first fifty years of European printing, Italy alone produced about 200 books on mathematics. During the next century the output was slightly over 1,500. The majority of course were elementary textbooks; but when a work with some real mathematics in it was printed, it became public property instead of the choice possession of a few who could afford a handmade copy. This was the second of the three major advances in the dissemination of mathematics. The first has already been noted in the transition from oriental secrecy to Greek free thought. The third was delayed for nearly four hundred years after the second, until 1826, when the first of scores of low-priced, high-grade periodicals devoted exclusively to mathematical research appeared. Printing also furthered mathematics through its economic insistence on a uniform, simplified symbolism.

Toward the close of this century, the discovery of America (1492) implied possibilities for mathematics that nobody could have predicted. The necessity for accurate navigation in mid ocean, and the determination of position at sea by tables based on dynamical astronomy, indicate the connection between 1492 and Laplace's celestial mechanics completed only in the first third of the nineteenth century. Some of the fundamental work (Euler's) of the eighteenth century in the lunar theory was undertaken to meet the need of the British Admiralty for reliable tables. The stimulus for these particular advances, originating in the voyages of Columbus and others, was about evenly divided among exploration, land grabbing, commerce, and the brutal struggle for naval supremacy. From Laplace's development of the Newtonian theory of gravitation in his dynamical astronomy, issued the modern theory of the potential and much of the analysis of the partial differential equations of physics in the nineteenth and twentieth centuries. Thus a modern mathematician, whether he lives in the United States or in China, who devotes his life to problems in potential theory with increasingly bizarre boundary conditions owes part of his livelihood indirectly to Columbus. So also does a mathematical physicist who computes perturbations in atomic physics; for the theory of perturbations was first elaborated in dynamical astronomy.

The sixteenth century was equally pregnant with great things for the future of mathematics. The names of Leonardo da Vinci (1452–1519), Michelangelo (1475–1564), and Raphael (1483–1520), three of the foremost among a host, will recall what this critical age, the century of Copernicus (1473–1543), was in art; while those of Torquemada (1420–1498), Luther (1483–1546), Loyola (1491–1556), and Calvin (1509–1564) may suggest what it was in the higher things of life. Cardan (1501–1576) published (1545) his *Ars magna*, the sum and crown of all algebra up to his time, only two years after Copernicus, on his deathbed, received the printer's proofs of his epoch-breaking *De revolutionibus orbium coelestium.*

The impact of Copernicus' work on all thought and on all social institutions is too familiar to require comment here. Mathematically, the Copernican theory was not a complete rejection of Ptolemy. The circular orbits of the Greeks remained, also thirty-four of Ptolemy's seventy-nine epicycles, and the sun itself had a small orbit. Although Aristarchus had anticipated the heliocentric theory of the solar system, Copernicus was profoundly original in his provision of a reasoned basis for what had been only a prophetic conjecture. If any one man is to be remembered as the precursor of modern mathematical-physical science, Copernicus has as good a claim as any.

Two further advances stand out as portents of the mathematics and science that were so shortly to become recognizably like our own. Stevinus (Simon Stevin, 1548–1620, of Bruges) is usually considered by physicists as the outstanding figure in mechanics between Archimedes and Galileo. Among other things, he stated (1586) the parallelogram of forces in the equivalent form of the triangle, and gave a complete theory of statical equilibrium. Modern statics is usually said to have originated with Stevinus. He also had as clear ideas of fluid pressure as was possible without the integral calculus. Incidentally, it may be noted that the development—such as it was—of mechanics by the predecessors of Stevinus is less readily evaluated than is the concurrent progress in mathematics. Reconsideration of medieval contributions to statics has usually deflated the first excessive claims in behalf of some more or less obscure writer to have anticipated Stevinus and even, on occasion, Galileo. Such was the case, for instance, with the suddenly inflated reputation of Jordanus Nemorarius (first half of thirteenth century) as a mechanist of high rank. It is now the opinion of mathematicians and physicists who have had the patience to sift his rhetoric that Jordanus was as unintelligent as his contemporaries in his conception of mechanics.

The other man of science in this period whose work was to influence mathematics indirectly but profoundly two centuries after his death was William Gilbert (1540–1603), physician to Queen Elizabeth of England. Except for some of its attempts at theory, Gilbert's *De magnete* (1600) was a thoroughly scientific treatise on the behavior of lodestones and other magnets. After the consequences of Newtonian gravitation had been elaborated, A. M. Ampère (1775–1836, French), Gauss, G. Green (1793–1841, English), and others in the first half of the nineteenth century created the mathematical theory of magnetism. Either the subject was inherently more difficult than gravitation, or less able mathematicians attacked it. Twice as long was required to breach it as had been needed for gravitation. Of course the immediate utility of the Newtonian theory may have enticed the leading mathematicians of the eighteenth century away from Gilbert's work; and there is the human possibility that the dynamics of inaccessible heavenly bodies appeared as a grander project than the attraction of magnets that could be weighed in the hand. Thus Laplace gave the sublimity of celestial mechanics as his chief reason for devoting his life to it, and he was by nature anything but sentimental. However, he did not always mean all he said.

All of these men of science of the sixteenth century are overshadowed in mathematical significance by one not usually reckoned nowadays with the professional mathematicians. Thirty-six years of Galileo's (1564–1642, Italian) life fell in this period of transition from ancient mathematics to modern. As the universally recognized founder of modern science, Galileo influenced all mathematics, pure and applied.

There was thus no lack of scientific daring in the sixteenth century, whatever the traditional custodians of consciences may have thought of it all. It may be said here once for all that some of the custodians did not always look kindly on young science struggling to free itself from the fetters of authoritative tradition, scholastic as well as ecclesiastic. Human nature being what it is, there is nothing remarkable in any of this. Both sides believed they were right; and the side which had all of the power save that of indomitable courage believed that it also had all the right. Conflict was inevitable. It was savage enough, but no more so than its belated echo in the third and fourth decades of the twentieth century, when science in some of the European states once more found itself fighting for its existence.

As for the hostility to science in the period of transition from ancient to modern thought, it is a mistake to blame one Christian sect rather than another. Anyone who cares to search the record may verify for himself that shades of creed were not the fundamental difference between those who welcomed science and those who sought to drive it out. The dissension lay deeper, in the ageless and irreconcilable antagonism between old minds and young, between those who can accept change and those who cannot. In the sixteenth and seventeenth centuries, the younger mind finally won its freedom and retained it for over two hundred years. During those brief centuries of free thought, science and mathematics prospered, and life for the majority on that account was less indecent than it was in the days of the Black Death and the Hundred Years' War.

It would be astonishing if mathematics had failed to respond to the crosscurrents of so tempestuous a transition from the old to the new as the four centuries from 1200 to 1600. But for the greater disasters for which human stupidity was only partly culpable, the response might have come two centuries earlier than it did.

A terminus in algebra

All through this period, as in the preceding, geometry continued to stagnate. Beyond translations of the Greek classics, such as the Latin editions of Euclid (1482) and Apollonius (1537), the work in geometry did not rise above the level of what would now be exercises in elementary textbooks. The Greek methods appeared to be exhausted, and mathematical progress was wholly in the divisions of arithmetic, algebra, and trigonometry. At the beginning of the period, arithmetic and algebra were still confused in a loosely coordinated alliance; at the end, they were satisfactorily divorced. Trigonometry also gained its liberty from astronomy in this period.

The *Liber abaci* (1202) of Leonardo of Pisa (c. 1175–c. 1250) has already been mentioned. This famous book by a man who was not by training a scholar at last converted Europe to the Hindu arithmetic. Leonardo himself is better known in mathematics by his other name, Fibonacci (son of Bonaccio). The son of a warehouse official, Fibonacci traveled for business and amusement in Europe and the Near East, observing and analyzing the arithmetical systems used in commerce.

The obvious superiority of the Hindu numerals and the Hindu-Moslem methods of computation inspired Fibonacci's book; and in spite of outraged protests from conservative merchants and the then-equivalents of chambers of commerce, the abacus and the counting board were finally (about 1280) relegated to the attic in European trade. Thus Fibonacci is indirectly responsible for the deluge of practical manuals on elementary computation and the flood of commercial arithmetics which have poured from the printing presses of the world ever since the fifteenth century. In spite of their great practical utility, none of these indispensable works has contributed anything of importance to the development of mathematics.[2]

Fibonacci also expounded the Eastern algebra with genuine understanding, but otherwise made no advance. His *Practica geometriae* (1220) gave an equally enlightened treatment of elementary geometry. His original work lay in the borderland between arithmetic and algebra. In his *Liber quadratorum*

(c. 1225), Fibonacci discussed some special diophantine systems of the second degree, such as $x^2 + 5 = y^2$, $x^2 - 5 = z^2$, which are harder than they look. Judged by the standard set by Euclid in his integer solution of $x^2 + y^2 = z^2$, Fibonacci's work is on a far lower level. It does not seem to have occurred to him that the real problem in diophantine analysis is to find all solutions, not merely some. This failure to sense the generality of a problem is characteristic of the distinction between ancient and modern algebra, also between mathematics and empiricism. Euclid was the one exception in about two thousand dreary years of a debased theory of numbers; even Diophantus, as we have seen, was content with special cases.

Although it is distinctly a minor issue, we must mention Fibonacci's famous recurring series defined by

$$u_{n+2} = u_{n+1} + u_n, \qquad n = 0, 1, \ldots, \qquad u_0 = 0, \qquad u_1 = 1,$$

which gives the sequence 0, 1, 1, 2, 3, 5, 8, 13, ... Fibonacci encountered this sequence in a problem (which the reader may recover for himself) concerning the progeny of rabbits. There is an extensive literature, some of it bordering on the eccentric, concerning these numbers and their simplest generalization, $u_{n+2} = au_{n+1} + bu_n$, a, b constant integers, the most interesting modern work being that inaugurated by E. A. Lucas (1842–1891, French) in 1878. Some professorial and dilettant esthetes have applied Fibonacci's numbers to the mathematical dissection of masterpieces in painting and sculpture with results not always agreeable, although sometimes ludicrous, to creative artists. Others have discovered these protean numbers in religion, phyllotaxis, and the convolutions of sea shells.

It would be interesting to know who first imagined anything transcendental in Fibonacci's numbers. Their simplest origin is in the Greek problem of dividing a line in extreme and mean ratio, the so-called golden section. It is said that some of the measurements of Greek vases, also the proportions of temples, exemplify the golden section; and one prominent psychologist even claimed to have proved that the pleasure experienced on viewing a masterpiece alleged to be constructed according to the golden section is a necessary consequence of the solid geometry of the rods and cones in the eye.

Fibonacci's quality as a mathematician emerges unmistakably in two isolated items, which also hint at the delay in the development of mathematics consequent on the social chaos that followed him. The first is his use of a single letter, in one instance, to denote a number in his algebra. Possibly this is the earliest definite trace of the generality of algebra as distinguished from mere syncopation and verbally expressed rules for numerical computation.

The second item marks Fibonacci as a true mathematician far ahead of his time. Being unable to give the algebraic solution of $x^3 + 2x^2 + 10x = 20$, Fibonacci attempted to prove that a geometrical construction of a root by straightedge and compass alone is impossible. He could not have succeeded with what was known at his time. He then proceeded to find a numerical approximation to a root. There was nothing in algebra like the inspiration for the attempted proof of impossibility till the nineteenth century.

All through the period of transition, algebra was concerned principally with the solution of equations. After quadratics had been solved within the limitations of the existing number system, the central problem was to find similar, that is, 'radical,' solutions for the cubic and quartic equations in one unknown.

There are two distinct problems in the solution of algebraic equations: to construct, by means of only a finite number of rational operations and root-extractions, performed on the literal coefficients of a given equation, all functions of those coefficients which shall reduce the equation to an identity; to construct a numerical approximation to a root of an equation with numerical coefficients. The first problem is called the solution by radicals, and is the one of greater interest in the development of algebra. The problem of approximating to a root is the one of importance in applications, for two reasons. As will appear considerably later, solution by radicals is impossible for the general equations whose degree exceeds four; the explicit radical solutions of the cubic and quartic are all but useless in numerical work.

The problem of approximation is said to have been effectively solved by the Chinese in the thirteenth and fourteenth centuries A.D. This work, if authentic, excelled most of what Europeans accomplished in numerical solutions[3] until almost the same procedure was reinvented by W. G. Horner (1773–1827, English) in 1819. Unfortunately, like nearly all oriental mathematics except Indian arithmetic and algebra, the Chinese method might as well never have been invented for any influence it had on the development of mathematics. Neither in the Orient nor in Europe did it start a forward movement, and it cannot be said to have passed into the living stream. Its human interest is the evidence it affords (if authentic) that mathematical talent is not the exclusive possession of any one race or of any one people.

The solution of the cubic and quartic by radicals was essentially completed by 1545. (The implied reservation refers to the lack of understanding at the time of negative and imaginary roots.) The history of this final triumph is spiced by a violence and chicanery that seem a trifle excessive even for the uninhibited sixteenth century. Cardan (1501–1576), whose name ornaments the solution of the cubic in every intermediate textbook on algebra, obtained the solution from Tartaglia under promise of secrecy and published it as his own in the *Ars magna* (1545). This ingenious algebraist is also renowned as an astrologer, his masterpiece in that direction being a horoscope of Christ. Among others of his excesses which offend our squeamish modern taste, Cardan is said to have disciplined a wayward son by cutting off his ears.

Tartaglia (Nicolo, 1500–1557, Italian; the nickname Tartaglia means 'the stutterer') had intended—so it is said—to crown a projected work of his own with his solution of the cubic. His name commemorates a split palate, which the traditional account credits to a saber slash inflicted by an inefficient soldier who, in doing his duty, was assisting in the massacre of the inhabitants of Tartaglia's native Brescia. They had taken refuge in the local cathedral. Tartaglia, a boy of twelve at the time, was left for dead but, owing to the devotion of his mother and of the dogs who licked his wounds, recovered. As a mature man, Tartaglia contributed to the obsolescence of sabers by pioneering work in exterior ballistics, investigating (1537) the range of a projectile and stating that the range is greatest when the angle of projection is 45°.

The solution of the quartic also harmonized with its social background. Its hero, Ferrari (1522–1565), was less fortunate than Tartaglia in his family affections. He is said to have been poisoned by his only sister. But Italy in the Renaissance without arsenic would be like veal without salt.

In recounting these traditional embellishments of Italian algebra in the sixteenth century, we have tried merely to show that mathematics can live and flourish in what purists might call barbarism, and to suggest that it may survive beyond 2000. There has been no intention of misjudging the protagonists in the drama of algebra by a misapplication of our own domestic ethics. As we live according to our lights, they live according to theirs; and if their domestic relations now seem strangely foreign to us, we have but to glance at modern international relations to feel thoroughly at home. On the domestic front, theft of scientific work is still practiced, unfortunately for struggling young men.[4]

The scientific history of the cubic and quartic is involved and possibly not yet thoroughly unraveled. It appears to be as follows. Scipio del Ferro (1465–1526) of the University of Bologna solved $x^3 + ax = b$ in 1515, and communicated the solution to his pupil, Antonio Fior, about 1535. Tartaglia then solved $x^3 + px^2 = q$ and rediscovered Fior's solution. Not believing that Tartaglia could solve cubics, Fior challenged him to a public contest, in which Tartaglia solved all of Fior's equations while Fior failed to solve any of Tartaglia's.

Cardan was not a lame copyist. He was the first to exhibit three (real) roots for any cubic. He advanced beyond the mere formal solution in recognizing the irreducible case (all roots real), when the radicals appearing are cube roots of (general) complex numbers. The first to recognize the reality of the roots in the irreducible case was R. Bombelli in 1572. Cardan also suspected that a cubic has three roots, although he was baffled by negatives and imaginaries. His most important advance, however, was the removal of the term of the second degree. This had an element of scientific generality which, apparently, he failed to appreciate fully.

Ferrari's solution of the quartic (c. 1540) also appeared in the *Ars magna*. The solution is substantially the same as that in textbooks on algebra, leading to a cubic resolvent.

These solutions of the cubic and quartic mark a definite end[5] in the algebraic tradition of Diophantus and the Hindus. They were sheer *tours de force* of ingenuity. Modern mathematics deprecates mere ingenuity and seeks underlying general principles. Proceeding from a minimum of assumptions, a modern mathematician exhibits the solutions of particular problems as instances of a general theory unified with respect to some concept or universally applicable method. An isolated solution obtained by ingenious artifices is more likely to be evidence of incomplete understanding than a testimonial to perspicacity. The ingenious solver of special problems may still be a useful member of mathematical society, in that he turns up mysterious phenomena for abler men to strip of their mystery, but he is no longer regarded as a mathematician; and to call a man who imagines he is a mathematician a problem solver is to offer him the unforgivable insult. This is one line of cleavage between ancient and modern mathematics.

It may be asked why these solutions of the cubic and quartic are still included in the second school course in algebra. Nobody with any sense attempts to use them for numerical computation, and as encouragements to skillful trickery they are positively detrimental. Instead of these relics of the sixteenth century it would be more to the purpose to offer the unified treatment of quadratics, cubics, and quartics which has been available since 1770–71, when Lagrange set himself the problem of seeing why the ingenuities of Tartaglia, Ferrari, and their successors worked. Such a treatment is considerably easier than that certified by historical tradition.

A beginning in algebra and trigonometry

Passing by an enormous mass of minor contributions, such as short tables of binomial coefficients, anticipations of Pascal's arithmetical triangle, improvements in algebraic notation, and the like, we come to the first notable progress toward generality in the methods of algebra and trigonometry, and hence in all mathematics. The slow accretion of details which ultimately passed into elementary mathematics in permanent if modified form is not comparable in importance for the whole of mathematics with the striving for uniformity in methods. Success in this direction created a science from what had been little better than a museum of tricks that worked occasionally, although nobody understood or seemed to care why.

The transition from the special to the general is first unmistakably discernible in the work of Vieta (François Viète, 1540–1603, French), who, like Fibonacci, was not a mathematician by training or profession. Vieta's activities ranged from cryptography in the military service to politics. At various times he was a member of parliament in Bretagne, a master of requests at Paris, and a king's privy councillor. His mathematics was his recreation. Some have stigmatized his work as prolix and obscured by a private jargon. But even if this were true, such superficial defects could not conceal the essential qualities of generality and uniformity in much of what Vieta invented.

There is no need to describe Vieta's contributions in detail.

His attack on quadratic, cubic, and quartic equations brings out the essentials. In each case he removed the second term of the equation by a linear transformation on the unknown. Cardan had already done this for the cubic; Vieta appreciated the importance of this step as a general procedure. For the cubic he made a second rational transformation, and finally a simple one involving a cubic irrationality. This produced a quadratic resolvent; and hence the solution of the cubic was reduced to the essential steps which the nineteenth-century theory of equations was to prove unavoidable by any device.

In this work we see also a germ of the theory of linear transformations, whose ramifications were to branch out all through later algebra and thence, in the concept of invariance, through mathematics as a whole. Here also we see a clear recognition of the art of reducing an unsolved problem to the successive solution of problems already solved, and the beginnings of tactical uniformity and generality. Vieta's solution of the quartic was similarly scientific, and led to the familiar cubic resolvent.

In Vieta's work we observe again the curious retardation of the number system. Negative roots appear to have been unintelligible to him, although he noted the simplest of the relations between the coefficients of a given equation and the symmetric functions of its roots. He also considered the possibility of resolving the polynomial $f(x)$ in an algebraic equation $f(x) = 0$ into linear factors. Anything approaching completeness or proof in this direction was far beyond the algebra of the time, and in fact was not attained till Gauss in 1799 settled the matter by giving a proof which would be admitted[6] today for the fundamental theorem of algebra.

Letters had been used before Vieta to denote numbers, but he introduced the practice (c. 1590) for both given and unknown numbers as a general procedure. He thus fully recognized that algebra is on a higher level of abstraction than arithmetic. This advance in generality was one of the most important steps ever taken in mathematics. The complete divorce of algebra and arithmetic was consummated only in the nineteenth century, when the postulational method freed the symbols of algebra from any necessary arithmetical connotation.

Improving on the devices of his European predecessors, Vieta gave a uniform method for the numerical solution of algebraic equations. Its nature is sufficiently recalled here by noting that it was essentially the same as Newton's (1669) given in textbooks. Although Vieta's method has been displaced by others, its historical significance is of more than antiquarian interest. The method applies to transcendental equations as readily as to algebraic when combined with expansions to a few terms by Taylor's or Maclaurin's series.

An algebraic equation of degree 45 which Vieta attacked in reply to a challenge indicates the quality of his work in trigonometry. Consistently seeking the generality underlying particulars, Vieta had found how to express sin $n\theta$ (n a positive integer) as a polynomial in sin θ, cos θ. He saw at once that the formidable equation of his rival had been manufactured from an equivalent of dividing the circumference of the unit circle into 45 equal parts. But for his lack of negative numbers, Vieta would have found all 45 roots instead of the 23 he did. More important than this spectacular feat was Vieta's suggestion that cubics can be solved trigonometrically. Indicative of the general haziness of algebra in the time of Vieta, his partial failure underlines the fact that even toward the close of the sixteenth century there was no clear conception of what is meant by the roots of an algebraic equation. Once more the obscurity arose from an incomplete understanding of the number system of algebra.

As evidence of Vieta's modern or Archimedean freedom, his application of both algebra and trigonometry to geometrical problems may be instanced, particularly in the use of algebra to replace geometric constructions wherever feasible. That he found nothing new in geometry of any lasting importance is immaterial; it was the boldness of his thought that mattered. But, systematic and general within its necessary limitations as this algebraized geometry was, it would be giving it infinitely more than its due to call it a precursor of analytic geometry in any but the strictly chronological sense. It nowhere even hints at the spirit of analytic geometry which made analysis and geometry complementary aspects of one mathematical discipline.

Vieta's principal advance in trigonometry was his systematic application of algebra. In both plane and spherical trigonometry he worked freely with all six of the usual functions, and in the former obtained many of the fundamental identities algebraically.

With Vieta, elementary (non-analytic) trigonometry was practically completed except on the computational side. All computation was greatly simplified early in the seventeenth century by the invention (1614) of logarithms. In pre-logarithmic computation Vieta extended (1579) the tables (1551) by G. J. Rhaeticus (1514–1576, German), giving the values to seven places of all six functions for every second of arc, instead of for every ten seconds as in Rhaeticus. The deliberate separation of trigonometry from astronomy is usually credited to Regiomontanus (Johannes Müller, 1436–1476, German) in his systematic *De triangulis* of 1464.

Elementary algebra at the close of the sixteenth century had still to receive many perfections, especially in notation, before it became the simple routine of our textbooks. But by 1600 the straight path for all future development had been clearly indicated. The way which Vieta had pointed out was followed with brilliant success by a host of workers, most of whom made incidental improvements in algebraic notation and technique while developing their major interests in the new mathematics. There was not another first-rank algebraist after Vieta till the eighteenth century,[5] when Lagrange sought and found deeper levels.

To conclude this account of Vieta, we quote the opinion of a first-rate mathematician and mathematical historian on Vieta's place in the history of mathematics. Writing in 1843, De Morgan expressed himself as follows.

> Vieta is a name to which it matters little that we have not dwelt on several points which would have made a character for a less person, such as his completion of the cases of solution of right-angled spherical triangles, his expressions for the approximate quadrature of the circle, his arithmetical extensions of the same approximations, and so on. The two great pedestals on which his fame rests are his improvements in the form of algebra, which he first made a purely symbolical science, and showed to be capable of wide and easy application *in ordinary hands;* his application of his new algebra to the extension of trigonometry, in which he first discovered the important relations of multiple angles; and his extension of the antient rules for division and extraction of the square and cube roots to the *exegetic* process for the solution of all equations. . . . If a Persian or an Hindu, instructed in the modern European algebra were to ask, "Who, of all individual men, made the step which most distinctly marks the separation of the science which you now return to us from that which we delivered to you by the hands of Mohammed Ben Musa [Al-Khowarizmi]?" the answer must be—Vieta. . . . When will the writer who asserts that Cardan was substantially in possession of Vieta's algebra attempt to substantiate his assertion by putting so much as half a page of the former side by side with one of the latter?

The development of symbolism

The importance of an easily manipulated symbolism, as implied by De Morgan, is that it enables those who are not great mathematicians in their generation to do without effort mathematics which would have baffled the greatest of their predecessors. The formulas in an engineers' handbook, for instance, if transposed into concise verbal equivalents, with a liberal use of abbreviations and conventional signs for the most frequently occurring words, might be intelligible to an Archimedes; to the average engineer they would probably be exasperating gibberish. And the prospect of having to combine several such verbalized formulas, in the hope of gaining useful information, might discourage even a modern Archimedes. In mathematics itself, as distinguished from its applications, the situation is the same. Unless elementary algebra had become "a purely symbolical science" by the end of the sixteenth century, it seems unlikely that analytic geometry, the differential and integral calculus, the theory of probability, the theory of numbers, and dynamics could have taken root and flourished as they did in the seventeenth century. As modern mathematics stems from these creations of Descartes, Newton and Leibniz, Pascal, Fermat, and Galileo, it may not be too much to claim that the perfection of algebraic symbolism was a major contributor to the unprecedented speed with which mathematics developed after the publication of Descartes' geometry in 1637. It is therefore of interest in following the evolution of mathematics to review the principal stages by which elementary algebraic symbolism reached its present maturity and to note how the lack of an effective symbolism hampered the progress of mathematics in some of its more productive periods. Two general observations may help to clarify the somewhat confused historical record.

In his analysis of Greek algebra (1842), G. H. F. Nesselmann (German) noted three historical phases of algebra, to which he gave the suggestive names rhetorical, syncopated, and symbolic. In the earliest phase, the rhetorical, the entire statement and solution of an algebraic problem were wholly verbal. It is not exactly clear why the outcome should be called algebra at all, unless it be that similar problems and their solutions reappeared in a later phase scantily clothed in at least the suggestion of a symbolism. The middle phase, the syncopated, was distinguished from the first only by the substitution of abbreviations for the more frequently occurring concepts and operations. Syncopated 'algebra' was thus an early instance of the quarter-truth that "mathematics is a shorthand." If algebra were nothing more than a shorthand, its contribution to the rudiments of mathematical thought would not be very impressive. The third phase, the symbolic, presents algebra as fully symbolized with respect to both its operations and its concepts. It also does much more than this.

Symbolic algebra replaces verbalized algebraic processes, which cost the practitioners of rhetorical and syncopated algebra much patient thought, by symbolic procedures summarizing chains of verbal reasoning in readily apprehended rules requiring only passive attention. The experience gained through centuries of laborious trial is condensed in mechanical processes which can be applied and manipulated with a minimum of thinking. If such manual dexterity as almost suffices for competence in solving linear equations, say, is condemned—as it frequently is—for its all but negligible educational value, it has had the merit of liberating the higher faculties of mathematicians to attack problems more difficult than any that taxed the devious ingenuity of the Greeks, the Indians, the Moslems, and the algebraists of the early Renaissance. In even elementary mathematics there is still opportunity enough for invigorating and profitable mental exercise. Finally, symbolic reasoning, as in the current phase of algebra, has suggested extensive generalizations and economical unifications. A typical example was the introduction (1655) of negative and fractional rational exponents, culminating about two centuries later in arbitrary complex exponents with a satisfactory theory to justify their use.

The second general remark regarding the evolution of mathematical symbolism is implicit in the recognition of the three phases of algebra. As algebra progressed, a multitude of individual names for members of what came to be recognized as one inclusive class were abandoned in favor of a uniform terminology significant for all members of the class. Further, in several instances, uniformity was possible only because the several members of the class were unified by some underlying property, usually simple when at last uncovered, of the rational numbers. When such was the case, an appropriate numerical character was imposed on the whole class, and an algoristic symbolism, amenable to the operations of rational arithmetic, brought the algebraically important characteristic of the class within the grasp of all but involuntary manipulative skill. For example, when it was finally perceived, after centuries of overlooking the elusive fact which now seems obvious, that the powers $x, x^2, x^3, x^4, x^5, x^6, \ldots$ are unified with respect to their exponents $1, 2, 3, 4, 5, 6, \ldots$, and that multiplication of powers of the unknown is effected by addition of exponents, an incredible mass of confusing terminology and inefficient rules was swept into the past, and with it, an equal or greater mass of tortuous thinking. Similar syntheses, again originating in some concealed but gradually perceived property of the rational numbers, accompanied the growth of the number concept. To cite a simple instance, the equations $ax^2 + bx = c$, $ax^2 = bx + c$, with a, b, c positive rational numbers, presented two distinct problems to algebraists before negative rational numbers were handled correctly and with (unjustified) confidence. The use of negatives reduced the solution of the two equations to that of the single equation $ax^2 + bx + c = 0$, with a, b, c rational numbers.

In passing, it is strange to find the two special quadratics treated independently in textbooks of less than a century ago. But perhaps this is not remarkable when we remember that Gauss consistently wrote x^2 as xx, for the curiously unmathematical reason that neither is more wasteful of space than the other. For ourselves, we still call x^2 and x^3 the square and the cube of x, possibly for easy diction, or perhaps because some of the ancients were mathematically fluent in the limpid jargon of areas and volumes. But there are already hints that squares and cubes may obsolesce from the vocabulary of algebra before many more centuries have passed. In the meantime, no tyro in algebra need be seriously discommoded by his inability to see immediately what squares and cubes have to do with x, provided only we do not afflict him with the corresponding names for $x^4, x^5, x^6, \ldots$ from the geometry of hyperspace, as we should do if we are to foster linguistic purity. But, as was discovered by too hasty penitents in the Middle Ages, even purity may sometimes cost too much for comfort.

The absence of symbolism in Babylonian algebra, already noted, poses the problem of what is to be recognized as algebra in the rhetorical and syncopated phases. Since it seems to be agreed among historians of mathematics that the Babylonians, the Egyptians, Diophantus, and the more rhetorical Moslems actually practiced a more or less rudimentary algebra, it is clear that the absence or presence of symbolism is not the historical criterion. More than a mere matter of words is involved.

From the frequently decisive impacts of symbolism on the general development of mathematical reasoning, it would seem that mathematics itself and not the pedantries of terminology is the important issue. An older conception of algebra than that now universal identified algebra with the solution of equations. If this antique be admitted, both the rhetorical and the syncopated phases are accepted as algebra without further qualification, and the entire historical development of the subject down to the beginning of the nineteenth century acquires a deceptive unity and a specious coherence. A somewhat similar conception of algebra indicates the use of unknowns, whether verbalized or symbolized, as the historical clue to be followed. This, however, comprises too much, as it includes such geometrical problems as the construction of a circle to satisfy prescribed conditions. To narrow the scope of the 'unknown' sufficiently, it may be restricted to the domain of numbers—a restriction which lost its stringency with the invention of analytic geometry. In an obvious sense all mathematics is a quest for the unknown. In addition to affording a clue to the development of symbolism, this inclusive definition has the signal advantage over its too numerous competitors of permitting the algebraists or the analysts to claim all mathematics as their province, as already done by some of the geometers.

In spite of all objections, the available data seem to show that for following the historical development of symbolism, either equations or unknowns offer a convenient directive. Again, the mere existence of equations all through the protracted evolution also foreshadowed a most important aspect of mathematical thought, which dominated much of the work of the recent period beginning in 1801, that of mathematics as a study of relations. In the earliest stages, the only relation considered was equality, and it required about three thousand years for this ubiquitous concept to reach full symbolic representation. This may serve as a typical example of the slowness with which the commonest paraphernalia of current mathematics evolved.

Operations seem to have been symbolized more readily than relations. If this is a correct statement of the facts, it accords with the order of increasing abstractness. But success in one department apparently did not appreciably stimulate inventiveness in another, and until the recent period the development of symbolism proceeded haphazardly.

In modern mathematics the creation of an efficient notation

may sometimes have been accidental, but usually it was the outcome of conscious effort. An example of the first is the notation $\frac{a}{b}$ (not a/b) for fractions, an invention whose full value may not have been appreciated by its author. Possibly the most striking instance of the second is Leibniz' $\frac{dy}{dx}$ (not dy/dx) for the derivative of y with respect to x. Nobody in the history of mathematics was more sensitive than Leibniz to the potentialities of a rationally devised symbolism, and nobody gave the 'philosophy' of mathematical notation more painstaking thought than he. More recently, the notations a/b for fractions and dy/dx for derivatives in some respects have illustrated progress in reverse. Centuries of easy habit were discarded to accommodate the incompetence of printers—the modern reinventor of a/b gave substantially this reason for his departure from custom. In our own century the incompetence of machines contributed in a similar manner to the delinquence of happily conceived notations, for example that of the tensor calculus, until printers learned that it pays to hire competent engineers to revise their machinery. This is one of the few instances where the economic motive has reacted simply and directly with the ideal of mathematical clarity to the benefit of both. In the matter of $\frac{a}{b}$, the motivation was partly a humane desire to moderate the repulsiveness of vulgar fractions to young children.

A small sample of the many notations for powers of the unknown will suffice to suggest the progression from rhetorical to symbolic algebra. Ahmes (seventeenth century B.C.) used a word, variously translated as 'heap,' 'amount,' 'mass,' to denote the unknown. Diophantus (third century A.D.) used a shorthand for the successive powers of the unknown: x^2 was the 'power,' x^3 the 'cube,' and 'power,' 'cube' were denoted by (what were probably) abbreviations of the corresponding Greek words. Say these abbreviations were P, C; then PP, PC, CC, denoted the fourth, fifth, sixth powers of the unknown, and so on. Here evidently there is a rationale behind the syncopation. Such a notation was but ill adapted to the simultaneous representation of several unknowns. Traces of operational symbolism also are attributed to Diophantus. Addition was indicated by juxtaposition, subtraction by a special symbol whose genesis is still in dispute—it may have been the first letter of a Greek word, or a genuine operational symbol in the sense that it was not derived from an abbreviation. If the latter, it was a significant step toward symbolic algebra. But Diophantus did not rise to relational symbolism, using the first two letters of the Greek word for 'equality' to denote 'is equal to.' It is doubted by some scholars whether Diophantus made any use of symbolism, his claim to having done so resting on a manuscript of his arithmetic written about a thousand years after his death.

Both the Indians and the Moslems followed Diophantus in what may be described as additive juxtaposition to denote successive powers of the unknown. Aryabhatta (fifth to sixth centuries A.D.) abbreviated the unknown to *ya*, its second, third, fourth, sixth powers to *va*, *gha*, *va va*, *va gha*, and so on. He also provided for several unknowns; *ya* (the first unknown), *ka*, *ni*, *pi* (the second, third, fourth unknowns) the abbreviations being those of the color names black, blue, yellow. Operations were indicated after the operands by the words *ghata*, *bha*, indicating addition, multiplication. Thus xy, is *ya ka bha*, where x, y are the unknowns. The substitute for a sign of equality was adequate, one of two equals being written under the other. The word for 'root' was *mula*. Combined with other words, as *varga mula*, *ghana mula* for 'square root,' 'cube root,' *mula* was hardly so much a mathematical symbol as a common noun. A closer approach to symbolism was the indication of the negative of a number by writing a dot or a small circle above the number. Of the Moslems, Al-Khowarazmi (first half of the ninth century), used *jidir* (root) for the unknown, and *mal* (power) for its square. Al-Karkhi (early eleventh century), with *kab* for the cube, composed the fourth, fifth, sixth, seventh, . . . powers by juxtaposition, as *mal mal*, *mal kab*, *kab kab*, *mal mal kab*. The Moslems generally followed Diophantus in simplifying equations by combining like terms. Both appear to have been led astray by this natural simplification. The significant classification of equations in one unknown is not according to number of terms, but by degree. However, the later Moslems, in spite of their ineffectual protosymbolism, recognized that the next problem after the quadratic in equations was the cubic, by no means an easy recognition from their point of view.

Late in the fifteenth century, the Moslems approached a purely symbolic representation of an operation in writing only the first letter of the Arabic for 'root' above a number to indicate square root. This might be considered an intermediate stage between syncopation and a matured symbolism in which operations are designated by specially devised signs whose verbal origin, if any, is no longer recognizable. An example of the last is the current sign for equality. Unless it were definitely known that Recorde invented this sign (*Whetstone of witte*, 1557), it might well be mistaken for a degenerated form of a word in medieval shorthand. But Recorde denoted equality by = because, it seemed to him, no two things could be "moare equalle" than "a paire of paralleles"— which is reminiscent of the remark that "William and John," twins, "are very much alike, especially William." But Recorde had the true symbolist's instinct for the ultimate perfection, by whatever conceit he chose to propitiate his syncopating contemporaries. The Egyptians had used the hieratic form of their hieroglyph for 'equality'; the Greeks, the first two letters of their word, the Moslems, the last letter of their word, till they reverted to total verbalism and wrote out 'equality' in full. It remained for Recorde to do the right thing.

In equations, the passage through the three phases was similar to the evolution of the progressively more algoristic notation for powers, which reached a climax in Wallis' (1655) x^{-n}, $x^{1/n}$ for $1/x^n$, $\sqrt[n]{x}$. Greek equations were partly rhetorical, partly syncopated, with little that would now be recognized as algebraic symbolism. Al-Khowarizmi's equations were purely rhetorical; a Latin translation of one is "*census et quinque radices equantur viginti quatuor*," or "the square of the unknown (*census*) and five unknowns (*radices*) are equal to twenty four," that is $x^2 + 5x = 24$. The Europeans of the sixteenth and seventeenth centuries gradually approached full symbolism in the writing of equations, as seen in the following specimens. Cardan (1545) wrote $x^3 + 6x = 20$ as "*cubus* $\bar{p}$ 6 *rebus aequalis* 20," in which there is nothing to indicate that *cubus* and *rebus* are powers (third, first) of the same unknown. The $\bar{p}$ is 'plus.' Vieta, with C, Q, N for 'cube,' 'square,' 'number' or 'unknown,' also left it to be inferred that these terms refer to one unknown in $IC - 8Q + 16N$ aequ. 40. At last Descartes (1637) settled the matter (except that he missed x^2 for xx) by writing x, xx, x^3, x^4, x^6, . . . for Vieta's N, Q, C, QQ, QC, . . . , putting all positive integral powers on the uniform notational basis familiar today. It all seemed so simple when it was finally done after centuries of effort.

Volumes might be (and have been) written about the evolution of mathematical symbolism. Probably almost anyone leafing through these will agree that lack of appropriate symbolism constrained the Greek arithmeticians and algebraists to consider special cases of what might have been their problems, and prevented the Indians and the Moslems from producing an elementary algebra within the capacities of ordinary adolescents.

CHAPTER 7

The Beginning of Modern Mathematics

1637–1687

Historical sketches may sometimes decoy us into artificial divisions of human progress by centuries or half-centuries demarked by precise dates. Having just passed through one such critical fifty years, we may well suspect that another is more imaginary than real. Be this as it may, the half-century from 1637 to 1687 is universally recognized as the fountainhead of modern mathematics. The first date marks the publication of Descartes' *Géométrie*, the second, that of Newton's *Principia.*[1]

From this prolific period, as from the Greek golden age, we shall select only those contributions which overtop a multitude of interesting details in their significance for the development of all mathematics. Some of the items omitted will be noted in later chapters, where they may be naturally included without interrupting the continuity of the main current. Thus, in this period infinite series advanced notably, but were of minor importance compared with the calculus. Again, the contribution of Leibniz to symbolic logic can be best described in the light of modern work, and will be noted in the final chapter. It may be remarked once for all that mathematics overshadows its creators; that we are primarily interested in mathematics; and that each of the men cited did far more than the few items described here, but that much of what is omitted has for long been of only antiquarian interest.

As some of the men whose work is to be reviewed were more directly responsible than others for the creation of modern mathematics, they will be given more extended notices than might be justified on purely impersonal grounds. Like Pythagoras, they too will doubtless vanish as personalities and live only in the body of mathematics as the centuries slip away; but at present they are close enough to us to be more than names attached to mathematical abstractions.

These outstanding originators of modern mathematics were not merely half a dozen eminent men in a crowd; they towered above the majority of those who preceded or came after them. Conspicuous eminence in mathematics was harder to achieve after these men had lived, simply because by the power of their methods they had quite suddenly raised the whole level of attainable mathematics. Geometers, for instance, were no longer condemned to crawl among the five conics and a handful of simple higher plane curves after Descartes had given them wings. It is arguable that even the most original of these men was indebted to the very humblest of those who preceded him. But their incomparable superiority in generality of outlook almost inclines us to regard all of them rather as sudden mutations touched off into explosive activity by accidents of their environment, than as orderly end products of a creeping evolution.

Five major advances

Modern mathematics originated in five major advances of the seventeenth century: the analytic geometry of Fermat (1629) and Descartes (1637); the differential and integral calculus of Newton (1666, 1684) and Leibniz (1673, 1675); the combinatorial analysis (1654), particularly the mathematical theory of probability, of Fermat and Pascal; the higher arithmetic (c. 1630–65) of Fermat; the dynamics of Galileo (1591, 1612) and Newton (1666, 1684), and the universal gravitation (1666, 1684–7) of Newton.

With these five, two further departures in new directions may be cited for their influence on subsequent advances: the synthetic projective geometry (1636–9) of Desargues and Pascal; the beginning of symbolic logic (1665–90) by Leibniz.

Throughout the first half of the century, reactionary hostility to science continued its losing fight, reaching its futile climax in the condemnation of Galileo by the Inquisition in 1633, only four years before Descartes, safe in Holland, permitted his masterpiece to be printed. Intolerance was partly offset by the scientific societies founded during this period or shortly after. Only the three most influential need be mentioned. The Royal Society of London was incorporated in 1662; Newton was its president from 1703 to 1727. The French Academy of Sciences (Académie des Sciences, Paris) crystallized in 1666 from the informal meetings of a group of savants, some of whom, including Mersenne, Descartes, and Mydorge, were primarily mathematicians. The Berlin Academy (Societät der Wissenschaften) was founded in 1700 at the instigation of Leibniz. He was its first president. The Paris and Berlin Academies have consistently been more cordial than the Royal Society to pure mathematics.

The importance of these and other academies for the advancement of science during the seventeenth and eighteenth centuries cannot be overestimated. Together, they did far more than the universities for science, one of their chief functions being the publication of research by their members. Even more important than this was the living example each of the scientific societies afforded of a nucleus of intelligent, influential men in a society still cowed by religious bigotry and scholarly intolerance. By the end of the seventeenth century, science had grown too sturdy for indiscriminate attack; and the forces of reaction, fighting among themselves, lacked the wit to combine against their common enemy.

A remarkable feature of the rapid development in mathematics was that the continuous and the discrete divisions advanced simultaneously. The advance in the continuous might have been expected; the other has the appearance of an accident. Neither more or less trivial arithmetic of permutations and combinations, nor unsystematized observations on games of chance, offer a sufficient explanation of the sudden and complete emergence of the fundamental principles of the theory of probability.

The most prolific of all the new acquisitions was the calculus; for when geometry became analytic, it derived most of its life from the analysis of functions continuous except at isolated singularities. There was thus provided an infinite store of curves and surfaces on which geometers might draw, and to which they applied the methods of the calculus to discover and investigate exceptional points, such as cusps and inflections, not intuitively evident from the equations.

It is not surprising that for more than a century after it became public property, the calculus and its applications to geometry, dynamical astronomy, and mechanics attracted all but a few of the ablest men, to the comparative neglect of combinatorial analysis, the theory of numbers, algebra (except improvements in notation and Descartes' work in equations), symbolic logic, and projective geometry. For more than twenty centuries, geometry and astronomy had dominated mathematical tradition in the works of the masters. Now here at last was the universal solvent for all the intractabilities of classical geometry and astronomy, and the philosopher's stone that changed everything it touched to gold. Difficulties that would have baffled Archimedes were easily overcome by men not worthy to strew the sand in which he traced his diagrams. Leibniz did not exaggerate when (1691) he boasted that "My new calculus [and Newton's] . . . offers truths by a kind of analysis, and without any effort of the imagination—which often succeeds only by accident—; and it gives us all the advantages over Archimedes that Vieta and Descartes have given us over Apollonius."

The calculus of Newton and Leibniz at last provided the long-sought method for investigating continuity in all of its manifestations, whether in the sciences or in pure mathematics. All continuous change, as in dynamics or in the flow of heat and electricity, is at present attackable mathematically only by the calculus and its modern developments. The most important equations of mechanics, astronomy, and the physical sciences are differential and integral equations, both outgrowths of the seventeenth-century calculus. In pure mathematics, the calculus at one sweep revealed unimagined continents to be explored and reduced to order, as in the creation of new functions to satisfy differential equations with or without prescribed initial conditions. One of the simplest of all such equations, $dy = f(x)dx$, in a sense defines the integral calculus; and the corresponding integral, $\int f(x)dx$, alone suggests an endless variety of functions according to the form of $f(x)$.

In the discrete division, continuity is of only secondary importance. Primarily, combinatorial analysis is concerned with the relations between subclasses of a given class of discrete objects, for example with the interrelations of the permutations and combinations of the members of a given countable class. Fermat's and Pascal's work of 1654 on probability lifted combinatorial analysis from the domain of mathematical recreations into that of severely practical mathematics; and only about fifty years elapsed between the creation of the mathematical theory of probability and the calculation of mortality tables by its use. In modern combinatorial analysis the calculus is indispensable in obtaining usable approximations to formulas beyond practicable exact computation.[2]

The other great advance in the discrete division, Fermat's creation of the modern higher arithmetic, was for long restricted to the study of relations between subclasses of the class of all rational integers. Since about 1850, numerous arithmeticians have extended the classical theory of Fermat and his successors to vastly wider classes of integers. The contribution to all mathematics of the higher arithmetic has been indirect, in the invention of new techniques, particularly in modern higher algebra and to a lesser extent in analysis, primarily for application to problems concerning the rational integers. Conversely, extensive tracts of the modern theory of numbers would not exist had not analysis made them possible.

The careers of synthetic projective geometry and symbolic logic afford an interesting contrast in mathematical obsolescence and survival. Both will be noted in subsequent chapters; here we remark only the striking difference between their fate and the uniform prosperity of the other creations of the seventeenth century. After its invention by Desargues and Pascal, synthetic projective geometry languished till the early nineteenth century, when it became extremely popular among geometers with a distaste for analysis. Leibniz' dream of a mathematical science of deduction lay dormant till the mid-nineteenth century, and even then it appealed to but very few, although Leibniz had foreseen the importance of symbolic logic for all mathematics, and had himself made notable progress toward an algebra of classes. Only in the second decade of the twentieth century did mathematical logic become a major division of mathematics. Concurrently, synthetic projective geometry was receding definitely into the past with the reluctant admission that an essentially Greek technique, even when revitalized, is hopelessly impotent in competition with the analytic methods of Descartes and his successors.

From all this it is clear that after the period of Archimedes, Euclid, and Apollonius, that of Descartes, Fermat, Newton, and Leibniz is the second great age of mathematics. If the fundamental distinction between the old and the new can be suggested in a word, it may be said that the spirit of the old was synthesis, that of the new, analysis.

'Anticipations'

Before proceeding to the individual advances, we must dispose of a purely historical matter which will not be further discussed. It concerns numerous aborted or sterile ideas which have not passed into living mathematics.

Behind each of the major advances of the seventeenth century were many short steps in the general direction of each, and some of these partial advances all but reached their unperceived goals. At least that is what we might be tempted to imagine now. Looking back on these efforts, some may be inclined in the generosity of their hearts to believe that without these halted steps final success would have been long delayed or unattained. In the specific instances of analytic geometry and the calculus, an examination of the mathematics—not the sentiments—involved has convinced a majority of professionals that the alleged anticipations are illusory. Especially is this the opinion of men who themselves create mathematics and who know from disconcerting experience that hindsight sees much to which foresight was blind.

In retrospect we can trace the evolution of analytic geometry, for example, back to Hipparchus, or even to the ancient Egyptians. Like every astronomer who has recorded the positions of the planets, Hipparchus used coordinates, in particular latitude and longitude. But the use of coordinates entitles nobody to priority in the invention of analytic geometry; nor does even an extensive use of graphs. As any intelligent beginner who has understood the first three weeks of a course in analytic geometry knows, analytic geometry and the use of coordinates in the plotting of graphs are a universe apart. Only in the sense that they preceded analytic geometry are such comparatively childish activities anticipations of that geometry. This also is the judgment of a majority of professional geometers, who probably are as competent as anyone in this matter.

We must refer the reader elsewhere[3] for a detailed evaluation and rejection of the romantic claims that several early mathematicians, and in particular Apollonius, Nicole Oresme (fourteenth century), and Kepler, 'anticipated' Descartes and Fermat in their independent invention of analytic geometry. To preserve the balance, and to exhibit a clean-cut instance of the absolute zero to which so many scholarly differences of opinion in the history of mathematics add up, we cite another evaluation of these 'anticipations' in which exactly the opposite conclusion is ably upheld.[4] Dozens more on either side might easily be mentioned; the two selected will suffice to orient whoever may be interested, and to start him on his own critical evaluations.

In each of the major advances of the seventeenth century some definite step led from confusion to a new method. Thus Newton himself states what gave him a hint for the differential calculus: "Fermat's way of drawing tangents."

But there is one 'anticipator' of the calculus, B. Cavalieri (1598–1647, Italian), who merits more than a passing citation for the lasting mischief his 'anticipation' has done. Cavalieri's method of indivisibles has endured, to the distraction of hundreds of teachers of the elementary calculus who must extirpate heretical notions of infinitesimals from their students' minds.

In the United States, much of this elementary confusion can be traced to a generation of college teachers who were thoroughly indoctrinated in their school course in solid geometry with Cavalieri's method of indivisibles. Their school geometries contained a seductive section on what some textbook writers called Cavalieri bodies; and these indivisibly-divisible nonentities were used, among other absurdities, to inculcate a disastrously nonsensical account of mensuration in three dimensions.

Cavalieri did not anticipate the calculus; he committed the unpardonable sin against it. But for his indivisibles and their absorption by scores of otherwise rational men who were to become college teachers, the common delusion that an infinitesimal is a 'little zero' would have been extinct two generations ago.

The historical appeal of Cavalieri's indivisibles is undeniable, and that, perhaps, is why some historians palliate their flagrant offenses. They were inspired by the scholastic lucubrations of Thomas Bradwardine, Archbishop of Canterbury (thirteenth century), and the submathematical analysis of Thomas Aquinas. As Cavalieri never defines his indivisibles explicitly, it is open to his apologists to read into them anything they know should be there but is not. But if his mystical exposition (1635) means anything at all, Cavalieri regarded a line as being composed of points like a string of countable but dimensionless beads, a surface as made up similarly of lines without breadth, and a solid as a stack of surfaces without ultimate thickness. These are the very notions which a conscientious teacher will purge out of his students if it takes four years.

A historical argument in favor of these indivisibles is that Leibniz was acquainted with them. But even this does not make Cavalieri an anticipator of the calculus. As will appear shortly, Newton clearly recognized the untenability of indivisibles; and although he did not fully succeed in clearing up his own difficulties, he did not mistake nonsense for sound reasoning. That is the fundamental distinction between one who imagined the calculus and one who did not. Contrary estimates of Cavalieri's work are readily available.

Descartes, Fermat, and analytic geometry

René Descartes (1596–1650, French) is more widely known as a philosopher than as a mathematician, although his philosophy has been controverted while his mathematics has not.

Descartes' family was of the lesser French nobility. His mother died shortly after her son's birth, but an unusually humane father and a capable nurse made up for this loss. After a broad education in the humanities at the Jesuit college of La Flêche, Descartes lived for two years in Paris, where he studied mathematics by himself, before joining Prince Maurice of Orange at Breda as a gentleman officer in 1617. In 1621, Descartes abandoned his military career, partly because he had seen enough service both active and passive, partly because, as he declared, three dreams on the night of November 10, 1619, suggested the germs of his philosophy and analytic geometry.

Much of the remainder of his life was spent in Holland, where he was safer from possible religious persecution than he would have been in France. These were his productive years; and in spite of his desire for tranquillity, he could not conceal the greatness of his thought. Rumors of what he was thinking were discussed wherever others with minds akin to his own dared to think. Largely through the efforts of Father M. Mersenne (1588–1648, French) of Paris, who acted as intermediary between the French intellectuals and the justly cautious Descartes, his fame spread over all Europe.

In 1637 Descartes published the work on which his greatness as a mathematician rests, the *Discours de la méthode pour bien conduire sa raison et chercher la vérité dans les sciences*, the third and last appendix of which, *La géométrie*, contains his subversive invention.

The closing months of his life were spent as tutor to the young and headstrong Queen Christina of Sweden. The rigors of a Stockholm winter and the inconsiderate demands of his royal pupil caused his death. In accordance with the ideals of his age, when experimental science was first seriously challenging arrogant speculation, Descartes set greater store by his philosophy than his mathematics. But he fully appreciated the power of his new method in geometry. In a letter of 1637 to Mersenne, after saying "I do not enjoy speaking in praise of myself," Descartes continues: " . . . what I have given in the second book on the nature and properties of curved lines, and the method of examining them, is, it seems to me, as far beyond the treatment in the ordinary geometry, as the rhetoric of Cicero is beyond the a, b, c of children."[5]

The famous appendix[5] on geometry consists of three books, of which the second is the most important. The third is devoted mostly to algebra. It will suffice to restate in current terminology the essential features of Descartes' advance.[6]

A plane curve is defined by some specific property which holds for each and every point on the curve. For example, a circle is the plane locus of a point whose distance from a fixed point is constant. Any point on a curve is uniquely determined by its coordinates x and y; and an equation $f(x, y) = 0$ between the coordinates completely represents the curve when the specific geometric property defining the curve is translated into a relation, denoted by the function f, between the coordinates x, y of the particular-general point on the curve.

There is thus established a one-one correspondence between plane curves and equations in two variables x, y: for each curve there is a definite equation $f(x, y) = 0$, and for each equation $f(x, y) = 0$ there is a definite curve.

Further, there is a similar correspondence between the algebraic and analytic properties of the equation $f(x, y) = 0$ and the geometric properties of the curve. Geometry is thus reduced to algebra and analysis.

Conversely, analysis may be spoken in the language of geometry, and this has been a fecund source of progress in analysis and mathematical physics.

The implications of Descartes' analytic reformulation of geometry are obvious. Not only did the new method make possible a systematic investigation of known curves, but, what is of infinitely deeper significance, it potentially created a whole universe of geometric forms beyond conception by the synthetic method.

Descartes also saw that his method applies equally well to surfaces, the correspondence here being between surfaces defined geometrically and equations in three variables. But he did not develop this. With the extension to surfaces, there was no reason why geometry should stop with equations in three variables; and the generalization to systems of equations in any finite number of variables was readily made in the nineteenth century. Finally, in the twentieth century, the farthest extension possible in this direction led to spaces of a non-denumerable infinity of dimensions. The last are not mere fantasies of the mathematical imagination; they are extremely useful frameworks for much of the intricate analysis of modern physics. The path from Descartes to the creators of higher space is straight and clear; the remarkable thing is that it was not traveled earlier than it was.

Another direct road from Descartes to the present may be noted in passing. The formula $(x_1 - x_2)^2 + (y_1 - y_2)^2$ for the square of the distance between any points (x_1, y_1), (x_2, y_2) in a plane (surface of zero curvature) suggested the corresponding formulas in differential geometry for the square of the line element joining neighboring points in any space, flat or curved, of any number of dimensions, as quadratic differential forms. The germ of this long evolution was the Pythagorean theorem.

In details, Descartes' presentation differs from that now current. Thus, he used only an x-axis and did not refer to a y-axis. For each value of x he computed the corresponding y from the equation, thus getting the coordinates x and y. The use of two axes obviously is not a necessity but a convenience. In our terminology, he used the equivalents of both rectangular and oblique axes. But in one important particular his procedure was needlessly restricted. He considered equations only in the first quadrant, as it was thence that he translated the geometry into algebra. This consistent but unnecessary limitation led to inexplicable anomalies in the translation back from algebra to geometry. As analytic geometry evolved and negative numbers were fearlessly used, the restriction was removed. By 1748, when Euler codified and extended the work of his predecessors, both plane and solid analytic geometry were practically perfected, except for the introduction of homogeneous coordinates in 1827.

The new method was not fully appreciated by Descartes' contemporaries, partly because he had deliberately adopted a rather crabbed style. When geometers did see what analytic geometry meant, it developed with great rapidity. But it was only with the invention of the calculus that analytic geometry came into its own. As early as 1704 Newton[7] was able to classify all plane cubic curves into seventy-eight species, of which he exhibited all but six. This comparatively early work in the geometry of higher plane curves is especially remarkable for its discussion of the nature of the curves at infinity, and for Newton's assertion, which he did not elucidate, that all species are obtainable as projections ('shadows') of the curves $y^2 = ax^3 + bx^2 + cx + d$. When we reflect that, only sixty-seven years before Newton published this work, geometers had been laboriously anatomizing those other 'shadows,' the conic sections, by the synthetic method of Apollonius and had not even imagined Newton's cubics, we begin to appreciate the magnitude of the revolution Descartes precipitated in geometry.

It is evident from Descartes' explanation of his method that he had an intuitive grasp of the elusive concepts 'variable' and 'function,' both of which are basic in analysis. Moreover, he intuited continuous variation. Vieta before him had used letters to denote arbitrary constant numbers; Descartes knew that the letters in his equations represented variables, and he clearly recognized the distinction between variables and arbitrary constants, although he defined neither formally. The significance of this advance for the calculus that was to follow only sixteen years after his death is plain.

Descartes' progress in generality is illustrated by two of his minor but geometrically important observations. He classified algebraic curves according to their degrees, and recognized that the points of intersection of two curves are given by solving their equations simultaneously. The last implies what actually is a major advance over all who had previously used coordinates: Descartes saw that an infinity of distinct curves can be referred to one system of coordinates. In this particular he was far ahead of Fermat, who, apparently, overlooked this crucial fact. Fermat may have taken it for granted, but nothing in his work shows unequivocally that he did.

Still seeking generality, Descartes separated all curves into two classes, the "geometrical" and the "mechanical." This is curious rather than illuminating. He defined a curve to be geometrical or mechanical according as (in our terminology) dy/dx is an algebraic or a transcendental function. Although this classification was abandoned long ago, it affords an interesting sidelight on the quality of Descartes' mind. The current definition of a transcendental curve as one which intersects some straight line in an infinity of points was given by Newton in his work on cubics.

Descartes' method for finding tangents and normals need not be described, as it was not a happy inspiration. It was quickly superseded by that of Fermat as amplified by Newton. Fermat's method amounts to obtaining a tangent as the limiting position of a secant, precisely as is done in the calculus today. The historical significance of this conception is evident when it is recalled that the tangent at (x, y) is drawn by a simple Euclidean construction once its slope dy/dx is known. Fermat's method of tangents is the basis of the claim that he anticipated Newton in the invention of the differential calculus. It was also the occasion of a protracted controversy with Descartes.

We pass on to Fermat and his part in the invention of analytic geometry. There is now no doubt that he preceded Descartes. But as his work of about 1629 was not communicated to others until 1636, and was published posthumously only in 1679, it could not possibly have influenced Descartes in his own invention, and Fermat never hinted that it had.

Fermat was one of those comparatively rare geniuses of the first rank, like Newton and Gauss, who find all their reward in scientific work itself and none in publicity. Under modern economic conditions, it is inexpedient for a scientist to hide his light under a bushel unless he wishes to starve to death in the dark, and few do. Of course it is impossible to say what effect the prospect of jobless starvation would have had on the more aloof scientists of the past. Some of them lived in economic security independently of any scientific work they might or might not do. Today men make the only livings they have at science or mathematics, and it seems like a misapplication of a warped yardstick to measure their professional ethics by those of a hypothetical past that may never have existed. For it has yet to be proved that a full mind can outargue an empty stomach. In Fermat's case, either lifelong security or excessive modesty made publication of very minor importance to him, and as a result his superb talents were all but buried in his own generation. Descartes, not Fermat, was the geometer whom others followed.

Pierre de Fermat (1601–1665, French, date of birth disputed) cultivated mathematics as a hobby. His profession, like Vieta's, was the law. As a counselor of the local parliament at Toulouse, he lived a quiet, orderly life which left him ample leisure for his favorite study. An accomplished linguist and classicist as well as a first-rank mathematician, Fermat knew the masterpieces of Greek mathematics at first hand.

Fermat did not discover his extraordinary powers in mathematics till he was about thirty, and even then he seems scarcely to have realized their magnitude. From his letters we get the impression that he regarded himself as a rather ingenious fellow, capable occasionally of doing a little better than Apollonius and Diophantus, but not very much after all in comparison with the ancient masters. Such sincere modesty would be engaging were it not exasperating: arithmeticians today would give a good deal for a glance at the methods Fermat must have devised but never published. In partial compensation for his indifference to publication, Fermat was a voluminous correspondent.

With the exception already noted concerning the use of one coordinate system for the representation of any number of curves, Fermat's analytic geometry[8] appears to be as general as that of Descartes. It is also more complete and systematic.[9] By 1629, according to Fermat's own dating, which there is no reason to question, he had found the general equation of a straight line, the equation of a circle with center at the origin, and equations of an ellipse, a parabola, and a rectangular hyperbola, the last referred to the asymptotes as axes.

The year (1638) after Descartes had published his geometry, Fermat communicated to him the accepted method of finding tangents. This originated in Fermat's investigation of maxima and minima, which he approached in substantially the same way as is done today in the calculus. What he did amounts to equating the derivative $f'(x)$ of $f(x)$ to zero to find the values of x which maximize or minimize $f(x)$. Geometrically, this is equivalent to finding the abscissas of the points on the curve $y = f(x)$ at which the tangent is parallel to the x-axis. He did not proceed to higher derivatives or their geometrical equivalent to determine whether $f'(x) = 0$ actually gives maxima or minima, as is necessary in a complete discussion. Nor did he isolate the calculation of the derivative from its implicit occurrence in problems of maxima and minima. Descartes either did not grasp the superiority of Fermat's method or was too chagrined to admit it, and his side of the controversy over tangents became somewhat acrimonious.

One positive gain has survived from this work on maxima and minima, Fermat's principle of least time[10] in optics.[11] This was the first (1657, 1661) of the great variational principles of the physical sciences.

As we shall pass on presently to Newton and his calculus, we may consider briefly here what the account given above of Fermat's tangents implies. If accepted at its full value, it makes Fermat an inventor of the differential calculus. The greatest mathematician of the eighteenth century, Lagrange, did so accept it. But the verdict is not unanimous.

The difference of opinion seems to hinge on Fermat's implicit conception of initially 'neighboring' but ultimately coincident points on a curve. To maximize or minimize $f(x)$ Fermat replaced x by $x + E$, where E differs but little from zero. He then equated $f(x)$ to $f(x + E)$, simplified the algebra, divided by E, and finally set E equal to zero.[12]

If this is legitimate differential calculus, then Fermat invented that calculus. If it is not, it seems no more illegitimate than its historical rivals. Thus Newton in his exposition of 1704, discussing the "fluxion" of x^n, n an arbitrary rational number, used his binomial formula (1676) to expand $(x + o)^n$, and formed the difference $(x + o)^n - x^n$. He then said, "Now let these augments [namely, $(x + o)^n - x^n$ and o] vanish, and their ultimate ratio will be 1 to nx^{n-1}." This was his method of "prime and ultimate ratios." In the Leibnizian notation used today, Newton thus finds $dx^n/dx = nx^{n-1}$.

The conclusion seems to be that either nobody in the seventeenth century invented the differential calculus, or Fermat was one of those who did. The matter is not settled by citing Newton's conception of a limit, because he did not develop a theory of limits in what he actually printed. But on this debatable difference of opinion everyone must form his own opinion after understanding the evidence as he may.

Before leaving the creators of analytic geometry, we may mention three further items from Descartes, although only the first is related to geometry. Descartes devised the notation $x, xx, x^3, x^4, \ldots$ for powers, and made the final break with the Greek tradition of admitting only first, second, and third powers ('lengths,' 'areas,' 'volumes') in geometry. After Descartes, geometers freely used powers higher than the third without a qualm, recognizing that representability as figures in Euclidean space for all of the terms in an equation is irrelevant to the geometrical interpretation of the analysis.[13]

The principle of undetermined coefficients was also stated by Descartes. Anything approaching what would now be admitted as a proof was about two centuries beyond the mathematics of his time. A second outstanding addition to algebra was the famous rule of signs given in every text on the theory of equations. This was the first universally applicable criterion for the nature of the roots of an algebraic equation. Even if it does not always yield any useful information, it admirably illustrates Descartes' flair for generality which made him the mathematician he was.

Newton, Leibniz, and the calculus

In the history of Newton's calculus, the temptation to read 'anticipation' into the works of his contemporaries and immediate predecessors is perhaps stronger than in that of any other major advance in mathematics. Knowing what we now do of the calculus and its implications in geometry and elementary kinematics, we can look back on many isolated discoveries in those domains and see in them what we *now* recognize as steps toward differentiation. But the discoverers, sometimes to our amazement, completely missed what we now perceive so plainly. They failed in each instance to take the last gigantic stride that now seems to us but a short step; and to credit them with strides they might have taken but did not is sheer sentimental romanticism.

As a relevant exercise in distinguishing between mathematical insight and facile prophecy after the fact, students of the calculus may wish to test their critical powers on the history of the "differential triangle" of Isaac Barrow (1630–1677, English). This was somewhat in the manner of Fermat. Ignoring this, we shall adhere to the generally accepted tradition and proceed on the hypothesis that Newton in his calculus did something new.

Isaac Newton (1642–1727, English), the posthumous son of a yeoman farmer, was born near Grantham, Lincolnshire, and passed his boyhood there. As a boy he was only passively interested in his schoolwork until he suddenly woke up at the age of adolescence. Earlier, he had shown unmistakable promise of experimental genius in the mechanical toys he invented and made to amuse himself and his young friends. It is interesting that both Newton and Descartes were delicate in childhood, and therefore had time to think and develop their own personalities while rougher boys were reducing one another to a very common denominator. Both matured into sturdy men, Descartes through military training, Newton by the inherited toughness of his farmer forebears.

After a desultory attempt to learn farming, Newton was sent to Trinity College, Cambridge, in 1661 (age nineteen). His undergraduate career, from all that is definitely known of it, was not particularly distinguished. Before going to Cambridge he had skimmed Euclid's *Elements*, and is said to have dubbed it "a trivial book." When, later, he understood Euclid's purpose, he revised his hasty judgment. In his own work he refers to Euclid with evident respect. His baffling encounters with "very little quantities" made him appreciate at least the tenth book of the *Elements*. It should encourage intelligent beginners to know that Newton found analytic geometry difficult at a first reading.

Perhaps fortunately for mathematics, Newton's studies were interrupted in 1665–6 by the Great Plague, when the university closed. Newton returned home, but not to farm. Before he was twenty-four years of age he had imagined the fundamental ideas of his fluxions (calculus) and his law of universal gravitation.

On returning in 1667 to Cambridge, Newton was elected a fellow of Trinity, and in 1669 succeeded Barrow, who resigned in his favor as Lucasian professor of mathematics. His first work to become known beyond the narrow circle of his intimate friends was in optics, beginning with his lectures of 1669.

As we are interested here mainly in Newton's calculus, we shall merely summarize the material circumstances of his career, full accounts of which are readily accessible. These also describe his epochal work in optics, which will not be discussed here as it belongs rather to physics than to mathematics.

In 1672 (age thirty), Newton was elected to the Royal Society, and from 1703 till his death was its president. His *Principia*, universally estimated by competent judges to be the greatest contribution to science ever made by one man, was composed in 1684–6 at the instigation of the astronomer E. Halley (1656–1742), at whose expense it was printed in 1687.

In 1689, and again in 1701, Newton was elected to represent Cambridge University in Parliament. He had no taste for debating, but he took his duties seriously, and showed a fine courage in championing the University's rights against the dictatorial meddling of King James the Second. At the age of fifty (1692) he suffered a severe illness and lost interest in scientific work, although he retained his unsurpassed intellectual powers to the end of his life.

Partly by his own desire, partly at the insistence of friends who wished to see him honored, Newton entered public life when he tired of science, and in 1696 was made warden of the mint. Having successfully directed the reform of the coinage, he was promoted to the mastership in 1699. In 1705 he was knighted by Queen Anne, and in 1727 he died. He is buried in Westminster Abbey.

Newton's excessive reluctance to publish his scientific work reflects certain aspects of his character. Although by no means a shy or timid man, Newton had a strong distaste for anything bordering on controversy. An unintelligent dispute over his work in optics at the beginning of his career taught him that scientific men are not always so objective as they might be, even in science, and he retired within himself in astonished disgust.

Nor was his notorious indifference to the survival of his scientific work affectation. But for the adroit coaxing and goading of Halley, the *Principia* would probably never have been written. Newton himself esteemed the theological writings to which he devoted the leisure of his later years far more highly than his science and mathematics. Again, in his work on light, Newton had proved himself one of the most acute experimentalists in the history of science; so it was but natural that he should spend much time and a considerable amount of money on what we should call alchemy, but what in his day was orthodox chemistry.

It was the ironic misfortune of this hater of profitless disputes to be embroiled in the most disastrous mathematical controversy in history, when some of his busy friends inveigled him into insinuating that Leibniz had plagiarized his own form of the calculus. We shall not discuss this, but merely state that the almost universal opinion now is that Leibniz invented his calculus later than Newton and independently. It must be gratifying to Englishmen to recall that it was another English mathematician, that born nonconformist Augustus De Morgan (1806–1871), who first undertook a judicial examination of the dispute and obtained some measure of justice for Leibniz.

The effects of this controversy on British mathematics for a century were deplorable. Patriotic loyalty to Newton blinded the English mathematicians to the obvious superiority of Leibniz' notation over Newton's dots, $\dot{y}$, $\ddot{y}$, . . . ; and early in the eighteenth century mathematical leadership passed to Switzerland and France. The Continental mathematicians, not Newton's countrymen, became his scientific heirs. At last, in the 1820's, the younger Cambridge mathematicians realized that their reactionary elders were not honoring Newton's memory by their obstinate nationalism, adopted Continental improvements in the calculus, and introduced analytic geometry and the Leibnizian notation into examinations. Cambridge again began to live mathematically.

Newton the mathematical astronomer, however, has continued to dominate English mathematics, at least in the popular estimate of what mathematics is. The most widely known English workers in the exact sciences are not England's leading mathematicians, but men who outside the British Empire would be classed as mathematical and theoretical physicists. There is more than an academic distinction between the two. A mathematician creates new mathematics or significantly improves and unifies what already exists. A mathematical physicist as a rule has but little time for purely mathematical pursuits, and uses mathematics merely as one of several convenient tools. A theoretical physicist has even less time to devote to mathematics.

Newton is almost unique in his triple supremacy as a pure mathematician, an applier of mathematics, and an experimentalist. Of the other two usually assigned to his class, Archimedes is generally considered his equal, and Gauss his superior in pure mathematics but his inferior in other respects.

Newton's competitor in the calculus was a man of strikingly different tastes and character. G. W. Leibniz (1646–1716, German) came of a scholarly family and had all the advantages (and disadvantages) of growing up with the Greek and Latin classics. Unlike Newton, he was remarkably precocious. His early mastery of languages, philosophy, theology, mathematics, and the law foreshadowed his later eminence in all fields of knowledge and, what seems regrettable, in diplomacy.

Leibniz is one of those rare characters in history of whom it can be said without exaggeration that he had a universal mind. His talents were purely intellectual. He had little or no artistic ability, and his lack of a feeling for tangible things occasionally betrayed him in science. Like Descartes,[14] who also went astray in science, Leibniz is probably most widely known today for his philosophy; but to a modern scientific mind his monads are as fantastically absurd[15] as Plato's eternal ideas. He thought incessantly. His unresting curiosity was attracted by everything and distracted by nothing. Perhaps the world is fortunate that much of his intelligence was dissipated in one way or other in the pursuit of money and fugitive honors.

As the reward for a revolutionary essay on the teaching of the law, Leibniz at the age of twenty-one was engaged by the Elector of Mainz as general agent and legal adviser. Most of his time thereafter was spent in travel on diplomatic missions for the Elector until the latter's death in 1673. Leibniz then became librarian, historian, and political factotum for the Brunswick family at Hanover.

During visits to France and England on political or diplomatic missions, Leibniz met the leading French and English men of science, and in exchange for some of their ideas disclosed his own. One such trade was to prove profoundly significant in the development of the calculus. If we seek the origin of modern work in the foundations not only of analysis but of all mathematics, we need look no farther than the following incident.

Until he met the great Dutch physicist and mathematician Christian Huygens (1629–1695) in Paris in 1672, Leibniz had but little if any competence in what was then modern mathematics. Such firsthand mathematical knowledge as he had was mostly Greek. Huygens enlightened him and undertook his mathematical education. Leibniz proved himself an exceedingly apt pupil. The two became good friends, corresponding till the death of Huygens in 1695. Leibniz begged Huygens for criticism of his projects and, naturally, got it.

It is only a speculation, but from Leibniz' ambitious character and his philosophic propensity for solving the universe, it is conceivable that his daring project for a universal symbolic reasoning was fostered by a determination to beat Descartes at his own game. The philosophic Frenchman had reduced all geometry to a universal method; the more philosophic German would similarly reduce all reasoning of whatever kind to a universal "characteristic" or, as would be said today, a symbolic mathematical science. Leibniz in 1679–80 confided his project to Huygens.[16] The physicist was not impressed.

By fatal mischance, Leibniz chose a trivial and singularly uninteresting geometrical problem to illustrate what he intended,[17] with the result that Huygens misunderstood the entire matter. He became somewhat polemical. This failure to see what Leibniz meant is the more remarkable as Huygens himself had a scientific vision which saw forests in spite of their innumerable trees. Possibly he was antagonized by Leibniz' boastful attitude. In his misunderstanding of what the ambitious philosopher-mathematician was trying to do, Huygens for once descended to the pedantries of captious criticism.

At first glance it may seem that Leibniz' attempts toward symbolic logic are irrelevant in the development of the calculus. Nothing could be farther from the fact. We shall see presently that Newton in his early encounters with continuity lost himself in the racecourse of Zeno, of whose paradoxes he perhaps had never heard. Subtly disguised but yet the same, these hoary difficulties have perplexed every mathematician, from Newton in the seventeenth century to Weierstrass in the nineteenth, who has sought not merely to obtain useful or interesting results by routine differentiations and integrations, but to understand the calculus itself. The calculus was difficult to Newton and Weierstrass; it is easy only to those who understand it too easily.

The modern attack on the fundamental problems of continuity has revealed the nature of the difficulties which baffled Newton, Leibniz, and the more thoughtful of their successors. It seems safe to say that without the mathematical logic which Leibniz advocated, and which he started to create, the critical work of the twentieth century on the foundations of analysis, and indeed of all mathematics, would have been humanly impossible.

Leibniz imagined the project of a 'calculus' of deductive reasoning; and if his own steps toward it were but short and hesitating, nevertheless it was his bold conception which encouraged others to proceed. It seems rather late in the day, therefore, to persist in seeing Leibniz the mathematician merely as a major satellite of Newton.

Historical tradition reiterates to weariness the undisputed fact that the *d*-notation of Leibniz is vastly superior to the dots of Newton. But if, with Gauss,[18] we believe that in mathematics notions are more important than notations, we must place the emphasis elsewhere. The greatest work of Leibniz, from the standpoint of modern mathematics, is not his improvement of the differential and integral calculus, great though that was, but his calculus of reasoning. He shines by his own light.

Little need be said here of Leibniz' career as a diplomat. To him is attributed that epitome of unstable equilibrium which later jugglers of destiny were to worship as the Balance of Power. In his diplomacy, Leibniz was neither more nor less unscrupulous than any of his famous successors in that dubious art. He was merely less incompetent than the majority. Unlike some of them, Leibniz did not succumb beneath a weight of honors heaped on him by grateful princes, but died neglected and forgotten by those whose petty fortunes he had made. When his employer departed from Hanover to become King George the First of England, Leibniz was discarded in the library to continue his history of the Brunswick family, surely a fitting occupation for one of the supreme intellects of all time. Only his secretary followed him to his grave.

Such were the two mortals who finally created the calculus.

Newton's version of the calculus

Newton's first calculus, of 1665–6, seems to have been abstracted from intuitive ideas of motion. A curve was imagined as traced by the motion of a 'flowing' point. The 'infinitely short' path traced by the point in an 'infinitely short' time was called the "momentum" and this momentum divided by the infinitely short time was the "fluxion." If the "flowing quantity" is x, its fluxion is denoted by $\dot{x}$. In our terminology, if x is the function $f(t)$ of the time t, $\dot{x}$ is dx/dt, the velocity at time t. Similarly, the fluxion of $\dot{x}$ is $\ddot{x}$, our d^2x/dt^2; $\dddot{x}$ is d^3x/dt^3, and so on.[10]

Newton regarded our dx/dt as the actual ratio of two "infinitely small quantities" in this first calculus. He had no approach to a limit that would be recognized today. The following extract from the *Principia* (1687) will indicate that Newton himself was dissatisfied with his own refinement of the method of fluxions.

> It is objected that there is no ultimate ratio of evanescent quantities because the proportion [ratio] before the quantities have vanished is not ultimate; and, when they have vanished, is none. But by the same argument, it might as well be maintained, that there is no ultimate velocity of a body arriving at a certain place, when its motion is ended: because the velocity, before the body arrives at the place, is not its ultimate velocity; when it has arrived, is none. But the answer is easy . . . There is a limit, which the velocity at the end of the motion may attain, but cannot exceed.

This is what Zeno and the tortoise knew, and what neither of them succeeded in clarifying. It is no disparagement of Newton to observe that the foregoing extract might have been written by Aristotle; indeed it bears a singular resemblance to Aristotle's discussion[20] of the infinite, the continuous, motion, and Zeno's paradoxes. A further observation of Newton's recalls Eudoxus:

"It may also be argued, that if the ultimate ratios of evanescent quantities are given, their ultimate magnitude will also be given; and so all quantities will consist of indivisibles, which is contrary to what Euclid has demonstrated concerning incommensurables, in the tenth book of his *Elements*." Again compare with Aristotle.

From the last it is clear that Newton's understanding of Euclid was sharper than Cavalieri's. It also suggests that the difficulties of intelligent beginners with limits and continuity are not mere willful perversity. In his third attempt (1704), Newton returns to the attack on continuity, and transfers the central difficulty to an unanalyzed "continued motion":

"I consider mathematical quantities in this place not as consisting of very small parts; but as described by a continued motion. Lines are described, and thereby generated not by the apposition of parts, but by the continued motion of points . . ."

It was considerations of this kind, among others, that drove the analysts of the nineteenth century to desperation and impelled them to attempt a meaningful foundation for the calculus. In spite of his resolute abandonment of "very small parts," Newton never quite circumvented the "very small quantity" which persistently annoyed him. In the *Principia* (Bk. I, Sec. I, Lemma I), he started toward a theory of limits and continuity:

"Quantities, and the ratios of quantities, which, in any finite time, tend continuously to equality; and before the end of that time, approach nearer to each other than by any given difference, become ultimately equal."

Leibniz' version

Leibniz for his part favored a species of differential, as that highly elusive concept is frequently misunderstood by practical engineers today. Thus, to find the differential of xy, he subtracted xy from $(x + dx)(y + dy)$ and rejected $dx\,dy$ because he considered it negligibly small in comparison with $x\,dy$ and

$y\,dx$—all without sound justification. This gave him the correct result, $d(xy) = x\,dy + y\,dx$.

On safer ground, he introduced the current notation for derivatives and the integral sign, $\int$, an elongated *s* from *summa* (sum). Both Leibniz and Newton were familiar with the fundamental theorem of the calculus connecting integrals as sums with integrals as anti-derivatives. They also established the elementary formulas of the calculus. It is of interest that the correct result for the derivative of a product eluded Leibniz on his first attempt.

Rigor; anticipations

It is generally agreed that reasonably sound but not necessarily final ideas of limits, continuity, differentiation, and integration came only in the nineteenth and twentieth centuries, beginning with Cauchy in 1821–3. This raises an extremely interesting question: how did the master analysts of the eighteenth century—the Bernoullis, Euler, Lagrange, Laplace—contrive to get consistently right results in by far the greater part of their work in both pure and applied mathematics? What these great mathematicians mistook for valid reasoning at the very beginning of the calculus is now universally regarded as unsound.

No short answer is possible; but history shows that frequently the essential, usable part of a mathematical doctrine is grasped intuitively long before any rational basis is provided for the doctrine itself. The creative mathematicians between Newton and Cauchy obtained mostly correct results—according to present standards—because, in spite of their ineffectual attempts to be logically rigorous, they had instinctively apprehended the self-consistent part of their mathematics.

Just as no short answer can dispose of our predecessors' good fortune, so none can dispose of ours. Like them, we consistently get meaningful results, although we realize that there is much obscurity in the foundations of our own analysis. It is now generally admitted that neither Cauchy nor his more rigorous successor Weierstrass said the last word, and we may confidently expect that it will not be uttered in our generation.

Whatever else may be said of Newton's calculus, it is still true that he endowed mathematics and the exact sciences with their most effective method of exploration and discovery. Linked to his own law of universal gravitation, the calculus in less than a century gave a more comprehensive understanding of the solar system than had accrued from thousands of years of pre-dynamical astronomy. And when differential equations, Newton's method of inverse tangents, were applied to the physical sciences, a new and unsuspected universe was revealed. The experimental method of Galileo combined with the calculus of Newton and Leibniz generated modern physical science and its applications to technology.

To conclude this account of the emergence of the calculus, we shall compensate for the deliberate neglect of pseudo anticipations by citing a real one in a subject which is basic for the modern attack on the foundations of analysis.

Galileo observed as early as 1638 that there are precisely as many squares 1, 4, 9, 16, 25, . . . as there are positive integers altogether. This is evident from the sequences

$$1,\ 2,\ 3,\ 4,\ 5,\ 6,\ \ldots,\ n,\ \ldots$$
$$1^2,\ 2^2,\ 3^2,\ 4^2,\ 5^2,\ 6^2,\ \ldots,\ n^2,\ \ldots$$

He thus recognized the fundamental distinction between finite and infinite classes that became current in the late nineteenth century. An infinite class is one in which there is a one-one correspondence between the whole class and a subclass of the whole. Or, what is equivalent, there are as many things in some part of an infinite class as there are in the whole class. The like is not true of finite classes.

A class whose elements can be put in one-one correspondence with the integers 1, 2, 3, . . . is said to be denumerable. All the points in any line segment, finite or infinite in length, form a non-denumerable set. A basic course in the calculus (usually the second) starts from the theory of point sets. The distinction between denumerable and non-denumerable classes was not stated by Galileo; it was observed about 1840 by Bolzano and in 1878 by Cantor. But Galileo's recognition of the cardinal property of all infinite classes makes him one of the genuine anticipators in the history of the calculus. The other was Archimedes.

Emergence of the mathematical theory of probability

Games of chance are probably as old as the human desire to get something for nothing; but their mathematical implications were appreciated only after Fermat and Pascal in 1654 reduced chance to law. Galileo had given a correct solution of a gaming problem by laborious tabulation of possible cases, but did not proceed to general principles. The "problem of points" which inspired the originators of the mathematical theory of probability was well known to Cardan who, among his other accomplishments, was a reckless gambler. He, however, did nothing of importance toward a science of chance; and it is customary without any quibbling to regard Pascal and Fermat as the founders of mathematical probability.

In the epoch-making problem, the first of two players who scores n points wins. If the game is abandoned when one has made a points and the other b, in what ratio shall the stakes be divided between them? This reduces to calculating the probability each has of winning when the game is stopped. It is assumed that the players have equal chances of making a point.

The problem was proposed to Pascal by a highly intelligent gentleman addicted to gaming, Antoine Gombaud, Chevalier de Méré, and Pascal communicated it to Fermat. Both solved it correctly, but by different reasoning. In some of his work, Pascal made a slip, which Fermat corrected. Thus originated the mathematics of chance which today is basic in all statistical analysis from stock-market trends and insurance to intelligence tests and biometrics.

As modern physics has become more certainly uncertain, the mathematics of probabilities has steadily increased in scientific importance. Newtonian mechanics is applicable to a completely determinate science in which differential equations imply the future history of a mechanistically determined universe. For the scientific interpretation of laboratory experiments, particularly in atomic physics, the strictly mechanical method of Newton, Lagrange, Laplace, and their successors, originating in Galilean mechanics and dynamical astronomy, is no longer adequate and is being increasingly supplemented by the mathematics of statistics and probability. The necessary mathematics all developed from the fundamental principles of mathematical probability laid down by Fermat and Pascal in about three months by a painstaking application of uncommon sense.[21]

Later applications of analysis to the theory of probability, chiefly to obtain usable approximations to the very large numbers occurring in even simple combinatorial problems, have made the modern theory highly technical. But with the exception of epistemological difficulties concerning the meaning of probability, the basic principles remain those of 1654 as stated in intermediate texts on algebra. In this connection it should be mentioned that the versatile Huygens got wind of what Pascal and Fermat were doing, and in 1654 published one of the earliest treatises on probability.[22] The concept of mathematical expectation was his.

The relative permanence of the mathematical foundations of probability as laid down in the seventeenth century is characteristic of the mathematics of the discrete, in which generally there has been less need for revision than in analysis.

The origin of modern arithmetic

We shall understand 'arithmetic' in the sense of the Greek arithmetica. Equivalents are 'the higher arithmetic' and 'the theory of numbers,' also, unfortunately, the hybrid 'number theory' with its Aryan adjective and adverb, 'number theoretic' and 'number theoretically.' Gauss, the foremost exponent of the classical theory after Fermat, preferred the simpler 'arithmetic' or, at longest, 'higher arithmetic.'

Modern arithmetic began with Fermat, roughly in the period 1630–65. Significant as was Fermat's work in other departments of mathematics, he is usually considered to have made his greatest and most personal contribution in arithmetic.

This extensive division of mathematics differs from others in its lack of general methods. Even comprehensive theorems appear to be more difficult to devise than, say, in algebra or analysis. Thus, in algebra there is a complete theory of the solution of algebraic equations in one unknown; in fact there are two complete theories. In arithmetic the simplest corresponding problem is the solution in integers of equations in two unknowns with integer coefficients, and for this there is nothing approaching a complete theory. Such progress as has been made since Fermat will be noted in a later chapter.

Many of Fermat's discoveries were either recorded as marginal notes in his books (the arithmetic in his copy of Bachet's Diophantus) or were communicated, usually without proof, to correspondents. Some of his theorems were proposed by him as challenge problems to the English mathematicians. For example, he demanded a proof that the only positive integer solution of $x^2 + 2 = y^3$ is $x = 5, y = 3$.

It will suffice to state those two of Fermat's discoveries which appear to have had the profoundest influence on arithmetic and algebra since his time, and the one general method in arithmetic due to him.

Fermat stated that if n is a positive integer not divisible by the positive prime p, then $n^{p-1} - 1$ is divisible by p. The Chinese "seem to have known as early as 500 B.C."[23] the special case $n = 2$. Any student of the theory of algebraic equations, or of modern algebra, or of arithmetic, will recall the frequent appearance of this fundamental theorem. The first published proof was Euler's in 1738, discovered in 1732; Leibniz had obtained a proof before 1683 but did not publish it. The rule of priority in mathematics is first publication.

The second famous assertion of Fermat, his celebrated 'Last Theorem,' states that $x^n + y^n = z^n$, $xyz \neq 0$, $n > 2$, is impossible in integers x, y, z, n. He claimed (1637) to have discovered a marvelous proof; and whether or not he had, no proof has yet (1945) been found. There seems to be but little point now in proving the theorem for special n's, enough in that direction being known to make it fairly plausible that the theorem is true. But, to take out insurance against a possible disproof tomorrow, it must be emphasized that arithmetic is the last place in mathematics where unsubstantiated guessing is either ethical or profitable. Numerical evidence counts for very little;[24] the only luxury a reputable arithmetician allows himself is proof.

It is generally agreed that the famous 'Last Theorem,' true or false, is of but slight interest today. But its importance in the development of arithmetic and modern algebra has been very great. This will be discussed in the proper place.

Fermat's general method, that of "infinite descent," is profoundly ingenious, but has the disadvantage that it is often extremely difficult to apply. In the particular theorem for which Fermat invented the method, it is required to prove that every positive prime of the form $4n + 1$ is a sum of two integer squares. From the assumption that the theorem is false for some such prime p, Fermat deduced that it is also false for a smaller prime of the same kind. Descending thus he proved, on the assumption of falsity, that 5 is not the sum of two squares. But $5 = 1^2 + 2^2$; hence the theorem.

The outstanding desideratum in arithmetic is the invention of general methods applicable to nontrivial types of problems. Further, "the arithmetical solution of a problem should consist in prescribing a finite number of purely arithmetical operations (exempt from all tentative processes), by which all the numbers satisfying the conditions of the problem, and those only are obtained."[25] Nobody after Euclid and before Lagrange in the eighteenth century even distantly approached this ideal.

Emergence of synthetic projective geometry

The sudden rise of synthetic projective geometry in the seventeenth century appears now as a belated resurrection of the Greek spirit. As already noted, Pappus in the fourth century A.D. anticipated a cardinal property of cross ratios; and even earlier Menelaus (first century A.D.) may have proved a theorem which can now be interpreted similarly. But it was only with G. Desargues' (1593–1662, French engineer and architect) eccentric *Brouillon project* (abbreviated title) of 1639 that synthetic projective geometry was developed as a new and independent division of geometry.

Doubtless the great advance in perspective drawing by the artists of the Renaissance made inevitable the emergence of a geometrical theory including perspective as a special case; and Desargues the architect was doubtless influenced by what in his day was surrealism. In any event, he composed more like an artist than a geometer, inventing the most outrageous technical jargon in mathematics for the enlightenment of himself and the mystification of his disciples. Fortunately, Desarguesian has long been a dead language. In current terminology, 'projective' means invariance under the group G of all general linear homogeneous transformations in the space (of 1, 2, 3, . . . dimensions) concerned, but not under all the transformations of any group containing G as a subgroup.

After his own fashion, Desargues discussed cross ratio; poles and polars; Kepler's principle (1604) of continuity, in which a straight line is closed at infinity and parallels meet there; involutions; asymptotes as tangents at infinity; his famous theorem on triangles in perspective; and some of the projective properties of quadrilaterals inscribed in conics. Descartes greatly admired Desargues' invention, but happily for the future of geometry did not hesitate on that account to advocate his own.

Desargues' most enthusiastic convert was the same Pascal who participated in the creation of the mathematical theory of probability. B. Pascal (1623–1662) was therefore a very considerable mathematician, even if, like Descartes and Leibniz, he is popularly remembered for other things. His magnitude as a religionist has overshadowed his accomplishments as a mathematician and physicist, and for one who has ever heard of Pascal's *Essay pour les coniques* there must be a million who have read at least a page of his *Pensées*. If anything, Pascal was more genuinely precocious than Leibniz. As a boy he was no mere sponge absorbing the learning of others, but a creative mathematician. At twelve he rediscovered and proved for himself several of the simpler theorems of elementary geometry. Four years later he had composed the famous essay on conics, in which he developed the consequences of his *hexagramma mysticum*—pairs of opposite sides of a hexagon inscribed in a conic intersect in collinear points. Combining his mathematics and his talent for physics, Pascal at nineteen (1642) invented an adding machine, the ancestor of all those in use today. This was greatly improved about thirty years later by Leibniz, whose machine did both addition and multiplication.

Pascal made grateful acknowledgment to Desargues for his skill in projective geometry. Perhaps in all of his mathematics Pascal was the brilliant commentator rather than the bold originator. Organically and spiritually ill for most of his thirty-nine years, he was unable, apparently, to concentrate his powers on the creation of a comprehensive method in anything, and his brilliance was dispersed and dissipated in the piecemeal illumination of other men's ideas. Much of his mentality was absorbed in the religious controversies of his time and in hopeless attempts to reconcile his own internal conflicts. Beyond his 'mystic hexagram' and his share in probability, it cannot be said that Pascal's contributions left more than a transient shadow on the surface of mathematics. The main stream flowed far deeper than he ever dreamed.

Before quitting the field for about a century and a quarter, synthetic projective geometry fought a terrific pitched battle to survive against its analytic antagonist, in the impressive *Sectiones conicae* (1685) of P. de la Hire (1640–1718, French). La Hire proved over three hundred projective theorems synthetically, and in an astounding appendix showed that all the theorems of Apollonius on conics are obtainable by the method of projection. But even these spectacular gymnastics failed to convince geometers that the synthetic method is as supple as the analytic. The conics no doubt were idealized from the Archetypal Circle by Plato's Geometer with the Eternal Idea of projective geometry at the back of his mind; but then, not all plane curves are conics. Nor were they even when La Hire fought his desperate rearguard action against Descartes' analytics. Synthetic projective geometry lapsed into temporary oblivion, and the treatises of Desargues and La Hire became collectors' rarities.

The other incidental advance of the seventeenth century that likewise was forgotten for a season, the universal characteristic of Leibniz, will be noted in a later chapter. We pass on to the origin of the applied mathematics which was to dominate the work of Newton's most prominent successors for a century after his death.

Origin of modern applied mathematics

The indebtedness of science and technology to pure mathematics was noted in the Prospectus. We shall now inspect the other side of the ledger at somewhat greater length than may seem necessary to those already conversant with the facts. We do this because hypersensitive mathematicians are sometimes inclined to exalt unduly the freedom and purely imaginative character of their creations, and to dwell exclusively on the admitted indebtedness of science to mathematics. The historical balance sheet indicates, as will be seen in greater detail later, that science and modern mathematics are so closely affiliated that neither owes the other anything, each borrowing freely from the other and repaying its debts a hundredfold.

Intermediate between pure and applied mathematics, improvements in numerical calculation are of more importance for the applications than for mathematics itself. Logarithms, for example, accelerated the practical developments of astronomy but were not a necessity even to that most efficient servant of civilization. The pertinacity of a Kepler cannot be thwarted by any amount of manual computation; and to claim that logarithms made modern astronomy or any other science possible is to forget that human zeal—or obstinacy—in pursuit of a fixed idea can withstand any finite punishment. But as logarithms undoubtedly hastened the sciences of the eighteenth and nineteenth centuries on their way to whatever is to be their ultimate contribution to civilization, or to its destruction, they must be included in any account of the origin of modern applied mathematics. The seventeenth-century invention of logarithms may therefore be properly assigned to applied mathematics.

Modern applied mathematics originated in Newton's theory of universal gravitation developed in his *Principia.* Astronomy before Newton was purely descriptive. The motions of the planets were described with increasing accuracy, and from the Babylonians to Ptolemy were fitted into geometrical frameworks of ever greater complexity. Copernicus simplified the geometry. But there was no physical hypothesis abstracted and consolidated in postulates from which the geometry could be deduced. Before such postulates could be stated profitably, accurate observations to determine the facts were necessary. These were provided in abundance by Tycho Brahe (1546–1601, Danish), whose industrious assistant for a short time, Johann Kepler, (1571–1630, German), subsumed the observations under the three laws of motion known by his name. The first two were published in 1609, the third in 1619: the orbit of a planet is an ellipse with the sun at one focus; the areas swept out in equal times by the line joining the sun to a planet are equal; the squares of the periodic times of the planets are proportional to the cubes of their mean distances from the sun.

Kepler's laws were the climax of thousands of years of an empirical geometry of the heavens. They were discovered as the result of about twenty-two years of incessant calculation, without logarithms, one promising guess after another being ruthlessly discarded as it failed to meet the exacting demands of observational accuracy. Only Kepler's Pythagorean faith in a discoverable mathematical harmony in nature sustained him. The story of his persistence in spite of persecution and domestic tragedies that would have broken an ordinary man is one of the most heroic in science.

The contemporaneous invention of logarithms was to reduce all such inhuman labor as Kepler's to more manageable proportions. The history of logarithms is another epic of perseverance second only to Kepler's. Baron Napier of Merchistoun (1550–1617, Scotch), in the leisure remaining from his duties as a landlord and his unavailing labors to prove that the reigning pope was Antichrist, invented logarithms.

When it is remembered that Napier died before Descartes introduced the notation n, nn, n,[3] . . . for powers, we cease to wonder that it took him all of twenty years to reason out the existence and properties of logarithms.

The fundamental idea of the correspondence between two series of numbers, one in arithmetic, the other in geometric progression, . . . was explained by Napier through the conception of two points moving on separate straight lines, the one with uniform, the other with accelerated velocity. If the reader, with all his acquired modern knowledge, will attempt to obtain for himself in this way a demonstration of the fundamental rules of logarithmic calculation, he will rise from the exercise with an adequate conception of the penetrating genius of the inventor of logarithms. (G. Chrystal.)

Add to this that Napier's logarithm of n would be our $10^7 \log_e (10^7 n^{-1})$, where e is the base of the natural system.

After the invention of the calculus, investigation of the logarithmic function, of greater significance in mathematics than the logarithms computed by its use, followed as a matter of course from the simple differential equation $dy = y\,dx$.

Napier gave Tycho a forecast of his invention in 1594, and in 1614 published his *Descriptio*. In 1624 a usable table by H. Briggs (1561–1631, English) was published, as also was one by Kepler. Other tables quickly appeared, and by 1630 logarithms were in the equipment of every computing astronomer.

For those interested in squabbles over priority, it may be recalled that logarithms are one of the most disorderly battlegrounds in mathematical history. It will suffice here to state the outcome of the fray as adjudicated in 1914. Napier's priority in publication is undisputed; J. Bürgi (1552–1632, Prague) independently invented logarithms and constructed a table between 1603 and 1611, while "Napier worked on logarithms probably as early as 1594 . . . ; therefore, Napier began working on logarithms probably much earlier than Bürgi."[26] The only facts concerning logarithms of any importance for the development of mathematics are those stated in the concluding sentence of the preceding paragraph.

Disputes like this and the other over the calculus have made more than one man of science envy his successors of ten thousand years hence, to whom Newton and Leibniz, Napier and Bürgi, and scores of lesser contestants for individual fame will be semimythical figures as indistinct as Pythagoras.

The harmonious geometry of Kepler's laws challenged mathematical ingenuity to devise a hypothesis from which they could be deduced. Among others, Newton's self-constituted rival and gadfly, the brilliantly original R. Hooke (1635–1703, English), had guessed and perhaps proved that Kepler's laws implied an inverse-square law of attraction, but could not determine the form of the orbit from this law. Newton, on being consulted in 1684, restored a proof, which he had discovered but mislaid, that the required orbit is an ellipse. This incident appears to have been the origin of the *Principia*. From his hypothesis of universal gravitation that any two particles of matter in the universe, of masses m_1, m_2 at a distance d apart, attract one another with a force proportional to m_1m_2/d^2 (m_1, m_2, d, and the force being measured in the appropriate units), Newton deduced Kepler's laws.

The deduction would have been impossible without a rational dynamics. This had been provided by Galileo[27] and by Newton himself. Just as the Pythagoreans had reduced the intuitive perception of form to geometry, and the great geometrical astronomers from Eudoxus and Hipparchus to Copernicus and Kepler had reduced the motions of the planets to geometry, so Galileo undertook the reduction of all motion to mathematics. He advanced beyond his predecessors chiefly because he aided reason by experiment, determining the facts in connection with falling bodies by accurate, controlled observation before venturing to mathematicize.

To some it seems incredible that any human being could ever have believed it possible to reason out the behavior of falling bodies without appeal to experiment. But one of the greatest intellects in history, Aristotle, had sufficient confidence in his logic to legislate for a universe which has but little respect for the unaided intellect. Others[28] see nothing questionable in attempts such as Aristotle's, substituting for the classical and medieval belief in Aristotelian logic an eager faith in the creative power of the intricate tautologies of mathematics. It may yet be too early to judge which side, if either, is right; but it is a fact that Galilean science, not Aristotelian logic and metaphysics, made our material civilization what it is.

Whether or not the legendary experiment in which Galileo confounded the Aristotelian scholastics by dropping shot from the Leaning Tower of Pisa ever took place,[29] Galileo knew by 1591 that a one-pound shot and a ten-pound shot dropped simultaneously from the same height strike the ground simultaneously. Experiments on motion down inclined planes gave him further data to be fitted into the mathematical theory of motion he sought to construct. As tentative hypotheses were subjected to experimental verification, the cardinal definitions and postulates of dynamics began to emerge.

In particular, Galileo mathematicized distance, time, velocity, and acceleration into the scientific (experimentally measurable) things they still are in classical dynamics. He sought to frame definitions that would respond to repeatable observations. He also understood an equivalent of Newton's first postulate of motion—inertia: every body will continue in its state of rest, or of uniform motion in a straight line, except in so far as it is compelled to change that state by impressed force. This postulate contradicted the naive intuitions of Galileo's predecessors and controverted the common sense of ages.

Galileo also understood at least special cases of Newton's second postulate: rate of change of momentum is proportional to the impressed force, and takes place in the direction in which the force acts. The mathematically important concept here is that of a rate; for rates are derivatives, and hence velocity, acceleration, and force are brought within range of the calculus. We have seen that Newton probably had velocity in mind when thinking about fluxions.

"In the two plague years of 1665 and 1666," as Newton states, he deduced from Kepler's third law that "the forces which keep the Planets in their Orbs must be reciprocally as the squares of their distances from the centers about which they revolve: and therefore [I] compared the force requisite to keep the Moon in her Orb with the force of gravity at the surface of the Earth, and found them answer pretty nearly."[30]

Further progress in the mathematical theory of gravitation was temporarily halted by Newton's lack at the time of a theorem in the integral calculus: the gravitational attraction between two homogeneous spheres can be calculated as if the masses of the spheres were concentrated in their centers. Once this theorem is proved, the Newtonian law of universal gravitation is applicable. If there is one master key to dynamical astronomy, this is it. With this key Newton proceeded in 1685 to unlock the heavens. He also, for the first time, gave a rational theory of the tides.

The Newtonian celestial mechanics was the first of the great syntheses of natural phenomena. From its very nature, celestial mechanics without the dynamics of Galileo and Newton, or without the calculus of Newton and Leibniz, is unthinkable. The Galilean method in science was to provide the model for even more recondite mathematical syntheses, as in the theories of heat, light, sound, and electricity. But the modern scientific method, invented by Galileo and Newton, of welding experiment and mathematics into a single implement of discovery and exploration was to remain fundamentally the same as it was in Galileo's *Discorsi* and Newton's *Principia.*

In a day when science is being discredited by messianic ignoramuses with enormous followings, it is well occasionally to recall the cliché, trite though it may be, that without this union of experiment and mathematics our civilization would not exist. Less trite is the more recent observation that because of this very union our civilization may cease to exist. And, while we are facing facts, we note the opinion of many observers that ever since the days of Aquinas science has been feared or secretly hated by nine human beings out of every ten who have sufficient animation to hate or fear anything. Science has been grudgingly tolerated since the days of Galileo and Newton only because it has increased material wealth. If science dies, mathematics dies with it.

To give an indication of how significantly dynamics and the Newtonian theory have influenced analysis, we may cite a few specific instances, some of which will be considered more fully in later chapters. Since the earth is not a sphere but a spheroid, its attraction on an exterior mass-particle cannot be calculated with the same precision as if its mass were concentrated at the center. When astronomy became more exact after Newton, the slight departure from perfect sphericity had to be included in the calculations, and this necessitated the invention of new functions, such as Legendre's in potential theory. So rudimentary a dynamical problem as Galileo's of the time of vibration of a simple pendulum of constant length leads at once in the general case to an elliptic integral. Such integrals, by inversion, generated the vast theory of doubly periodic functions. These in their turn were recognized in the late nineteenth century as but special cases of automorphic functions, whose theory still is far from complete.

All of the earlier functions together suggested to Lagrange, Cauchy, and others in the late eighteenth and early nineteenth centuries general theories of functions, culminating in the theory of functions of a complex variable. Fourier's analytical theory of heat (final form 1822), devised in the Galileo-Newton tradition of controlled observation plus mathematics, is the ultimate source of much modern work in the theory of functions of a real variable and in the critical examination of the foundations of mathematics. Finally, the gravitational interactions of a system of mass-particles, in particular of three, generated the theory of perturbations and all its intricate analysis; and the problem of three bodies, partly topologized in the late nineteenth century, is the source of the modern theory of periodic orbits from which a qualitative, topologized dynamics is rapidly developing.

Geometry also has enriched itself by successive alliances with mechanics. In the seventeenth century the astronomical need for the accurate measurement of time inspired Huygens to construct the first pendulum clock (1656). Incidentally, he was compelled to investigate the (small) oscillations of a compound pendulum, the first dynamical problem beyond the dynamics of particles to be discussed mathematically. From practical clock making Huygens was led to his great work[31] in horology (1673), in which he defined and investigated evolutes and involutes. The cycloid, sometimes called the Helen of geometry, partly on account of its graceful form and beautiful properties, figures prominently in this science. Huygens proved the remarkable theorem that the cycloid is the tautochrone. In more recent times, the four-dimensional geometry of Plücker (nineteenth century) in which straight lines instead of points are taken as the irreducible elements of space, found a ready interpretation in the dynamics of rigid bodies. Conversely, this dynamics suggested much to be done in line geometry. But what is perhaps the greatest service a physical science has ever rendered geometry was the sudden acceleration imparted to differential geometry by Einstein's general relativity and his relativistic theory of gravitation in the second decade of the twentieth century. In the curiously sanguine months after the buzzards of Versailles had completed their labors, it was frequently said that Einstein's work would outlast the memory of the world war, as the science and mathematics of Archimedes has outlasted the Punic wars in the consciousness of all but professional historians. Twenty years later the human race had made a complete recovery from its attack of optimism.

Here we leave the mathematics of the seventeenth century and commit ourselves to the turbulent stream that gushed from that inexhaustible source.

CHAPTER 8

Extensions of Number

In following the development of mathematics since the death of Newton (1727), we might start from any one of arithmetic, algebra, geometry, or applied mathematics. As arithmetic preceded the others in the historical order from Babylon to Göttingen, we shall discuss it first. Those who are more interested in one of the other topics may pass at once to it.

The detailed growth of the number concept to be described in this and the following five chapters being quite intricate, we shall indicate first the principal features to be observed.

Four critical periods

About four centuries of generalization, confused and hesitating at first, produced the number systems of analysis, algebra, mathematical physics, and the higher arithmetic of the twentieth century. The final gain left mathematics with three major acquisitions: the ordinary complex numbers of algebra and analysis, and their subclasses of algebraic integers; the hypercomplex number systems of algebra, geometry, and physics; the continuum of real numbers as it appears in the modern theories of functions of real and complex variables. The five periods of most radical change were the decade bisected by the year 1800, the late 1830's and early 1840's, the 1870's, and the twenty years bisected by 1900.

With the first period is associated one beginning of modern abstract arithmetic and algebra, in the use by Gauss (1801) of a particular equivalence relation, which he called congruence, to map an infinite class of integers on a finite subclass. The general method of mapping (homomorphism) implicit in this early work was not clearly formulated and isolated for independent study until the twentieth century, when it became basic in abstract algebra, topology, and elsewhere.

In the 1830's, the British algebraists clearly recognized the purely abstract and formal character of elementary algebra. This was followed, in the 1840's, by Hamilton's quaternions and the vastly more general algebras of Grassmann, from which the vector algebras of mathematical physics evolved. From the standpoint of pure mathematics, the lasting residue of this period was a widely generalized conception of number.

The 1870's saw the inception of the modern attack on the real number system in the work of Cantor, Dedekind, Méray, and Weierstrass. The outcome, in the late nineteenth century, was the arithmetization of analysis and the beginning of the modern critical movement. What now appears as the most enduring residue of this stormy period is the enormous expansion of mathematical logic during the first four decades of the twentieth century.

The third period passed into the fourth about 1897 with the first appearance of the modern paradoxes of the infinite. The latter were largely responsible for the sudden growth of mathematical logic, which has reacted strongly on all mathematics and in particular on the number concept.

We shall have occasion to refer frequently in anticipation to problems as yet unsettled concerning the nature of number and the continuum of real numbers. Because a problem defies solution is no ground for believing it to be unsolvable. The outstanding obstacles that have hitherto blocked clear perception of the nature of number may be removed tomorrow. In any event, none of the unsolved problems of number has stopped progress in both pure and applied mathematics. On the contrary, in pure mathematics these unresolved difficulties have inspired much valuable work; and for applied mathematics, even the most serious doubts have as yet proved wholly irrelevant in obtaining scientific conclusions which can be checked against experience in the laboratory.

Having followed the development of number in its purely mathematical aspects, we shall return in a later chapter to the impact of science on mathematics. It will be seen that applied mathematicians have been justified in their bold use of an analysis which may not yet meet all the demands of logical rigor.

From time to time we shall call attention to points of special significance to be observed while following details. A general observation may be emphasized here. As mathematics passed the year 1800 and entered the recent period, there was a steady trend toward increasing abstractness and generality. By the middle of the nineteenth century, the spirit of mathematics had changed so profoundly that even the leading mathematicians of the eighteenth century, could they have witnessed the outcome of half a century's progress, would scarcely have recognized it as mathematics. The older point of view of course persisted, but it was no longer that of the men who were creating new mathematics. Another quarter of a century, and it had become almost a disgrace for a first-rank mathematician to attack a special problem of the kind that would have engaged Euler in much of his work. Abstractness and generality, directed to the creation of universal methods and inclusive theories, became the order of the day. There had been one precedent in the eighteenth century for such work, the dynamics of Lagrange. There is a second clue through the intricate development. This leads back to Pythagoras, and is so suggestive that we shall describe it next by itself in relation to the preceding sketch.

The Pythagorean adventure

Perhaps the feature of greatest general interest in the entire development is the wide departure from the Pythagorean program of basing all mathematics on the 'natural' numbers 1, 2, 3, . . . in the periods of intensest creativity, and the final return to Pythagoras for a brief interval after the natural numbers had been extended to meet the demands of analysis, geometry, physics, algebra, and the higher arithmetic. What was probably the golden age of the Pythagorean program lasted through the second half of the nineteenth century. Thereafter, the modern critical movement sought to base the natural numbers, and hence all their acquired extensions, on mathematical logic. This later program was already strongly hinted in that of the nineteenth century, which had attempted to derive number from the theory of infinite classes.

To be appreciated as the great adventure in thought that it is, this circling movement away from Pythagoras and back to him, with the subsequent flying off at a tangent he was incapable of imagining, must be inspected in some detail. This we shall do in the following five chapters. We shall see modern mathematics as a whole becoming increasingly self-conscious and critical of its own naive behavior in the eighteenth century, and passing safely through adolescence in the decade 1820–30. Mathematics then became less interested in the uncritical analysis that produced surprisingly accurate results in the heavens with Laplace's celestial mechanics (1799, 1802, 1805), and on earth in the intuitive calculations of Fourier's analytic theory of heat conduction (1822). Most of the great mathematicians of the eighteenth and early nineteenth centuries were more like engineers than modern mathematicians in their thinking; a formula revealed in a flash of intuition, or hastily inferred from loose reasoning, was as good as any other provided it worked. Their formulas worked admirably. Gauss (1777–1855) was the first great mathematician to rebel successfully against intuition in analysis. Lagrange (1736–1813) had tried and failed.

The focus of the last serious trouble was found most unexpectedly in the speciously innocuous natural numbers 1, 2, 3, . . . that, since the days of Pythagoras, had been eagerly accepted by mathematics as manna from heaven. Indeed L. Kronecker (1823–1891, German), himself a confirmed Pythagorean and one of the leading algebraists and arithmeticians of the nineteenth century, confidently asserted that "God made the integers; all the rest is the work of man." By 1910, some of the more wary mathematicians were inclined to regard the natural numbers as the most effective net ever invented by the devil to snare unsuspecting men. Others, of a yet more mystical sect, maintained that the natural numbers have nothing supernatural of either kind about them, asserting that the 'unending sequence' 1, 2, 3, . . . is the one trustworthy 'intuition' vouchsafed to Rousseau's natural man. The tribes of the Amazon Basin were not consulted.

Clashes between these and other opposing factions of mathematical orthodoxy temporarily relegated the Pythagorean program to limbo shortly after 1900. All parties to the many-sided dispute united in torturing logic into new and fantastic shapes to make it reveal at last what meaning, if any, there may be in the natural numbers and in the dream of Pythagoras concerning them. Do these numbers tell the truth about mathematics and nature, or do they not, 'truth' being merely a self-consistent description? If they do not, is it necessary for human needs that the mathematics constructed on the natural numbers be 'true' in this sense? Whatever may be the answer to the first question, that to the second seems to be an emphatic No. A vast amount of mathematical reasoning now known to be unsound led in the past to extremely useful consequences. However, our concern here is not with such profound questions, all of which may be meaningless, but with the technical mathematics which has bred these inquiries and many more like them. But we may note in passing that mathematics is not the static and lumpish graven image of changeless perfection that some adoring worshippers have proclaimed it to be.

The extensions of the number system since the sixteenth century are one of the outstanding accessions of all mathematics. In the opinion of those competent to estimate the technical evidence, these extensions are likely to be of value for many years to come. The 'crisis' of the early twentieth century in all mathematics, induced by an uncritical acceptance of the formalism that had generated the successive extensions of the number system, was precipitated by too bold a use of infinite classes in an attempt to be logically rigorous. Infinite classes penetrated the domain of number, as will appear later, from two diametrically opposite points, namely, the finite cardinal numbers of common arithmetic and the continuum of analysis. In the higher arithmetic also, Dedekind's generalization (about 1870) of the rational integers and their unique decomposition into primes to the algebraic integers with a corresponding unique decomposition into prime ideals introduced denumerably infinite classes of algebraic integers. The concurrent entry of non-denumerably infinite classes of rational numbers came with the theories of Cantor and Dedekind, devised to provide the continuum of real numbers in analysis with a self-consistent foundation. Thus the central obstacle, the mathematics of the infinite, that had stopped Pythagoras halted his successors over two thousand years after he had become a legend. Eudoxus seemed to have discovered a way round the obstacle, or perhaps through the thickest part of it; and the builders of the modern continuum, following essentially the same path, sought only to clear it of obstructions and give it a firmer foundation. What at first looked like security appeared on closer inspection to be an illusion. The road had yet to be built.

To anticipate the report in a later chapter and bring this forecast to a close, we recall the singular conversion of H. Poincaré (1854–1912, French) in 1908. This, in a way, sums up the progress of 2,300 years. At the close of the nineteenth century, Poincaré was a major prophet of a self-confident mathematics. In 1900 he declared that all obscurity had at last been dispelled from the continuum of analysis by the nineteenth-century philosophies of number based on the theory of infinite classes (*Mengenlehre*). All mathematics, he declared, had finally been referred to the natural numbers and the syllogisms of traditional logic; the Pythagorean dream had been realized. Henceforth, reassured by Poincaré, timid mathematicians might proceed boldly, confident that the foundation under their feet was absolutely sound.

Eight stormy and eventful years changed the prophet's vision: "Later generations will regard the Mengenlehre as a disease from which one has recovered."

Thirty years after Poincaré's somewhat caustic prognosis was pronounced, the theory from which mathematics was to have recovered was still flourishing. This of course disproves nothing; Euclid's geometry lasted unmodified in the minds of generation after generation of mathematicians who for two thousand years believed it to be flawless. Both the prognosis and its possibly retarded realization are recalled here merely to exhibit in a just perspective the great acquisitions of number since the sixteenth century. If in its continual development mathematics seldom if ever attains a finality, the constant growth does mature some residue that persists. But it is idle to pretend that what was good enough for our fathers in mathematics is good enough for us, or to insist that what satisfies our generation must satisfy the next.

Extension by inversion and formalism

The earliest extensions of the system of natural numbers were the Babylonian and Egyptian fractions. These illustrate one prolific method of generating new numbers from those already accepted as understood, namely, inversion. To solve the problem 'by what must 6 be multiplied to produce 2?', a new kind of 'number,' the fraction $\frac{1}{3}$, must be invented. Here the direct operation is multiplication, and the inverse, division. The other pairs of elementary inverses are addition and subtraction; raising to powers and extracting roots.

All of these elementary operations were known to the ancients. The inverses, division and subtraction, of the rational operations, multiplication and addition, necessitated the invention of common fractions and negative numbers; the operation inverse to powering was in part responsible for the invention of irrationals, including the pure imaginaries and the ordinary complex numbers. The solution of an algebraic equation, or of a system of such equations in several unknowns, can be restated as a problem in inversion with respect to iterations of addition and multiplication. Up to about 1840, algebraic equations were probably the most prolific source of extensions of the natural numbers.

Considering the extensions up to and including the acquisition of ordinary complex numbers, we shall take a point of view which may be indefensible historically but which can be justified mathematically: mere accidental encounters with, say, negative numbers do not constitute mathematical discovery. Nor does a rejection of imaginary roots of equations entitle anyone to priority in the invention of complex numbers. Until a conscious attempt was made to understand negative and complex numbers, and to state rules, however crude, for their use wherever they might occur, neither had any more right to be considered a mathematical entity than has an unconceived child to be considered a human being. Mathematically, these numbers did not exist until the conditions indicated were met.

Professional historians are in substantial agreement on the following details in the development of negatives. Diophantus, in the third century A.D., encountering -4 as the formal solution of a linear equation, rejected it as absurd. In the first third of the seventh century, Brahmagupta is said to have stated the rules of signs in multiplication; he discarded a negative root of a quadratic. The rules of signs became common in India after their restatement by Mahavira in the ninth century. Al-Khowarizmi, of about the same time, made no advance except that he appears to have exhibited a positive and a negative root for a quadratic without explicitly rejecting the negative.

Of the Europeans, Fibonacci in the early thirteenth century rejected negative roots, but took a step forward when he interpreted a negative number in a problem concerning money as a loss instead of a gain. It has been claimed that the Indians did likewise. L. Pacioli (1445?–1514, Tuscan) in the second half of the fifteenth century is credited with a knowledge of the rule of signs on such evidence as $(7 - 4)(4 - 2) = 3 \times 2 = 6$. M. Stifel[1] (1487?–1567, German) a fine algebraist for his time, called negative numbers absurd in the middle of the sixteenth century. Cardan, in his *Ars magna* (1545), stated the rule 'minus times minus gives plus' as an independent proposition; he also is said to have recognized negative numbers as 'existent,' but on evidence which seems doubtful. In fact, he called negatives 'fictitious.'

Bombelli in 1572 showed that he understood the rules of addition in such instances as $m - n$, where m, n are positive integers. Vieta, about the same time, rejected negative roots. Finally J. Hudde (1628–1704, Dutch) in 1659 used a letter to denote a positive or a negative number indifferently. As an historical curiosity, it may be mentioned that T. Harriot (1560–1621, English) was one of the first Europeans to duplicate the feat of the ancient Babylonians in permitting a negative number to function as one member of an equation. But he refused to admit negative roots.

With one exception, all items in the foregoing list may be classified as partial extension by formalism. The extension was incomplete because no free use of negatives was made until the seventeenth century. The extension was formal because it had no basis other than the mechanical application of rules of calculation that were known to produce consistent results when applied to positive numbers, and were assumed to be legitimate in the manipulation of negatives. This unbased assumption was to be elevated in the 1830's to the dignity of a general dogma in the notorious and discredited 'principle of permanence of form.' By the middle of the seventeenth century, untrammeled use of negatives had given mathematicians a pragmatic demonstration that the rules of common algebra lead to consistent results. But there was no attempt to go any deeper and put a substratum of postulates under the rickety formalism.

The one glimmer of mathematical intelligence in the early history of negatives is the suggestion of Fibonacci that a negative sum of money may be interpreted as a loss. This appears to have been the first step toward the second stage in the evolution of negatives, that of interpreting the results of formalism in terms of something which is accepted as consistent. It marks the beginning of two distinct but complementary philosophies of mathematics: the products of mathematical formalism are to be admitted only if they can be put in correspondence with some already established system accepted as self-consistent; all mathematics is a formalism without meaning beyond that implied by the postulates defining the formalism. For example, if Euclidean geometry is accepted as self-consistent, and if the formal algebraic operations with complex numbers can be interpreted in terms of that geometry, the formalism of complex numbers is admissible. This is according to the first philosophy, which was that instinctively and subconsciously adopted by Fibonacci in his encounter with negatives. The second philosophy is illustrated by the rules of algebra in any modern elementary text, where $a, b, c, \cdots, +, \times, =$ are displayed, and it is postulated that $a = a$, $a + b = b + a$, etc.

Each philosophy has greatly enriched mathematics. The first, seeking interpretations, may be called synthetic; the second, beginning and ending in a formalism within its own postulated universe,[2] may be termed analytic. The designations are merely for convenience, and are not intended to recall Kant's terminology, although the parallel may be suggestive. The development of the number system is the record of a continual interplay between the synthetic and the analytic approaches. For applied mathematics, as in quaternions and the vector algebras that evolved from the geometrical interpretation of ordinary complex numbers, it is the synthetic philosophy that dominates; in pure mathematics the analytic alone is relevant.

The history of mathematics holds no greater surprise than the fact that complex numbers were understood, both synthetically and analytically, before negative numbers. Accordingly, we shall retrace first the principal steps by which complex numbers arrived at mathematical maturity. The negatives will then enter incidentally.

From manipulation to interpretation

The early history of complex numbers is much like that of negatives, a record of blind manipulations unrelieved by any serious attempt at interpretation or understanding. The first clear recognition of imaginaries was Mahavira's extremely intelligent remark in the ninth century that, in the nature of things, a negative number has no square root. He had mathematical insight enough to leave the matter there, and not to proceed to meaningless manipulations of unintelligible symbols. It is of more than historical interest that Cauchy[3] made the same observation a little less than a thousand years later (1847): "[we discard] the symbolic sign $\sqrt{-1}$, which we repudiate completely, and which we may abandon without regret, because one does not know what this alleged sign signifies, nor what meaning one should attribute to it." These sentiments were the origin of Kronecker's project in 1882–7 for a unified derivation of all extensions of the natural numbers.

The next step forward after Mahavira was toward the analytic philosophy of number. Cardan in 1545 regarded imaginaries as fictitious, but used them formally, as in the resolution of 40 into conjugate complex factors $5 \pm \sqrt{-15}$, without raising any question as to the legitimacy of the formalism. A more vicious species of pure formalism appeared in a totally unwarranted conjecture of A. Girard (?1590–?1633, Dutch). Having noticed that some equations of low degrees n have n real roots and that some quadratics have two imaginary roots, Girard inferred that any equation of degree n has n roots, supplying any awkward lack of real roots by guessing that the deficiency would be exactly met by complex roots.

Leibniz[4] by 1676 had progressed no farther than Cardan. He gave a formal factorization of $x^4 + a^4$, and succeeded in convincing himself that he had done something remarkable when he verified by actual substitution that Cardan's solution of the general cubic in the irreducible case satisfies the equation. He was equally astonished by a similar verification of the expression of a special real radical as a sum of conjugate complexes. The truly astonishing thing historically about Leibniz' performances with complex numbers is that less than three centuries ago one of the greatest mathematicians in history should have thought that any of these meaningless manipulations were mathematics, or that their outcome was more unexpected than is that of turning a tumbler upside down twice in succession. That a mathematician, logician, and philosopher of the caliber of Leibniz could so delude himself, substantiates Gauss' observation that "the true metaphysics of $\sqrt{-1}$" is hard. It also suggests that mathematics really has progressed since the ever-memorable seventeenth century.

In the eighteenth century, blind formalism at last produced a formula of the first magnitude. About 1710, R. Cotes (1682–1716, English), the man whose death moved Newton to lament, "If Cotes had lived, we might have known something," stated an equivalent of the result usually called DeMoivre's theorem in trigonometry. In current notation, i denoting[4a] $\sqrt{-1}$, Cotes' formula[5] is $i\phi = \log_e (\cos \phi + i \sin \phi)$. DeMoivre's theorem (1730), $\cos n\phi + i \sin n\phi = (\cos \phi + i \sin \phi)^n$ $[= e^{ni\phi}]$, n an integer > 0, is an immediate *formal* consequence. Euler (1743, 1748) extended the last to any n; he also gave the exponential forms of sin ϕ, cos ϕ—evident from Cotes' result. Thus, by 1750 trigonometry had become a province of analysis, and all that remained was to derive the analytic formulas with due attention to convergence and to create a self-consistent theory of complex numbers. The first desideratum was met in the third decade of the nineteenth century, by Cauchy; the second, in the last decade of the eighteenth century, by Wessel. So, after about a thousand years of meaningless mystery, the so-called 'imaginary' numbers were incorporated into unmystical mathematics.

Before proceeding to Wessel and his successors, we recall two noteworthy steps toward what has just been described. John Bernoulli observed the connection between inverse tangents and natural logarithms. This was in Cotes' direction. Even more significant was the long stride toward a geometrical interpretation of complex numbers taken in 1673 by J. Wallis (1616–1703, English), an original mathematician and at one time a fashionable preacher. Wallis missed by a hairsbreadth the usual geometrical interpretation of complex numbers. But in mathematics a hair may be as thick as a ship's cable; and Wallis is not usually credited with Wessel's invention. In effect, Wallis represented the complex number $x + iy$ by the point (x, y) in the plane of Cartesian coordinates; what he missed was the use of the y-axis as the axis of imaginaries.

It remained for a Norwegian surveyor, C. Wessel[6] (1745–1818), to take the final step and produce a consistent, useful interpretation of complex numbers in 1797. He modestly called his completely successful effort "an Attempt." It fully explained what is customarily misnamed the Argand diagram in texts, and mapped the formal algebra of complex numbers on properties of the diagram. J. R. Argand (1768–1822, French) independently arrived at similar conclusions in 1806.

Wessel's decisive contribution suffered the misfortune of being published (1799) in a scholarly journal that mathematicians were not likely to read. A French translation in 1897, exactly one hundred years after Wessel had communicated his paper to the Royal Danish Academy, secured its author whatever reward there may be in posthumous fame. Thus a work that could have hastened the development of the number system might as well never have been written for all the influence it had; and it remained for the great authority of Gauss (1831) to get complex numbers accepted as respectable members of mathematical society.

Two possible generalizations were suggested by Wessel's interpretation. The geometry of complex numbers was obviously translatable into a description of rotations and dilatations in a plane. Were further extensions of the number system possible by which rotations in space of three dimensions could be described? Or were the complex numbers themselves adequate for the purpose? The answer to the first was to be affirmative, to the second, negative; but this could scarcely have been predicted in 1799 when Wessel's interpretation was published.

The geometrical approach was not the 'natural' way to the heart of the problem, although W. R. Hamilton (1805–1865, Irish) was to follow it successfully. But Hamilton, as will be seen, rested content with an algebra adapted to space of three dimensions. In mathematics, however, three is no more sacred than any other cardinal number, and the real problem was to extend complex numbers to 'space' of n dimensions.

The Euclidean program

Gauss took as the subject for his doctor's dissertation (1799) a proof of the fundamental theorem of algebra: an algebraic equation has a root of the form $a + bi$, a, b real. (For the precise statement of the theorem, we refer to any text on the theory of equations. The foregoing statement, like others in this account, is intended merely to recall the theorem.) After Girard's conjecture, there had been attempts at proof, including essays by D'Alembert (1746) and Euler (1749). All were faulty, as were the first and fourth attempts (1799) by Gauss.[7] It may be stated in passing that the fundamental theorem in its classic form, as proved in the theory of functions of a complex variable, is no longer regarded as belonging to algebra. It is supplanted in modern algebra by a statement which is almost a triviality.[8] The basic ideas of the modern treatment go back to Galois (1811–1832), Dedekind (1831–1916), and Kronecker (1823–1891), not to Gauss.

This first serious work of the greatest mathematician since Newton convinced him that a satisfactory theory of complex numbers had yet to be created. Unaware of Wessel's work, Gauss himself arrived at a geometrical representation.[9] But the man who gave it as his mature opinion that "Mathematics is the Queen of the Sciences, and Arithmetic the Queen of Mathematics," could not be satisfied with a helpful but irrelevant geometrical picture of what, he believed, was purely a question of number. By 1811, Gauss had convinced himself that a 'formal' treatment alone could provide a sound theory of complex numbers; and he came within an ace of committing himself to the mysterious principle of permanence which was to guide others to the desired end about a quarter of a century later. But in 1825 he confessed that "the true metaphysics of $\sqrt{-1}$" was elusive.

By a formal treatment, Gauss meant the deduction of the properties of complex numbers from the accepted postulates of common arithmetic. He sought proofs in the manner of Euclid from definitions and explicit assumptions. We shall return later to the principle of permanence.

What Gauss regarded as "the true metaphysics" of complex numbers was invented by him in 1831, six years before Hamilton communicated his independent discovery of the same method to the Royal Irish Academy. The 'true metaphysics' banished geometric intuition entirely, defining $a + bi$, a, b real, as the number-couple (a, b) subjected to postulates necessary and sufficient to yield the desired properties of complex numbers as given by algebraic manipulations. For example, equality $(a, b) = (c, d)$ is defined to mean $a = c$, $b = d$; addition,

$$(a, b) + (c, d),$$

by definition, is $(a + c, b + d)$; multiplication, $(a, b) \times (c, d)$ is $(ac - bd, ad + bc)$. The mysterious i has vanished, and the algebra of complex numbers is replaced by what De Morgan and others called a "double algebra" of couples of real numbers a,b, c,d, . . . subject only to the accepted laws of arithmetic and common algebra, as $a + b = b + a$, $ab = ba$,

$$a(b + c) = ab + ac, \text{ etc.}$$

The occasion for Gauss' disclosing his anticipation of Hamilton's method was a letter of 1837 from his old university friend W. Bolyai (1775–1856, Hungarian) in which Bolyai reproached Gauss for having propagated a geometrical theory of complex numbers. Bolyai argued that geometry has no place in the foundations of arithmetic, and that complex numbers should be referred to the real numbers whose arithmetic was assumed to be known. Gauss replied that he was of exactly the same opinion and that, in 1831, he had done what Bolyai demanded. He remained of this opinion, and only five years before his death emphasized that the "abstract," postulational method is the desirable approach to complex numbers. This method is now fairly common in texts on college algebra.

Anyone seeing the algebra or arithmetic of number-couples for the first time might be pardoned for thinking it a sly subterfuge, or at best a beating of the imaginary devil round a supposedly real bush. Familiarity corrects misapprehension; and when the suggestive notation (a, b) for couples is extended to triples (a, b, c), and beyond, to sets of n ordered real numbers or elements of a field, with appropriately defined laws of addition and multiplication, the creative power of Hamilton's simple invention becomes evident. Multiple algebra, with its innumerable applications to the sciences, was already in sight when Hamilton replaced $a + bi$ by (a, b). He himself elaborated the algebra and geometry of number-quadruples (a, b, c, d) in his quaternions; Grassmann almost simultaneously took a more general point of view and created the algebra of number-n-ples, $(a_1, a_2, \ldots, a_n)$. We shall resume this in a later chapter; for the moment, we follow the consequences of the resumption of the Euclidean methodology by Gauss and Hamilton. After about twenty-three centuries of sightless wandering, arithmeticians and algebraists opened their eyes and saw what Euclid had done: definitions, postulates, deduction, theorems. They then took a long stride ahead.

It may have been clear to Euclid that his geometry was that of a postulated, ideal universe having no necessary connection with an intuitively perceived 'real world' of common experience; but if so, he did not convey the full import of his philosophy to his successors. Euclid's geometry, or any other mathematical system constructed on the deductive pattern, is now almost universally regarded as a free and arbitrary creation of the mathematician constructing the system, whether the initial impulse came from experiences of the material world sublimated into abstractions, or whether it originated in formal extensions of algebraic symbolism, as in the passage from number-couples to ordered sets of n real numbers. The philosophy behind the Euclidean program as now conceived is analytic.

It seems singularly appropriate that the conception of algebra as pure formalism should have first appeared in that country which above all others has revered Euclid. It was an Englishman, G. Peacock (1791–1858), at one time Lowndean professor in Cambridge University, later a Dean of Ely, who first[10] (1834,

1845) perceived common algebra[11] as an abstract hypothetico-deductive science of the Euclidean pattern.

Peacock was not an 'important' mathematician in the accepted sense of wide reputation; so possibly the following is a just estimate of his place in mathematics: "He was one of the prime movers in all mathematical reforms in England during the first half of the 19th century, although contributing no original work of particular value."[12] He was merely one of the first to revolutionize the whole conception of algebra and general arithmetic.

The Euclidean program advocated by Peacock was developed by the British school, notably by D. F. Gregory (1813–1844, Scotch), and A. De Morgan; but it did not become widely known until H. Hankel (1839–1873) in 1867 expounded it with insight and massive German thoroughness. Hankel also reformulated the principle of permanence of formal operations, which had been stated in less comprehensive terms by Peacock: "Equal expressions couched in the general terms of universal arithmetic are to remain equal if the letters cease to denote simple 'quantities,' and hence also if the interpretation of the operations is altered." For example, $ab = ba$ is to remain valid when a, b are complex.

It is difficult to see what the principle means, or what possible value it could have even as a heuristic guide. If taken at what appears to be its face value, it would seem to forbid $ab = -ba$, one of the most suggestive breaches of elementary mathematical etiquette ever imagined, as every student of physics knows from his vector analysis. As a parting tribute to the discredited principle of permanence, we note that since $2 \times 3 = 3 \times 2$, it follows at once from the principle that $\sqrt{2} \times \sqrt{3} = \sqrt{3} \times \sqrt{2}$. But the necessity for proving such simple statements as the last was one of the spurs that induced Dedekind in the 1870's to create his theory of the real number system. According to that peerless extender of the natural numbers, "Whatever is provable, should not be believed in science without proof."[13]

The device of number-couples, invented to exorcise the imaginary and reduce the theory of complex numbers to that of pairs of real numbers, also banished rational fractions and negative numbers. Thus, for negatives, $-n$ is replaced by $[m, m + n]$, where m is an arbitrary positive number; zero is $[m, m]$, and n is $[m + n, m]$. As the details are available in standard texts, we pass on. The last and most difficult step in reducing all 'number' to the natural numbers 1, 2, 3, . . . concerned the real irrationals. This arithmetized analysis.

Between the final step and the formalizing of algebra and arithmetic by Peacock, De Morgan, Hamilton, and others, the natural numbers were vastly extended in another direction, that of algebraic numbers, beginning with Gauss in 1831 and continuing into the twentieth century. Concurrently, the generalizations of number-couples to multiple algebra were developed. Another type of arithmetization, originating in 1801 in the work of Gauss and reaching one of its climaxes in Kronecker's work of 1882–7, incidentally provided another means of reducing all numbers to the natural numbers. This will be described in the following chapter.

The Euclidean program, which would ultimately reduce all mathematics to a pure formalism, had its opponents as well as its partisans in the nineteenth century as it has had since. To illustrate the ironies of prophecy, we recall the vigorous attack delivered in 1882 by a distinguished analyst, P. du Bois-Reymond (1831–1889, German), whose penetrating researches contributed much to the progress of analysis in its second heroic age—that of Newton and Leibniz being the first—in the nineteenth century. The program of formalism, du Bois-Reymond declared with considerable passion, would replace mathematics by "a mere play with symbols, in which arbitrary meanings would be attached to the signs as if they were the pieces on a chessboard or playing cards." He went on to prophesy that such a 'meaningless' outcome would fritter out in barren efforts and be the death of mathematics as Gauss had pictured the Queen of the Sciences. Since 1920, mathematics for one highly productive school has become exactly what the prophet feared it might. Those who call themselves formalists revel in their endless game of chess and exult that it has no meaning whatever beyond the rules of the game. Ultimate realities and eternal truths, at least in mathematics and science, suffered an eclipse in the twentieth century.

Thus has ended one quest after the meaning of number; and this conclusion, disconcerting to some, was reached by following the same road that Euclid took. It was D. Hilbert's (1862–1943, German) close scrutiny of the postulates of elementary geometry, in an endeavor to put a solid foundation under that venerable if somewhat palsied science, that led him to a similar inspection of the bases of common arithmetic. Addressing the second international congress of mathematicians in 1900, Hilbert observed[15] that the noncontradiction of the postulates of geometry is demonstrated by constructing a suitable domain of numbers such that, to the geometrical postulates, there correspond analogous relations between the numbers of the domain. Consequently, any contradiction in the conclusions drawn from the geometrical postulates would necessarily be recognizable in the arithmetic of the domain. Thus the self-consistency (noncontradiction) of the postulates of geometry is referred to the self-consistency of the postulates of arithmetic. Consequently Hilbert emphasized as one of the outstanding unsolved problems of mathematics in 1900 the proof that, by proceeding from the postulates of arithmetic, it is impossible to reach contradictory results by means of a finite number of logical deductions. The problem was still open in 1945. Attempts to solve this seemingly elementary problem were in part responsible for the formalistic—chess-playing—school of mathematical philosophy led by Hilbert, the foremost mathematician of his generation.

Enough has been said to indicate the fundamental importance of the natural numbers for all mathematics, not merely for arithmetic and its algebraic extensions, and to suggest that Euclid's methodology is as vital in modern mathematics as it was in ancient. It will be well before proceeding to further extensions of number to cast up the account of Pythagoras with mathematics thus far.

Pythagoras to 1900

Glancing back over the confused effort to incorporate imaginaries with the reals in one self-consistent number system, we note the curious fluctuations in mathematical creed accompanying the struggle. The abrupt check experienced by the Pythagoreans in their encounter with irrationals practically abolished mensuration in orthodox Greek mathematics, and the investigation of number independently of its geometrical representation all but ceased, Euclid's summary of arithmetica being a partial exception. The academic Greek mathematicians were at ease with number only when it was geometrized into 'magnitude'—a vague concept whose tenability they seem never to have questioned. Thus number was supposed to be apprehensible through form, the opposite of what the Pythagoreans first held and of what a majority of mathematicians have believed since Descartes.

The diagrams of Wessel and Argand were an unseasonable reversion to pre-Cartesian mathematics. Exactly nothing is provable by the geometrical representation of complex numbers unless it be assumed that the underlying geometry is founded consistently. Gauss also, we have seen, at first sought to 'justify' imaginaries geometrically, but later decided that this was a mistake. In all these earlier developments, geometry was accepted without question as the irreversible court of last appeal. But with increasing sophistication, it was perceived that the geometrical justification is merely disguised arithmetic, real numbers entering with the coordinates of a point in the plane of complex numbers. The geometrical interpretation is thus left without a foundation until the real number system is firmly based in self-consistency. Going deeper, Hilbert in 1899 resumed the Pythagorean program for all geometry, referring form to number and demanding a proof of noncontradiction for the real number system, or even for its subset of rational integers.

Hamilton also was a Pythagorean in his escape from geometry into number-couples. His was the more suggestive method for future extensions to hypercomplex numbers. But he was less critical than Hilbert, in that he took the self-consistency of the real number system for granted.

The modern attack, as in abstract algebra, attempts to strip Hamilton's number-couples (a, b) of all arithmetical connotations by postulating that the 'coordinates' a, b are defined by the postulates of an abstract field. The last vestige of number as the Pythagoreans conceived it has been sublimated from the 'meaningless marks' a, b and their equally 'meaningless rules of combination.' But the problem of proving that the rules will never produce a contradiction is not eliminated by manipulating a set of postulates which, by assumption, completely define the mathematical system deducible from them.

In escaping from form to number, and back from number to form, thence again to number, and finally into complete abstraction, mathematicians from Pythagoras to Hilbert have sought to validate their creations by deductive reasoning. Hilbert was the first to recognize the futility of such vacillation until deductive reasoning as applied in all mathematics should itself be shown to be incapable of producing contradictions. This put the whole Pythagorean program on trial; for the cardinal hypothesis of the Pythagoreans assumed that number and form may be consistently described by deductive reasoning. The final question, then, is, How far can mathematical deduction be trusted if it is not to produce contradictions such as '*A* is equal to *B*, and *A* is not equal to *B*'? This is debated in the symbolic language of reasoning foreseen by Leibniz. Some of the conclusions will be noted in the last chapter.

For the moment it is sufficient to observe that the ability of modern mathematics to discuss such questions profitably is one of its titles to superiority in power over its predecessors. And, whatever the outcome is to be, the admitted utility of complex numbers, whether conceived as affixes of points in a plane or as number-couples, in both pure and applied mathematics, will doubtless remain substantially unaffected.

CHAPTER 9

Toward Mathematical Structure

1801-1910

Three new approaches to number, in 1801 and in the 1830's, were to hint at the general concept of mathematical structure and reveal unsuspected horizons in the whole of mathematics. That of 1801 was the concept of congruence, introduced by Gauss in what many consider his masterpiece, the *Disquisitiones arithmeticae*, published when its author was twenty-four. To this and the revolutionary work (1830–2) of E. Galois (1811–1832, French) in the theory of algebraic equations can be traced the partial execution of L. Kronecker's (1823–1891, German) revolutionary program in the 1880's for basing all mathematics on the natural numbers.

The same sources are one origin of the modern abstract development of algebraic and geometric theories, in which the structure of mathematical systems[1] is the subject of investigation, and it is sought to obtain the interrelations of the mathematical objects concerned with a minimum of calculation. 'Structure' may be thought of for the present in any of its intuitive meanings; it was precisely defined in 1910 by the mathematical logicians. It might be compared to morphology and comparative anatomy. We shall approach mathematical structure through the union effected in the nineteenth century between algebra and arithmetic.

Abstraction and the recent period

From the standpoint of mathematics as a whole, the methodology of deliberate generalization and abstraction, culminating in the twentieth century in a rapidly growing mathematics of structure, is doubtless the most significant contribution of all the successive attempts to extend the number concept. But at every stage of the progression from the natural numbers 1, 2, 3, . . . to other types of numbers, each of several fields of mathematics adjacent to arithmetic was broadened and enriched.

New acquisitions in other fields reacted reciprocally on arithmetic. For example, the first satisfactory theory of ordinary complex numbers to become widely known was that of Gauss (1831), devised to provide a concise solution for a special problem in diophantine analysis: If p, q are primes, what conditions must p, q satisfy in order that at least one of the equations $x^4 = qy + p$, $z^4 = pw + q$, shall be solvable in integers x, y, z, w? The theory of complex numbers necessitated a radical revision and generalization of the concept of arithmetical divisibility, which in turn suggested a reformulation of certain parts (intersections of varieties) of algebraic geometry. The latter in its turn was partly responsible for further generalizations (modular systems) in the algebraic arithmetic—or arithmetical algebra—of the twentieth century.

The like may be observed in the creation of the numerous vector algebras invented during and after the 1840's for application to the physical sciences. The first of these evolved directly from the vectorial interpretation of ordinary complex numbers. The extension in the 1840's of vector algebra in a plane to space of more than two dimensions was one origin of the hypercomplex number systems of algebra, and these again supplied arithmetic with new species of integers. The development of the corresponding arithmetic in turn reacted, particularly in the twentieth century, on the algebra in which it had originated. It would seem to be incorrect, therefore, to say that any one division of mathematics was alone responsible for the steady progression since 1800 from the special and detailed to the abstract and general. The forward movement was universal, and each major advance in one department induced progress in others.

In following this development, one misapprehension above all others that might be possible is to be particularly guarded against. Those who are not mathematicians by trade are sometimes inclined to confuse generality with vagueness, and abstraction with emptiness. The exact opposite is the case in the mathematical generalizations and abstractions with which we shall be concerned here. Each, on appropriate and definitely prescribed specialization, yielded the specific instances from which it had evolved. The theory of hypercomplex numbers, for example, contains as a mere detail that of ordinary complex numbers; and once the general theory of hypercomplex number systems has been elaborated, the special theory of ordinary complex numbers follows automatically. Moreover, each generalization gives in addition a whole universe of mathematical facts distinct from those in the special instances from which the generalization proceeded.

It was remarked in the Prospectus that the separation of all mathematical history into a remote period, to 1637, a middle, 1638–1801, and a recent, 1801–, distinguishes three well-marked epochs in the development of mathematics. Following the rapid growth of arithmetic and algebra, we shall see that in the passage from the middle period to the recent there was a profound change in the quality of mathematical thought and its objectives. This change is most simply observed, perhaps, in the evolution of the number concept. It is the item of greatest interest to be noticed in this chapter and the following. Geometry might have been considered instead of arithmetic to exhibit the same change. But as the transformed algebra and arithmetic played important parts in the expansion of geometry, it seems more natural to consider them first. It is to be borne in mind, however, that while arithmetic and algebra were being transformed into shapes the mathematicians of the eighteenth century would not have recognized as mathematics, geometry and analysis were undergoing corresponding transformations.

Prospect

The abstract approach of the 1830's to algebra parallels the epochal advance in geometry made simultaneously with the publication in 1829 of N. I. Lobachewsky's (1793–1856, Russian) non-Euclidean geometry. This also stems from 1800 or earlier in the preparatory work of Gauss and others. As this properly belongs to geometry, it will be discussed in that connection. The relevant detail here is that geometers and algebraists perceived almost simultaneously that mathematical systems are not supernaturally imposed on human beings from without, but are free creations of imaginative mathematicians. Lobachewsky's new geometry was the earliest mathematical system to be recognized as such a free creation. It provided the first proof of the complete independence of a particular postulate (Euclid's postulate of parallels) in a system which tradition and common sense had agreed must contain that postulate. The significance of this radical step in methodology was only slowly appreciated; and it would seem that the almost simultaneous advance of algebra and arithmetic in a parallel direction was more directly responsible than geometry for the modern abstract view of mathematics.

The explicit recognition, by the British school in the 1830's, of common algebra as a purely formal mathematical system shortly led to a revolution in arithmetic and algebra of significance comparable to that precipitated by non-Euclidean geometry.

Hamilton's rejection (1843) of the commutative 'law' (postulate) of multiplication, in his invention of quaternions, opened the gates to a flood of algebras, in which one after another of the supposedly immutable 'laws' of rational arithmetic and common algebra was either modified or discarded outright as too restrictive. By 1850 it was clear to a majority of creative mathematicians that none of the postulates of common algebra, which up to 1843 had been thought necessary for the self-consistency of symbolic reasoning, was any more a necessity for a noncontradictory algebra than is Euclid's parallel postulate for a self-consistent elementary geometry. To the astonishment of some, it was found that the modified algebras, such as Hamilton's quaternions, were adaptable to mechanics, geometry, and mathematical physics. The dead hand of authoritative tradition had been brushed aside; mathematics was free. As G. Cantor (1845–1918, Germany), one of the boldest extenders of the number concept, was to say about three-quarters of a century later, "the essence of mathematics is its freedom." No mathematician, not even Gauss, could have conceived such a thought in 1801. The accomplished facts of the revolutions in geometry and algebra of the 1830's and 1840's made freedom conceivable.

From supernaturalism to naturalism

The change during the nineteenth century from what may be called Platonic supernaturalism to modern naturalism in mathematics is reflected in three aphorisms, the first of which may have expressed the Greek reverence for synthetic geometry; the second, the early nineteenth-century worship of arithmetic and analysis; and the third, the final admission that mathematics is made by men. The second and third, following the example of the first, were phrased in classic Greek.

Plato is said to have asserted that "God ever geometrizes"; C. G. J. Jacobi (1804–1851, German), a great arithmetician and analyst, declared that "God ever arithmetizes"; while J. W. R. Dedekind (1831–1916, German), first and last an arithmetician, wrote as the motto for his famous essay on the nature of number (*Was sind und was sollen die Zahlen?*, 1888), "Man ever arithmetizes."

A proposition less open to objection than the last today would be "Man attempted to arithmetize during the second half of the nineteenth century, and came to grief so doing early in the twentieth." Although this is less elegant than the original, it is closer to historical fact. Nevertheless, man's failure as yet to complete the Pythagorean program of arithmetizing mathematics and the universe has been, and is, a potent stimulant to the continual creation of new and interesting or useful mathematics.

During the nineteenth century the physical sciences were given to solving the universe by dissolving it in vast generalizations distilled from inadequate data. So powerful were some of these solvents that they dissolved themselves. Having learned by disconcerting experience that the universe is not to be solved between breakfast and lunch, the physical sciences took a more modest view of their function, and in the early twentieth century, after critical introspection, contented themselves with consistent descriptions intelligible to instructed human beings. Universe-solving went out of fashion temporarily about the time of the first world war.

Mathematics in the meantime was experiencing similar difficulties in its abortive struggle to comprehend its own vast empire in the all-inclusive generalization of Pythagoras. What is to be the outcome is not yet predictable. But two things may be reasonably conjectured.

The Pythagorean project of deriving all mathematics from number will continue for many years to suggest new accessions to mathematics, and it will remain essentially as the Pythagoreans imagined it in their unmystical moments. If form, for example, is to be better described by our successors in terms of something other than number, we have no inkling at present of what that something may be, unless it be symbolic logic or analysis situs, themselves in partial process of arithmetization.

We may also conjecture that mathematics, like the physical sciences, will take a less inflated view of itself as a result of its critical self-analysis. There will be less mysticism in the mathematics of the future than there has been in that of the past, and fewer grandiose claims to immortality and eternal truth. Mathematics will become less self-conscious, less introspectively critical, and more boldly creative. It will resign its soul to the metaphysicians for such tortures as they may choose to inflict, feeling nothing; for it will continue to serve with its living body the purposes of the men who create it to meet human needs rather than to be the plaything of sterile philosophies. The very implements of torture—symbolic logic, for example—were devised in mathematics itself as by-products of more immediately useful inventions.

It is the latter that are of vital significance in a scientific civilization; the by-products too often carry with them the cold smell of a mildewed scholasticism. The spirit of the Middle Ages, which the successors of Galileo and Newton imagined they had laid forever in science, stirs again in the twentieth-century disputes concerning the nature and meaning of number. Ignoring these for the present, we shall continue with the more profitable arithmetic of 1801 which, a century later, was to deliquesce in metaphysics. But history will compel us to return to these disputes. Our attitude in what follows will be that of Molière's despised "average sensual man," who seeks through science merely to make life less barbarous for himself and his fellows, and who is content to leave what professed humanists call "the really important questions" to God and the philosophers.

Congruence from 1801 *to* 1887

'Congruence,' like 'analysis,' 'formal,' 'ideal,' 'functional,' 'analytic,' 'normal,' 'conjugate,' 'modulus,' 'integral,' and a dozen others, is one of those overworked technical terms in mathematics which appear to have been invented to confuse the uninitiated by a multitude of meanings having no connection with one another. Congruences in higher geometry, as in congruences of lines or circles, are unrelated to congruence in elementary geometry, as in congruent triangles; and congruence in arithmetic, with which we are concerned here, has nothing in common with congruence elsewhere. Nor are the ideal elements of projective geometry significantly connected with ideals as in arithmetic and algebra.

Gauss in 1801 defined two rational integers a, b to be *congruent* with respect to the rational integer *modulus* m if, and only if, a, b leave the same remainder on division by m; and he expressed this by writing $a \equiv b \bmod m$. Otherwise stated, if $a \equiv b \bmod m$, then $a - b$ (or $b - a$) is a multiple of m, and conversely; and $x \equiv 0 \bmod m$ asserts that x is exactly divisible by m.

This simple but profound invention is one of the finest illustrations of Laplace's remark that a well-devised notation is sometimes half the battle in mathematics. Writing 'x is divisible by m' as $x \equiv 0 \bmod m$ at once suggested to Gauss extremely fruitful analogies between algebraic equations and arithmetical divisibility. The last is one of the central and most elusive concepts of all arithmetic. It is not this technical aspect of congruence, however, that is of primary importance for our immediate purpose, but another, of far deeper significance, which was perceived only by the successors of Gauss. If Gauss did foresee this, he seems to have left no record of the fact.

To bring out the point, of the first importance for an understanding of modern mathematical thought, we must return for a moment to a hypothetical prehistory, long before Pythagoras and even prior to Sargon. Abstraction, to judge by the behavior of contemporary primitives, is not by any means 'natural' to Rousseau's carefree savage. Numbers first were nouns as concrete as father and mother, an early instance, perhaps, of 'one,' 'two.'

No trace survives of the actual passage from concreteness to abstractness, when 'two' was realized as applicable to a couple of parents, a stick and a stone, or any other of its innumerable manifestations; and we can only imagine the dismay of human beings when they were first overwhelmed by the appalling revelation that the natural numbers have no end. Traces of the attempt to cope with that first deluge of knowledge survive in the symbolism, meaningless to us, of number mysticism. Sympathetically viewed, all that prehistoric nonsense was the outcome of men's first groping efforts to regiment the generative freedom, 'n into $n + 1$,' of the numbers in the unending sequence $1, 2, 3, \ldots, n, n + 1, \ldots$. If only some finite restraint could dominate the endless generations of numbers, they would be less terrifying.

It must have given the mind that first perceived that 'odd' and 'even' suffice to comprehend all the natural numbers a sense of almost supernatural power. The endless sequence after all was no more mysterious than humanity itself, which could be subsumed under 'male' and 'female.' Accordingly, the mathematically useful separation of the natural numbers into only two classes was made more concretely satisfying to the primitive mind by calling odd numbers male and even numbers female. Arithmetic and numerology thereafter flourished together in happy and fruitful symbiosis. But however nonsensical the numerological fruits of that early union may now appear, its occasion was the urge to comprehend an infinite totality in finite terms, and hence to bring the infinite within the grasp of a finite syntax.

Gaussian congruence has proved the most fruitful of all classifications of the rational integers $0, \pm 1, \pm 2, \pm 3, \ldots$ into a finite number of classes, as may be appreciated on inspecting any elementary text on the theory of numbers. What Gauss could not have foreseen was that his invention of mapping one assemblage, finite or infinite, of individuals on another, by classifying the individuals in the first set with respect to some relation having the abstract properties of reflexiveness, symmetry, and transitivity, shared by his relation of congruence, was to become a guiding principle to the structure of algebraic theories. With the gradual evolution of this and similar ideas, mathematics transcended the Pythagorean dream and, as in the theories of groups, fields, point sets, symbolic logic, etc., escaped from the natural numbers into a domain where number is irrelevant and the structure of relations is the subject of investigation.

The concepts mentioned above being fundamental in modern mathematics, we shall recall their definitions. A relation, denoted by $\sim$, is said to be binary with respect to the members $a, b, c, \ldots$ of a given class of things (which need not be 'numbers' of any kind) if $a \sim b$ is either true or false for any a, b in the class. If $a \sim a$ for every a in the class, $\sim$ is reflexive; if $a \sim b$ implies $b \sim a$, $\sim$ is symmetric; and finally, if $a \sim b$ and $b \sim c$ together imply $a \sim c$, $\sim$ is transitive. A relation such as $\sim$ is called an 'equivalence relation' for the given class. Equality, $=$, is a simple instance of $\sim$. If $m, a, b, c, \ldots$ are rational integers, and $m \neq 0$, and if $a \sim b$ is interpreted as $a \equiv b \bmod m$, it is easily verified that this Gaussian congruence is an equivalence relation. Further, congruence is preserved under addition and multiplication: if $x \equiv a \bmod m$, $y \equiv b \bmod m$, then $x + y \equiv a + b \bmod m$, and $xy \equiv ab \bmod m$.

Any equivalence relation separates its class, whether finite or infinite, into subclasses, all those members, and only those, of the whole class that are equivalent to a particular member (and hence by transitivity to one another) being included in a particular subclass. Any member of a subclass may be taken as representing the entire subclass. Congruence with respect to the positive integer modulus m separates all the rational integers into precisely m classes, whose representatives may be taken as $0, 1, 2, \ldots, m - 1$. Congruence is a typical example, and historically the first, of the modern methodology of mapping an infinite totality on a comprehensible finite set. The theory of arithmetical congruence as developed by Gauss and his successors belongs to the higher arithmetic, and will be noted in that connection. Our present interest in congruence is in yet another direction, of deeper significance than the technical applications to arithmetic for mathematical thought as a whole.

In the preceding chapter we noted Cauchy's objections (1847) to the symbol i $(\equiv\sqrt{-1})$. An immediate extension of Gaussian congruence to congruences between polynomials in one variable (more properly, 'indeterminate') x provided Cauchy with the escape into the illusory 'reality' he so ardently desired. If

$$F, = \sum_{r=0}^{m} A_r x^{m-r}, \qquad \text{and} \qquad M, = \sum_{s=0}^{n} B_s x^{n-s},$$

are polynomials, with $m \geqq n$ and $A_0 B_0 \neq 0$, there is exactly one polynomial R of degree $\leqq n - 1$, and exactly one polynomial Q, such that $B_0^{m-n+1} F = QM + R$. Cauchy wrote the particular case of this in which $B_0 = 1$ as a congruence, $F \equiv R \bmod M$, and imitated the Gaussian theory in his easy development of such congruences for polynomials.

For the particular modulus $x^2 + 1(=M)$, Cauchy found that his 'residues' R had all the formal properties of complex numbers, his 'x' taking the place of 'i.' He was thus enabled to construct a wholly 'real' algebra abstractly identical with (having the same structure as) that of complex numbers. A moment's reflection will show why his ingenious device succeeded. It offered an alternative to Hamilton's number-couples.

It seems rather surprising that Cauchy, having gone so far, should not have continued to the expulsion of negatives from his paradise of 'real,' 'existent' numbers, for they surely are as 'unreal' and as 'inexistent' as i to the Pythagorean mind. Naturally enough, the Cauchy who in 1821 had given the first satisfactory definitions of limits and continuity in the calculus noticed nothing demanding reform in 1847 in the continuum of real numbers with their non-denumerable infinity of irrationals. A thoroughgoing Pythagorean would have expelled the real irrationals along with i.

Cauchy extended his invention to what he called algebraic keys; but, as these were very special instances[2] of some of the algebras[3] already implicit in the work (1844) of H. G. Grassmann (1809–1877, German), they have rather missed fire. The prolific Cauchy passed on to new creations more conformable to his passion for analysis. His ingenious suggestion went unnoticed for forty years, when (1887) it reappeared, greatly amplified, in the arithmetical program of Kronecker.[4]

Here at last was the modern Pythagoras. Gauss is said to have ascribed an 'external reality' to 'space' and 'time,' while reserving for number the ideal purity of a 'creation of the mind.' Kronecker denied this philosophy, insisting that geometry and mechanics are expressible wholly in terms of relations between numbers, and by numbers he meant the positive integers 1, 2, 3, ... Thus, for him, the continuities 'space' and 'time,' fused in the concepts of kinematics, had meaning only in terms of the ineradicable discontinuities of these same God-given natural numbers. Continuity had no meaning; all was discrete.

To show how his subversive program might be carried out, Kronecker expelled negative numbers by means of congruences to the modulus $j + 1$, precisely as Cauchy had banished imaginaries $a + bi$ with his modulus $i^2 + 1$. Since only the natural numbers existed for Kronecker, he exorcised rational fractions by a similar magic, introducing (in effect) a new symbol, or 'indeterminate,' for each objectionable fraction. To dissipate $\frac{3}{4}$, for example, it sufficed to use congruences to the compound modulus $4k + 3j$. An irrational, say $+\sqrt{-2}$, could be dispensed with by a new indeterminate t and an additional modulus $t^2 + 2$. Arithmetic, algebra, and analysis began to grow somewhat complicated. But that was beside the point.

In the work cited and in an extensive earlier memoir (1882), Kronecker outlined in some detail how the program of Pythagoras could be realized in modern mathematics. Whether such a project was worth doing is irrelevant. Kronecker was interested primarily in showing that the Pythagorean vision could be materialized. Provided that were once demonstrated, careless mortals presumably were to be permitted to use negatives and irrationals in the customary manner and in the usual notations, on the understanding, however, that they admitted their workable mathematics to be merely a convenient shorthand for the only true mathematics, that of Kronecker's modular systems.

It would be interesting to know what Gauss would have thought of this devastating outcome of his simple device of writing 'n is divisible by m' as $n = 0 \bmod m$. Eminent mathematicians have called it anything from anarchy to hocus-pocus. Kronecker, however, might have recompensed himself at the expense of the nineteenth-century analysts, had he lived to participate in the debates of the twentieth century on the consistency of classical analysis. For few in the 1940's would have written with the resolute conservatism of E. W. Hobson (1856–1933, English) in 1921:

> Kronecker's ideal . . . that every theorem in analysis shall be stated as a relation between positive integral numbers only, . . . , if it were possible to attain it, would amount to a reversal of the actual historical course which the science has pursued; for all actual progress has depended upon successive generalizations of the notion of number, although these generalizations are now regarded as ultimately dependent on the whole number for their foundation. The abandonment of the inestimable advantages of the formal use in Analysis of the extensions of the notion of number could only be characterized as a species of Mathematical Nihilism.

Apart from its Pythagoreanism, Kronecker's effort left a useful residue, his theory of modular systems. This provides an alternative approach to algebraic numbers, Dedekind's being that usually followed.

One of the elementary by-products of Kronecker's algebra (1882) provided a rigorous theory of elimination for systems of polynomials in any number of variables. This rendered obsolete many unsatisfactory attempts, particularly by algebraic geometers, to give sound proofs for the speciously simple formalism of such methods as grow out of J. J. Sylvester's (1814–1897, English) of 1840, and E. Bézout's (1730–1783, French) of 1764. The latter was also invented independently by L. Euler (1707–1783, Swiss). The usual textbook discussion is still in the spirit of 1764, although there are honorable exceptions.

The same cynical fate awaited Kronecker's reduction of all mathematics to the natural numbers that seems sooner or later to nullify all human attempts to solve the universe at one stroke. It does not appear to have occurred to him that the natural numbers themselves might some day be put on trial as he had tried all other numbers and found them wanting in meaning.

Any savage might have suggested such a possibility; but it remained for the mathematical logicians of the early twentieth century to demonstrate the possibility up to the hilt and beyond.

A period of transition

We must briefly indicate the involuntary participation of E. Galois (1811–1832, French) and N. H. Abel (1802–1829, Norwegian) in the development of Kronecker's Pythagoreanism. Galois himself adhered to no such creed. Nor did Abel. But it was in the attempt to understand and elucidate the Galois theory of equations, left (1832) by its young author in a rather fragmentary and unapproachable condition, that Kronecker acquired some of his skill. Both Kronecker and Dedekind, two of the founders (E. E. Kummer [1810–1893, German] being a third) of the theory of algebraic numbers, were inspired partly by their scrutiny of the Galois theory to begin their own revolutionary work in algebra and arithmetic. Kronecker also began some of his researches in the arithmetization of algebra with a profound study of abelian equations.

Galois and Abel mark the beginning of one modern approach to algebra. The transition from highly finished individual theorems to abstract and widely inclusive theories is plainly evident in the algebra of Gauss contrasted with that of Abel and Galois. The like is seen in other fields, as will appear presently. This transition took place about 1830, contemporaneously with the abstract approach of the British algebraists.

Younger than Gauss by thirty-four years, and dying twenty-three years before him, Galois now, curiously enough, seems more modern than Gauss. A single example will suffice to substantiate the radical distinction between the two minds.

The occasion for Gauss' making mathematics his lifework was his spectacular discovery at the age of nineteen concerning the construction of regular polygons by means of straightedge and compass alone. Gauss proved that such a construction is possible if, and only if, the polygon has n sides where n is an integer of the form $2^s p_1 p_2 \ldots p_r$, $s \geqq 0$, in which $p_1, p_2, \ldots, p_r$ are r different primes, each of the form a power of 2 plus 1. The algebraic equivalent of this theorem, concerning binomial equations, is partly developed in the seventh and last section of the *Disquisitiones arithmeticae.* This particular work marks the end of its era in mathematical outlook.

Gauss terminated his investigations on the nature of the solutions of algebraic equations with the binomial equation

$$x^n - 1 = 0.$$

Galois grasped and solved (1830) the general problem, proving, among other things, necessary and sufficient conditions for the solution by radicals of any algebraic equation. Mathematics after Gauss, and partly during his own lifetime, became more general and more abstract than he conceived it. Interest in special problems sharply declined if there was a general problem including the special instances to be attacked. Or, what amounts to the same, mathematics after Gauss turned to the construction of inclusive theories and general methods which, theoretically at least, implied the detailed solutions of infinities of special problems. In this sense Galois was more modern than Gauss. In the same sense, Gauss was less modern than Abel, his junior by a quarter of a century, whom he outlived by thirty-six years.

The reasons in Abel's case are similar to those for Galois. The abelian equations in which Kronecker began his most individual work were named after Abel; they are a generalization of the equations discussed by Gauss in his problem of regular polygons. If it ever occurred to Gauss that there was a generalization, he has left no record of any attempt to attack the more inclusive problem. Confronted with binomial equations, Abel immediately saw through the special case to the abstract generality behind it, and elaborated the general theory. There is a similar contrast in the approaches of the two men to elliptic functions.

In the respects indicated by these comparisons, Gauss was closer to the eighteenth century than he is to the twentieth. His Athenian motto, "Few but ripe," was doubtless responsible for the classic perfection of the masterpieces he published himself. But the very perfection, reminiscent of the Greeks at their most rigid, repelled younger and less patient workers to whom time was the essence of their contract with genius, and they sought smoother ways round the obstacles in their path. Though they spoke of the master with respect and unsuccessfully courted his approbation, they seldom followed in his footsteps.

The liberation of algebra

Algebra attained its first freedom in the 1830's and 1840's in the hypercomplex number systems of Hamilton and Grassmann.

These two liberators of algebra are among the nineteenth century's major mathematical prophets. Both were richly gifted in many things besides mathematics. Hamilton at the age of thirteen was an accomplished classicist and a proficient linguist in the oriental as well as the European tongues; Grassmann was a profound scholar of Sanskrit. At twenty-seven Hamilton was famous as the result of his mathematical prediction of conical refraction, a deduction from his comprehensive theory of systems of rays in optics; and by thirty he had practically completed his fundamental work in dynamics, an advance beyond Lagrange comparable to Lagrange's beyond Euler. At the age of thirty-eight (1843) he overcame the difficulty which had prevented him from extending the algebra of coplanar vectors to a theory of vectors and rotations in space of three dimensons. He discovered that the commutative law of multiplication is not necessary for a self-consistent algebra. Thereafter Hamilton's scientific life was devoted to the elaboration of his theory of quaternions, in the mistaken hope that the new algebra would prove the most useful addition to mathematics after the differential and integral calculus.

Honors were showered on Hamilton; none fell on the less fortunate Grassmann. Neither had a particularly happy life. Hamilton was afflicted by domestic troubles and personal weaknesses; Grassmann supported himself, his wife, and nine children by elementary teaching, a profession for which he was eminently unsuited. A steadfastly pious man, Grassman trusted that if his contemporaries failed to reward his signal merits, the Lord would. He never complained of the torments he endured from the young savages he was meagerly paid to civilize. His avocations were his life the Sanskrit classics, philosophy, phonetics, harmony, philology, physics, theology, politics being among the extraordinary miscellany. But with the possible exception of theology, Grassmann's creation of "a new branch of mathematics"[5] in 1840–4 gave him the most abiding satisfaction. Here his inventive imagination and his perverse originality had free play. His theory of extension (*Ausdehnungslehre*), in which Hamilton's quaternions are a potential detail, was first published in 1844, about a year after Hamilton had found the clue to his problem of rotations in the equations $i^2 = j^2 = k^2 = ijk = -1$ defining the quaternion units i, j, k.

It has often been observed that it is not healthful for a mathematician to be a philosopher. Whether or not this is a general theorem, it was certainly true in the unfortunate Grassmann's case. Endowing his theory with the utmost generality it could support, he all but smothered it in philosophical abstractions. This was one of the greater tragedies of mathematics. Gauss looked the *Ausdehnungslehre* over, and blessed it with his qualified approval. It was partly in the same directon, he said as he himself had taken almost half a century earlier. But it was too philosophical with its "peculiar terminology" even for Gauss, himself no inconsiderable amateur of philosophy.

Gauss in the meantime had recorded his own independent discovery of Hamilton's quaternions. In a brief abstract which he never published,[6] ascribed to the year 1819, Gauss wrote out the fundamental equations of what he called mutations in space, essentially quaternions.

Grassmann continued his efforts to gain recognition for his own incomparably more general theory. Eighteen years (1862) after the first publication of his book, he brought out a completely revised, greatly amplified, and somewhat less incomprehensible version.[5] But a mathematician who has once been seriously called a philosopher might as well have been hanged for all the hearing he is likely to get from his fellow technicians. The second edition followed the first into temporary oblivion. Grassmann abandoned mathematics. The scope of his theory was perhaps not fully appreciated until the twentieth century. As one implicit detail, Grassmann's work included the algebra of the tensor calculus that became widely known only after its application (1915–16) in general relativity.

The central difficulty that had blocked Hamilton in his attempt to create an algebra of vectors in space of three dimensions was the commutative law of multiplication. His own graphic account of how he saw his way round the obstruction in a flash of certainty after much fruitless work being readily accessible, we need not repeat it here. But it is well worth thoughtful consideration by all students, especially by those who imagine that mathematical inventions fall into people's laps from heaven.

Before Hamilton succeeded, able men had failed to find the clue to a consistent algebra of rotations and vectors in space. For one, A. F. Möbius (1790–1868, German), who in 1823 had been a pupil[7] of Gauss, took a considerable step toward the desired algebra of four fundamental units in his barycentric calculus of 1827, a work which Gauss complimented as being composed in the true mathematical spirit. But Möbius was balked by the commutative law of multiplication, which he lacked the daring to reject. However, his new algorithm was of importance in the development of analytic projective geometry, particularly in the use of homogeneous equations, and he was an independent discoverer of the geometric principle of duality.[8] So his effort was anything but wasted.

The four fundamental units 1, i, j, k of Hamilton's quaternions $a + bi + cj + dk$, (a, b, c, d real numbers), do for rotations and stretches in space what 1, i do for the like in a plane. But whereas multiplication of complex numbers is commutative, that of quaternions is not. Familiar as mathematicians are today with swarms of algebras in which the postulates of common algebra are severally violated, they can still appreciate the magnitude of Hamilton's success when, in a flash, he transcended the tradition of centuries. His insight is comparable to that of the founders of non-Euclidean geometry, or to that of the arithmeticians who restored the fundamental law of arithmetic to the seemingly lawless algebraic integers.

It is radical departures from traditional orthodoxy such as these that carry mathematics forward what seems like a century or more at one stride. The painstaking, detailed cultivation of a newly discovered territory is necessary if it is to be fruitful; but such work can be well done by mere competence, while radically new discovery (or invention) is possible only to men who may imagine they are conservatives, but who at heart are rebels. Their boldness may cost them their scientific reputations or the comforts of a decent livelihood; for the way of the transgressor—who may be only a harmless innovator with the courage to step out in front of the rabble of respectable mediocrity—is sometimes as hard in science as it is elsewhere. Grassmann paid for his rashness with eighteen years of obscurity and final scientific extinction for the remainder of his life. Gauss, long in possession of non-Euclidean geometry, preferred his peace of mind to what he called "the clamor of the Boeotians," and kept his treasure to himself. Hamilton, having won an imperishable success in optics and dynamics, courted the indifference of his contemporaries when he devoted all of his superb talents to quaternions and, during his lifetime, acquired exactly one competent disciple in algebra. P. G. Tait (1831–1901, Scotch) gave up his all in mathematics to follow quaternions.

Ten years (1853) after his initial discovery, Hamilton published his *Lectures on quaternions* (64 + 736 + lxxii pages), in which he showed the utility of quaternions in geometry and spherical trigonometry. But the geometry was Euclidean and of three dimensions. The massive *Elements of quaternions* (lvii + 762 closely printed pages) followed in 1866, the year after Hamilton's death. If anything could have convinced geometers and physicists that quaternions were the master key to geometry, mechanics, and mathematical physics that Hamilton anticipated, his *Elements* should have done so. Literally hundreds of applications to these subjects were made by Hamilton in this elaborate work, which he considered his masterpiece.

Many reasons have been suggested for the failure of quaternions to fulfill Hamilton's expectations. A sufficient explanation, which includes many of the others, is that the calculus of quaternions was simply too hard for the busy scientists whom Hamilton would have helped. It took too long to master the tricks. But the possibility of an algebra specifically adapted to Newtonian mechanics and some parts of mathematical physics had been more than merely suggested, and it was reasonably certain that such an algebra would be forthcoming when the need for it became acute. Whatever form the desired algebra might assume, it was also a fair guess that it would follow the example of quaternions and reject the commutative law of multiplication.

On the long view, then, the permanent residue of Hamilton's tremendous labor was the demonstrated existence of a self-consistent algebra in which the commutative law of multiplication does not hold. This in turn, like the invention of non-Euclidean geometry, encouraged mathematicians to break the iron law of custom elsewhere and to create new mathematics in defiance of venerated traditions. A striking instance, which was to prove of cardinal importance in the development of algebra and the number system, was the construction of algebras in which $ab = 0$ without either a or b being zero, or in which $a^n \neq 0 (n = 0, 1, \dots, m)$, but $a^{m+1} = 0$. A simple instance of the former occurs in Boolean algebra (belonging to the algebra of logic), in which the fact stated is the symbolic expression of Aristotle's law of contradiction. Linear associative algebra furnishes any desired number of algebras containing 'divisors of zero'—such as a, b described above—also any number of algebras of the second species. The origin of all these modifications of common algebra is in the work of Hamilton and Grassmann in the 1840's.

Grassmann's outlook was much broader than Hamilton's. To appreciate how much broader it was, we must remember that in 1844, when Grassmann published his first *Ausdehnungslehre*, 'space,' for all but A. Cayley (1821–1895, English), was still imprisoned in Euclid's three dimensions. Cayley's sketch of a geometry of n dimensions is dated 1843; it could not possibly have influenced Grassmann's theory of 'extended magnitude,' which also can be phrased in the language of n-dimensional space. A 'real' space or 'manifold' of n dimensions is the set, or class, of all ordered n-ples $(x_1, x_2, \ldots, x_n)$ of n real numbers $x_1, x_2, \ldots, x_n$, each of which ranges over a prescribed class of real numbers. It is sufficient for purposes of illustration to let each of $x_1, x_2, \ldots, x_n$ range independently over all real numbers. The class of all $(x_1, x_2, \ldots, x_n)$ is also called an n-dimensional real number manifold.

In effect, Grassmann associated with $(x_1, x_2, \ldots, x_n)$ the hypercomplex number $x_1e_1 + x_2e_2 + \cdots + x_ne_n$, where $e_1, e_2, \ldots, e_n$ are the fundamental units of the algebra, which he proceeded to construct, of such hypercomplex numbers. Two such numbers, $x_1e_1 + \cdots + x_ne_n$ and $y_1e_1 + \cdots + y_ne_n$, are defined to be equal if, and only if, $x_1 = y_1, \ldots, x_n = y_n$.

Addition was defined by

$$(x_1e_1 + \cdots + x_ne_n) + (y_1e_1 + \cdots y_ne_n) \\ = (x_1 + y_1)e_1 + \cdots + (x_n + y_n)e_n,$$

of which an instance is common vector addition if $n = 2$ or if $n = 3$. The various special kinds of multiplication that can be defined at will give the general algebra its chief interest.

To demand a definition of multiplication without stating what properties the product is to have is meaningless. If, for example, the associative law $a(bc) = (ab)c$ is to be preserved, this is equivalent to imposing certain conditions on the fundamental units $e_1, \ldots, e_n$; if either of the distributive laws, $a(b + c) = ab + ac$, $(b + c)a = ba + ca$, is to hold, this must be expressed in terms of relations between $e_1, \ldots, e_n$; and similarly for the commutative law of multiplication, $ab = ba$. Thinking partly in terms of geometrical imagery, Grassmann defined several types of multiplication. In particular, multiplying out $(a_1e_1 + \cdots + a_ne_n)(b_1e_1 + \cdots + b_ne_n)$, and assuming that the 'coordinates' $a_1, \ldots, a_n, b_1, \ldots, b_n$ commute with the units $e_1, \ldots, e_n$, so that $a_1e_1b_1e_2 = a_1b_1e_1e_2$, etc., Grassmann called $e_1e_1, e_1e_2, e_2e_1, \ldots, e_{n-1}e_n, e_ne_{n-1}$, in the distributed product $a_1b_1e_1e_1 + a_1b_2e_1e_2 + a_2b_1e_2e_1 + \cdots$ units of the second order; and he first imposed conditions on these new units. For example, the product of $a_1e_1 + \cdots$ and $b_1e_1 + \cdots$ was called an inner product if $e_re_s = 1$ or 0 according as $r = s$ or $r \neq s$; and an outer product if $e_re_s = -e_se_r$, for $r, s = 1, \ldots, n$. From these two kinds of products, Grassmann constructed others for more than two factors. For example, if $e_r|e_s$ denotes the inner product of e_r, e_s, and $[e_re_s]$ the outer product, there are the possibilities $[e_r|e_s]e_t$, $e_r|[e_se_t]$, among others, for the definition of products of three factors. A type of particular importance is that in which each of the n^2 products e_re_s is a linear homogeneous function of the fundamental units $e_1, \ldots, e_n$, and multiplication is postulated to be associative. The linear associative algebras of B. Peirce (1809–1880, U.S.A.), developed in the 1860's but first printed in 1881, are of this type.[9]

A third type of product, called 'open' or 'indeterminate,' was to prove of central importance in the creation (1881–84) of a practical vector analysis by J. W. Gibbs (1839–1903, U.S.A.). The modern name for such a product is a matrix.[10] Gibbs, one of the most powerful mathematical physicists of the nineteenth century, was perhaps better qualified than Grassmann or Hamilton to sense the kind of algebra that would appeal to students of the physical sciences. His most original mathematical contributions in this direction were in dyadics and the linear vector function.

These hints must suffice to suggest that as early as 1844, Grassmann was in possession of an extensive theory capable of almost endless developments by specialization in various directions. As elaborated by its creator, this theory of 'extended magnitudes' might be interpreted as a greatly generalized vector analysis for space of n dimensions. It incidentally accomplished for any finite number of dimensions what Hamilton's quaternions were designed to do for Euclidean space of three dimensions. We have already noted that Grassmann's algebra includes quaternions as a very special case. As a general kind of algebra it also includes the theories of determinants, matrices, and tensor algebra. In short, Grassmann's theory of 1844–62 was anywhere from ten to fifty years ahead of its epoch.

Our present interest in Grassmann's work is the wide generalization it afforded of complex numbers $x_1 + ix_2$ as number-couples (x_1, x_2) to hypercomplex numbers $(x_1, \ldots, x_n)$. We must now relate this extension of the number concept to another, made explicitly in 1858 by Cayley but already implicit in the work of Grassmann, namely, matrices. The elements of the theory of matrices are now included in the usual college course in algebra; and since their appearance (1925) in the quantum theory, matrices have become familiar to mathematical physicists.

The invention of matrices illustrates once more the power and suggestiveness of a well-devised notation; it also exemplifies the fact, which some mathematicians are reluctant to admit, that a trivial notational device may be the germ of a vast theory having innumerable applications. Cayley himself told Tait[11] in 1894 what led him to matrices. "I certainly did not get the notion of a matrix in any way through quaternions: it was either directly from that of a determinant; or as a convenient mode of expression of the equations

$$x' = ax + by$$
$$y' = cx + dy."$$

Symbolizing this linear transformation on two independent variables by the square array $\begin{pmatrix} a & b \\ c & d \end{pmatrix}$ of its coefficients or 'elements,' Cayley was led to his algebra of matrices of n^2 elements by the properties of linear homogeneous transformations of n independent variables.

Behind this invention there is a relevant bit of history. Cayley had shown (1858) that quaternions can be represented as matrices $\begin{pmatrix} a & b \\ c & d \end{pmatrix}$ with a, b, c, d certain complex numbers. To Tait, the pugnacious champion of quaternions ever since he had elected himself Hamilton's disciple in 1854,[11] this discovery of Cayley's was conclusive evidence that Cayley had been inspired to matrices by his master's quaternions. Because the multiplication of matrices is in general not commutative, and since the like is true of quaternions, therefore, etc. This illustrates the unreliability of circumstantial evidence in mathematics as elsewhere. But for Cayley's testimony, critics might even now be asserting that Hamilton anticipated Cayley in the invention of matrices, or at least that Cayley got the notion of a matrix from quaternions.

Applications, or developments, of these extensions of number followed two main directions. The first, in the geometrical tradition of Hamilton and Grassmann, led to the extremely useful vector algebras of classical mechanics and mathematical physics, and later to the tensor algebra and calculus of relativity with its modifications and generalizations in modern differential geometry, also to the matrix mechanics of the quantum theory. The second, in the arithmetical spirit of Gauss, guided in part by the abstract algebraic outlook of Galois, led to a partial but extensive arithmetization of algebra. The course of both was highly intricate and blocked by innumerable details, many of which still promise to be of some enduring significance. But to see the principal trends at all, special and strictly limited developments must be ignored, for the present at least; and we shall attend only to the shortest paths from the past to the gains just indicated.

From vectors to tensors

The line of descent of vector algebra in general is fairly clear. The composition of velocities or of forces in the corresponding parallelogram laws suggested the addition of 'directed magnitudes.' Wessel's or Argand's diagram for depicting complex numbers was equally suggestive visually, geometrically, and kinematically. Hamilton and De Morgan's 'double algebra' of number-couples, replacing that of complex numbers, naturally suggested a generalization to number-triples, -quadruples, and so on. As we have seen, the central difficulty was the purely algebraic obstacle of commutative multiplication. Thus, in at least the early stages, geometrical and mechanical intuition shared about equally with formal algebra in the creation of a workable mathematics of vectors.

The famous *Treatise on natural philosophy* (1879) of Thomson and Tait offered a magnificent opportunity to display the power of quaternions as an implement of exposition and research in mechanics. Tait exhorted Thomson to repent of his Cartesian sins and embrace the true faith of quaternions. But W. Thomson (Lord Kelvin, 1824–1907, Scotch), declaring that Hamilton's good mathematics had ended with the masterpieces on optics and dynamics, hardened his heart and persisted in his iniquitous coordinates. The great opportunity was missed.

Tait had a somewhat better success with J. C. Maxwell (1831–1879, Scotch). In his epoch-making *Treatise on electricity and magnetism* (1873, Art. 11), Maxwell made a slightly damning concession: "I am convinced . . . that the introduction of the ideas, as distinguished from the operations and methods of

Quaternions, will be of great use . . . especially in electrodynamics . . . " And, with one exception, Maxwell studiously avoided quaternions. The exception (Art. 618) is a summary in quaternion notation of the electromagnetic equations. No use is made of this summary. But Maxwell did use "the ideas," not of quaternions, but of his own conception of vector analysis. His convergence is the negative of the divergence in use today, and he introduced (Art. 25) what is now called the curl of a vector. These innovations have lasted.

The most profitable departure from quaternionic orthodoxy was that of J. W. Gibbs in his vector analysis of the 1880's. This will be noted presently. The next was by O. Heaviside (1850–1925, English), in his profoundly individualistic *Electromagnetic theory* of 1893. In a chapter of 173 pages, Heaviside elaborated his own vector notation. His methods resembled those of Gibbs; but of Gibbs' notation, Heaviside confessed, "I do not like it." Germany provided the next (1897) considerable variation on the now familiar theme, in A. Föppel's *Geometrie der Wirbelfelder*—geometry of vortex fields. By 1900 the contest between rival claimants to physical favor had narrowed down in English-speaking countries to Gibbs versus Heaviside. Quaternions appeared to have been knocked out. Tait, their most formidable champion, died in 1901; the vector analysis of Gibbs or some modification of it prevailed in the U.S.A.

Much of this tortuous development was enlivened by one of the most spirited mathematical controversies of modern times. Unlike the numerous squabbles over priority, the quaternions-versus-vectors war was refreshingly scientific. The casus belli was a purely mathematical difference of opinion: were quaternions a good medicine for applied mathematics, or was some one of several diluted substitutes a better? The uninitiated might think that so abstract a bone of contention would provoke only dry academic discourse, with at worst an occasional growl of dissent. It did nothing of the kind. The language of the disputants even bordered on quite un-Victorian indelicacy at times, as when Hamilton's devoted Tait[12] in 1890 called the vector analysis of Gibbs "a sort of hermaphrodite monster, compounded of the notations of Hamilton and Grassmann." That was Scotch and Irish against American. Gibbs, being a New Englander to the marrow and a confirmed bachelor cherished only by his married sister, was but slightly acquainted with the inexhaustible resources of the American language. Tait got away with his abnormal physiology, but Gibbs got the better of the mathematical argument.

Frenchmen, Germans, and Italians, urging their respective substitutes for quaternions, added to the din. By the second decade of the twentieth century there was a babel of conflicting vector algebras, each fluently spoken only by its inventor and his few chosen disciples. If, at any time in the brawling half-century after 1862, the bickering sects had stopped quarreling for half an hour to listen attentively to what Grassmann was doing his philosophical best to tell them, the noisy battle would have ended as abruptly as a thunderclap. Such, at any rate, seems to have been the opinion of Gibbs. In retrospect, the fifty-year war between quaternions and its rivals for scientific favor appears as an interminable sequence of duels fought with stuffed clubs in a vacuum over nothing.

The disputes ceased to have any but a mathematically trivial significance almost as soon as they began. As Gibbs[10] emphasized in 1886, in his account of the development of multiple algebra, the mathematical root of the matter is in Grassmann's indeterminate product, that is, in the theory of matrices. Gibbs also remarked the superior generality of Grassmann's many possible kinds of product in multiple algebra over the unique product insisted upon by Hamilton:

> Given only the purely formal law of the distributive character of multiplication—that is sufficient for the foundation of a science. Nor will such a science be merely a pastime for an ingenious mind. It will serve a thousand purposes in the formation of particular algebras. Perhaps we shall find that in the most important cases the particular algebra is little more than an application or interpretation of the general.

The whole of Gibbs' judicial and profound evaluation (1886) of multiple algebra in relation to its applications might be studied with profit at any time by those interested in the continued improvement of applied algebra. Vector analysis and even the infinitely more inclusive *Ausdehnungslehren* of Grassmann are after all only provinces, although highly cultivated ones, of algebra, which itself is but a territory of modern mathematics. Those interested in the advancement of mathematics, rather than in the perpetuation of individuals as dictators of provinces, will not be dismayed when particular theories to which they may be personally attached are supplanted by others. Obsolescence is a necessary adjunct of progress; and any effort such as Tait's to keep quaternions unsullied and perpetually fresh is likely to be as futile as an attempt to stop the earth in its orbit. The vector analysis of Gibbs gradually displaced quaternions as a practical applied algebra in spite of the utmost efforts of the quaternionists; and after 1916 it seemed that the several special brands of vector analysis were about to be supplanted in their turn by the tensor algebra and analysis that became popular in 1915–16 with the advent of general relativity.

As in the struggle of vector analysis against quaternions, the advance to tensors generated its own opposition. Vector analysis, like some human beings, needed above all else to be delivered from the good intentions of its partisan friends. Progress here, as elsewhere in the past of mathematics, appeared to be possible only when all the friends and former pupils of some great and justly famous master should have died. Then only might it be possible to see the mathematics rather than the man.

Such retardations due to misdirected enthusiasm are frequent enough in mathematics. The master founds a 'school'; the pupils, remembering perhaps among other things an encouraging pat on the head from their first competent teacher, graduate into a world that does not stop dead no matter who dies, to keep on repeating for the rest of their lives the only lesson they ever really learned. The school itself expires, leaving its useful contribution encrusted with an accumulation of artificially stimulated growths that must be cut away before the creative idea of the originator can begin to live and function freely. Aware of these possibilities, some mathematicians, including one of the first rank, have refrained from propagandizing their own ideas or those of their teacher, and have made no attempt to gather a following of bigoted disciples. Kronecker took pride in the fact that he had never tried to found a school or to acquire a host of students. He believed, as did Gibbs, that "the world is too large, and the current of modern thought is too broad, to be confined by the *ipse dixit* even of a Hamilton."

There seems to be but little doubt that applied algebra was held back by the partisans of jealous schools. The road to unity can be traced back from about 1940, when the rudiments of the tensor calculus had become fairly common in undergraduate instruction, to Grassmann's n-dimensional manifolds of 1844. Three dimensions are inadequate for modern physics, or even for classical mechanics with its generalized coordinates. G. F. B. Riemann (1826–1866, German) in 1854 took the next long step forward after Grassmann when he introduced Gaussian (intrinsic) coordinates and made n-dimensional manifolds basic in his revolutionary work on the foundations of geometry. Another work of Riemann's, published after his death, contained what is now known as the Riemann-Christoffel tensor in the relativistic theory of gravitation. Riemann encountered this tensor in a problem on the conduction of heat. E. B. Christoffel (1829–1900, German) was the next to make significant progress toward a general tensor calculus, in his work of 1869 on the transformation (equivalence) of quadratic differential forms. Finally, in the 1880's, the Italian geometer M. M. G. Ricci combined and added to all the work of his predecessors. The result, published in 1888,[13] was the tensor calculus. Thus the mathematical machinery demanded by the theory of general relativity was available a year after the Michelson-Morley experiment, which was partly responsible for the special theory of relativity in 1905; without the tensor calculus the general theory of 1915–16 would have been impossible. The above assertion about the Michelson-Morley experiment does not imply that Einstein was motivated by the experiment in his construction of special relativity. In fact he has stated explicitly that he knew of neither the experiment nor its outcome when he had already convinced himself that the special theory was valid.

The new method attracted very little attention. On the invitation of F. Klein (1849–1925, German), Ricci and his former pupil, T. Levi-Civita 1873–1942, Italian), prepared an article on the tensor calculus and its applications to mathematical physics for publication in a journal read by mathematicians of all nationalities. The article, in French, appeared in 1901. It fell rather flat. However, a few curious geometers outside of Italy became aware of the new calculus, and at least one, M. Grossmann of Zurich, mastered it and taught it to Einstein. The tensor calculus was the particular kind of generalized vector algebra appropriate for expressing the differential equations of relativity in covariant form as demanded by a postulate of the theory.

The debt of algebra and geometry to general relativity is as great as that of relativity to algebra and geometry. Although Ricci and Levi-Civita in their expository article of 1901 had offered abundant evidence of the utility of tensor analysis in applied mathematics, the new calculus was seriously taken up by mathematical physicists only after their curiosity had been roused by the experimentally verified mathematical predictions of relativity. The tensor method quickly induced a vast development of differential geometry.

Gibbs had predicted in 1886 that vector analysis would someday greatly simplify what in his time was modern higher algebra —the theory of algebraic covariants and invariants. He had in mind the possibilities of Grassmann's theory. His prediction was verified in the 1930's. Another prediction of Gibbs of the same kind was verified in 1925, when W. Heisenberg found in the algebra of matrices the implement he needed for the non-commutative mathematics of his quantum mechanics. Physicists took less kindly to $ab \neq ba$ than they had to tensors; and it was a great relief to many when C. Eckart (U.S.A.) and E. Schrödinger (Austria) in 1926 showed independently and simultaneously that matrix mechanics could be replaced by wave mechanics, in which the theory of boundary-value problems, already familiar in classical mathematical physics, is the key to the mathematics.

It seems probable that Grassmann did not anticipate any such outcome for his extremely general 'geometrical algebra.' Two of his successors, Riemann and W. K. Clifford (1845–1879, English), both more physical-minded than Grassmann, ventured to predict the twentieth-century geometrization of some parts of mathematical physics. This was in the middle stage from Grassmann to tensors, and it was as remarkable a prophecy as any that mathematicians have ever made. But it must not be forgotten that mathematicians no less than scientists and others have made many false prophecies. The successes are remembered.

Toward structure

"Mathematics," according to Gauss in 1831, "is concerned only with the enumeration and comparison of relations." B. Peirce (1809–1880, U.S.A.), one of the creators of linear associative algebra, asserted[9] in 1870 that "Mathematics is the science which draws necessary conclusions." Peirce also remarked that "all relations are either qualitative or quantitative," and that the algebra of either kind of relation may be considered independently of the other, or that, in certain algebras, the two may be combined.

These opinions, from what is now a remote past mathematically, might be admitted by some formalists as anticipations of their own conception of mathematics as the theory of structure. In particular, the Pythagorean program is superseded. Euclid's postulational method remains. Large tracts of mathematics have become entirely formal and abstract; the content of a mathematical theory is the structure of the system of postulates from which the theory is developed by the rules of mathematical logic, and from which are derived its various interpretations.

This excessively abstract view of mathematics evolved from the formalization of elementary algebra in the 1830's, which has already been described; the work of Abel and Galois in the theory of algebraic equations, of about the same time; the development of linear algebra throughout the nineteenth and early twentieth centuries; the creation of mathematical logic, beginning with Boole in 1847–54 but vigorously pursued only in the twentieth century; and finally, from the free invention of non-Euclidean geometries after 1825, and the renewed interest in postulational methods following Hilbert's work of 1899 on the foundations of geometry.

Of all these influences, two in particular are germane here: the development of linear algebra; and the infiltration of the ideas of Abel and Galois into algebra as a whole. The Galois theory of equations was acknowledged by both Dedekind and Kronecker to be the inspiration for their own general and semi-arithmetical approach to algebra. Two of the basic concepts of the Galois theory, domains of rationality, or fields, and groups, were the point of departure. Both groups and fields will be described presently. For the moment we observe the underlying methodology which might have been followed and which would be followed today (1945), but which was not followed historically, in the generation of linear algebras, groups, and other systems in modern algebra.

The methodology is that of generalization by suppression of certain postulates defining a given system. The system defined by the curtailed set of postulates is then developed. Linear algebra is obtainable in this way from the algebra of a field. Vector algebras, as we have seen, received their initial impulse from Hamilton's suppression of the postulate that multiplication is commutative in common algebra. Common algebra is the most familiar example of a field.

Groups also may be derived from common algebra by the same technique of generalization. But they were not so obtained originally; and it is doubtful whether they would ever have attracted the attention they did, had not the momentum of history thrust them forward. There are 4,096 (perhaps more) possible generalizations of a field. To develop them all without some definite object in view would be slightly silly. Only those that experience has suggested have been worked out in any detail. The rest will keep till they are needed; the apparatus for developing them is available. Nevertheless, the postulational technique has been one of the most suggestive of twentieth-century mathematics; and we shall have occasion to recur to it frequently as we proceed.

Fields being the most familiar of all mathematical systems, we shall define them first. A field[14] (*Körper*, corpus, *corps*, domain of rationality) F is a system consisting of a set S of elements $a, b, c, \ldots, u, z, \ldots$ and two operations, $\oplus$, $\odot$, which may be performed upon any two (identical or distinct) elements a, b of S, in this order, to produce uniquely determined elements $a \oplus b$ and $a \odot b$ of S, such that the postulates (1) to (5) are satisfied. Elements of S will be called elements of F. For simplicity, $a \oplus b$, $a \odot b$ will be written $a + b$, ab.

(1) For any a, b of F, $a + b$ and ab are uniquely determined elements of F, and $b + a = a + b$, $ba = ab$.

(2) For any a, b, c of F, $(a + b) + c = a + (b + c)$, $(ab)c = a(bc)$, $a(b + c) = ab + ac$.

(3) There exist in F two distinct elements z, u such that if a is any element, $a + z = a$, $au = a$.

(4) For any element a of F, there is in F an element x such that $a + x = z$.

(5) For any element a, other than z, of F, there is in F an element y such that $ay = u$.

It should be noticed that equality, $=$, has been assumed as a known relation. For completeness: equality is an equivalence relation (as defined earlier in connection with congruence). That is, if a, b are any elements of F, $a = b$ or $a \neq b$, $\neq$ meaning 'not equal to'; $a = a$; if $a = b$, then $b = a$; if $a = b$ and $b = c$, then $a = c$.

This familiar and somewhat elaborate abstraction of common algebra and rational arithmetic will serve to illustrate the meaning of structure and the history of its development. We note first that these precise postulates date only from 1903; and that in the postulates as given in 1903 (and 1923), the precise meaning of equality is not stated, being taken for granted. In texts of 1930 or later, it became customary to define equality as an equivalence relation before using equality in the postulates of a field. This is typical of the continually increasing precision in elementary mathematics since the first explicit definition of a number field in 1879 by Dedekind. As a final instance of the same tendency, it was only in the 1920's that it became customary to state explicitly that $a = b$ or $a \neq b$. There is therefore little reason to suppose that even these precisely stated postulates have explicated all the assumptions underlying our habitual use of common arithmetic.

If, in the postulates (1) to (5), the elements, $a, b, c, \ldots$ be interpreted as rational numbers, and u, z as 1, 0, with $a + b$, ab, the sum, product of a, b, it is seen that the rational numbers are an instance of a field with respect to addition and multiplication. Subtraction and division follow from (4), (5). Similarly, ordinary complex numbers $x + iy$ furnish another instance; as also do Hamilton's number-couples (x, y) with the appropriate definitions of u, z, addition, and multiplication, which the reader may easily recover. The rational integers $0, \pm 1, \pm 2, \ldots$ do not furnish an instance, on account of (5). If F is any field, and $x_1, \ldots, x_n$ are independent variables (or indeterminates), the set of all rational functions of $x_1, \ldots, x_n$, with coefficients in F, is another field.

With 'structure' still not defined formally, it is intuitively evident what is meant by the statement that all instances of a field have the same structure, and that this structure is as in (1) to (5). Further, it is clear that if the logical consequences of (1) to (5) are developed, the body of theorems so obtained will be valid for each instance of a field. The last is indeed 'clear,' although a proof of it might be difficult and, as a matter of fact, no generally accepted proof had been devised up to 1945. A thoroughly satisfactory proof must demonstrate that the rules of mathematical logic applied to (1) to (5) will never produce a contradiction, such as "$a = b$ and $a \neq b$." It seems as if this must be the case; but seeming in mathematics is not the same as being. 'Existence,' for one school, is indeed identified with proof.

The earliest recognitions of fields, but without explicit definition, appear to be in the researches of Abel[15] (1828) and Galois[16] (1830–1) on the solution of equations by radicals. The first formal lectures on the Galois theory were those of Dedekind to two students in the early 1850's. Kronecker also at that time[17] began his studies on abelian equations. It appears that the concept of a field passed into mathematics through the arithmetical, works of Dedekind and Kronecker. Both, especially Dedekind,[18] early recognized the fundamental importance of groups for algebra and arithmetic. With Dedekind's famous *Eleventh supplement* to the third edition (1879) of P. G. L. Dirichlet's (1805–1859, German) *Vorlesungen über Zahlentheorie*, the concept of a number field was firmly established in mathematics. We note, however, that Dedekind in this work was interested only in algebraic numbers—roots of algebraic equations with rational number coefficients. The fields he defined were therefore those of real and complex numbers. Kronecker followed in 1881 with his domains of rationality, that is, fields. Although Kronecker's definition was more general than Dedekind's, it did not attain the complete generality of the postulate system quoted above.

The passage to final abstractness took about a quarter of a century. This need not be traced in detail here; the references given are sufficient to orient anyone who wishes to elaborate the history. The turning point was Hilbert's work on the foundations of geometry in 1899. Although this did not concern algebra or arithmetic directly, it set a new and high standard of definiteness and completeness in the statement of all mathematical definitions or, what is equivalent, in the construction of postulate systems. Compared to what came after 1900 in this basic kind of work, that before 1900 now seems incredibly slack. With abundant resources at hand to continue the Euclidean program of stating explicitly what a mathematical argument is to be about, a majority of nineteenth-century mathematicians left their readers to guess exactly what was postulated. Neglect to state all the intended assumptions incurred its own penalties in faulty proofs and false propositions. The change for the better after 1900 was most marked, but there is still room for improvement, especially in mathematics of the intuitive kind—such as the repeated appeal thus far to intuition for the meaning of structure.

Passing to groups, we shall state in full a set of postulates for a group, as 'group' in the technical sense defined by these postulates will occur repeatedly in the sequel. We shall then be in a position to define structure.

A group G is a set S of elements $a, b, c, \ldots, x, y, \ldots$ and an operation O, which may be performed upon any two (identical or distinct) elements a, b of S, in this order, to produce a uniquely determined element aOb of S, such that the postulates (1) to (3) are satisfied.

(1) aOb is in S for every a, b in S.

(2) $aO(bOc) = (aOb)Oc$ for every a, b, c in S.

(3) For every a, b in S there exist x, y in S such that $aOx = b$, $yOa = b$.

These postulates may appear strange to those acquainted with others for a group; but they are simpler than some, and all are equivalent. Historical notes on groups will be given later; our present interest is in mathematics. We proceed to structure,[19] which seems to have been first recognized, but not defined, in groups.

Consider two groups, with the respective elements a_1, b_1, c_1, . . . and a_2, b_2, c_2, . . . and the respective operations O_1, O_2. These groups are said to be simply isomorphic, or to have the same structure, if it is possible to set up a one-one correspondence between the elements such that, if $x_1O_1y_1 = z_1$, then $x_2O_2y_2 = z_2$, and conversely, where x_1, y_1, z_1 are the respective correspondents of x_2, y_2, z_2. For further details we must refer to the texts.

This definition is probably the simplest example of what is meant by 'same structure.' Note that 'structure' is not defined, but that 'same structure' is. For the purposes of algebra this is sufficient. If 'same structure' seems at first glance to define absolute identity, an example to the contrary is supplied by all the normal men in a community, all of whom have the same shape—two arms, one head, etc.—but no two of whom are identical except perhaps topologically.

A general theory of structure was developed by A. N. Whitehead (1861–, English) and B. Russell[21] (1872–, English) in 1910. It will suffice here to recall a cardinal definition: A relation P between the members of a set x_p has the same structure as a relation Q between the members of a set y_q if there is a one-one correspondence between the elements of x_p and y_q such that, whenever two elements of x_p are in the relation P to each other, their correlates (by the correspondence) in y_q are in the relation Q to each other, and vice versa.

If in any division of mathematics there are relations P, Q, . . . having the same structure, it suffices to elaborate the implications of one, say P, when those of Q, . . . follow on translating from P, x_p, . . . to Q, y_q, . . . by means of the relevant correspondence in each case. Each of the postulates of a mathematical system can be restated as a relation between the data ('elements' and 'operations') of the system.

If it is possible to establish a one-one correspondence between the postulates of two systems such that correlated postulates have the same structure, then the systems are said to have the same structure. Instead of saying that two systems have the same structure, it is customary in the U.S.A., following E. H. Moore (1862–1932, U.S.A.) who used the concept in his lectures and writings from about 1893 on, to say that the systems are abstractly identical. Abstract identity is itself an equivalence relation. If several systems are abstractly identical, obviously it is sufficient to develop the mathematics of one in order to have that of all. The systems so developed will differ in the interpretations assigned to the abstract elements and operations; each assignment provides an 'instance' of the theory. For example, the algebras of real and of complex numbers, or of Hamiltonian number-couples, are instances of the theory of an abstract field.

Tracing the evolution of algebra since the 1830's, we note a constant but largely subconscious striving toward abstractness. Concomitantly, abstract identity was sought, sometimes deliberately, as in the theories of groups and fields. Most classification is an effort in the same direction preparatory to comparison of different theories and the detection of abstract identities. Klein's unification of diverse geometries by the theory of groups in 1872, which will be described in connection with invariance, was a conspicuous example of the advantages accruing from a recognized abstract identity. But it seldom happens that anything so simple as a group unifies apparently unrelated divisions of mathematics with respect to anything deeper than superficialities.

Enough has been said about structure to indicate what Gauss may have had in mind when he observed that "mathematics is concerned with the enumeration and comparison of relations." He made this statement in connection with complex numbers. On another occasion he expressed a doubt that any 'numbers' other than the real and complex—such as quaternions, for example—would ever be of any use in the higher arithmetic. In our further pursuit of algebra and arithmetic we shall be guided by this hint of Gauss', and endeavor to see what he might have had in mind. This, of course, is not the only road by which number might be followed from the 1830's to the twentieth century. But following it, we shall have a definite object in view by which to orient some of the major trends of algebra and arithmetic on the way.

CHAPTER 10

Arithmetic Generalized

Continuing with the modern developments of number and their influence on the emergence of structure, we shall observe next the expansion of modern arithmetic—the Greek arithmetica—from its origin in 1831 in the work of Gauss on the law of biquadratic reciprocity to its end in mathematical logic. Our immediate interest in this chapter is the greatly generalized concept of whole number, or integer, which distinguished the higher arithmetic of the late nineteenth century from all that had preceded it. In a subsequent chapter we shall follow some of the main lines of descent of the classical arithmetic from Fermat, Euler, Lagrange, and Gauss to the present. Historically, many of these older developments preceded the work to be described here. But their interest, great though it may be intrinsically, is as yet comparatively negligible for mathematics as a whole.

There are six major episodes to be observed, four of which will be described in this chapter and the following. The four are the definition by Gauss, E. E. Kummer (1810–1893, German), and Dedekind of algebraic integers; the restoration of the fundamental theorem of arithmetic in algebraic number fields by Dedekind's introduction of ideals; the definitive work of Galois on the solution of algebraic equations by radicals, and the theory of finite groups and the modern theory of fields that followed; the partial application of arithmetical concepts to certain linear algebras by R. Lipschitz (1831–1903, German), A. Hurwitz (1859–1919, Swiss), L. E. Dickson (1874–, U.S.A.), Emmy Noether (1882–1935, German), and others. All of these developments are closely interrelated. The last marks the farthest extension of classical arithmetic up to 1945, and is either the climax or the beginning of a structural arithmetization of algebra foreseen as early as 1860 by Kronecker, but only partly achieved by him in the 1880's. As if in preparation for the climax, the algebra of hypercomplex numbers rapidly outgrew its classificatory adolescence of the 1870's, represented by the work of B. Peirce and his successors, and became progressively more concerned with general methods, reaching a certain maturity early in the twentieth century.

The fifth major episode, which logically would seem to be a necessary prelude to the others, strangely enough came last. Not until the closing years of the nineteenth century was anyone greatly perturbed about the natural numbers 1, 2, 3, ... All mathematics, from the classical arithmetic of Fermat, Euler, Lagrange, A. M. Legendre (1752–1833, French), Gauss, and their numerous imitators, to geometry and analysis, had accepted these speciously simple numbers as 'given.' Without them, none of the major advances of modern arithmetic would ever have happened. Yet no arithmetician asked, "By whom are the natural numbers 'given'?" Kronecker ascribed them to God, but this was hardly a mathematical solution. The question arose, not in arithmetic, but in analysis. It was answered by the modern definition of cardinal and ordinal numbers. This finally united arithmetic and analysis at their common source.

The sixth and last major episode in the evolution of the number concept was the application of arithmetic to the differential and integral calculus. It is a point of great interest, as will be seen in a later chapter, that one of the strongest initial impulses for the final application of arithmetic to analysis came from mathematical physics. Fourier's theory of heat conduction (1822) disclosed so many unforeseen subtleties in the concepts of limit and continuity that a thorough overhauling of the basic ideas of the calculus was indicated. Many toiled at this for the rest of the nineteenth century. It was gradually perceived that the cardinals and ordinals 1, 2, 3, . . . demanded clarification. By 1902 the last subtle obscurity then uncovered had been removed, only to make room for a yet more subtle. The arithmetic of 1, 2, 3, . . . , and with it mathematical analysis, resigned its soul to the searching mercies of mathematical logic.

About twenty-five centuries of struggle to understand number thus ended where it had begun with Pythagoras. The modern program is his, but with a difference. Pythagoras trusted 1, 2, 3, . . . to 'explain' the universe, including mathematics; and the spirit animating his 'explanation' was strict deductive reasoning. The natural numbers are still trusted by mathematicians and scientists in their technical mathematics and its applications. But mathematical reasoning itself, vastly broadened and deepened in the twentieth century beyond the utmost ever imagined by any Greek, supplanted the natural numbers in mathematical interest.

When, if ever, mathematical logic shall have surmounted its obscurities, the natural numbers may be clearly seen for what they 'are.' But there will always remain the possibility that any unscaled range may conceal a higher just beyond; and if the past is any guide to the future, arithmeticians will come upon many things to keep them busy and incompletely satisfied for the next five thousand years. After that, perhaps, it will not matter to anyone that 1, 2, 3, . . . 'are.'

Generalized divisibility

The class of positive rational integers 1, 2, 3, . . . was first extended, as a class of *integers*, by the adjunction of zero and the negative rational integers -1, -2, -3, ... We recall that Euclid in the fourth century B.C. proved one of the cardinal theorems concerning positive rational primes: If a prime p divides the product of two positive rational integers, p necessarily divides one of them. A rational prime admits as divisors only itself and the units 1, -1. The extension of Euclid's theorem to all the rational integers is immediate and need not be recalled. But to emphasize the non-trivial character of the generalizations of the rational integers by Gauss, Kummer, Dedekind, and others, the preceding definitions must be reformulated so as to apply to the generalized 'integers' in question. It may be remarked that this simple recasting of the definitions of rational arithmetic was one of the three most difficult steps toward the desired generalization. The other two were a redefinition of arithmetical divisibility, as distinguished from division in algebra, and the closely related problem of selecting from a given class of numbers those which are to be defined as integers.

First, as to units. With 'integer' as yet unspecified, a unit in a given set of integers is an integer that divides each integer in the set. An integer α 'divides' an integer β if there is an integer γ such that $\beta = \alpha\gamma$.

Second, as to 'irreducibles.' An integer α is said to be irreducible if '$\alpha = \beta\gamma$,' with β, γ integers, implies that one of β, γ is a unit and the other is α.

Third, as to primes. An integer α is called a prime if it is irreducible, and if further the assertion 'α divides $\beta\gamma$' implies at least one of the assertions 'α divides β,' 'α divides γ.'

These definitions accord with those for the rational integers. But whereas rational primes and rational irreducibles coincide, the like is not true for all of the generalized integers to be described.

The manufacture of definitions is likely to be a profitless pursuit unless there is a definite end in view. The goal here is the fundamental theorem of arithmetic: the 'integers' defined are to be resolvable into powers of distinct 'primes' in one way only, apart from 'unit' factors and permutations of the factors. This requirement is too drastic for the 'arithmetic' of most linear algebras; it is that at which the founders of the theory of algebraic numbers aimed. It was to prove unattainable.

The means by which the original program was replaced by another, which accomplished the essentials of what had been sought originally, is one of the finest examples of generalization in the history of mathematics. The generalization concerned the fundamental concepts of common arithmetic, particularly 'integer' and 'divisibility.'

To be of more than trivial significance, any generalization in mathematics must yield on appropriate specialization all the instances from which the generalization proceeded, and must give in addition more than is contained in all of those special instances. The profoundest generalizations appear to be those in which the interpretations of all the symbols in the structure (postulates) of a given system are changed. The passage from rational integers to algebraic integers was of this kind.

For example, in the theorem of rational arithmetic, "if a divides b, then b does not divide a unless a, b are units," a, b, $(b \neq a)$, and the division-relation are all assigned interpretations in the generalization differing from those of rational arithmetic. But these interpretations are such that the statement "if a, etc." remains true for the new interpretations.

The extension of rational arithmetic to an arithmetic of algebraic numbers and, considerably later, to a partial arithmetization of linear algebra, originated in two distinct sources: the proof by Gauss in 1828–32, or earlier, of the law of biquadratic reciprocity; Kummer's attempt in the 1840's to prove Fermat's last theorem. We begin with Gauss.

If there is a rational integer x such that, when n, p, q are given positive integers, $x^n - q$ is divisible (without remainder) by p, q is called an n-ic residue of p. Restated as a Gaussian congruence, q is an n-ic residue of the given p if and only if $x^n \equiv q \bmod p$ is solvable for x. For simplicity, we describe only the case in which p, q are positive odd primes. Gauss was particularly concerned with $n = 4$, $n = 3$. For $n = 2$, Legendre's law of quadratic reciprocity, which Gauss called "the gem of arithmetic," is

$$(p|q)(q|p) = (-1)^{\frac{1}{4}(p-1)(q-1)},$$

where $(p|q)$ denotes 1 or -1 according as $x^2 \equiv p \bmod q$ is, or is not, solvable for x, and similarly for $(q|p)$ and $x^2 \equiv q \bmod p$. Gauss long sought a reciprocity law for $n = 4$ as simple as that for $n = 2$. He found it only when he passed beyond rational integers to complex integers. A Gaussian complex integer is a number of the form $a + bi$, where a, b are rational integers. Defining units, primes, and divisibility for his complex integers in the straightforward way suggested by analogy with rational integers, Gauss proved that the fundamental theorem of arithmetic holds for integers $a + bi$. By means of these integers he was enabled to state the law of biquadratic ($n = 4$) reciprocity concisely. For $n = 3$ he found an equally simple theory, based on 'integers' $a + b\rho$, where ρ is a root of $y^2 + y + 1 = 0$ and a, b are rational integers; but he did not publish his results.

The history of reciprocity laws for $n > 4$ would fill a large book. This highly developed subject has been cultivated by scores of arithmeticians, and it has had a considerable influence on the evolution of modern algebra. But as this specialty, rich though it may be intrinsically, is rather to one side of the principal advance, we must leave it here with a remark.

What is essentially the law of quadratic reciprocity was known to L. Euler (1707–1783, Swiss) in 1744–6 but was not proved by him.[1] He discussed the law more fully in 1783. Legendre in 1785 attempted a proof, but slipped in assuming as obvious a theorem which is as difficult to prove as the law itself. Gauss first published a proof in 1801, and gave six in all. For $n > 2$, the reciprocity laws depend upon the algebraic number fields entering through binomial equations of degree n. This brings us to the next stage in the development of algebraic numbers. A particular algebraic number field of degree n is the set of all rational functions of a root of a given irreducible algebraic equation of degree n with rational integer coefficients.[2]

In his attempt to prove the impossibility of $z^p = x^p + y^p$, in which x, y, z, p are rational integers, $xyz \neq 0$, and p is a prime

>2, Kummer in 1849 resolved $x^p + y^p$ into its p linear factors,

$$(x + y)(x + \alpha y) \cdots (x + \alpha^{p-1}y),$$

where α is an imaginary pth root of 1. This led him to extend the theory of Gaussian complex integers to the algebraic number field defined by $\alpha^{p-1} + \alpha^{p-2} + \cdots + \alpha + 1 = 0$. With appropriate definitions of integers, primes, etc., in this field, Kummer persuaded himself for a time that he had proved Fermat's last theorem. But, as P. G. L. Dirichlet (1805–1859, German) pointed out to him, he had assumed that the fundamental theorem of arithmetic holds for these integers constructed from α. For certain primes p the fundamental theorem is valid in the corresponding α-field; for others it is not. The complete proof (or a disproof) of Fermat's theorem was still open.

Undaunted by this totally unforeseen failure, Kummer invented a new kind of number, which he called 'ideal'—not to be confused with Dedekind's ideals. There would be no point in describing these here,[3] as they are too far off the main road. They apply to the particular number fields considered by Kummer in connection with Fermat's last theorem.

Making a completely fresh start in the early 1870's, J. W. R. Dedekind (1831–1916, German) created a theory of algebraic integers applicable to the general case of an algebraic number field defined by a root of an irreducible equation

$$a_0x^n + a_1x^{n-1} + \cdots + a_n = 0$$

of any degree n with rational integer coefficients $a_0, \ldots, a_n$. A root of this equation is called an algebraic number of degree n; if $a_0 = 1$, this number is an algebraic integer; if in addition $a_n = 1$ (or -1), the algebraic integer is a unit. Note that any rational integer r is an algebraic integer of degree 1, since r is the root of $x - r = 0$.

All this detail has been recalled to indicate that the generalization from rational integers and rational units to algebraic integers of any degree demanded unusual insight. At first glance it seems impossible that a number such as $(-13 + \sqrt{-115})/2$ should have any of the divisibility properties of a common whole number. This specimen, being a root of the irreducible equation $x^2 + 13x + 71 = 0$, is in fact an algebraic integer of the second degree.

Algebraic number fields in which there is unique decomposition of algebraic integers into primes are the exceptions. To restore the fundamental theorem of arithmetic to the integers of any algebraic number field, Dedekind reexamined divisibility for the rational integers. This was the critical step, leading to the invention of what Dedekind called ideals.

An (integral) ideal of an algebraic number field F is a subset, say $\mathfrak{a}$, of all the integers of F such that, if α, β are in $\mathfrak{a}$, and ξ is any integer in F, then $\alpha - \beta$ and $\alpha\xi$ are in $\mathfrak{a}$. The ideal $\mathfrak{a}$ is said to divide the ideal $\mathfrak{b}$ if every integer in $\mathfrak{b}$ is also in $\mathfrak{a}$, that is, if $\mathfrak{a}$, considered as a class, contains $\mathfrak{b}$. The unit ideal is the set of all integers of F; it divides every ideal. An ideal $\mathfrak{p}$ is prime if, and only if, $\mathfrak{p}$ and the unit ideal are the only ideals dividing $\mathfrak{p}$.

Unique factorization was restored thus—'replaced' would be more strictly accurate. If α is any integer of F, the set of all $\alpha\xi$, where ξ runs through all integers of F, is easily seen to be an ideal. This ideal, denoted by (α), is called the principal ideal corresponding to α; and it follows immediately from the definitions that, if α, β are any integers of F, α divides β when, and only when, (α) divides (β). 'Divides,' in 'α divides β,' means that there is an integer γ of F such that $\beta = \alpha\gamma$; 'divides,' in '(α) divides (β),' means that the principal ideal (α) contains the principal ideal (β); and the theorem asserts that each of these division-relations implies the other.

In rational arithmetic, for example, '3 divides 12' is equivalent to 'the class of all integer multiples of 3 contains the class of all integer multiples of 12.' Again, if a, b are given integers, the class of all integer multiples of a contains the class of all integer multiples of b if and only if a divides b.

Decomposition of an algebraic integer into a product of algebraic integers is now mapped onto a decomposition of an ideal into a product of ideals. The fundamental theorem of arithmetic is valid in the map. The mapping is as follows.

The integers α, β of F are replaced by their corresponding principal ideals (α), (β), ... Since the capital theorem of Dedekind's theory establishes the unique decomposition of any ideal (in F) into a product of powers of prime ideals, each of (α), (β), . . . has such a unique decomposition.[4]

Roughly, the crux of the matter is the replacement of the relation of arithmetical divisibility by the relation of class-inclusion as in either classical or symbolic logic. And, still roughly, this replacement is in part responsible for the appearance of ideals as linear sets of particular kinds in modern algebra and in algebraic geometry.

The invention of ideals has been given what may seem more than its legitimate share of space because it is an admirable and easily described example of the modern tendency to generalization. The most characteristic detail, possibly, is that of replacing the familiar concept of arithmetical divisibility by another that includes it. A central relation is replaced by another bearing no superficial resemblance to the first. Nevertheless, after the replacement, a cardinal theorem (unique decomposition into primes) is restored, by mapping or one-one correspondence, to a domain in which, before the replacement, the theorem did not hold generally. And further, the replacement leaves essentially unaltered those cases in which the theorem held before the replacement.

From another point of view, the replacement of the set of algebraic integers by the set of correlated principal ideals introduces uniformity and brings apparent anomalies under a new and wider law. An earlier instance of the same procedure occurred in the introduction of ideal elements (points, lines, planes, . . . at infinity) into projective geometry during the first half of the nineteenth century. Such elements have only a remote connection with algebraic number ideals, but in both cases the methodology of generalization by extension to regularize exceptions is the same.

A feature of this theory that strikes those approaching it for the first time as rather peculiar is characteristic of much of Dedekind's thinking about number: a strictly finite problem is solved in terms of infinite classes. The problem for algebraic integers is that of unique decomposition; Dedekind's solution is through the particular infinite classes of algebraic integers which he called ideals. His theory (1872) of the real number system is based on a similar escape from the finite to the infinite by means of what he called cuts. To define $\sqrt{3}$, for example, Dedekind imagined all rational numbers to be separated into two classes, say L, U; L contains all those rational numbers, and only those, whose squares are less than 3, and U all those, and only those, whose squares are greater than 3; L, U are said to define a 'cut' in the system of all real numbers, and this particular cut is said to define $\sqrt{3}$.

The arithmetic of Dedekind cuts is a map of the usual properties of real numbers, such as $\sqrt{2} \times \sqrt{3} = \sqrt{2 \times 3}$, familiar to analysts and algebraists through centuries of formal manipulations. The purely formal work with irrationals produced consistent numerical approximations, and was sufficient for scientific applications of analysis. Dedekind aimed to provide a sound logical basis for the traditional formalism of number. The outcome of his efforts was a deeper formalism of the infinite. In 1926, the leading mathematician of his age, D. Hilbert (1862–1943, German) asserted that "The significance of the *infinite* in mathematics has not been completely clarified." Nor had it by 1945.

Further developments

Dedekind's theory of ideals was but one of several constructed for the purpose of restoring the fundamental theorem of rational arithmetic to algebraic numbers. The other which has survived[5] is Kronecker's theory of 1881, already mentioned in connection with complex numbers. Both Kronecker's and Dedekind's theories have extensive ramifications in other departments of mathematics, and both exerted a decisive influence on the development of modern abstract algebra.

A third theory has become prominent since its creation in the 1900's by K. Hensel, in which numbers are represented by power series. This theory originated in the remark that any rational integer can be developed into a series of positive integral powers of a given prime p, with coefficients chosen from $0, 1, \ldots, p - 1$. It may be considered as the ultimate extension of the Babylonian, Mayan, and Hindu place-systems of numeration in common arithmetic. Analogies with the theory of functions of a complex variable, also with the theory of algebraic functions of one variable and their representation on Riemann surfaces, appear to have guided this arithmetical theory in its rapid development. 'Algebraic function' is used here in its customary technical sense: if $P(w, z) = 0$, where P is a polynomial, w is called an algebraic function of z. We shall see later that the detailed study of such functions and their integrals was a major activity of nineteenth-century mathematics.

It may be of interest to indicate very briefly how the concept of arithmetical divisibility as generalized to class-inclusion by Dedekind and Kronecker became significant in departments of mathematics far distant from arithmetic. Any detailed description soon becomes highly technical, and we can give only enough to suggest that far-reaching applications might have been anticipated from the finished form of Dedekind's theory and the broad outline which Kronecker left of his.

A familiar example from elementary analytic geometry offers the plainest hint. If $C_n(x, y) = 0 (n = 1, 2, \ldots, m)$ are the equations of m given plane curves, then

$$f_1(x, y)C_1(x, y) + \cdots + f_m(x, y)C_m(x, y) = 0,$$

in which the f's are functions of x, y (or constants), not identically zero, is the equation of a curve passing through the points common to the m given curves. For simplicity, let all the C's and f's be polynomials in x, y. Then the system of all polynomials $f_1(x, y)C_1(x, y) + \cdots + f_m(x, y)C_m(x, y)$, in which the C's are held the same and the f's are constants or range over all polynomials in x, y, contains, or 'divides,' any particular polynomial in the system. 'Divides' here is as in Dedekind's ideals or Kronecker's modular systems.

A modular system is a set M of all polynomials in s variables $x_1, \ldots, x_s$ defined by the property that if P, P_1, P_2 belong to the system, then so do $P_1 + P_2$ and QP, where Q is any polynomial in $x_1, \ldots, x_s$. One further definition enables us to state a capital theorem of modern algebra. A basis of a modular system M is any set of polynomials $B_1, B_2, \ldots$ of M such that every polynomial of M is expressible in the form

$$R_1B_1 + R_2B_2 + \cdots,$$

where $R_1, R_2, \ldots$ are constants or polynomials (not necessarily belonging to M). Hilbert's basis theorem of 1890 states that every modular system has a basis consisting of a finite number of polynomials or, equivalently, a polynomial ideal has a finite basis.

Anyone might be excused for doubting this theorem until he had followed its remarkably simple proof. In fact, when Hilbert applied it to prove the fundamental theorems for algebraic forms, P. Gordan (1837–1912, German), who had previously obtained the same theorems by laborious calculations, exclaimed "This is not mathematics; it is theology!"

There was a double-edged truth in Gordan's protest. Hilbert's theorem marks a major turning point in algebra. It was the first example to attract universal attention to the modern abstract non-calculating method. Gordan's proofs were by highly ingenious algorithms; Hilbert's attacked the structure of the systems concerned—algebraic forms and their covariants and invariants. The algoristic method was incapable of revealing the general underlying principle of which Gordan's theorems are but special manifestations. We shall return to this when we consider invariance.

The sharper edge of Gordan's protest was felt only in the late 1920's. A proof in theology, it may be recalled, usually demonstrates the existence of some entity without exhibiting the entity or providing any method for doing so in a finite number of humanly performable operations. Mathematics, particularly analysis, abounds in proofs theological in this sense. To Kronecker, all theological proofs in mathematics were anathema. He insisted that those existence proofs are invalid and therefore worthless in mathematics which do not provide a method for exhibiting, or constructing, in a finite number of humanly performable operations, the mathematical object whose existence is alleged to be proved. To most algebraists it is intuitive that a polynomial $P(x)$ with rational coefficients either is or is not rationally reducible—the product of two polynomials in x with rational coefficients. Kronecker would not admit this statement until he had devised a method for deciding in a finite number of steps whether a polynomial is actually reducible or irreducible.

Since Kronecker first demanded constructive existence proofs, it has been suspected by some that the free use of 'theological' existence proofs may lead to inconsistencies. In particular, the admissibility of Hilbert's non-constructive existence proof for his basis theorem was questioned in the 1930's, although neither he nor the majority of working mathematicians sensed anything objectionable or dangerous in the continued application of the theorem. Without it, a vast tract of modern abstract algebra and a considerable amount of algebraic geometry would evaporate into nothing. A finitely constructive existence proof of the basis theorem had not been given up to 1945. The implied doubts in this lack are of a piece with those arising from the work of the nineteenth century on the real number system.

None of these deep uncertainties deters mathematicians in their technical labors, any more than an occasional eruption discourages the vineyardists on the slopes of Etna and Vesuvius. The periodic upheavals and submersions under rivers of incandescent lava are indeed regarded as blessings, except by the generations who must endure them. The decomposing lava revitalizes the exhausted soil, and the grapes produce a richer wine. But it is rather unpleasant for those who must be suffocated or incinerated in order that their successors may prosper. Much of the mathematics of the nineteenth and twentieth centuries seems now to be significant chiefly because it may contribute to a sounder mathematical prosperity in the twenty-first century. But we have no assurance that it will. In the meantime, our generation endures or enjoys metamathematics and continues to create new mathematics. And so has it been since existence proofs were first questioned.

The general gain to 1910

What may be called the second heroic age of the theory of algebraic numbers ended in the 1870's–1880's in the work of Dedekind and Kronecker. The first great age was that of Gauss and Kummer in the 1830's–1840's. The principal innovations in each of these periods naturally inspired a considerable number of technical developments. But it does not appear that any fundamentally new concept, or any novel approach comparable in general significance to those indicated, was suggested earlier than the third great epoch beginning in 1910 and the 1920's. It must be recalled here that we are interested primarily in the development of mathematical thought as a whole, rather than in the detailed exploitation of special fields. Before passing on to the third period and its relation to the general progress, we may glance back at the first two and note once more their origin, in order to see their principal residue. Perhaps the most significant contribution is the methodological approach in both periods.

The theory of algebraic numbers originated in two definite problems concerning the rational integers: the laws of n-ic reciprocity, designed to yield criteria for the solvability of binomial congruences $x^n \equiv r \bmod m$; the proof or disproof of Fermat's last theorem. A solution[c] of the first problem for n prime was given in what long remained its classic form by F. M. G. Eisenstein (1823–1852, German) in 1844–50, and by Kummer in 1850–61. Thus, in this direction, algebraic numbers accomplished the purpose for which they were invented. The underlying structure of the modern theory of reciprocity laws, dating from about 1908, is that of the modernized Galois theory of fields and finite groups.

The second problem—Fermat's last theorem—responsible for the theory of algebraic numbers has resisted the best efforts of three generations of arithmeticians since Kummer made the first notable progress. In this direction, then, the theory has not attained its goal, although it has found much on the way. Of both problems it seems fair to say that they, as definite ends, have waned in interest, while the methods devised for their solution have steadily waxed in importance for modern mathematics. Algebraists, for example, who have but a slight interest in either reciprocity laws or Fermat's theorem constantly use the machinery (fields, ideals, rings, etc.) devised in the first instance to handle these problems.

The like is true of the Galois theory of equations. Galois himself made the terminal contribution so far as algebraic equations are concerned, and subsequent reworkings of his initial theory have added nothing basically new to his criteria for solvability by radicals. Even the modernized presentation of the Galois theory, as in the streamlined model of E. Artin (Germany, U.S.A.), is a tribute to the mathematical creed of Galois, in its elimination of all superfluous machinery. For this modern release from algebraic calculation, the direct approach of A. E. 'Emmy' Noether (1882–1935, Germany, U.S.A.) in the 1920's was primarily responsible. Much of her mathematics was in the spirit of Galois. But his methods, sharpened and generalized by his successors, have transcended the problem for which they were invented, and have rejuvenated much of living pure mathematics.

It is to be noticed concerning the vital residue of the theory of algebraic numbers that it, like the Galois theory, can be traced to definite, highly special problems. Neither Galois nor the creators of the theory of algebraic number fields set out deliberately to revolutionize a mathematical technique; their comprehensive methods were invented to solve specific problems.

Such appears to have been the usual path to abstractness, generality, and increased power. Some difficult problem that has appeared in the historical development of a particular subject is taken as the point of departure without any conscious effort to create a comprehensive theory; repeated failures to achieve a solution by known procedures force the invention of new methods; and finally, the new methods, having been necessitated by a problem which appeared in the historical development, themselves pass into the main stream.

Both Fermat's last theorem and the arithmetical theory of reciprocity are but very special cases of a central problem in diophantine analysis. It is required to devise criteria to decide in a finite number of non-tentative steps whether or not a given diophantine equation is solvable. The extreme complexity of the theories invented for the two special cases suggests that only insignificant progress toward a solution of the general problem is likely without the invention of radically new methods.

The contribution from algebraic equations

The third great epoch in the extension of arithmetic is that of the twentieth century after 1910. To anticipate, the introduction of general methods into linear algebra, beginning in the first decade of the twentieth century, prepared that vast field of mathematics, first opened up by Hamilton and Grassmann in the 1840's, for partial arithmetization in the second and third decades of the century. In 1910, E. Steinitz (1871–?, Germany), proceeding from, and partly generalizing, Kronecker's theory (1881) of "algebraic magnitudes," made a fundamental contribution to the modern theory of (commutative) fields. His work was one of the strongest impulses to the abstract algebra of the 1920's and 1930's, with its accompanying generalized arithmetic. The outstanding figure in the later phase of this development is usually considered to have been Emmy Noether[7] (1882–1935, Germany) who, with her numerous pupils, laid down the broad foundations of the modern abstract theory of ideals, also a great deal more in the domain of modern algebra. The application of this work to the 'integers' of linear associative algebras affords the ultimate extension up to 1945 of common arithmetic.

One of the main clues threading this intricate maze is the Galois theory of fields as it has developed since 1830. The Galois theory of equations itself was the concluding episode in about three centuries of effort to penetrate the arithmetical nature of the roots of algebraic equations. Accordingly we shall consider this first. It is of interest in this connection to recall an opinion expressed by Hilbert[8] in 1893 which still retains its force:

> With Gauss, Jacobi and L. Dirichlet frequently and forcefully expressed their astonishment at the close connection between arithmetical questions and certain algebraic problems, in particular with the problem of cyclotomy. The basic reason for these connections is now completely disclosed. The theory of algebraic numbers and the Galois theory of equations have their common root in the theory of algebraic fields. . . .

After the solution of the general cubic and quartic in the sixteenth century, there appears to have been only one contribution of lasting significance to the algebraic solution of equations before the late eighteenth century. E. W. Tschirnhausen[9] (or Tschirnhaus, 1651–1708, German) in 1683 applied a rational substitution—reducible to a polynomial substitution—to remove certain terms from a given equation. This generalized the removal of the second term from cubics and quartics by Cardan, Vieta, and others. About a century later (1786), E. S. Bring (1736–1798, Swedish) reduced[10] the general quintic to one of its trinomial forms, $x^5 + ax + b = 0$, by a Tschirnhaus transformation with coefficients involving one cube root and three square roots, a result of capital importance in the transcendental solution of the quintic.

Euler, about 1770, solved the general quartic by a method differing from that of Ferrari. This unexpected success led him to believe that the general equation is solvable by radicals. As remarked in connection with the Greek problem of trisecting an angle, it demanded originality of a high order to doubt the possibility of a solution by radicals in the general case. Were such a solution for the general quintic possible, Euler no doubt would have found it; for he was without a superior on the manipulative side of algebra. But the quintic called for a different kind of mathematics. As Abel pointed out, failure to solve the general quintic by radicals might indicate only incapacity on the part of the would-be solver; and no number of failures could be of any value as an indication whether the problem was solvable.

A long stride forward was taken by Lagrange[11] in 1770–1. Instead of trying to solve the general quintic by ingenious tricks, Lagrange critically examined the extant solutions of the equations of degrees 2, 3, 4 in a successful attempt to discover why the particular devices used by his predecessors had succeeded. He found that in each instance the solution is reducible to that of an equation of lower degree, whose roots are linear functions of the roots of the given equation and roots of unity. Here at last was a seemingly universal method. But on applying his reduction to the general quintic, Lagrange obtained a sextic. The degree of the resolvent equation, instead of being reduced as before, was raised. We see now that this was a strong hint of the impossibility of a solution by radicals; but Lagrange apparently missed it. He had, however, found the germ of the theory of permutation groups.

In this discovery, Lagrange took the first step toward the general theory of groups, a step of immeasurably greater significance for mathematics as a whole than a complete disposal of the theory of algebraic equations. Permutation groups suggested abstract finite groups. These in turn suggested infinite discontinuous groups, and finally the group concept entered analysis and geometry with the invention by M. S. Lie (1842–1899, Norwegian) of continuous groups in the 1870's. The reaction upon both the discrete and the continuous divisions of mathematics was far reaching and profound. With invariance, closely related to the group concept, the theory of groups in the nineteenth century transformed and unified widely separated tracts of mathematics by revealing unsuspected similarities of structure in diverse theories. This, however, belongs to the subsequent development of Lagrange's discovery, and will be considered in the proper connections. For the moment we are concerned only with the application of groups to algebraic equations.

To recall briefly the nature of what Lagrange found, let $x_1, \ldots, x_n$ denote the roots of the general equation of degree n. Then, if a rational function f of $x_1, \ldots, x_n$ is left unaltered by all those permutations on $x_1, \ldots, x_n$ that leave unaltered another rational function g of $x_1, \ldots, x_n$, f is a rational function of g and the coefficients of the general equation.

A set of permutations $S_1, \ldots, S_r$ on a given set of letters (as $x_1, \ldots, x_n$ above) form a group in the technical sense already defined, when a product, such as S_iS_j, of two permutations S_i, S_j, is interpreted as the permutation which results when S_i is applied first, and S_j is then applied to the new arrangement of $x_1, \ldots, x_n$ generated by S_i. For example, if $n = 4$, and the 4 letters are a, b, c, d, the symbol $(abcd)$ means the permutation which takes each letter into its immediate successor, a being considered the successor of d in this cycle: a into b, b into c, c into d, d into a. The permutation (acd) takes a into c, c into d, d into a. Hence $(abcd)(acd)$ takes a into b, b into d, d into c, and c into a. Thus $(abcd)(acd) = (abdc)$. The identical permutation I, or the 'identity,' takes each letter into itself or, what is the same, leaves each arrangement of the letters unaltered. The set of all possible $n!$ permutations of $x_1, \ldots, x_n$ is called the symmetric group on $x_1, \ldots, x_n$. If $x_1, \ldots, x_n$ denote the roots of an irreducible equation of degree n, the properties of the symmetric group on $x_1, \ldots, x_n$ are the clue to necessary and sufficient conditions that the equation be solvable by radicals. It is impossible to go into details here, and we must refer to any modern text on the theory of equations or higher algebra. The fundamental concept for applications of finite groups to algebraic equations is that of a solvable group. The meaning of this term will be explained presently.

Lagrange did not explicitly recognize groups. Nevertheless, he obtained equivalents for some of the simpler properties of permutation groups. For example, one of his results, in modern terminology, states that the order of a subgroup of a finite group divides the order of the group. Normal (self-conjugate, invariant) subgroups, basic in the theory of algebraic equations and in that of group structure, were introduced by Galois, who also invented the term 'group.'

Both Abel and Galois were indebted to Lagrange in their own profounder work on algebraic equations. Before Abel set himself (1824) the problem of proving the impossibility of solving by radicals the general equation of degree greater than four, an Italian physician, P. Ruffini (1765–1822), beginning in 1799, had attempted to do the same. Ruffini's definitive effort (1813) is said by some who have examined it to be essentially the same as Wantzel's simplification of Abel's proof. Abel published this proof at his own expense in 1824; it was reprinted in 1826 by A. L. Crelle (1780–1855, German) in the initial volume of his great journal. Remediable defects are said by some competent algebraists to mar the final proofs of both Ruffini and Abel. But as the oversights are not fatal, it is customary to say that each of these two proved the impossibility of solving by radicals the general equation of degree greater than four. Their work was entirely independent. The unique importance of Abel's proof is that it inspired Galois to seek a deeper source of solvability, which he found in the theorem that an algebraic equation is solvable by radicals if and only if its group, for the field of its coefficients, is solvable.

We cannot enter into the technicalities of Galois' theorem. But assuming some acquaintance with the modern theory of algebraic equations, which, after all, is well over a century old as this is written, we shall use a few of its concepts to illustrate the meaning of structure as exemplified in this capital theorem of algebra. The simple isomorphism of any two groups was defined in connection with the postulates for a group. Galois considered simply isomorphic groups as the same group which, abstractly, they are. A. Cayley[12] (1821–1895, English) in 1878 expressed this by saying that the properties of a group are defined by its multiplication table.

A subgroup H_1 of a group G is said to be a normal divisor of

G if for every s in G, $sH_1 = H_1s$, where sH_1 denotes the set of all products sh, h in H_1, and similarly for H_1s and the set of products hs; equality here means that the two sets contain the same elements. A subgroup of G other than G itself is called a proper subgroup. A maximal normal divisor of G is a proper normal divisor of G that is not a proper subgroup of any proper normal divisor of G. The maximal normal divisors of the group of $n!$ permutations of the roots of the general equation of degree n appear in the criteria for solvability by radicals. To state the connection, we require the definition of quotient (or factor) groups.

The order of a group is the number of distinct elements in the group; and Lagrange proved (1770–1) in effect that the order of a subgroup divides the order of the whole group. If H_1 is a normal divisor of order m_1 of a group G of order n, then must $n = m_1q_1$, q_1 an integer; and it can be shown that

$$G = H_1 + s_1H_1 + \cdots + s_{q_1-1}H_1,$$

where no two of the sets $H_1, s_1H_1, \ldots, s_{q_1-1}H_1$ have an element in common, and the plus signs mean that all the elements of G are separated out into these q_1 mutually exclusive sets. That is, the $+$ is logical addition, as in the Boolean algebra of classes. Let $K_1, K_2, \ldots, K_{q_1}$ denote these q_1 sets (in any order). Then, if K_iK_j denotes the set of all products formed by multiplying an element of K_i by an element of K_j, it can be shown that precisely m_1 of these products are distinct, and that these m_1 are all the elements of some one of the K's. Moreover, with multiplication K_iK_j as just described, $K_1, K_2, \ldots, K_{q_1}$ form a group, called the quotient (or factor) group of G with respect to the maximal normal divisor H_1 of G. This quotient group is denoted by G/H_1; its order is q_1, and the order $n(= m_1q_1)$ of G, divided by the order (m_1) of H_1, is called the index of H_1 under G. Thus the index of H_1 under G is here q_1.

Now there may be more than one proper maximal normal divisor of G. If there is, its quotient group can be formed as above, and its index under G is known. Our concern here is with all of these possibilities at each stage of the process next described.

Proceeding with H_1 as we did with G, we find its quotient group H_1/H_2 with respect to any maximal proper normal divisor H_2 of H_1. This divisor may be only the identity-group consisting of the single element I (the identity) of G. The process is now repeated with H_2, and so on, until it stops automatically with I. In this way, starting with G and ending with I, we get the sequence of groups G, H_1, H_2, . . . , $H_t(= I)$, each of which (after G) is a proper maximal normal divisor of its immediate predecessor. There is also determined the sequence of quotient groups G/H_1, H_1/H_2, H_2/H_3, . . . , H_{t-1}/H_t, and the corresponding indices, say q_1, q_2, . . . , q_t. Two final definitions, and we can state several striking consequences of this iterated process. A group having no normal divisors except itself and the identity-group I, formed of the single element I (the identity of the group), is said to be simple. If all the indices q_1, q_2, . . . , q_t are prime numbers, the group G is said to be solvable.

Remembering that there may be several ways of proceeding at each step, we state the following conclusions. First,[13] in whatever way we proceed, we get the same number of groups G, H_1, H_2, . . . H_t. Second, all the factor groups displayed above are simple. Third,[14] in whatever way we proceed, the factor groups are the same, although not necessarily in the same order, and hence similarly for the indices q_1, q_2, . . . , q_t. In a sense which need not be elaborated, these theorems of C. Jordan (1838–1922, French) (1870) and O. Hölder (1859–1937, German) (1889) are a remarkable revelation of the structure of any finite discrete group. Since 1930 they have been refined and extended in what may be likened to the minute anatomy of any articulated organism.[15] Recalling the capital theorem of Galois for the solvability of an algebraic equation by radicals, we see that these theorems go to the root of the matter. To anticipate slightly, it may be noted here that the Jordan-Hölder theorems have themselves been structurally analyzed as phenomena of the theory of 'lattices' or 'structures.' This theory will be discussed in the next chapter.

The further development of the theory of groups will be described presently. For the moment we note that the question of solvability by radicals received a conclusive answer in the theory of finite groups.

After the impossibility of solving the general equation of degree higher than the fourth by radicals had been proved, the next problem was to find what kind of functions would suffice to solve the general quintic. The general cubic had long been known to be solvable by circular (trigonometric) functions. The circular functions are uniform (single-valued) singly periodic functions of one variable. They are degenerate forms of the elliptic functions, which are uniform doubly periodic functions of one variable, or 'argument' x. If $f(x)$ is an elliptic function, and p_1, p_2 are its two periods, p_1/p_2 is necessarily imaginary, and $f(x + n_1p_1 + n_2p_2) = f(x)$ for all choices of the integers n_1, n_2. As will be seen when we consider analysis, Abel and Jacobi in the 1820's discovered the elliptic functions through the inversion of elliptic integrals.[16] An extensive department of the theory of elliptic functions is the problem of the division of the periods: if there is an integer n such that nx is a period, the problem of division by n is to find elliptic functions having x as argument. This problem leads to certain algebraic equations which, for $n = 2, 3, 4, 3.2^s$, are solvable by radicals. The degree of the equation for division by any odd n is $(n^2 - 1)/2$. Thus for $n = 5$ the degree is 12. But if n is prime, the equation is obtainable in a much simpler form, being only of degree $n + 1$. We recall that Lagrange was led to a resolvent equation of degree 6 in his attempt to solve the general equation of degree 5 by radicals. The problem of division of elliptic functions for $n = 5$ incidentally provided the functions of a, b which reduce the trinomial form $x^5 + ax + b = 0$ of the general quintic to an identity. The transcendental functions necessary to solve the general equation of the fifth degree had therefore been constructed.

This unexpected result was found by C. Hermite (1822–1905, French) in 1858. Hermite was led to it by his intimate knowledge of elliptic functions. He observed that an equation occurring in the problem of quinquisection of elliptic functions could be transformed into Bring's form of the general quintic. Simultaneously, Kronecker was nearing the same goal by another road. Kronecker's method differed profoundly from Hermite's. It was closer to what Galois might have done, had he lived. In 1853 Kronecker set himself the task of solving a fundamental problem encountered by Abel in his attack on algebraic equations: to find the most general function of $x_1, \ldots, x_n$ that can be a root of an algebraic equation with coefficient in a given field. He proved that the equations arising from the theory of division of certain transcendents suffice to solve the general equations of certain degrees, and in this way obtained transcendental solutions of the general cubic and quartic. He then attacked the general quintic without previous reduction of the equation by a Tschirnhaus transformation to remove certain terms. His object was to find a method capable of extension to equations of any degree.

Interest in such problems flagged during the second half of

the nineteenth century. So far as the general quintic is concerned, F. Klein (1849–1925, German) in 1884 reviewed[16] all the labors of his predecessors, and unified them with respect to the group of rotations of a regular icosahedron about its axes of symmetry. The earliest discussion from the standpoint of groups of the (modular) equations arising in the division of elliptic functions was by Galois.

As a specimen of later results in this same general field, a theorem of Hilbert may be cited: the general equation of degree nine requires for its solution functions of four arguments.

In summary, the chief contribution of the theory of algebraic equations to the number concept appears to have been the characterization of the irrationalities required for explicit general solutions. The attempt to solve the general algebraic equations of degrees higher than the fourth in terms of functions constructed from the given coefficients by a finite number of additions, subtractions, multiplications, divisions, and root extractions ended in the proof by Abel and Ruffini that such solutions do not exist. But such solutions—as for the general equations of degrees 2, 3, 4—define certain species of irrationalities. It therefore became necessary to seek totally different kinds of irrationalities to effect the solution of equations of degree higher than the fourth. These were found for degree 5 in the elliptic modular functions.

The work of Galois and his successors showed that the nature, or explicit definition, of the roots of an algebraic equation is reflected in the structure of the group of the equation for the field of its coefficients. This group can be determined nontentatively in a finite number of steps, although, as Galois himself emphasized, his theory is not intended to be a practical method for solving equations. But, as stated by Hilbert, the Galois theory and the theory of algebraic numbers have their common root in that of algebraic fields. The last was initiated by Galois, developed by Dedekind and Kronecker in the mid-nineteenth century, refined and extended in the late nineteenth century by Hilbert and others, and finally, in the twentieth century, given a new direction by the work of Steinitz in 1910, and in that of E. Noether and her school since 1920.

Between these later developments and the work of the nineteenth century on algebraic numbers and algebraic number fields, the elaboration of hypercomplex number systems intervened. This will be our concern in the following chapter.

It is to be noted that the principal clues to the final generalization of arithmetic have been Dedekind's theory of algebraic number fields, Kronecker's parallel theory of what he called "algebraic magnitudes," and the theories of general fields, groups, and rings. A ring differs from a field in that an inverse for multiplication is not postulated. The rational integers, 0, ±1, ±2, . . . are the simplest instance of a ring; the class of all rational integers is a group with respect to addition and is closed under multiplication. In a general ring, multiplication is not assumed to be commutative. Linear associative algebras are instances of rings.

Changing outlooks, 1870–1920

Allusions to groups will appear frequently as we proceed. From 1870 to the 1920's, groups dominated an extensive sector of mathematical thought, and were occasionally rather rashly touted as the long-sought master key to *all* mathematics. The first date, 1870, marks the publication of C. Jordan's (1838–1922, French) classic *Traité des substitutions et des équations algébriques*, which contained a great deal more than its modest title indicates. This is emphasized because Jordan was one of the few leading specialists in groups who made signal contributions to other departments of mathematics, including analysis. Others in the same category were Klein, Lie, Poincaré, G. Frobenius (1849–1917, German), W. Burnside (1852–1927, English), and L. E. Dickson (1874–, U.S.A.). But the great majority of those who labored in groups were specialists in the narrowest sense; and some of them after 1920 were content to parrot the informed but obsolete opinions of well-rounded mathematicians on the place of groups in mathematics as a whole, without acquiring the knowledge necessary to enable them to form a reasonable personal judgment.

During the fifty years from 1870 to 1920, mathematics did not stagnate; and although groups remained one of the seemingly permanent additions to mathematical thought, informed opinion after 1920 was less immoderate than it had been in the 1890's in its claims for the domination of groups over *all* mathematics. It was therefore inexcusably misleading to retain in the 1930's inflated estimates of groups that may have been valid in the 1910's, as if mathematics had remained stationary since the death (1912) of Poincaré, even when chapter and verse for the estimates in question were cited in the works of some of the greatest mathematicians of a bygone generation.

As a specific instance, a prominent specialist in finite groups reproduced in 1935 the long-since-superseded dictum of Poincaré that "The theory of groups is, as it were, *the whole* [our italics] of Mathematics stripped of its matter and reduced to 'pure form,'" as if this extravagance were the considered verdict of competent opinion in 1935. That it was not, will be seen in detail when we follow certain developments of geometry since 1916, to be described in connection with that other outstanding addition of the nineteenth century to all mathematical thought, the concept of invariance. Poincaré's dictum was a gross exaggeration even when it was first uttered. At best, it was an understandable overstatement, possibly for emphasis, that deceived nobody above mathematical illiteracy. While we may continue to remember the achievements of the great creators in mathematics with gratitude, we do the masters of the past a left-hand honor when we perpetuate their outmoded opinions for the misguidance of oncoming generations.

Some of the more conspicuous landmarks in the development of groups after Galois may be noted here, exclusive of the continuous groups discussed in a later chapter. Even before Galois coined the term 'group,' A. L. Cauchy (1789–1857, French) made (1815) extensive investigations in what are now called permutation groups, and discovered some of the simpler basic theorems. He returned to the subject in 1844–6, and just missed the fundamental theorem (1872) of L. Sylow (1832–1918, Norwegian), proved in most texts on the theory of finite groups. Cayley (1854) stated the earliest set of postulates for a group, thereby defining groups in the accepted technical sense. This definition sank out of sight and, as will be seen in connection with continuous groups, some of the leading experts, including Lie and Klein, occasionally used the term 'group' for systems which are not groups in the technical sense now universal. Consequently, the statements of certain theorems in older work require amendment. Another set of postulates was given (1882) by H. Weber (1842–1913, German), whose *Algebra* (3 vols., ed. 2, 1898–9) presented a masterly synopsis of algebra as it was at the close of the nineteenth century.

In passing, there is no more instructive demonstration of the change in outlook and objectives that distinguished the algebra of the early twentieth century from that of the late nineteenth than a comparison of Weber's classic with an advanced treatise of the 1930's. The transition from the old to the new began in 1910 with the work of Steinitz. It used to be said before 1910 that a thorough mastery of three famous classics, Weber's *Algebra*, J. G. Darboux' (1842–1917, French) *Leçons sur la théorie générale des surfaces et les applications géométriques du calcul infinitésimal* (2 vols., 1887–8; ed. 2, 1913–15), and É. Picard's (1856–1941, French), *Traité d'analyse* (3 vols., 1891–6; ed. 3, 1922–7), would suffice for a liberal education in mathematics, and enable a competent student to begin creation in what were then topics of living interest to research workers in mathematics. Less than a third of a century sufficed to render this particular liberal education hopelessly antiquated for anyone seeking to orient himself quickly for a creative career in vital mathematics. Those who in 1940 would arrive at the front of progress were, for the most part, taking short cuts that did not exist before 1910, or even before 1920.

The first decade of the twentieth century witnessed a somewhat feverish activity in the postulational analysis of groups, in which American algebraists produced numerous sets of postulates for groups, with full discussions of complete independence. By 1910, nobody could possibly misunderstand what a group is.

In another department of finite groups, also, American algebraists were incontinently prolific: the determination of all finite groups of a given order, especially all permutation groups on a small number of letters. One of the earliest attempts at a complete census was that (1858) of T. P. Kirkman (1806–1895), an English clergyman in a muggy parish, who claimed that his methods sufficed for an exhaustive enumeration. Kirkman will appear again in connection with topology. Not a very well-known mathematician, although rather a notorious one in his own day for the perfection of his sarcasm, Kirkman appears in retrospect to have been one of the born combinatorialists in mathematical history. For various reasons, he received practically no encouragement and about as much recognition. Of Americans who made notable contributions to what may be called Kirkman's program, F. N. Cole (1861–1927) and G. A. Miller (1863–) were among the most prolific.

Groups of linear homogeneous substitutions on n variables, also groups of such substitutions with integer coefficients, including congruence groups, were studied by many after Jordan (1870) had shown their importance in several departments of algebra and analysis, including hyperelliptic functions and the geometry of plane quartics. After Jordan's own work in this division, that of E. H. Moore[17] (1862–1932), Dickson, and H. F. Blichfeldt (1874–), all of the U.S.A., from the late 1890's to the second decade of the twentieth century, accomplished most. Interest in this specialty collapsed after about 1918, and the more imaginative algebraists turned their efforts in other directions.

By 1920, the painstaking collection and detailed analysis of special groups had become a thing of the past. Should the fruits of all this devoted labor of about half a century ever be required in either pure or applied mathematics, the toilers of the future will be spared decades of some of the hardest and most thankless drudgery ever successfully carried through in the history of algebraic computation. Finite groups are an episode in modern combinatorial analysis and, as such, are as difficult to civilize as any other phenomena in that inchoate science.

A brilliant exception to purely combinatorial methods appeared in 1896–9 in the algorithm of group-characters invented by Frobenius, and applied by him and others[18] with conspicuous success to several difficult problems in finite groups. The necessary computations, although often tedious, are non-tentative and non-combinatorial. They may therefore presage a more intelligently reasoned, less grubbing attack on the problem of group structure than that of the taxonomic period. With the appearance of groups in the early 1930's in quantum mechanics, the somewhat neglected algorithm of Frobenius became of possible scientific significance, and the heavy labor of applying it in detail to the permutation groups required in physics was undertaken. So possibly science may stimulate the algebra of the future to devise more practicable methods of calculation and enumeration in finite groups than those of the heroic age of uninspired hard labor.

Mathematics and society

In reviewing the contribution of algebraic equations to the development of the number system, any mathematician today must be impressed by the apparent permanence of the ideas introduced by Abel (1803–1829) and Galois (1811–1832), and the profound difference between their approach to mathematics and that of their predecessors including, in some respects, Gauss

(1777–1855). To these young men, perhaps more than to any other two mathematicians, can be traced the pursuit of generality whch distinguishes the mathematics of the recent period, beginning with Gauss in 1801, from that of the middle period. They initiated for the whole of mathematics the deliberate search for inclusive methods and comprehensive theories. Their forerunners in the middle period were Descartes with his general method in geometry; Newton and Leibniz with the differential and integral calculus created to attack the mathematics of continuity by a uniform procedure; and Lagrange, with his universal method in mechanics. Their contemporary in recent mathematics was Gauss, who in his arithmetic sought to unify much of the uncorrelated work of the leading arithmeticians from Fermat to Euler, Lagrange, and Legendre. Both Abel and Galois acknowledged their indebtedness to the theory of cyclotomy created by Gauss; and although they went far beyond him in their own algebra (Abel in analysis also), it is at least conceivable that neither Abel nor Galois would have chosen the road he followed had it not been for the hints in the Gaussian theory of binomial equations.

Both Abel and Galois died long before their time, Abel at the age of twenty-seven from tuberculosis induced by poverty, Galois at twenty-one of a pistol shot received in a meaningless duel. When Abel's genius was recognized, he was subsidized by friends and the Norwegian government. By nature he was genial and optimistic. Galois spent a considerable part of his five or six productive years in a hopeless fight against the stupidities and malicious jealousy of teachers and the smug indifference of academicians. Not at first quarrelsome or perverse, he became both.

Whoever, if anybody, was responsible for the colossal waste represented by these two premature deaths, it seems probable that mathematics was needlessly deprived of the natural successors of Gauss. What Abel and Galois might have accomplished in a normal lifetime cannot be even conjectured. That it would have been much and of the highest quality seems probable. Early maturity and sustained productivity are the rule, not the exception, for the greatest mathematicians. It may be true that the most original ideas come early; but it takes time to work them out. Gauss spent about fifty years developing the inspirations that came to him (this is substantially his own description) before he was twenty-one, and even with half a century of continual labor, he brought only a fraction of his ideas to maturity.

All this raises the question of what 'society' would do for a Gauss, an Abel, or a Galois today. Statesmen, including Disraeli, have said that society is an ass; closer inspection reveals it as a nebulous abstraction. Nevertheless, we shall use the term, because nearly everyone has a clear image of what he thinks it means.

Gauss, the penniless son of a day laborer, was educated by society as represented by the Duke of Brunswick. Today he would be educated at public expense, at least in the United States.

An Abel, no doubt, would be sent by the municipal health authorities to a sanitarium, where he might recover.

A Galois almost certainly would find himself at outs with respectability, or in the protective custody of the police on some trumped-up charge or other, or in a concentration camp. For there is but little evidence that teachers are less helpless in the disturbing presence of a mind of the very highest intelligence than they were in Galois' day, or that the guardians of law and order are less nervous than they were when they sentenced Galois to six months in jail on a legal technicality. Aesop's fable of the peacock and the crows has an element of permanence in it: you are different from us; get out or be plucked.

Society, however, has learned something it did not know when it permitted Galois to throw his life away in a duel. Galois was not considered "a dangerous radical" on account of his mathematics, but because of his politics—now strangely respectable. He was a Republican; and a reward for his capture was posted on that account. Royalist society was gravely concerned for the continued security of its protracted decadence. Quite sensibly, when we consider the interests of the individuals composing it, society in the early 1830's regarded Galois' revolutionary ideas in mathematics with complete indifference. But in the early 1920's, society discovered that a purely mathematical theory—relativity, or biometrics, to be specific—may conceal dangerous threats to right political thinking. Human biometrics was frowned upon in Russia, relativity in Germany. So it would seem unjust to say that society has remained stationary since it discarded Galois in 1832.

CHAPTER 11

Emergence of Structural Analysis

With the material already described as a background, we shall now observe in some detail the trend toward ever greater generality and more refined abstraction which distinguished much mathematics of the recent period from nearly all that preceded 1840. Structure, in a sense to be noted and described, was the final outcome of this accelerated progression from the particular to the general. The entire movement may be seen in geometry as clearly as in algebra and arithmetic, and will be remarked in that connection in later chapters. The course followed here continues that already indicated merely for convenience. It is typical of that for all divisions.

We saw that Gauss in 1831 invented his complex integers $x_1 + ix_2$ to solve a specific problem in rational arithmetic. His $x_1 + ix_2$ written as a number-couple (x_1, x_2) suggested hypercomplex numbers $(x_1, \ldots, x_n)$ with n coordinates $x_1, \ldots, x_n$; and it was natural to ask whether any of these extended numbers, with real or ordinary complex coordinates, might be useful in rational arithmetic. More generally, how may 'integers' be defined in a system of hypercomplex numbers, and what is their 'arithmetic'?

Before either problem could be attacked, or even formulated precisely, the algebra of hypercomplex number systems had to be developed. But this in its turn was not a definite problem. Once the algebraic problem had been made precise, its solution followed rapidly.

There appear to have been three principal phases after the problem was first posed by the invention of quaternions.

Three phases in linear algebra

The first phase was represented by such work as that of B. Peirce in 1870, which sought to find and exhibit all the linear associative algebras in a given (finite) number of fundamental units.[1]

The second phase, merging into the third, began in the first decade of the twentieth century and continued to about 1920. In this period, general theorems applicable to all linear associative algebras were the objective.

The third phase was distinguished for its restatement in abstract form of much that was already known, and the introduction of arithmetical concepts, such as ideals and valuations, into the resulting abstract algebra. The outcome[2] was an extensive and intricate theory assignable to either algebra or arithmetic, according to taste. The algebraic number rings and fields long familiar in the theories of algebraic equations and algebraic numbers; the ideals and dual groups of Dedekind; the relative fields of Hilbert; the modular systems of Kronecker; and the Galois theory of fields, all contributed to the final abstract theory. The finished product exhibits the broad outlines of the theories from which it evolved as but different, particularized aspects of a unified whole, like the varying projections of an intricate geometrical configuration on a moving plane. In addition, the abstract theory gives a wealth of results not obtainable from its classical instances.

The abstract method

The entire development required about a century. Its progress is typical of the evolution of any major mathematical discipline of the recent period; first the discovery of isolated phenomena; then the recognition of certain features common to all; next the search for further instances, their detailed calculation and classification; then the emergence of general principles making further calculations, unless needed for some definite application, superfluous; and last, the formulation of postulates crystallizing in abstract form the structure of the system investigated. The detailed elaboration of the abstract system implicit in the postulates then proceeds undistracted by what may be adventitious circumstances in any special instance. Incidentally, this is the reason that our extremely practical decadic Hindu-Arabic numerals are a positive detriment, except for numerical checks, in investigating the properties of numbers. The p-adic and g-adic numbers of Hensel are closer to arithmetic.

The full import of the abstract formulation appears only when it is taken as the point of departure for the deliberate creation of new mathematics. Certain postulates in the original set are suppressed or contradicted, and the consequences of the modified set are then worked out as were those of the original. For example, in a field as first defined, multiplication is commutative. This raises the question whether it is possible to construct consistent 'algebras' subject to all the postulates of common algebra except the commutativity of multiplication. Again, in common algebra (a commutative field), if $a \neq 0$, and $ab = ac$, it follows by division that $b = c$. But division by a presupposes that a has an inverse with respect to multiplication. Division is not defined in a ring; nevertheless there are rings in which, if

$$a \neq 0$$

and $ab = ac$, then $b = c$. Hence the existence of inverses as in a field is not postulated, but is replaced by the weaker condition just stated taken as a postulate. The result is a type of algebra more general than a field, in that it is based on some but not all of the field postulates or their consequences.

While viewing the abstract method we need not let our sincere admiration for its undeniable beauties betray us to the fate of Narcissus. It is just possible that our descendants may record that we perished of hunger while staring at the seductive reflection of our own superficial ideas—

"What thou seest,

What there thou seest, fair creature, is thyself—",[3]

or that we fell in and were drowned in something we never suspected just below the entrancing surface. To illuminate this heresy with an even more heretical example, what reason is there for supposing that because Dedekind's ideals did what was required for algebraic number rings, something very much like them should be introduced into other rings? Is it even evident that the arithmetic of a non-commutative ring should be based on ideals at all? Or that the usual definition of integral elements of a ring by close analogy with the like—by means of the rank equation—for an algebraic number ring is the most promising clue? The obvious retort is to demand something better of the doubter. Admitting the justice of this, we may nevertheless consider the other possibility.

The root of these doubts seems to be the unimaginative lack of a clearly recognized objective. If the aim is merely to create new theories which many find intensely interesting and even beautiful, then the abstract method keeps on reaching its goal. In this respect it somewhat resembles Stephen Leacock's hero, who leapt on his horse and dashed furiously off in all directions. But the taste of another generation may find our abstractions boring and our beauties vapid. They will have a hard way to travel before coming upon something different. The tangled masses of our theories will impede their every step. If they are to progress, their only possible road will circumvent our work almost entirely. The like happened once before, when Descartes walked clear round the synthetic labors of Euclid and Apollonius. To the skeptically inclined, viewing the vast accumulations in abstract geometry, abstract algebra, and abstract analysis of the twentieth century, another Descartes seems about due. Unless he arrives within the next two thousand years, no two mathematicians in the world twenty centuries hence will understand each other's words. In the meantime we may appreciate the tremendous accomplishments of the modern abstract method and trace the main steps by which it finally arrived since Hilbert gave it the strongest impulse since Euclid in his geometry of 1899.

The first phase in the development of abstract algebra, that of calculation and tabulation, is on the same scientific level as systematic botany. All sciences of the past seem to have been condemned to creep through this Linnaean stage of development. If only Linnaeus would stay as dead in mathematics as he appears to be in biology, mathematics might be far leaner and more virile than it is. But the irrepressible botanist keeps on rising from the dead; it is impossible to keep him down. While the vanguard of mathematical progress is advancing to genera, a host of straggling camp followers busies itself with the collection and classification of trivial or discarded subspecies. It was so in the theories of algebraic invariants and finite groups. An exasperating instance from recent analysis is the introduction of two new technical terms to distinguish $x > 0$, $x \geqq 0$. After enough classification has been done to indicate some inclusive characteristic, there would seem to be no point in collecting further specimens unless they are to be used. Linear algebra fortunately escaped the intensest fury of the taxonomists in its rapid passage to the second phase.

Toward structure in algebra

We shall select a few typical episodes from the history of linear algebra[4] to illustrate the general trend since about 1870, in nearly all mathematics, from the detailed elaboration of special theories to the investigation of interrelations between the theories themselves. Technicalities are unavoidable, but they may be ignored by those unfamiliar with the subject; the important items are not the facts but the relations between them. Even without technical knowledge, it is possible to appreciate distinctions in scope and generality between different theorems. The technicalities, however, are of interest on their own account. Some of them are landmarks in their own province. Only what appear to have been the principal steps leading from the enumerative stage of hypercomplex number systems to the abstract theory of all such 'algebras' will be considered here.

The work of B. Peirce (1870) aimed at principles for the exhaustive tabulation of linear associative algebras in a given finite number of fundamental units, with real or complex number coefficients. His methods were adequate, and his partial failure to find all the algebras he sought in less than seven units was due to mere oversights.

Peirce's problem was equivalent to that of exhibiting all sets of n linearly independent symbols (the basal, or fundamental, units) $e_1, \ldots, e_n$ forming closed systems under associative multiplication, on the assumption that any product e_re_s is a linear function of $e_1, \ldots, e_n$. For n given, the problem can be solved brutally by actually constructing all possible multiplication tables, applying the associativity condition

$$e_r(e_se_t) = (e_re_s)e_t,$$

and retaining only those which actually close. The labor of such an undertaking quickly becomes prohibitive with increasing n, and Peirce proceeded otherwise. Two of his guiding principles depended on the presence or absence of nilpotent and idempotent units: if there is a positive integer $r > 1$ such that $e^r = 0$, Peirce called e nilpotent; and similarly for $e^2 = e$ and idempotent. These pervasive concepts of linear algebra were the foundation of Peirce's classification. His work was continued by his son, C. S. Peirce (1839–1914, U.S.A.), who also made outstanding contributions to mathematical logic and is said to have invented the peculiarly Yankee philosophy known as pragmatism.

In an appendix to his father's memoir, C. S. Peirce proved a famous theorem,[5] which we restate in its customary form: The only linear associative algebras in which the coordinates are real numbers, and in which a product vanishes only if one factor is zero, are the field of real numbers, the field of ordinary complex numbers, and the algebra of quaternions with real coefficients. The advance here beyond tabulation is evident.

The theorem also indicates one possible kind of answer to the question asked by Gauss. It suggests that quaternions might have a useful arithmetic, as they are so closely related to ordinary complex numbers. Extensive arithmetics of ordinary quaternions were constructed by R. Lipschitz (1832–1903, German) in 1886, and by A. Hurwitz in 1896. Dickson in 1922 simplified these arithmetics. The historical order here is that of increasing simplicity.

From the late 1870's through the 1890's, linear algebra took several new directions which looked extremely promising at the time, but which do not appear to have influenced the main advance significantly. Thus G. Frobenius (1849–1917, German) in 1877 developed an interesting connection between hypercomplex number systems and bilinear forms. This was followed in 1884 by H. Poincaré's discovery of a similar connection with the continuous groups due to M. S. Lie, (1842–1899, Norwegian). Lie's outstanding contributions to nineteenth-century mathematics will be noted in connection with invariance. Poincaré replaced the problem of classification by an approachable equivalent for a wide class of linear algebras: to find all continuous groups of linear substitutions whose coefficients are linear functions of *n* arbitrary parameters. This line of attack was developed with considerable success by G. W. Scheffers (1866–, German), a scientific legatee of Lie, who in 1891 undertook to show that Lie's theory of (finite) continuous groups contains the theory of hypercomplex number systems. The theory of such groups therefore afforded principles for the classification of linear associative algebras.

In the 1890's Lie's theory was perhaps more assiduously cultivated than it was to be for a generation, when it was revived in a renovated differential geometry. Reflecting this widespread interest, E. Cartan (1869–, French) in 1898 applied the Lie theory to obtain a classification of hypercomplex number systems which followed 'naturally' from that theory. But such apparently promising leads were practically abandoned shortly after 1900, and the main advance proceeded in another direction, beginning in 1907 with the work of J. H. M. Wedderburn (1882–, Scotch, U.S.A.).

Frobenius and Cartan had obtained numerous special results for hypercomplex number systems in which the coordinates are rational numbers, whereas in a general theory the coordinates should be elements of nothing more restricted than an abstract field. The postulates for such a situation were formulated in 1905 by Dickson. As much of Cartan's development depended on the characteristic equation of a linear associative algebra, it was not always extensible to the general case. In any event, algebra took a new turn in 1907, heading directly toward a theory of structure.

The theorem of Frobenius and Peirce on quaternions suggested the search for all linear associative[6] algebras satisfying certain preassigned conditions. The most important of these are the division algebras, in which, if $a(\neq 0)$ and b are any elements of the algebra, each of the equations $ax = b$, $ya = b$ has a unique solution. Two of the earlier results on division algebras may be recalled, to provide concrete examples of structure.

Galois initiated the study of fields containing only a finite number of distinct elements. It was proved in 1893 by E. H. Moore (1862–1932, U.S.A.) that every finite commutative field is of the type considered by Galois; that such a field is uniquely determined by a pair of positive integers p, n, of which p is prime; and that the corresponding field contains p^n distinct elements. This theorem exhibits one of the many variants of structure: an exhaustive characterization is prescribed for *all* (commutative) fields containing only a finite number of distinct elements. Another variant appears in Wedderburn's theorem of 1905, that if the coordinates of a linear associative division algebra are elements of a finite field, then necessarily multiplication in the algebra is commutative.

As will be seen presently, division algebras play a dominant part in the theory of algebraic structure. The determination of all division algebras, or the discovery of comprehensive classes of such algebras, thus became a central problem in the theory. Dickson in 1914 constructed division algebras in n fundamental units with coefficients in any field F.

At this point we may repeat that we are primarily interested here in the emergence of the theory of structure, particularly as exemplified in linear algebra. To bring out the essentials, it will be necessary next to state a rather formidable-looking theorem containing several technical terms whose meaning has not been explained. Those already acquainted with the subject will recognize the statement as one of the fundamental theorems[7] (Wedderburn's, 1907) on the structure of algebras. Those seeing it for the first time may substitute letters S, X, Y, . . . , for the technical terms 'sum,' 'semi-simple,' 'direct sum,' . . . , as we shall do in a moment, and attend solely to the construction of the sentences constituting the theorem. The structural character, which is our concern here, is then evident. As a mere verbal convenience, a linear associative algebra whose coordinates are in a field F is said to be over F. The theorem states that:

(1) Any linear associative algebra over a field F is the sum of a semi-simple algebra and a nilpotent invariant subalgebra, each over F;

(2) A semi-simple algebra over F is either simple or the direct sum of simple algebras over F.

(3) Any simple algebra over F is the direct product of a division algebra and a simple matric algebra, each over F, including the possibility that the modulus is the only unit of one factor.[8]

Eliminating the technicalities, irrelevant for our purpose, we restate this with S, X, Y, ...

(1) Any linear associative algebra over a field F is the S of an X-algebra and a Y-algebra, each over F;

(2) An X-algebra over F is either a Z-algebra or the DS of Z-algebras over F;

(3) Any Z-algebra over F is the DP of a W-algebra and a U-algebra, each over F.

The theorem exhibits the structure of *any* linear associative algebra over *any* (commutative) field F, in terms of three kinds of operations, S, DS, DP, and five species of algebras X, Y, Z, W, U. Thus any linear associative algebra over any field F may be dissected into algebras of the five kinds specified, and always in the same way, namely, by S, DS, DP. Without further elaboration, this is what is meant by saying that all linear associative algebras over any field F have the same structure with respect to certain specified kinds of subalgebras. Attention may therefore be confined to the five specified kinds of subalgebras. One of these, W, comprises the division algebras.

The radical distinction between general structure theorems of this kind and the cataloguing type of algebra which preceded it is obvious. The shift of objective is typical of modern abstract mathematics. Specimens are no longer prized for their own curious sake as they were in the nineteenth century. It is as if some industrious company of fossil collectors who had never heard of Darwin were suddenly enlightened by an evolutionist. Their interesting but somewhat meaningless collections would simplify themselves in an unsuspected coherence.

Toward abstraction in analysis and geometry

Three further advances of 1906–20 toward abstractness and generality may be mentioned here, as they are connected with the real numbers. None originated in algebra; yet at least one—J. Kürschák's (1864–1933, Hungary)—was to suggest far-reaching consequences in both algebra and arithmetic. All were in the direction of a theory of valuation generalizing that of real numbers and ordinary complex numbers.

With any ordinary complex number $x + iy$, written as a Hamiltonian number-couple (x, y), is associated the unique real number $|(x, y)|$, the 'absolute value' of (x, y), which is the positive square root of $x^2 + y^2$. But if x, y are elements of an abstract field F, $x^2 + y^2$ is not a real number. To distinguish the zero element of F from the zero, 0, of the real number field, we shall write it $0'$.

Extending the properties of absolute values for the field of ordinary complex numbers to the elements $0', x, y, \ldots$ of an abstract field, F, Kürschák in 1913 associated with any element z of F a unique real number, its 'absolute value,' denoted by $\|z\|$, subject to the postulates $\|0'\| = 0$; $\|x\| > 0$ if $x \neq 0'$; $\|zw\| = \|z\| \|w\|$; $\|z + w\| \leqq \|z\| + \|w\|$, for any z, w in F.

The last is sometimes called the triangular inequality. It is the analogue of the theorem in plane Euclidean geometry that any side of a triangle is less than, or equal to (when the vertices are collinear), the sum of the other two sides. If the 'distance' between any two elements x, y be defined by $\|x - y\|$, the postulates for these absolute values reproduce the properties usually associated with the concept of distance. They were so defined in 1906 by (R.) M. Fréchet (1878–, French) in his thesis for the doctorate at Paris. This work is one of the sources of modern general or abstract analysis, the theory of abstract spaces, and topology (all to be noted in other connections). A further advance in this direction was made by S. Banach (1892–1941, Polish) in 1920, who removed the restriction that the elements $x, y, \ldots$ be in a field; his $x, y, \ldots$ are elements of any class whatever.

It might be thought that nothing not already known could issue from such faithful copies of the simplest properties of real and complex numbers. To substantiate the contrary, we need cite only one rather unexpected outcome of this process of abstraction. It has been found that much of the analysis based on the real or complex numbers has its image in general analysis. It is not necessary to assume that $x, y, \ldots$ are real or complex numbers to obtain many results which formerly were supposed to be consequences of that assumption. In Fréchet's analysis, for example, limit points and convergent series are definable, and the theorems on convergence for this abstract analysis are applicable to the special case of the analysis in which the basic elements are real or complex numbers. Again, it might be imagined that for the rational numbers $r, \ldots$ the only possible $\|r\|$ is the familiar $|r|$. But it was shown (1918) by O. Ostrowski (1893–, Russian) that there are in fact precisely two distinct types. Thus, in this instance at least, abstraction led to something new and unexpected.

A terminus in arithmetic

Gauss' question concerning the possible utility of hypercomplex numbers in the higher arithmetic is of peculiar interest, both historically and mathematically. We have seen that Gauss was acquainted with quaternions, and we have referred to applications of quaternions to the classical theory of numbers. It is unlikely that Gauss had developed quaternions far enough to suspect that they have interesting arithmetical properties. Nevertheless, he asked[9] "whether the relations between things, which furnish a manifold of more than two dimensions, may not also furnish permissible kinds of magnitudes [numbers] in general arithmetic?"

There has been much speculation as to what Gauss considered 'permissible.' Each guess generated its own answer in an enumeration of all algebras satisfying the 'permissible' condition. We shall report only one of several closely similar answers, as the direction taken by general arithmetic since the time of Gauss certainly was not foreseen by him.

Weierstrass is said to have proved the following theorem in his lectures of 1863; at any rate he published it, with several more of a like kind, in 1884. The only hypercomplex number systems with real coordinates, in which a product vanishes only if at least one of its factors does, and in which multiplication is commutative, are the algebra with one fundamental unit e such that $e^2 = e$, and the two-unit system of ordinary complex numbers. If Gauss did not permit divisors of zero, and if he insisted on commutative multiplication, this answered his question. The algebra with $e^2 = e$ is of little interest; the other gave Gauss his own arithmetic of complex integers $a + bi$, a, b real integers. In short, he himself had reached the end of the road he may have imagined but did not explicitly indicate.

Further progress was in other directions, starting in the work of Dedekind on algebraic number fields and ideals, and in Kronecker's theory of modular systems as developed by himself and others, notably E. Lasker (1868–1941, German, former world chess champion) in 1905, J. König (1849–1913, Hungary) in 1903, and F. S. Macaulay (1862–1937, English) in 1916. We pass on to a short enumeration of what appear to have been the principal steps toward these vast developments of modern algebra and arithmetic.

Newer directions

From the great mass of work that has been done since 1900 on the arithmetization of algebra—or vice versa—we shall select only three items, to indicate the trend toward abstractness and the analysis of structure.

In his algebraic theory of fields (1910), E. Steinitz sought all possible types of fields and the relations between them. Proceeding from the simplest types, explicitly defined, he extended these by either algebraic or transcendental adjunctions. In algebraic adjunctions, Steinitz followed the method of Cauchy as exploited by Kronecker, described in an earlier chapter. The concepts of characteristic, prime field, complete field, and others now familiar in modern texts on higher algebra were fully treated in this profound reworking and extension of Kronecker's theory of algebraic magnitudes. The final outcome may be roughly described as an analysis of the structure of fields with respect to their possible subfields and superfields.

The next item, dating from about 1920, marks a distinct advance. It is represented by a host of vigorous workers who, in the twentieth century, undertook to do for an abstract ring what Dedekind had done for any ring of algebraic numbers, and to extend the Galois theory to abstract fields. Thus the Dedekind theory of ideals was abstracted and generalized, as was also the Galois theory. The first of these may properly be assigned to arithmetic, as one of the chief objectives is the discovery, for any ring, of unique decomposition theorems analogous to the fundamental theorem of arithmetic, or to the unique representation of a Dedekind ideal as a product of prime ideals. It was not to be expected that there would be a single type of decomposition obviously preferable above all others for a general ring; nor was it reasonable to suppose that rational arithmetic or the theory of algebraic numbers would be translatable into the new domain with only minor modifications. Only the arithmetic of rings with commutative multiplication had been discussed with anything approaching completeness up to 1945.

In spite of radical differences between the arithmetic of commutative rings (usually without divisors of zero, the so-called domains of integrity) and that of algebraic numbers, the theory of Dedekind ideals proved a valuable clue. For example, in the Dedekind theory, an ideal has a finite basis; that is, any number of the ideal is representable as $n_1b_1 + \cdots + n_rb_r$, where $b_1, \ldots, b_r$ are fixed integers of the algebraic number field concerned, and $n_1, \ldots, n_r$ range independently over all integers of the field. But this fact does not directly suggest a profitable generalization to rings. However, it is equivalent to the theorem that any sequence $A_1, A_2, A_3, \ldots$ of ideals which is such that A_{j+1} is a proper divisor of A_j, for $j = 1, 2, \ldots,$ ends after a finite number of terms. This 'chain theorem,' valid in Dedekind's theory, is generalizable.

Two basic but rather inconspicuous-looking items of the classical theory of algebraic number ideals passed unchanged into the abstract theory, the G.C.D. ('greatest' common divisor) and L.C.M. ('least' common multiple). Although at the first glance these are mere details, experience has shown that they are the framework of much algebraic structure and that, when their simplest properties are restated abstractly as postulates, the resulting system unifies widely separated and apparently distinct theories of algebra and arithmetic. They lead, in fact, to what seemed the most promising theory of algebraic-arithmetic structure devised up to 1945. We shall therefore describe their properties in some detail.

The relevant phenomena appeared first in mathematical logic, specifically in the algebra of classes, now called Boolean algebra after its founder, G. Boole (1815–1864, English). If the letters $A, B, C, \ldots$ denote any whatever classes, and if the symbol $>$ is read 'includes,' or 'contains,' it is true that from

$A > B$ and $B > C$ follows $A > C$. The symbol $<$ is read 'is included in,' or 'is contained in.'

If A, B are any two classes, their 'intersection,' denoted by $[A, B]$, is the most inclusive—largest—class whose members are in both A and B. For example, if A is the class of all animals with red hair on their heads, and B is the class of all girls, $[A, B]$ is the class of all red-headed girls. If A is as before and B is the class of all vegetables, $[A, B]$ is the null class—the class with no members. Again, if A, B are any classes, their 'union,' denoted by (A, B), is the least inclusive class whose members are in A, or in B, or in both. In the second of the above examples, (A, B) is the smallest collection each of whose members is either an animal with red hair on its head, or a vegetable.

We may restate these definitions as follows. If A, B are any two classes, they have a unique most inclusive common subclass, $[A, B]$, and a unique least inclusive common superclass, (A, B). By definition, A is equal to B, written $A = B$, if, and only if, $A > B$ and $B > A$; and by convention, $A > A$. It is now a simple exercise in language to verify the following statements concerning any set (or class) $\mathfrak{S}$ of classes $A, B, C, D, D_1, \ldots,$ $M, M_1, \ldots$

(1) If A, B, C are any three members of $\mathfrak{S}$, such that $A > B$ and $B > C$, then $A > C$.

(2) If A, B are any members of $\mathfrak{S}$, there is a member of $\mathfrak{S}$, say D, such that $D \leqq A$, $D \leqq B$; and such that, if also $D_1 \leqq A$, $D_1 \leqq B$, then $D_1 \leqq D$. There is also a member of $\mathfrak{S}$, say M, such that $M \geqq A$, $M \geqq B$; and such that, if also $M_1 \geqq A$, $M_1 \geqq B$, then $M_1 \geqq M$.

The assertions in (2) are true when D is the intersection $[A, B]$ of A, B, and M is their union (A, B). Reading $(A, (B, C))$ as the union of A and (B, C), and $[A, [B, C]]$ as the intersection of A and $[B, C]$, we have the following theorems as immediate consequences of (1), (2):

$[A, B]$, (A, B) are uniquely defined; $[A, B] = [B, A]$, $(A, B) = (B, A)$; $[A, A] = A$, $(A, A) = A$; $[A, [B, C]] = [[A, B], C]$, $(A, (B, C)) = ((A, B), C)$; $(A, [A, B]) = A$, $[A, (A, B)] = A$.

It may be left to the reader's ingenuity to decide whether or not the next is true for classes,

(3) If $A < C < (A, B)$, then $C = (A, [B, C])$.

It is important for our purpose to verify that (1), (2), and the simple theorems above are satisfied for the following wholly different interpretations of A, B, C, . . . , $>$, $=$, $<$: A, B, C, . . . is any class of positive rational integers; ‘ $=$ ’ is equality as in common arithmetic; ‘ $>$ ’ means ‘divides’; ‘$<$’ means ‘is divisible by’; $[A, B]$ is the L.C.M. of A, B, and (A, B) their G.C.D. Recalling that ‘divides’ in the theory of Dedekind ideals means ‘contains’ or ‘includes,’ as in inclusion for classes, we see why the interpretation in terms of classes should be relevant for the theory of ideals.

We now empty A, B, C, . . . , $>$, $<$, $[A, B]$, (A, B) of all interpretation and of all meaning, and take (1), (2) as postulates defining the meaningless marks A, . . . , $>$, $<$, etc. Denoting the D, M in (2) by $[A, B]$, and (A, B) respectively—a mere convenience of notation—we can deduce from our postulates the same theorems as before.

The abstract system so defined is not vacuous; for we have actually exhibited two of its instances. One would have sufficed. There are many more. The abstract system defined by (1), (2) has been called by various names, including ‘structure’ and ‘lattice.’ The second is to be preferred here, to avoid confusion with ‘structure’ as previously defined in mathematical logic.

Here we reach the ultimate in arithmetized algebra or algebraized arithmetic up to 1945. It may turn out to be a very bad guess but, as this is written, many of the younger generation of algebraists and arithmeticians believe that in this theory of lattices they have at last unified a welter of theories inherited from their prolific predecessors. The theory has been most vigorously developed in the United States. To vary the historical monotony of dwelling almost exclusively on the dead, we may mention the names of two of the most active contributors to this rapidly expanding domain of abstract mathematics, G. Birkhoff[10] and O. Ore.

The assertion (3), which was abandoned to the ingenuity of the reader, is not a consequence of (1), (2) in the abstract theory. When (3) is included with (1), (2) among the postulates, the resulting system defines a special type of lattice which is named after Dedekind, because he was the first (1897) to investigate such systems. That he did so, is but another instance of his penetrating and prophetic genius. His work passed practically unnoticed for a third of a century, when its significance was realized in the theory of lattices.

To leave this comprehensive theory of lattices here with no indication of its scope would do it but scant justice. A very brief quotation[11] from one of the creators of the theory contains the meat of the matter. "In the discussion of the structure of algebraic domains, one is not primarily interested in the *elements* of these domains, but in the relation of certain distinguished *subdomains*, like invariant [normal] subgroups in groups,[12] ideals in rings, and characteristic moduli in modular systems. For all of these are defined the two operations of union and intersection, satisfying the ordinary axioms"—the postulates (1), (2).

The rapid expansion of this theory of structures or lattices, after a quiescence of about a third of a century following Dedekind's introduction of dual groups, is typical of much in the recent development of mathematics. Among other common features is the apparently inevitable slowness with which a basic simplicity underlying a multitude of diversities finally emerges. Usually the unifying concept implicit in all of its different manifestations is almost disconcertingly obvious once it is perceived. But there is as yet no recognized technique for perceiving the obvious or for not confusing significance with triviality, and each instance, it seems, must wait its own more or less random occasion. Nor are mathematicians always reliable prophets of what mathematics is to retain its vital interest or acquire a new importance. An example which deserves to become a historical classic is the sudden rise to popularity of the tensor calculus which, until the relativists adopted it, was the neglected waif of the mathematicians. In all such revaluations it is easy after the fact to see that they might have occurred much earlier than they did. Yet nobody can foresee where the next is to happen. Mere neglect, however, is not necessarily an assurance of immortality to any who may be overanxious about their reputations or the permanence of their work.

In the matter of lattices or structures, one of the determining characteristics of the theory might have been anticipated—but was not—in 1854 when Boole published his *Laws of thought*. The irrelevance of the nature of the individual elements of the classes whose unions and intersections give Boolean algebra its distinctive character was recognized by Boole himself. Apparently without fully realizing what he had done, and what now seems so plain, Boole had taken the first and the decisive step toward the abstract algebra and some of the geometry of the 1930's–1940's. In these newer developments the primary interest is not in the elements of certain domains, but in the inclusions, the intersections, and the unions of distinguished subdomains formed of classes of elements of the original domain. Sets of elements rather than the elements themselves become the basic data, and the original level of abstraction rises to the next above it in what experience has shown to be a natural hierarchy of abstractions.

The Boolean algebra of classes attracted but little attention from mathematicians during the nineteenth century, and the hints it might have offered projective geometers and algebraists passed unnoticed. If Boole had assimilated the controversy of the 1820's over the validity of the geometric principle of duality, he might well have anticipated the algebraic interpretation of projective geometry of the 1930's. The controversy will be noted in a later chapter; for the moment it suffices to state that the 'real space' some of the contestants imagined they were discussing evaporated. We shall return to this presently.

Another clue that might have been noticed appeared in the various decomposition theorems of arithmetic and algebra. Decomposition, as in unique factorization in rational arithmetic and the theory of algebraic integers, reduces a given system to a system of simpler parts, the parts being combined according to prescribed rules to produce the elements of the given system. The 'simpler parts' in rational arithmetic are the rational primes; in algebraic numbers, the prime ideals; and in both, the rules are those of multiplication as in an abelian group. The decomposition may not be unique, as for instance the basis of an abelian group. Further suggestive examples of decomposition were the Jordan-Hölder theorem, described earlier, for finite groups, Wedderburn's theorems on the structure of linear algebras, and A. E. Noether's determination of all commutative rings in which there is unique prime ideal factorization. For the Jordan-Hölder theorem it is the subsystem of normal (invariant, self-conjugate) subgroups of a given finite group with respect to which the decomposition is effected that matter; the elements of the group itself are of only minor importance. For algebraic integers the distinguished subsets are the ideals, and so on; in each instance the original elements are subsidiary. With these and other examples before them, it gradually became evident to algebraists that some common characteristic must be the ultimate source of at least a part of the phenomena of decomposition in the several theories. Axiomatic formulations reveal the common characteristic to be an underlying lattice. In his work of 1900 on the dual groups generated by three moduli, Dedekind had noticed the affiliation with what is now called Boolean algebra. Once the unifying feature was recognized, the natural next step was to develop the algebra of lattices as an independent theory on its own merits. The outcome was an abstract theory of structures or lattices.

Abstraction for its own sake may prove fruitful or barren. If it also suggests new theorems or expresses what is already known more simply and more clearly, abstraction rises to the level of creative mathematics. As the theory of lattices advanced, it unified and simplified much of existing algebra and the fundamental parts of certain other disciplines; it also aided in the discovery of new results. It was not to be expected that the general lattice by itself would clarify all the phenomena of decomposition, and as the theory progressed, specialized lattices of various types were defined to accommodate particular theories. One significant clue was the invariance of the chain length—the number of factors of composition—in the Jordan-Hölder theorem for finite groups. Other equally patent clues might have been followed initially. But all this is in retrospect, and it will be in closer conformity with time if we merely mention some of the items which the theory of lattices illuminated. For further details we may refer to G. Birkhoff, *Lattice theory*, 1940.

Boolean algebra, the historical source of lattice theory, found its natural place in the theory as a special type of lattice (complemented, distributive). Distributive lattices may be realized by rings of sets, an observation—in other terminologies—which goes back substantially to Euler. From this follows, on appropriate specialization, the representation theory of Boolean algebras by fields of sets. Partly suggested by the last, there are applications of lattice algebra to the classic theories of sets and measure. In a later chapter we shall note the wide generalizations of the geometric 'space' of the nineteenth century which evolved into the twentieth-century geometry of abstract space. Among these generalizations, the function spaces of the twentieth century readily accommodated themselves to abstraction. L. Kantorovich (Russian) essentially defined (1937) a partially ordered linear space as a (real) linear space having non-negative elements f, symbolized $f \geq 0$, subject to the following three postulates: If $f \geq 0$ and $\lambda \geq 0$, then $\lambda f \geq 0$; if $f \geq 0$ and $-f > 0$, then $f = 0$; if $f \geq 0$ and $g \geq 0$, then $f + g \geq 0$. It was shown by G. Birkhoff in 1940 that every function space then known forms a lattice with respect to this partial ordering. Vector lattices were also defined and shown to be decomposable into their suitably defined positive and negative components. In connection with abstractions of the absolute values of real and complex numbers, we shall note in a later chapter the type of abstract space named for the Polish mathematician S. Banach. Banach lattices were defined as vector lattices with a suitably specialized norm (generalized absolute value), and it was shown that all the examples given by Banach of his space are such lattices. As a last result in this direction, the decompositions of partially ordered function spaces when characterized abstractly yield components forming a Boolean algebra.

From its historical origin, lattice theory might have been expected to have applications to mathematical logic and the mathematical theory of probability, and such was found to be the fact. A curious application (1936) provided a model of quantum mechanics. We recall that the concept of observables is central in this mechanics, also that the indeterminacy principle imports probability into the metaphysics of all physical observation. A basically different application (1944) of logic and probability to the quantum theory will be noted in the concluding chapter.

A revealing application of lattices was the restatement of projective and affine geometries by K. Menger in 1928, *Bemerkungen zu Grundlagenfragen IV*, as instances of what was later to be called lattice algebra. Menger considered a system of abstract elements for which two associative, commutative operations, denoted by $+$, $\cdot$, are defined. The operations admit neutral elements, the 'vacuum' V and the 'universe' U, such that $A + V = A = A \cdot U$ for all A in the system, and it is postulated that $A + A = A = A \cdot A$. The characteristic feature of the algebra is the postulate of 'absorption': if $A + B = B$, then $A \cdot B = A$, and conversely for all A, B in the system. Thus the algebra is essentially what G. Birkhoff (1934) called a lattice. Birkhoff also (1934) independently reduced projective geometry to a topic in lattice algebra, and Menger (1935) published a somewhat amplified account of his theory of 1928, in which it had been stated that the algebra is applicable to the theories of measure and probability. So far as the reduction of projective geometry to lattice algebra is concerned, this now seems an inevitable but unanticipated climax of the thoroughgoing axiomatization of projective geometry—not very well known in Europe, apparently—by O. Veblen and J. W. Young in 1910.

The first two decades of the twentieth century witnessed an unprecedented activity in axiomatics, especially in the U.S.A., consequent on the work of D. Hilbert (1862–1943, German) in the foundations of geometry. The first volume of the *Projective geometry* (1910) at Veblen and Young subjected projective geometry to a logical rigor it had not experienced before, reconstructing this department of geometry as a hypothetico-deductive abstract system in accordance with Hilbert's general formalistic program for all mathematics. Although geometric intuition may not have been abandoned in the rigorous treatment, it was not recognized, either officially or unofficially. The opening sentences of the work assert that

> Geometry deals with the properties of figures in space. Every such figure is made up of various elements (points, lines, curves, planes, surfaces, etc.), and these elements bear certain relations to each other (a point lies on a line, a line passes through a point, two planes intersect, etc.). The propositions stating these properties are logically interdependent, and it is the object of geometry to discover such propositions and to exhibit their logical interdependence.

A more inclusive but somewhat vaguer description of geometry by one of the authors cited will be reported in a later chapter. The above passage is sufficient for the present; it almost begs to be translated into the language of lattices. In the expressive slang of 1945, it is a natural for such translation by anyone who has ever glanced at the postulates for a lattice. Yet essentially these postulates were accessible ten years before the above manifesto of the aims of geometry was printed, and it was not until a quarter of a century after it appeared that the connection between lattices, or Dedekind's dual groups, and geometry was perceived. But again this overlooking of what is now plain almost to the verge of truism is no reflection on the perspicacity of creative geometers. Rather is it merely another instance of the historical commonplace, emphasized long ago by W. Bolyai in connection with the final emergence of non-Euclidean geometry after centuries of seemingly unnecessary struggle, that mathematical discoveries, like the springtime violets in the woods, have their season which no human effort can retard or hasten.

A further quotation from the *Projective geometry* of 1910 disposes, in the current mathematical manner, of certain debates on the nature of 'space' that have exercised metaphysicans for many centuries: "Since any defined element or relation must be defined in terms of other elements and relations, it is necessary that one or more of the relations between them remain entirely undefined; otherwise a vicious circle is unavoidable." Although, as will be seen in connection with 'space' of any finite number of dimensions, it is not necessary to choose 'points' as the ultimate undefined elements of geometry, they usually were chosen and frequently still are. Lines, etc., then became classes of points. Geometry, even of the elementary-school kind, was never primarily concerned with the elements in these classes, but with the intersections and unions of the classes. The lines, surfaces, etc., were the distinguished subdomains of the initially structureless chaos of points with which geometry actually dealt. However, points were usually in the background, if not in the mind, of the geometer doing geometry—implicitly Veblen's definitions of 'geometer' and 'geometry'—ever since the time of Euclid. According to Euclid, "a point is that which has no parts and has no magnitude." But this nihilistic attempt at a definition was not incorporated into the deductive development of geometry until, as noted by Menger (1935), the lattice reformulation of both projective and affine geometry made logical use of Euclid's definition in the deductive treatment of geometry. Euclid and his successors omitted to state what they meant by 'part'; Menger supplied the deficiency by a precise logical definition. By corresponding explicit statement and inclusion of all hypotheses underlying his proofs, Menger showed that "great parts of projective and affine geometry may be developed upon the basis of 'trivial' axioms such as, for example, the postulates of [intersection and union as defined in Boolean algebra]."

The outcome amounted to the reduction of projective geometry to the algebra of a suitably specialized lattice. The appropriate lattice elements represent the several geometric objects or configurations of historical or mathematical interest, such as points, lines, planes; the intersection of two configurations, in the Boolean sense, is the geometric intersection, or the configuration common to the two; the union of two configurations is the least inclusive configuration containing both. Dedekind's axiom (for lattices, noted earlier), with a suitable finiteness condition classifies the elements dimensionally by the corresponding Jordan-Hölder chains. In contrast to the previous axiomatizations by Hilbert and by Veblen and Young, the lattice representation distinguishes no configuration as basic; all enter the theory symmetrically. To include affine geometry, Menger (1935) imposed on lattices a reasonable axiom of parallelism. He was thereby enabled to develop projective and affine geometry together.

In G. Birkhoff's reduction (1934), projective geometries were correlated with Boolean algebras, fields and rings of point-sets, systems of normal subgroups, systems of ideals, modules of a modular space, and systems of subalgebras of abstract algebras. There were also further correlations with the reduction of group representations, semi-simple hypercomplex algebras, and compact Lie groups. All of these were to be expected from the reduction of projective geometries to lattices of certain types, and from the earlier instances of lattices given mostly by G. Birkhoff himself.

From a philosophical point of view, perhaps the most interesting consequence of the lattice (or structure) representation of projective and affine geometries was a possible repercussion on the perennial speculations of metaphysicians concerning the nature of space. The debate of the 1820's on the real spacial existence of the ideal elements of projective geometry has been mentioned, and it was suggested that the disputed topic had no meaning in the sense intended by the debaters. The intimate connections between the Boolean algebra of logic, projective and affine geometries, and lattices at least hint that the debaters may have been discussing merely the ingrained habits of their own reasoning processes, inherited for thousands of years as a legacy from the two-valued logic in which, it appears, human beings reason with a minimum of thought. 'Space' may be nothing more mysterious than a triviality of a rudimentary logic.

Retrospect and prospect

Having rapidly traveled one main highway of modern mathematics as far as 1945, we may now glance back over our route and note certain landmarks of more than local—arithmetical or algebraic—interest. This has indeed been our objective from the start; to observe those changes in spirit and point of view which distinguish the whole of mathematics since 1900 from that of the nineteenth century.

Vast empires of mathematics, where hundreds of assiduous

workers have toiled and where many still labor, have not been even noticed. A detailed survey of the territory traversed would fill many volumes but, so far as may be judged at present, a complete account would neither add to nor subtract from two conclusions, evidence for only one of which has so far been presented. Both may be falsified tomorrow by the advent of another Descartes or a modern Gauss, or by a successor of Galois or Abel.

The mathematics of the twentieth century differs chiefly from that of the nineteenth in two significant respects. The first is the deliberate pursuit of abstractness, in which relations, not things related, are the important elements. The second is an intense preoccupation with the foundations on which the whole intricate superstructure of modern mathematics rests. It may be hazarded as a very problematical guess that when the history of mathematics is written a century hence, if mathematics is to last that long, the early twentieth century will be remembered chiefly as the first great age of healthy skepticism in mathematics as in much else.

The nineteenth century in mathematics seems in retrospect to be of a piece with the rest of that smugly optimistic age. God was in his Heaven and all was right with the world. European civilization shipped its blessings wholesale to the heathen of all continents. There seemed to be no limit to the quantity of flimsy goods that could be manufactured and dumped at a handsome profit on the unenlightened who were unable to distinguish tin plate from sterling silver, or brass nose rings from solid gold. Nor was there any restraint in the output and consumption of mathematics. Nearly everybody appears to have believed that nearly everything was sound beyond all doubt. With the turn of the century a period of criticism and revaluation set in, and all but unteachable reactionaries agreed that the change was a decade or more overdue.

The origins of both the abstract method and the critical approach can be traced definitely to the 1880's. Neither attracted much attention till Hilbert in 1899 published his work on the foundations of geometry, and about the same time pointed out the basic importance for all mathematics of proving the self-consistency of common arithmetic. But it seems just to attribute the initial impulse to G. Peano (1858–1932, Italian) in his postulates for arithmetic (1889). Resuming the Euclidean program, Peano undertook the deduction of common arithmetic from an explicitly stated set of postulates which were as free of concealed assumptions as he could make them. The postulational method is the source of both the modern critical movement and abstractness.

In the least flattering light, both criticism and abstraction reflect the leaden hue of decadence. Viewed thus, the mathematics of the twentieth century is a modernized version of the Alexandrian age of criticism and sterile commentary that were the lingering death of Greek mathematics. Even if this should prove to be a correct diagnosis of twentieth-century mathematics, it does not necessarily follow that mathematics is about to expire. Although they were long in coming, Archimedes had successors.

Looking more sympathetically at the mathematics of the twentieth century, as nearly all professionals do, we see it full of life and more vigorous than ever. Criticism is necessary to see exactly what is sound so that the next step may be taken with reasonable safety. The abstract, postulational method is not mere cataloguing and pigeon-holing. It also is creation, but of a kind more basic than the disorderly luxuriance of the nineteenth century. Unless the enormous accumulations from that most prolific century in mathematical history are sorted out and reduced to manageable proportions, mathematics will be smothered in its own riches. In the process of putting order into the huge mass by the abstract method, it is seen that much can be ignored. Should any of these neglected acquisitions ever be required, they are now obtainable with much less labor than formerly. On the creative side, the postulational analysis of mathematical systems suggests innumerable new problems, some of which may be worth detailed investigation.

Probably few engaged in this work of revaluation, simplification, and generalization imagine that an end has been reached in any direction. If Dedekind's dual groups passed all but unnoticed for over thirty years, is it credible that all the promising approaches have been explored, or that others will not be unexpectedly come upon? A historical detail which was omitted in our rapid survey may be recalled here to suggest at least the possibility of progress in directions not yet followed.

One of the clues that led Dedekind to his creation of ideals was the theory of composition of binary quadratic forms, which is reflected in the multiplication of ideals in a quadratic field. Gauss systematized this theory of composition, and had essentially completed it in 1801, but only for forms of degree 2 in 2 variables. Dedekind's ideals give an immediate generalization to certain very special forms (norms of algebraic integers of degree n) of degree n in n variables.

The point of particular interest here is that all this originated in Fibonacci's identity. There are similar identities for sums of 4 or of 8 squares, but not for any other number of squares. Fibonacci's identity follows at once from the properties of ordinary complex numbers; Euler's four-square identity is a property of quaternions; Degen's eight-square identity is similarly connected with Cayley's algebra in eight basal units. In the algebra of Fibonacci's identity, all the postulates of a field are valid; in that of Euler's identity, commutativity of multiplication no longer holds; in Cayley's algebra, multiplication is neither commutative nor associative. In passing, it seems rather remarkable that such a truncated algebra as Cayley's could have any physical significance, but it has been applied to the quantum theory.

Now, the classical arithmetic associated with these identities and their related algebras is a detail in the arithmetical theory of quadratic forms in n independent variables. Reduced to its simplest terms, this theory is concerned with the following problem, in which all integers are rational: $a_{ij}(i, j = 1, \ldots, n)$ and m are constant integers: $a_{ij} = a_{ji}$; $x_1, \ldots, x_n$ are variable integers; the summation, Σ, refers to $i, j, = 1, \ldots, n$; it is required to state criteria for the solvability of $\Sigma a_{ij}x_ix_j = m$, and, if the equation is solvable, to find all solutions $x_1, \ldots, x_n$. A 'natural' generalization of this problem is the similar one obtained on replacing the quadratic form $\Sigma a_{ij}x_ix_j$ by any form of degree greater than two; and this is in fact the problem of representation by arithmetical forms. If a further generalization is required, the coefficients and variables may be integers in any algebraic number field.

The algebras with their corresponding theories of ideals, etc., which so far have been created are wholly inadequate for an attack on the general theory of arithmetical forms. So simple an equation as $x_1^3 + x_2^3 + x_3^3 = n$ apparently defies them. It seems reasonable to suppose that if significant progress is ever made in the theory of arithmetical forms of degree higher than the second, algebras of a character totally different from those whose abstract theory has been developed will come into being. Indicative of the rapidity with which a situation in diophantine analysis may change, since the first edition (1940) of this book,

L. J. Mordell (1888–, U.S.A., England) proved (1942) that the preceding equation has no solutions with x, y, z, quartic polynomials in one parameter with rational coefficients unless $n = a^3$ or $n = 2a^3$, where a is a rational number; and B. Segre (Italian, England) made (1943) extensive applications of algebraic geometry to polynomial diophantine equations.

There is the possibility, however, that the problem of arithmetical forms will lose its interest for mathematicians. In other words, it will cease to be considered important. But importance sometimes, humanly enough, is only a tribute to the self-esteem of an egocentric mathematician. The problems which he can solve are, by definition, important; those which baffle him are unimportant. To say that a particular problem has lost its importance for modern mathematics may therefore be merely a rationalized confession of incapacity. Until a method has been devised for solving a problem, or for proving that it is unsolvable if such be the case, common professional pride would seem to demand that it be considered.

If this point of view is justified, the conclusion is that neither algebra nor arithmetic has reached an end in the modern abstract method, although that method may be a significant prelude to what our successors will create. We shall see that physics suggests the like in analysis. In any event, the beautiful achievements of the method would have delighted Euclid, who first of all mathematicians produced a rounded example of the postulational technique. He may not have realized what he was doing, as a modern mathematician sees his work, but he did it. Pythagoras, on the other hand, would have stood bewildered before the modern concept of number. He would want to know what had become of the natural numbers 1, 2, 3, ... All this time we have been taking them for granted, not questioning their specious simplicity.

We must now return to these so-called natural numbers—Kronecker's gift from God—and see what happened to them while number was disporting itself in a heaven Pythagoras never dreamed of. This will supply us with the link connecting algebra and arithmetic with analysis. Having examined this, we shall be in a position to proceed in later chapters to geometry and applied mathematics. Finally we shall be led back to the foundations of the whole structure, to see in them what our successors may designate as the characteristic distinguishing the mathematics of the twentieth century from all that preceded it—critical, constructive doubt.

CHAPTER 12

Cardinal and Ordinal to 1902

Lecturing in 1934–5 to his Chinese students in Peiping on the theory of functions of real variables, a distinguished American analyst[1] observed that "The student has thus far taken the system of real numbers for granted, and worked with them. He may continue to do so to the end of his life without detriment to his mathematical thought. . . . On the other hand, most mathematicians are curious, at one time or other in their lives, to see how the system of real numbers can be evolved from the natural numbers." The natural numbers are the positive rational integers 1, 2, 3, ... A little earlier, a distinguished German analyst,[2] writing for his beginners in analysis, made an unusual request: "Please forget all that you have learned in school; because you have not learned it." He was referring to such simple matters as $1 + 1 = 2$.

Subscribing to all of these sentiments, we shall indicate the main steps by which mathematicians in the second half of the nineteenth century reached the modern concept of real numbers. The real numbers are the soil in which the classical theory of functions grows and flourishes and, as remarked above, are usually taken for granted by students. Also by some others.

To anticipate, theories were constructed deriving the real numbers from the natural numbers. It was then sought to derive the natural numbers from something yet more basic, the theory of classes as in mathematical logic. By another of those curious coincidences regarding dates, what seemed like finality in this direction was reached in the closing years of the nineteenth century. Those years, in more senses than one, were the end of a great epoch.

It is still no doubt true that students may take the real numbers for granted, as they did in 1902, without detriment to their mathematical thought. But it is no longer true that the more basic natural numbers can be taken for granted by anybody as they were by nearly everybody in 1902. Drowsy intuition has been shocked awake since the close of the nineteenth century; and the program of Eudoxus, resumed in the late nineteenth century by the founders of the modern real number system, gave way in the twentieth century to another, more fundamental than any imagined by the Greek mathematicians. The center of interest shifted here as it is always shifting in mathematics. Indeed, it might have been mathematics and not an insignificant minor planet of a second-rate sun that Galileo had in mind when he muttered—according to a legend which should be true even if it may not be—"And still it moves," as he rose from his knees and bowed to the Grand Inquisitor. Nobody yet has succeeded in stopping mathematical progress, in 1902 or in any other climactic year, as Joshua stopped the motion of a heavenly body at the battle of Gibeon.

Equivalence and similarity

The modern number concept is the link connecting the arithmetic and algebra of the past with the analysis, geometry, and mathematical logic of the present. Like the typical student of analysis today, we have been taking for granted the system of natural numbers from which, by successive generalizations, evolved the system of complex numbers, which in its turn suggested the hypercomplex numbers of modern algebra. Our immediate concern in this chapter is to indicate the main steps by which mathematicians in the latter half of the nineteenth century sought to 'arithmetize' analysis. In the following chapter, we shall reach the same goal by a different route, that of the calculus from Newton and Leibniz to the year 1900; and we shall see again that the close of the nineteenth century marks a definite terminus in one direction of mathematical thought.

This terminus, as will appear in the concluding chapter, was a turning point in the evolution of mathematics comparable in significance to that in the fourth century B.C., when Eudoxus parted company with the Pythagoreans. Once more the occasion for the new departure was the nature of irrational numbers. But in the modern pursuit of number, a far deeper spring of mathematical knowledge than any that refreshed the Greeks was tapped. The natural numbers had been taken for granted, and it was found that they too presented subtle obstacles to clear understanding. The point of cardinal importance to be noted in the following summary account is the precise nature of the subtlest of these obstacles, which was perceived with dramatic suddenness only in 1902.

The modern attack on number was directed against two closely related objectives: that of rigorizing the concepts of function, variable, limit, and continuity in analysis; that of penetrating the logical disguise of number. The first eventuated in the retreat from intuitive ideas of the calculus sublimated from unanalyzed conceptions of motion and continuous curves; the second culminated in the identification of cardinal numbers with classes. In both, the concepts of equivalence (or similarity) of classes, especially for infinite classes, played a dominant part. It is a matter of great historical interest (as indicated in an earlier chapter), that equivalence for classes was firmly grasped as early as 1638 by Galileo,[3] just a year after Descartes published his geometry. Galileo's work was translated into English in 1665, the year that young Newton, rusticating at Woolsthorpe, thought out his first calculus.

To us it seems strange that so plain an indication as Galileo's of a feasible attack on all matters pertaining to the infinite was not pursued sooner than it was. But there is the earlier parallel of the Greek indifference to Babylonian algebra to suggest that mathematics does not always follow the straightest road to its future.

As it would be difficult to find a clearer, more graphic statement than Galileo's of the critical points, we quote[4] what he puts into the mouths of his characters, the sagacious Salviatus (*Salv.*), and the questioning Simplicius (*Simp.*). The talk has been about the "Continuum of Indivisibles."

Salv. . . . an Indivisible, added to another Indivisible, produceth not a thing divisible; for if that were so, it would follow, that even the Indivisibles were divisible. . . .

Simp. Here already riseth a doubt, which I think unresolvable. . . . Now this assigning an Infinite bigger than an Infinite is, in my opinion, a conceit that can never by any means be apprehended.

To make the infinite plain even to Simplicius, Salviatus patiently explains what a square integer is before proceeding as follows.

Salv. Farther questioning, if I ask how many are the Numbers Square, you can answer me truly, that they be as many, as are their propper roots; since every Square hath its Root, and every Root its Square, nor hath any Square more than one sole Root, or any Root more than one sole Square.

This is the kernel of the matter: the one-one correspondence between a part of an infinite class, here that of all the natural numbers, and one of its subclasses, here that of all the integer squares. Continuing the argument, Salviatus compels Simplicius to surrender.

Simp. What is to be resolved on this occasion?
Salv. I see no other decision that it may admit, but to say, that all Numbers are infinite; Squares are infinite; and that neither is the multitude of Squares less than all Numbers, nor this greater than that: and in conclusion, that the Attributes of Equality, Majority, and Minority have no place in Infinities, but only in terminate quantities. . . .

In modern terminology, two classes which can be placed in one-one correspondence are said to be equivalent[5] or similar.[5] In Galileo's example, the class of all square integers is equivalent to the class of all positive integers. Again, a part (strictly, proper part) of a class C is any class which contains some but not all members of C, and nothing else. Galileo's example shows that a class may be equivalent to a part of itself. A class is defined to be infinite if it is equivalent to a part of itself; and a class which is not infinite is defined to be finite. This means of distinguishing between finite and infinite classes was postulated[6] by B. Bolzano (1781–1848, of Prague), philosopher and theologian, and is basic in the modern theory of classes, both finite and infinite. Without it, Cantor's theory of sets of points, fundamental in modern analysis, would not exist.

It is interesting to note that Leibniz pointed out the similarity of the classes of all natural numbers and all even natural numbers, but drew the incorrect conclusion, rectified in Cantor's theory, that "the number of all [natural] numbers implies a contradiction."[7]

Arithmetized analysis

This is not the place to expound the theories of real numbers constructed by Cantor, Dedekind, and Weierstrass; and we shall merely recall the few fundamental concepts necessary to give point to the historical climax of 1902. We observe first that the concept of a class (set, aggregate, assemblage, ensemble, Menge) was taken as intuitive in the preceding section. That 'class' is by no means an intuitive notion was recognized by Cantor, who in 1895 defined it thus: "By a class (Menge) we understand any summary (Zusammenfassung) into a single whole of definite well-distinguished objects of our intuition (Anschauung) or of our thought (Denkens)." Possibly the exquisitely modulated philosophical German[8] is untranslatable into any blunter language, or incomprehensible to any but the initiated. With hesitation, then, we offer the following substitute in crude American:[9] "A class is said to be determined by any test or condition which every entity (in the universe considered) must either satisfy or not satisfy." It seems clear to the unphilosophic mind that either of these definitions somewhat rashly invites philosophers to philosophize; and indeed, the invitation was accepted with alacrity. Whether this was a happy issue for the analysis of the nineteenth century out of all its afflictions seems to be in some doubt among professional analysts.[10]

Another fundamental point in Cantor's theory is the radical distinction between cardinal and ordinal numbers. For finite classes and numbers the distinction is almost trivial. Finite classes have the same cardinal number if and only if they are similar. Note that this does not define 'cardinal number'; it defines 'same cardinal number,' a significant distinction. It is quite possible to know that two criminals have the same name without knowing what the name is. The symbol 1, or 2, or 3, . . . denoting a cardinal number (not yet defined!) of a class is a mere mark or tag which is characteristic of the class without reference to the order in which its members are arranged. When the members of a finite class are counted in a given order, the mark or label 1 being assigned to the first, the mark 2 to the next, and so on, an ordinal number is correlated with each element of the ordered class; and if n is assigned to the last, n is also the mark denoting the cardinal number of the class. But for infinite classes, as shown by Cantor, the like is no longer true; the marks for (transfinite) cardinals and ordinals differ, and the distinction between cardinal and ordinal is not trivial.

The cardinal number of any given class, finite or infinite, was defined by F. L. G. Frege (1848–1925, German) to be itself a class, namely, the class of all those classes similar to the given class. Thus the familiar cardinals 1, 2, 3, . . . of our unlearned youth have vanished in the all-nesses of infinities of classes containing respectively 'one' thing, 'two' things, and so on for as many things as there may be in our *Anschauung* or in our

Denkens. This outcome may seem rather disappointing at first. But on prolonged reflection we are forced to agree with E. Landau (1877–1938, German) that what we learned at school we did not learn. Any class similar to the class of all the natural numbers is said to be denumerable or countable.

Resolving Simplicius' doubt about the conceit of "assigning an Infinite bigger than an Infinite," Cantor proceeded to describe any desired number of such bigger Infinites. First, there is said to be no difficulty in imagining an ordered infinite class; the natural numbers 1, 2, 3, . . . themselves suffice. Beyond all these, in ordinal numeration, lies ω; beyond ω lies $\omega + 1$; then $\omega + 2$, and so on, until $\omega 2$ is reached, when $\omega 2 + 1$, $\omega 2 + 2$, . . . are attained; beyond all these lies ω^2, and beyond this, $\omega^2 + 1$, and so on, it is said, indefinitely and forever. If the first step—after which all the rest seems to follow of itself—offers any difficulty, we have but to grasp the scheme 1, 3, 5, . . . , $2n + 1$, . . . |2, in which, after all the odd natural numbers have been counted off, 2, which is not one of them, is imagined as the next in order. One purpose of Cantor in constructing these transfinite ordinals, ω, $\omega + 1$, . . . was to provide a means for the counting of well-ordered classes, a class being well-ordered if its members are ordered and each has a unique 'successor.'

For cardinal numbers also Cantor described "an Infinite bigger than an Infinite" to confound the Simpliciuses of mathematics and enchant the Salviatuses. He proved (1874) that the class of all algebraic numbers is denumerable, and gave (1878) a rule for constructing an infinite non-denumerable class of real numbers. Were we to make a list of the spectacularly unexpected discoveries in mathematics, these two might head our list. Cantor's proof is strictly one of existence. Providing no means for constructing any of the infinity of transcendental numbers whose existence is demonstrated, Cantor's proof is in the medieval tradition of submathematical analysis. It would have convinced and delighted Aquinas. J. Liouville (1809–1882, French), on the other hand, invented a method (1844) for constructing any one of an extensive class of transcendental numbers. His numbers were the first to be proved transcendental; Hermite's proof of the transcendence of $e(= 2.718 \ldots)$ followed in 1873; F. Lindemann's (1852–1939, German) for π in 1882. Hinting at the controversies to come in the twentieth century, Kronecker demanded of Lindemann, "Of what value is your beautiful proof, since *irrational numbers do not exist?*" We shall return to

Kronecker's program of arithmetization presently. It was quite different in both aim and scope from that of Cantor, Dedekind, and Weierstrass in their project of arithmetizing analysis. In passing, we note that A. Gelfond in 1934 proved the transcendence of a^b, where a is any algebraic number $\neq 0$, 1, and b is any irrational algebraic number.

In the program of arithmetizing analysis, the rational numbers presented no difficulty. By the device of number-couples subjected to appropriate postulates, the properties of positive rationals were referred to those of positive integers, and negative rationals were driven back with equal ease to positive rationals. Thus all the rationals were derived by a simple routine from the natural numbers. Proceeding to infinitely the greater part of the continuum of real numbers, Cantor defined irrationals by infinite sequences of rationals; $\sqrt{2}$, for example, may be defined by the sequence 1, 14/10, 141/100, 1,414/1,000, 14,142/10,000, ... Generally, if $a_1, a_2, a_3, \ldots$ is any infinite sequence of rational numbers which is such that, for each rational $\epsilon > 0$, however small, there is an index m such that $|a_n - a_\nu| < \epsilon$ for every $n, \nu \geqq m$, the sequence is said to be regular. It is postulated that every regular sequence defines a number; the class of all so-defined numbers is the real number system. With suitable definitions of equality, greater, less, sum, difference, product, and quotient, it was shown that these numbers satisfy the requirements of experience. In particular, a meaning was given to the formalism of such useful statements as $\sqrt{2} \times \sqrt{3} = \sqrt{2 \times 3}$, $\sqrt{2} \times \sqrt{3} = \sqrt{3} \times \sqrt{2}$. Cantor had arithmetized the continuum of real numbers.

Geometry also shared in the benefits which arithmetization had conferred on analysis. A one-one correspondence between all the points on any segment of a straight line and the continuum of real numbers was established. This done, C. Jordan (1838–1922, French) banished intuition from the conception of curved lines by giving a strictly arithmetical definition[11] of a curve as a plane set of points which can be put in one-one correspondence with the points of a closed segment $[a, b]$. This seems like restating a platitude in pedantic obscurity when we cite its simplest example, the parametric equations $x = r \cos t$, $y = r \sin t$ of the circle $x^2 + y^2 = r^2$. It seems rather less platitudinous when we recall that Peano (1890) constructed a real continuous plane curve, as the locus of a point (x, y) whose coordinates are given by $x = f(t)$, $y = g(t)$ with f, g uniform, continuous functions of the real variable t in the range $0 \leqq t \leqq 1$, completely filling the square $0 \leqq x \leqq 1$, $0 \leq y \leq 1$. In fact, he described two such curves passing through every point of the unit square. Many more examples of such 'space-filling curves' have been constructed since Peano exhibited the first; and what in 1890 appeared as a collapse of the geometrical heavens has reappeared as a commonplace phenomenon in Ph.D. dissertations. "And still it moves."

Equally unexpected miracles began illuminating the continuum itself. By a fairly immediate generalization, classes (sets) of points in a continuum ('space') of any finite or of a denumerably infinite number of dimensions were invented. Cantor proved that in each instance all the points in the whole space can be put in one-one correspondence with all the points on any straight-line segment. In a plane, for example, there are precisely as many points on a segment an inch long as there are in the entire plane. This, of course, is contrary to common sense; but common sense exists chiefly in order that reason may have its Simpliciuses to contradict and enlighten. However, some Simplicius occasionally interjects a shrewd objection, upsetting the even progress of the discussion; and if he but seldom gets the better of an argument, he can at least cause it to stumble badly. Kronecker elected himself the Simplicius of Cantor, Dedekind, and Weierstrass. His objections will be noted presently.

A profound question which exercised Cantor's utmost powers was this: Can the continuum of real numbers be well ordered? In 1883 he thought he had answered this affirmatively. Objections to his attempted proof were largely responsible for the descent of mathematics after 1900 to deeper levels in its efforts to escape deceptive intuition. Another problem which baffled Cantor was to prove or disprove that there exists a class whose cardinal number exceeds that of the class of natural numbers and is exceeded by that of the class of real numbers. This problem seemed to be still open[12] in 1945.

Whatever may be the ultimate fate of Cantor's theories of the infinite, continuity, and the number system, it appears likely that he will be remembered with Eudoxus as one of those who breached what is, after all, the central fortress of mathematical analysis. So also will Dedekind and Weierstrass. Like Cantor, these two also derived the number system of analysis from the natural numbers, Dedekind by his device of cuts, Weierstrass by classes of rationals. Another analyst who reached the same goal was C. Méray (1835–1911, French); but possibly the difficulty of his exposition deprived him of his just share of fame. One or other of these theories is so familiar today to every student of an advanced course in the calculus that there is no need to describe them here. They are open to precisely the same objections that mathematical logicians have raised to Cantor's Mengenlehre. But in stating this plain matter of fact, we do not imply that these theories have been rejected as totally wrong or barren. They still afford the most promising approach to an understanding of numbers and their part in analysis. If a theory is imperfect, that is perhaps only because it is not yet either dead or useless.

In the following chapter we shall see that the analysis of trigonometric (Fourier) series was partly responsible for the attempt to put a firm logical foundation under the continuum of real numbers. Many contributed significantly to the attempt; but the four whose work has been noted were the earliest to see clearly what needed doing and the first to attempt it.

Of the many who prepared for the final success, one may be mentioned here, P. du Bois-Reymond (1831–1889, German), partly for his own subtle researches in analysis, partly because it was owing to his insistence that Weierstrass permitted a most disconcerting invention of his own to become public. Intuitively, a continuous arc of a curve has a tangent at every point on the arc; Weierstrass constructed the equation of a continuous curve having no tangent at any of its points. He is said to have communicated this to his circle in 1861, but for some reason withheld it until du Bois-Reymond in 1874 asked him whether such a curve was possible. This example alone showed the necessity for some rigorous theory of the real number system to replace the pernicious intuitions which had seeped into analysis from geometry and kinematics.

Existence and constructibility

Kronecker has already been mentioned for his technical contributions to algebra and the higher arithmetic. These probably represent his choicest creations to those who have taken the pains to appreciate them; but Kronecker is more widely known to the mathematical public for his philosophy of mathematics. At one time he was looked upon by some of the analysts, including Weierstrass, as a sort of personal devil. It was feared that Kronecker's philosophy was wholly destructive; and it cannot be gainsaid that he hated the highly speculative analysis of his famous contemporaries. If Kronecker spelled Satan to Cantor, Cantor signified the personification of all mathematical evil to Kronecker.

Dedekind's definition of irrationals as cuts in infinite classes of rationals, Cantor's sequences of rationals defining irrationals, and Weierstrass' irrationals as classes of rationals, all ultimately referred the continuum of real numbers to the natural numbers. The 'magnitudes' of Eudoxus were replaced by hypothetical constructions performed upon the numbers 1, 2, 3, ... Thus the arithmetization of analysis was a return to the program of Pythagoras. Mathematical mechanics having been reduced to a department of analysis, it too was potentially arithmetized, at least by implication, and likewise for geometry. All at last was reduced to number as the Pythagoreans imagined number, but at a cost which they would never have attempted to meet, infinities upon infinities.

As thoroughgoing a Pythagorean as Pythagoras himself, Kronecker insisted that the infinities be banished, and that *all* mathematics be built up by finite constructions from the natural numbers. Unless a mathematical object were constructible in a finite number of non-tentative steps, it did not exist for Kronecker, no matter how many and how rigidly logical the transcendental proofs of its existence.[13] Such a philosophy of mathematics did not make nonsense of analysis as Kronecker's rivals had re-created it; analysis was simply abolished.

As if to lend some plausibility to the destructive part of Kronecker's program, flagrant contradictions began appearing in the late 1890's in reasoning of apparently the same general character as that used by the arithmetizers of analysis. Twenty-four centuries after Zeno's runner had lost his race, his heirs appeared on another course, fresher and fleeter of foot than that ancient had ever been. The new antinomies of the infinite sprang from the protean 'all' by which the irrationals were generated in the arithmetization of analysis: 'all' natural numbers; the class of 'all' rational numbers whose squares are less than 2, and the class of 'all' those whose squares are greater than 2 as the 'cut' defining $\sqrt{2}$, and an infinity more.

The first of the new and more vigorous paradoxes was fathered in 1897 by the Italian mathematician C. Burali-Forti: the well-ordered series of *all* ordinal numbers defines a *new* ordinal number which is not one of the *all*.

A less technical paradox on 'all' was B. A. W. Russell's (1872–, English) of 1902: Is the class of all those classes which are not members of themselves a member of itself? Either Yes or No leads to a contradiction. The same irrepressible successor of Zeno recorded a yet simpler 'all' paradox: A barber in a certain village shaves all those, and only those, who do not shave themselves; does the barber shave himself? There are many more. All—if we may use the word without danger of engendering another exasperating paradox—or, if not all, then many conceal a dubious 'all.' One root of the technical mathematical difficulties has been traced to the concept of 'class' itself, Cantor's 'Menge' as defined by him. The attempt to be logically precise ended in hopeless confusion.

The problems of the nineteenth century met those of the twentieth in the extraordinarily subtle mind of F. L. G. Frege (1848–1925, German), part of whose lifework[14] was an endeavor to put a self-consistent foundation under the number concept. In 1884, Frege was led to his famous definition of the cardinal number of a given class as the class of all those classes similar to the given class. From this definition, Frege derived the usual properties of numbers familiar in common arithmetic. Unfortunately, to develop the subtlety of his reasoning with precision, he had found it necessary to clothe his proofs in a complicated diagrammatic symbolism which repelled all but the hardiest and most obstinate readers. As a result, the epoch-making definition embedded in his work passed unnoticed by the mathematical public till Russell, by different reasoning, independently reached (1902?) the same definition and expounded it in English.

Frege had used the theory of classes. The second volume of his masterpiece[14] appeared in 1903. It closes with the following confession. "A scientist can hardly encounter anything more undesirable than to have the foundation collapse just as the work is completed. A letter from Mr. Bertrand Russell put me in this position as the work was all but through the press." Russell's letter contained his paradox on the class of all classes that are not members of themselves.

Frege's pessimism is understandable enough. But it was to prove unjustifiable in the long view of mathematical progress. The attempt to found the number system on a theory of classes seemed to have failed, and no doubt it had, at least temporarily. By the collapse of the class theory of the number system, analysis was left without a foundation, and hung suspended in mid-air like Mahomet's coffin, sustained only by a miracle of faith. But the very failure revealed the nature of the fundamental weakness. A younger, more vigorous generation attacked the problem of bringing analysis down to reason again. Profiting by the experience of the nineteenth-century arithmetizers, the mathematical logicians of the twentieth century set themselves the task of putting a self-consistent foundation under all mathematics, not merely under analysis. Their efforts quickly carried the program of Leibniz for a strict symbolic reasoning far beyond anything he ever conceived, and in so doing created much new mathematics.

In the meantime, analysts, geometers, arithmeticians, and algebraists continued their technical labors as if there were no 'crisis' in the foundations, creating interesting and useful things as their forerunners had done for centuries. Their confidence in the security of the essentials of their creations is justified by experience. Changing philosophies of mathematics may transform proofs and even theorems out of easy recognition as mathematics develops, and much is thrown away. But if history is a reliable prophet,[15] there will remain of the analysis of the nineteenth century as much, relatively, as remains of Euclid's proposition I, 47.

CHAPTER 13

From Intuition to Absolute Rigor

1700–1900

In following the development of the number concept to its final phase in modern arithmetic and abstract algebra, we have caught occasional glimpses of the spirit of mathematics as it has become since the close of the eighteenth century. Similar profound changes appear when we observe analysis. We now return to the eighteenth century and note the first attempts to construct a logically sound differential and integral calculus.

The contrast between what passed for valid reasoning then and what is now demanded is violent. Passing back to the eighteenth century, we find ourselves in a dead world, almost in another universe. Some of Newton's successors who strove to make sense out of the calculus are among the greatest mathematicians of all time. Yet, as we follow their reasoning, we can only wonder whether our own will seem as puerile to our successors a century and a half hence. It is not a question here of the enduring things these famous men did with their analysis in applied mathematics, or even of the basic algorithms which they invented and which have also lasted. We are concerned solely with their avowed attempts to put consistent meaning into the analysis itself.

Two decisive turning points

We saw that Newton himself was dissatisfied with his account of the fundamental concepts of the calculus. The like holds for

Leibniz, who half promised Huygens that some day he would return to the beginning and set everything right. But he never did. After the passing of Newton and Leibniz, critics of both approaches made themselves heard, and conscientious analysts, responding to legitimate objections, attempted to put a firm foundation under the calculus. Their efforts gradually disclosed the depth of the difficulties, and in the nineteenth century were partly responsible for the creation of vast new departments of mathematics, such as the theories of Dedekind and Cantor. We shall indicate the main stages in this extremely complex evolution by which the calculus of 1700 developed into that of 1900. As wide a gulf separated the great analysts of the late eighteenth century from those of the early nineteenth as severed the Pythagoreans from Eudoxus. After 1929—the historic year in which the great depression began in the United States—another deep fissure opened up, cutting off retreat to the nineteenth century, apparently forever, when K. Gödel re-examined the possibility of a consistency-proof for rational arithmetic.

In following any account of the evolution of rigor in the calculus, it must be remembered that opinions on many unsettled points differ, sometimes widely. Further, it has been difficult for some to avoid reading their own more exact knowledge into work of their predecessors which, if taken at what seems its intended value, gives no hint that its authors were ever conscious of what later stood out as fatal defects. For example, the generous J. le R. d'Alembert (1717–1783, French) in 1770 ascribed to Newton a fully developed theory of limits that but few analysts today can detect in what Newton published. And last, before proceeding to details, we emphasize once more that in exhibiting the shortcomings in the work of the older analysts, there is no implication that perfection has been attained in our own. The mistakes and unresolved difficulties of the past in mathematics have always been the opportunities of its future; and should analysis ever appear to be without flaw or blemish, its perfection might only be that of death.

Five stages

The general trend from 1700 to 1900 was toward a stricter arithmetization of three basic concepts of the calculus: number, function, limit. More subtle questions concerning the meaning of 'variable' scarcely entered before the twentieth century. In the period under discussion there were five well-marked stages, which may be easily retained by the names and dates of certain leaders in each. With the first are associated Thomas Simpson (1710–1761, English) in England, and G. F. A. l'Hospital (1661–1704, French) on the Continent. Euler (1707–1783, Swiss) represents the second stage; Lagrange (1736–1813) the third; Gauss (1777–1855) and Cauchy (1789–1857) the fourth; and Weierstrass (1815–1897, German) the fifth. Euler is the culmination of the almost wholly uncritical schools of Newton and Leibniz; Lagrange marks the earliest recognition by a mathematician of the first rank that the calculus was in a thoroughly unsatisfactory condition; Gauss is the modern originator of rigorous mathematics; Cauchy is the first modern rigorist to gather any considerable following; and Weierstrass, dying in 1897, epitomizes the progress made in exactly one hundred years from the first publication (1797) of Lagrange attempting to rigorize the calculus.

The golden age of 'nothing'

'Formalism' in analysis means manipulation of formulas involving infinite processes without sufficient attention to convergence and mathematical existence. Thus the formal binomial theorem applied to $(1 - 2)^{-1}$ gives

$$-1 = 1 + 2 + 4 + 8 + 16 + \dots,$$

a meaningless result which did not astonish Euler, the greatest but not the last of the formalists. 'Intuition' in analysis, as we shall use it here, means an unreasoning faith in the universal validity of what the senses report to the intellect concerning motion and geometrical diagrams. Newton was the greatest of the intuitionists in analysis, with the far more philosophical Leibniz a distant and therefore honorable second. (Both formalism and intuitionism in the mathematics of the twentieth century have different meanings, to be noted in the concluding chapter.) The direction of evolution in the calculus has been constantly away from formalism and intuition, although neither is yet extinct.

The first and crudest stage is represented in England by the two editions (1737–1776) of Simpson's classic *Treatise on fluxions*, in which[1] intuition flourishes freely and rankly. Attempting to clarify Newton's intuitive approach to fluxions through the generation of "magnitudes" by "continued motion," Simpson succeeded only in adding a deeper obscurity of his own. This can scarcely be considered an advance. The Continental mathematicians, following the Leibnizian tradition as handed down by John Bernoulli[2] (1667–1748, Swiss) in 1691–2 to l'Hospital[3] in 1693, proceeded from the mystical doctrine that "a quantity which is increased or decreased by an infinitely small quantity is neither increased nor decreased." This was the golden age of the 'little zero,' the happy morning of innocence in which the axiom of Archimedes was indefinitely suspended. The old Greek who had died all of two thousand years before they were born could have taught the analysts of the early eighteenth century more real calculus than any of them ever dreamed of.

Taylor's contribution

Intuition and formalism were to reach their climax in the masterpieces of Euler. An irresistible temptation to Euler's unrestrained and almost wholly uncritical use of infinite processes was provided by B. Taylor (1685–1731, English) with the publication (1715–17) of his *Methodus incrementorum directa et inversa,* in which 'Taylor's theorem' of the calculus, discovered as early as 1712, as also its easy consequence 'Maclaurin's theorem,' appeared in print for the first time. These expansions were an irresistible incitement to license and later, in the attempt of Lagrange to rigorize analysis, a hint that order might be imposed on the luxuriant confusion.

Taylor's work also expounded the calculus of finite differences, of which he is commonly said to have been the inventor. This too reacted on the evolution of the calculus, as will appear shortly. Taylor offered nothing that would have passed for a proof of his theorem at any time after Cauchy's work of 1821. Anything that has been nonsense for over a century may fairly be called such without qualification as to its epoch. Taylor's attempted proof being in this category, it is rather surprising to find its equivalent in widely used elementary texts on the calculus as late as 1945.

Attack by an amateur

While the analysts were pouring out new and correct formulas in lavish profusion with scarcely a qualm as to the legitimacy of their formalism, uncompromising critics were protesting against the deluge of what they considered metaphysical nonsense. We need not explore the human motives behind a goodly part of these savage assaults on the work of eminent mathematicians. The shrewdest attack of all was delivered by a man who was not a mathematician and who made no claim to be, G. Berkeley (1685–1753, Irish): half-heir of Jonathan Swift's Vanessa; at one time self-appointed apostle of culture to Bermuda but shipped by mistake to Newport, Rhode Island, where for three stagnant years (1728–31) he rusticated; later Bishop of Cloyne in his native Ireland; famed for his subjective idealism that out-idealized Plato, and immortal for his advocacy of tar water as a remedy for spiritual disorders and smallpox. A mind as keen as Berkeley's was needed to expose once for all the subtle fallacies in Newtonian fluxions, and the sagacious bishop spared no logic in his withering attack. A philosophical amateur did what professional mathematicians had shown themselves either too partisan or too tender-minded to do. Although few professionals would admit that either corpse was dead, Berkeley slew both fluxions and 'prime and ultimate ratios.'

Berkeley's assault in his *Analyst* (1734) was not just another of the vulgar wrangles, like the controversy over priority in the calculus, that disfigure the career of the Queen of the Sciences. It was one of the ablest critiques the leading mathematicians of any period have ignored, possibly because it came from one who was not a member of their somewhat exclusive guild. For once a philosopher turned the tables on mathematicians by convicting the fluxionists of changing their hypothesis in the middle of an argument. Until Berkeley's time, it had been supposed that this effective tactic in logomachy was an exclusive prerogative of dialecticians. Berkeley contended that substituting $x + o$ for x in x^n and letting o vanish in the final step, to get the fluxion of x^n, is a shift in the hypothesis: " . . . for when it is said let the increments be nothing,[4] or let there be no increments, the former supposition that the increments were something, or that there were increments, is destroyed, and yet a consequence of that supposition, *i.e.*, an expression got by virtue thereof, is retained." There were replies to this; but that which is unanswerable cannot be answered, and the controversy blew over, leaving scarcely a ripple on the muddy waters of mathematical analysis as they were in 1734.

Berkeley's criticisms were well grounded, but neither he nor they were taken seriously by the leading analysts of his time, and mathematics sought salvation in its own way. It is amusing to recall in passing that another question of salvation inspired Berkeley to his attack on fluxions. The full title of his work is *The analyst: or, a discourse addressed to an infidel mathematician. Wherein it is examined whether the object, principles, and inferences of the modern analysis are more distinctly conceived, or more evidently deduced, than religious mysteries and points of faith.* Only an Irish bishop who was also an idealistic philosopher could have conceived such a heroic project. It seems that Newton's friend Halley, posing as a great mathematician, had proved conclusively to some deluded wretch the inconceivability of the dogmas of Christian theology. The converted one, a friend of Berkeley's, refused the latter's spiritual offices on his deathbed. This was in the very year that Berkeley became a bishop. Profoundly shocked by the soul-destroying savagery of "the modern analysis," and mindful of his education in semicivilized Rhode Island, the good bishop went after the scalp of fluxions. He secured it; and the wretch who had been converted to infidelity by a nonsensical argument was avenged, although it may have been too late to save his soul.

The triumph of formalism

Euler's almost total capitulation to the seductions of formalism is one of the unexplained mysteries in mathematics. Like Newton, Euler was aware that series must 'in general'[5] converge if they are to be practically useful, as in astronomy; but unlike Newton, he was unable to restrain himself this side of absurdity. Euler appears to have believed that formulas can do no evil; and so long as they continued to furnish their parent with ever new and more prolific variations of themselves, he encouraged them to increase and multiply, trusting no doubt that someday all their offspring would somehow be legitimized. Many of them have, and flourish today as lusty theories whose first bold steps were taken in several editions of three masterpieces of this most prolific mathematician in history: *Introductio in analysin infinitorum* (1748); *Institutiones calculi differentialis* (1755); *Institutiones calculi integralis* (1768–1794).

The aim of the *Introductio* is to obtain by elementary means the utmost of what is so obtainable, but which is usually derived by the differential and integral calculus. The work is in two parts, an analytic and a geometric. Among the host of results are the expansions of the circular (singly periodic) functions, transformations of infinite products into infinite series, and developments into series of partial fractions. The last suggested one approach in the nineteenth century to elliptic (doubly periodic) functions. One chapter derives the basic formulas in the analytic-algebraic theory of partition of numbers. There are two heroes in this great drama of formalism, the expansion of e^x from the limit (Euler's style) of $(1 + x/n)^n$ as n tends through positive values to infinity, and the cardinal formula of analytic trigonometry, $e^{ix} = \cos x + i \sin x, i \equiv (-1)^{\frac{1}{2}}$. Creative formalism such as this is responsible for the impatient criticism of extreme mathematical rigor as *rigor mortis*.

The geometric part handles analytic geometry, both plane and solid, with equal freedom and complete mastery. The material includes special curves and surfaces, tangents and tangent planes, normals, areas, and volumes.

Turning partly away from intuitionism, Euler abandoned geometry in the *Institutiones*. This work is remarkable for its exhibition of analogies between the infinitesimal calculus and the calculus of finite differences, and the use of the latter to approximate to results in the former. There is no hint of convergence, but slowly convergent series are converted in masterly fashion into others more rapidly convergent. Here also the usual formal parts of the differential and integral calculus are developed in minute detail. One prophetic triumph of manipulative skill may be specially cited: Euler obtains the addition theorem for elliptic integrals as an exercise in differential equations.

A function to Euler became a congeries of formal representations transformable into one another by ingenious devices ranging from elementary algebra to the calculus. Glorying in the pragmatic power of his methods, Euler needed to see nothing absurd in his conception of the differential calculus as a process of determining the ratio of vanished increments. His differentials are first and last absolute zeros whose ratios by some incomprehensible spiritualism materialize in finite, determinate numbers. As the usually courteous Lagrange observed, Euler's calculus does not make sense.

If the end ever justifies the means in analysis, Euler was justified. He sought beautiful formulas, and he found them in overwhelming abundance. But obviously the calculus could not continue indefinitely on the primrose path so happily followed by this boldest and most successful formalist in history. Even Euler caught an occasional whiff of the everlasting bonfire in the absurdities that floated up now and then from the pit just ahead of him. Others scented damnation more keenly than he, among them his friend d'Alembert.

Best known for his principle in mechanics (1743), d'Alembert should be remembered also for having been the first (1754) to state[6] that "the theory of limits is the true metaphysics of the differential calculus." That nobody in the eighteenth century carried out the implied program, or was capable of doing so, is beside the point; d'Alembert saw clearly that what the calculus needed was not more formulas but a foundation. He regarded Newton's calculus of prime and ultimate ratios as a method of limits. Newton might have agreed with this had it been pointed out to him.

Lagrange's remedy

A new direction was taken by Lagrange in his ambitious *Théorie des fonctions analytiques* (1797, 1813), and his *Calcul des fonctions* (1799, 1806). These were conscious attempts to escape from Euler's conception of a function as a mere formula or algorithm, although Lagrange himself substituted another kind of formula, the power series, for the representation of all functions. His escape took him from one kind of formalism to another. Dissatisfied[7] with the efforts of all his predecessors and contemporaries, he rejected both infinitesimals and limits as being unsound, too difficult for neophytes, and, in the least complimentary sense, metaphysical.

Lagrange was the leading mathematician of the eighteenth century and one of the greatest in history. He also was the first to restate Taylor's theorem with a remainder term. Keeping all this in mind we shall, if we have the least grain of caution in us, be extremely conservative in our estimates of current rigor when we remember what finally satisfied Lagrange.

He based his calculus on the expansion of a function in a Taylor series, assuming that, "by the theory of series,"

$$f(x + h) = f(x) + ah + bh^2 + ch^3 + \dots .$$

From this he convinced himself that if $a = f'(x)$, "the derived function of $f(x)$," then $2b = f''(x)$, where $f''(x) = (f'(x))'$, and so on, all, as he imagined, without benefit of limits. He points out[8] that anyone familiar with the usual form of the calculus will see that $f'(x)$ is really $df(x)/dx$. But by what he has just deduced from "the theory of series," it is clear that $df(x)/dx$ does not in any way depend upon limits, prime and ultimate ratios, or infinitesimals as formerly; $f'(x)$ is merely the coefficient of h in the expansion of $f(x + h)$ in ascending powers of h. Need more be said?[9]

Gains to 1800

The net gains in the eighteenth century appear to have been four. Berkeley disposed of fluxions and of prime and ultimate ratios. Euler produced a vast wealth of results by purely formal uses of the calculus; and so sure was his instinct for what was to remain valid that his work was the point of departure from which many of his more productive successors made some of their most significant advances. To cite only two instances, Gauss, Abel, Jacobi, and Hermite were indebted directly to Euler in their more rigorous work on the theta and elliptic functions; the Eulerian integrals suggested to Legendre, Gauss, and Weierstrass extensive developments in the theory of the gamma function.

The third outstanding gain was d'Alembert's demand that the calculus be founded on the method of limits. The execution of this program had to wait for Cauchy (1821). The fourth gain was the hint implicit in Lagrange's abortive attempt to generate the calculus from power series. Weierstrass, in his theory of analytic functions, carried out what might have been an eighteenth-century program had Lagrange seen just a little more clearly what he was actually doing.

Ridiculous interlude

Formalism of a narrower sort than Euler's reached its absurd climax in the period between Lagrange and Cauchy. Combinatorial analysis, in the trivial sense of manipulating binomial and multinomial coefficients, and formally expanding powers of infinite series by applications ad libitum ad nauseamque of the multinomial theorem, represented the best that academic mathematics could do in the Germany of the late eighteenth century. The combinatorial school headed by C. F. Hindenburg (1741–1808, German) was the unlovely offspring of two human failings, neither of which is popularly supposed to have any relevance for the sublimities of pure mathematics: blind hero worship and national jealousy.

The German Leibniz in his combinatorial analysis of the discrete had created a rival for the English Newton's infinitesimal analysis of the continuous. Therefore, abandoning the calculus and its astronomical applications to the British, the Swiss, and the French, who had the lead already, the German mathematicians would loyally follow their national hero. Completely missing the deeper significance of Leibniz' program as a step toward the 'universal characteristic,' his patriotic disciples elaborated its superficialities in a wealth of useless formulas. The title of the masterpiece in this ambitious futility brazenly proclaimed the multinomial theorem to be the most important truth in the whole of analysis.[10]

Still more grandiose pretensions to omnipotence were urged by an egocentric Pole, H. Wronski (1778–1853), ardent envier of Lagrange, also a disciple of the combinatorial school, although his transcendent conceit[11] denied any progenitor but Wronski for himself and his "Supreme Law" which, he insisted, contained *all* analysis, past, present, and future. Both Wronski's claims and those of the combinatorialists have been disallowed by the supreme court of mathematical progress, from which there is no appeal. His criticism of Lagrange's attempt at rigor was justified; but his own substitute was no better.

The labors of this almost forgotten combinatorial school, however, were not without lasting benefit for the calculus. They filled young Gauss with such an intense disgust for formalism and all its works that he resolved to go his own lonely way and put some meaning into analysis, even if it cost him the patronage of every academician in Germany. He even favored the mighty Hindenburg with an extremely sarcastic letter.

In 1812 Gauss published his classic memoir[12] on the hypergeometric series, in which, for the first time in the history of mathematics, the convergence of an infinite series was adequately[13] investigated. Others before Gauss had gone as far as stating tests for convergence, notably Leibniz, for alternating series, and E. Waring (1734–1798, English), who had given what is usually called Cauchy's ratio test as early as 1776; but Gauss was the first to carry through a rigorous treatment.

Intuition transformed

From the calculus of Newton and Leibniz to that of Lagrange there is no indication that analysts were aware of the necessity for an understanding of the real number system. Nor is there in the next stage, that of Cauchy. Even as late as 1945 'quantities' occurred frequently in the writings of professional analysts with no explanation of what a 'quantity' may signify.

Imagining, perhaps, that he was banishing deceptive intuition forever from analysis, Cauchy succeeded in driving it down to a far deeper level where it might continue its subtle mischief unobserved. The crude visual and geometrical intuition of the early analysts was transformed into an uncritical faith in the logical possibility of the continuum of real numbers. Cauchy, Abel, and possibly Gauss[15]—for he seems to have left no record of his beliefs on this matter—adhered to this faith.

The definitions of limit and continuity current today in thoughtfully written texts on the elementary calculus are substantially those expounded and applied by Cauchy in his lectures and in his *Cours d'analyse* (1821), his *Résumé des leçons données à l'école polytechnique* (1823), and his *Applications du calcul infinitésimal à la géométrie* (1826). The differential quotient, or derivative, is defined as the limit of a difference quotient, the definite integral as the limit of a sum, and differentials as arbitrary real numbers. The continuity of a function and the convergence of an infinite series are referred to the concept of a limit. Thus Cauchy in effect created the elements of the classical theory of functions of a real variable. It was Cauchy's rigor that inspired Abel on his visit to Paris in 1826 to make the banishment of formalism from analysis a major effort of his projected lifework.

But, indicative of the subtleties inherent in consistent thinking about the infinite and the continuum, even so cautious a mind as Cauchy's went astray when it surrendered itself to intuition. He believed for a time that the sum of any convergent series of continuous functions is continuous, and that the integral of the sum is always obtainable by termwise integration. Later (1853, 1857) he recognized uniform convergence, discovered independently by the mathematical physicist G. G. Stokes (1819–1903, Irish) in 1847 and P. L. v. Seidel (1821–1896, German) in 1848. Cauchy also fell foul of the traps guarding interchange of limits in double-limit processes, as also did Gauss,[16] another plain hint that the real number system is less innocuous than it appears to naive intuition.

A suggestion from physics

It is rather surprising to find a main source of modern rigor in the work of a mathematical physicist who had almost a contempt[17] for mathematics except as a drudge of the sciences. J. B. J. Fourier (1758–1830, French) published his masterpiece,

La théorie analytique de la chaleur, in 1822, the year after Cauchy had rigorized the calculus. But if it had appeared twenty years after Cauchy's lectures, it probably would not have differed materially from what it was. Fourier had obstinately refused for fifteen years to heed the objections of Lagrange and others that vital parts of his analysis were unsound. In his famous classic on the conduction of heat,[17] Fourier proved himself the Euler of mathematical physics. Leaving convergence to take care of itself, he trusted his physical intuition to lead him to correct results, as it usually did.

The sixth section of Fourier's *Théorie*[17] is the one which concerns us here. It is devoted to the solution "of a more general problem, which consists in developing any function whatever in an infinite series of sines or cosines of multiple arcs. . . . We proceed to explain the solution."[18] Having done so for a special case, Fourier continues,[19] "We can extend the same results to any functions, even to those which are discontinuous and entirely arbitrary. To establish clearly the truth of this proposition, we must examine the foregoing equation," which he does, in the manner of Eulerian formalism.[20] The outcome is the expansion of an 'arbitrary' odd function in a sine series. Lagrange in 1766 had constructed by a process of interpolation a finite summation formula from which Fourier's result is obtainable by a leap into infinity, but he "abstained from the transition from this summation formula to the integration formula given by Fourier."[21] Lagrange's difficulty was that he had a mathematical conscience. Physical intuition supplied Fourier's lack of mathematical inhibitions and guided him to the general statement of his famous theorem.

The mathematical physicist's boldness taught pure mathematicians several things of the first importance for the future of analysis. Purists gradually came to realize that their intuitions of 'arbitrary' function, real number, and continuity needed clarification. P. G. L. Dirichlet's (1805–1857, German) definition[22] (1837) of a (numerical-valued) function of a (real, numerical-valued) variable as a table, or correspondence, or correlation, between two sets of numbers hinted at a theory of equivalence of point sets. When G. F. B. Riemann (1826–1866, German) in 1854 investigated[23] the representation of a function by a trigonometric (Fourier) series, he discovered that Cauchy had been too restrictive in his definition of an integral, and showed that definite integrals as limits of sums exist even when the integrand is discontinuous. Later (date uncertain) he invented a function, defined by a trigonometric series, which is continuous for irrational values of the variable and discontinuous for rational values.[24] It was clear that the continuum of real numbers had not been thoroughly understood. With our present knowledge we see once more what Cantor was the first to perceive, the necessity for a theory of sets of points. Cantor's investigations, like Riemann's, began in Fourier series.

The demand for clearer understanding of limits, continuity, and derivatives was further emphasized in 1874 by the publicity given to Weierstrass' example of a continuous function having no derivative or, what is equivalent, of a continuous curve admitting no tangent at any point. Intuition all but expired.

Such appear to have been the principal impulses behind the creation of the modern continuum. The unforeseen phenomena cited, and many others almost equally unexpected but of the same general character, seemed to indicate that all the difficulties were ultimately rooted in the real number system. Urged by this conviction, Dedekind, Cantor, and Weierstrass, by different methods but with a common aim, returned to the problem of Eudoxus and stripped it of its disguised intuitive geometry. 'Magnitudes,' as we have seen, were replaced by 'numbers,' and geometrical intuitions were driven out to make room for those of traditional logic. Nebulous 'quantities' persisted in the analysis of some. In the numerical epsilons and deltas of rigorous Weierstrassian analysis the calculus of the nineteenth century attained its classic perfection. The ϵ, δ technique became part of the standard equipment of every working analyst, and an advanced course in the calculus toward the end of the century usually included the rudiments of Cantor's theory of sets.

Finality in 1900

In the preceding chapter we followed the development of the real number system to the close of the nineteenth century, and we have just seen that one origin of the modern concept of real numbers was analytic necessity. In the retreat of geometric and kinematic intuition to the classical logic which validated the work of Dedekind, Weierstrass, and Cantor, the calculus returned at the close of the nineteenth century to the paradoxes of the infinite that had exercised generation after generation of logicians from Zeno to Russell. Before further progress was possible, a more subtle logical technique had to be developed in the twentieth century, and this was forthcoming only when the symbolic logic prophesied by Leibniz was extended and refined far beyond his utmost imaginings. Thus, after two centuries, the calculus returned for new strength and health to one of the minds from which it had sprung. What it received will be our concern after we have reviewed some of the triumphs of analysis in the two centuries following Newton and Leibniz.

For the moment, we recall the benediction pronounced by Henri Poincaré (1854–1912, French) at the second international congress of mathematicians in 1900. On this historic and somewhat solemn occasion, Poincaré, the outstanding mathematician of his epoch and the Lagrange of the nineteenth century, contrasted the roles of intuition and logic in mathematics. In particular he reviewed the movement which has just been sketched and which, in the late nineteenth century, was called the arithmetization of analysis. The comforting assurances of this bold master of analysis induced a warm glow of security and pride in all who heard him, or who read his memorable address, and who, at least temporarily, had forgotten all they knew of mathematical history.

Having recalled[25] that mathematicians had once been content with the ill-defined and rough images of things mathematical as they appear to the senses or the imagination, Poincaré credited the logicians, for whom he had a dislike sharpening occasionally into acid ridicule, with having remedied this unsatisfactory state of affairs. Likewise, he continued, for irrational numbers and "the vague idea of continuity that we owe to intuition," now (1900) resolved into "a complicated system of inequalities concerning integers." By such means, he declared, all difficulties concerning limits and infinitesimals had been clarified.

> Today [1900] there remain in analysis only integers and finite or infinite systems of integers, inter-related by a net of relations of equality or inequality. Mathematics, as we say, has been arithmetized.
>
> . . . Is this evolution ended? Have we at last attained absolute rigor? At each stage of the evolution our fathers believed that they too had attained it. If they deceived themselves, do not we deceive ourselves as they did?
>
> We believe that we no longer appeal to intuition in our reasoning. The philosophers tell us that this is an illusion . . .
>
> Now, in analysis today, if we care to take the pains to be rigorous, there are only syllogisms or appeals to the intuition of pure number that could possibly deceive us. We may say today [1900] that absolute rigor has been attained.

Here we may refer to the last section of the preceding chapter. Some of the concrete achievements of analysis in applied mathematics will be discussed later.

CHAPTER 14

Rational Arithmetic after Fermat

We shall conclude our account of number since the seventeenth century with a few typical items from the vast domain of the classical theory of numbers. Arithmetic in the tradition of Fermat, Euler, Lagrange, Legendre, and Gauss has been concerned mainly with the rational integers $0, \pm 1, \pm 2, \ldots$. Although it has attracted several of the greatest mathematicians since the seventeenth century, rational arithmetic has had far less influence than its nineteenth-century offshoot, the theory of algebraic numbers, on the rest of mathematics. Intensively cultivated for its own fascinations by hundreds of mathematicians of very different tastes, rational arithmetic has developed into an ever-growing expanse of loosely coordinated results with fewer general methods than any other major division of modern mathematics.

From all this heterogeneous miscellany we shall select only three topics in which there is some coherence of method and an approach to completeness in certain details. The rest is largely a wilderness of dislocated facts offering a strange and disconcerting contrast to the modernized generality of algebra, geometry, and analysis. Much of it is hopelessly archaic in both aim and results. Rational arithmetic appears to be the one remaining major department of mathematics where generalizing a problem makes it harder instead of easier. Consequently it has attracted fewer merely able young mathematicians than any other.

The subject falls naturally into the complementary divisions of multiplicative and additive arithmetic. The multiplicative theory develops the consequences of unique factorization into primes; the additive division is concerned with the composition of integers as sums of prescribed types. A capital project in both divisions is that of enumeration: how many integers of a specified kind satisfy given conditions? For example, how many primes are there between given limits? Or in how many ways may any integer be represented as the sum of a fixed number of positive cubes?

A problem in rational arithmetic is said to have been solved when a process is described whereby the required information is obtainable by a finite number of non-tentative operations. Time certainly is not the essence of the contract between the rational integers and the human intellect. The problem of resolving a number into its prime factors is solvable; yet the finite number of operations at present required for a number of a few thousand digits might consume more ages than our race is likely to have at its disposal.

The problem of finding the prime factors of a number must strike an amateur as a natural one. To say that it has been solved in any respect that would satisfy common sense is a flattering exaggeration, and the like is true of many other arithmetical problems that seem natural to the inexperienced. The professional ignores these natural problems in favor of others which he or his predecessors have constructed, and for which he may hope to find at least partial solutions. Complete solutions, even of manufactured problems, are comparatively rare; and it would seem that rational arithmetic in the twentieth century is still relatively in the same position as geometry was before Descartes. Compared with what we should like to know in each of several directions, such progress as has been made is almost negligible. Yet all the resources of algebra and analysis have been hurled into the assault on this most elementary of all divisions of mathematics.

Outgrowths of diophantine analysis

The nature of diophantine analysis has already been described in connection with Fermat. Its most extensive outgrowth, the arithmetical theory of quadratic forms, slowly took shape during the eighteenth century, principally in the prodigious output of Euler and the more restrained contributions of Lagrange and Legendre. Finally, in 1801, with the publication by Gauss of his *Disquisitiones arithmeticae*, diophantine analysis[1] in the sense of Fermat and Euler suffered an eclipse that was to last a century, until arithmeticians began to realize that the

Gaussian theory of quadratic forms does not exhaust the subject of indeterminate equations.

The second great branch of modern arithmetic that sprang from diophantine analysis was the theory of congruences. This also originated in the *Disquisitiones*. The suggestiveness of Gaussian congruence for modern algebra and the development of structural theories was noted in an earlier chapter.

After Diophantus and Fermat, Euler was the great master of indeterminate analysis. But, like nearly all the predecessors of Lagrange, Euler contented himself with special sets of integers or of rational numbers satisfying his equations. The only interest such work has had at any time since Lagrange's discussion of $x^2 - Ay^2 = 1$ in 1766–9 is in showing that a particular equation, or set of equations, with integer coefficients is in fact rationally or integrally solvable when the existence of a solution has been doubted. Thus a single numerical instance would dispose of the doubt (1945) concerning the solvability of $x^4 + y^4 + z^4 = w^4$, $xyzw \neq 0$, in integers. In a modern setting, this problem is equivalent to determining the number of representations of zero in the quartic form $x^4 + y^4 + z^4 - w^4$ and, if there are any, finding all. Euler (1772) conjectured that there are no solutions.

The daring of this baseless conjecture typifies the cardinal distinction between indeterminate analysis before Lagrange and after him. Euler and others in the older tradition did not hesitate to suggest problems of great difficulty without offering the slightest suggestion for a method of attack. And when ingenuity furnished special solutions of an equation, the solver dropped the matter. Lagrange was the first to impose some common mathematical morality on diophantine analysis. He refrained from facile guessing; and when he did propose a problem, he also invented methods for obtaining its solutions.

The turning point is marked by Fermat's equation[2]

$$x^2 - Ay^2 = 1,$$

where A is any positive non-square integer, and all integer solutions x, y are sought. Fermat (1657) asserted that there are an infinity of solutions, a fact which Lord Brouncker and J. Wallis were unable to prove, although they gave a tentative method, improved in 1765 by Euler, for finding solutions. Euler proceeded from the conversion of $\sqrt{A}$ into a continued fraction. But he was unable to prove the existence of a solution with $y \neq 0$. Lagrange (1766–9) supplied the crucial proof, and in 1769–70 gave a non-tentative method for obtaining all integer solutions of $x^2 - Ay^2 = B$, where A, B are any given integers.

It was noted in earlier chapters that the Pythagoreans approximated to quadratic irrationalities by what amounts to solving special cases of Fermat's equation by continued fractions, and that Brahmagupta in the seventh century gave a tentative method for solving $x^2 = Ay^2 + B$ in integers. But the mathematical distance between such empirical work as this and Lagrange's proofs of necessity and sufficiency is immeasurable; and it is fantastic to claim that the Hindu mathematicians anticipated Lagrange. There is honor enough for Brahmagupta and Bhaskara in having imagined a problem that, centuries after they were dead, was to prove of cardinal importance in modern arithmetic. But in this they may have been merely lucky, for they devoted much time to numerous other problems that are essentially trivial. Fermat's equation and its solution by Lagrange are indispensable in the Gaussian theory of binary quadratic forms, also in that of algebraic number fields of the second degree. Lagrange's solution was the first determination of the units in an algebraic number field other than the rational.

In addition to being haphazard, Euler's attack on diophantine equations was absurdly ambitious. If a single equation of the second degree in two unknowns proved unexciting, Euler increased the degree to three or four. If this failed to provide an attractive equation, he simultaneously increased the number of unknowns. As a last resort, he increased the number of equations and exercised his unequaled ingenuity on simultaneous systems. It is not surprising that he made but little progress toward either general methods or general theorems. Nor did any of his hundreds of successors who equaled or excelled him in ambition, but who fell far short of him in ingenuity.

Advances toward real mathematics began when unambitious men like Lagrange and A. M. Legendre (1752–1833, French) confined their main efforts to the humble task of systematically investigating a single equation of the second degree in not more than three unknowns. Their work smoothed the way for Gauss, who also set himself a program which, compared to the rank opulence of the pre-Lagrangian period, is poverty itself. And without the pioneering work of Lagrange and Legendre, it is at least doubtful whether even Gauss would have been able to compose the *Disquisitiones*.

Arithmetical forms

The basic technique of the arithmetic of forms originated with Lagrange's theory of binary quadratics, in 1773, four years before Gauss was born. To describe it we shall use the standard terminology introduced in 1801 by Gauss and modified by later arithmeticians. Several of the definitions given next will be useful in slightly modified shape when we come to invariance.

A form in rational arithmetic is a homogeneous polynomial $P, \equiv P(x_1, \ldots, x_n)$, in the n indeterminates (or variables) $x_1, \ldots, x_n$, with integer coefficients. If the degree is m, the form is called an n-ary m-ic. For $n = 2, 3, 4, 5, \ldots$ the forms are called binary, ternary, quaternary, quinary, . . . , respectively. In what follows, 'form,' unqualified, shall mean an n-ary m-ic. The fundamental concepts are equivalence and reduction of forms, and representation by a form.

The form $P(x_1, \ldots, x_n)$ is said to contain the form

$$Q(x_1', \ldots, x_n')$$

if Q is derived from P by a linear homogeneous substitution $T{:}x_i = a_{i1}x_1' + \cdots + a_{in}x_n' (i = 1, \ldots, n)$ with integer coefficients a_{ij} whose determinant $|a_{ij}|$ is not zero. If $|a_{ij}| = \pm 1$ the inverse, T^{-1}, of T, expressing $x_1', \ldots, x_n'$ as linear homogeneous functions of $x_1, \ldots, x_n$, will have integer coefficients, and Q will contain P. When each of two forms thus contains the other, the forms are said to be equivalent; the equivalence of P, Q is written $P \sim Q$. It follows readily that this $\sim$ is an instance of the abstract equivalence described in connection with Gaussian congruence. For P is either equivalent or not equivalent to Q; $P \sim P$; and if $P \sim Q$, then $Q \sim P$; also, $P \sim Q$ and $Q \sim R$ together imply $P \sim R$. Hence all forms equivalent to a given form are equivalent to one another; and therefore all forms may be separated into classes with respect to equivalence, two forms being put into the same class if and only if they are equivalent.

The link with diophantine analysis is supplied by the concept of representation: an integer r is said to be represented by (or in) the form $P(x_1, \ldots, x_n)$ if and only if the equation

$$P(x_1, \ldots, x_n) = r$$

is solvable in integers $x_1, \ldots, x_n$. If $x_1 = s_1, \ldots, x_n = s_n$ is such a solution, $(s_1, \ldots, s_n)$ is called a representation of r in P. The diophantine problem, as reformulated in 1773 by Lagrange, is to decide whether or not a given r is represented in P and, if it is, to find all representations.

It follows at once from the definitions that if r is represented by a particular form in a given class, it is represented by every form in that class; and that if it is not represented by a particular form in the class, it is represented by no form in the class. The diophantine problem of finding all integer solutions of

$$P(x_1, \ldots, x_n) = r$$

is thus reduced to two others: to assign criteria, expressed in terms of the given coefficients of P, sufficient to decide whether or not r is represented by P; to find all the forms equivalent to P. The second of these suggests as a preliminary a third: given the coefficients of two forms, to determine whether the forms are equivalent and, if so, to transform one into the other. This in turn requires the automorphs of a given form, namely, those transformations which leave a form unaltered. Once the automorphs of P and one transformation taking P into Q are known, all such transformations are known. For binary quadratic forms, the automorphs are obtained by solving certain of Lagrange's equations $x^2 - Ay^2 = B$.

The remaining problem of modernized diophantine analysis, that of the reduction of forms, is on a different level. Suppose that in each class of forms it is possible to isolate a unique form by imposing appropriate conditions on the coefficients of all the forms in the class. Then this so-called reduced form, being equivalent to every form in the class, may be taken as a representative of its entire class in the problems of equivalence and representation of numbers. Thus attention may be concentrated on individual forms instead of being dispersed over possible infinities of forms in the different classes. Incidentally, the problem of determining the number of classes of forms whose invariants have any preassigned integer values is suggested. Lagrange solved the problem of reduction for binary quadratics in 1773; a solution for ternary quadratics was first obtained by L. A. Seeber (German) in 1831.

A little trial and a great deal of error will readily convince any experimenter that a complete solution of these basic problems is not to be anticipated in the immediate future. Nevertheless, their mere formulation was a notable achievement. If nothing else, they stripped the ancient diophantine analysis of its specious simplicity and revealed the nature of its inherent difficulties. In this respect they are an outstanding example of mathematical strategy as practiced by masters.

These modern, clearly defined problems may prove to be so intractable in the general case that they will be abandoned. The entire program of a frontal attack on diophantine analysis has been questioned. Our successors may be forced to resume the classical technique of manufacturing problems which they can solve. We recall that diophantine analysis originated in the Pythagorean equation $x^2 + y^2 = z^2$; and it is conceivable that the generalized problem which sprang from this equation is artificial. The Pythagorean equation entered mathematics through geometry, not through arithmetic. A generation less respectful of tradition than ours may succeed in formulating and solving problems closer in some as yet unimagined sense to the nature of rational arithmetic, whatever that may be. In any event, interest in the modernized problems of diophantine analysis described above declined rapidly toward the close of the nineteenth century. The problems were simply too hard; and all the impressive machinery of modernized algebra and analysis succeeded only in making a great clatter which failed to silence the insistent questionings of arithmetic.

By far the major part of all the advances was in the theory of quadratic forms. A rough estimate of the amount of work done on the several kinds of forms gives eighty per cent in quadratic and twenty per cent in all others. Of the work on quadratics, about eighty per cent was devoted to binaries, eight per cent to ternaries, three per cent to quaternaries, and three per cent to n-aries. The remaining six per cent on quadratics was accounted for by binaries with coefficients in a few special quadratic fields. These statistics suggest that the general program of an arithmetical theory of forms was still largely a hope after over a century and a half of industrious exploitation by several hundred arithmeticians, including such men as Lagrange, Legendre, Gauss, Eisenstein, Dirichlet, Hermite, H. J. S. Smith, Minkowski, and Siegel.

We shall now indicate briefly a few of the outstanding landmarks in the theory of forms of low degree. By his general treatment of binary quadratics (1773), Lagrange obtained incidentally and uniformly many of the special results of his predecessors, such as Euler's theorem (1761) that every prime $6n + 1$ is represented by $x^2 + 3y^2$. Lagrange's principal achievement, however, was the introduction of universally applicable methods into the theory of binary quadratic forms.

Legendre in 1798 published his *Théorie des nombres*, the first treatise devoted exclusively to the higher arithmetic, in which Lagrange's theory was simplified and extended. This work contains the earliest systematic attack on ternary quadratics. Much use was made of the law of quadratic reciprocity, of which the first complete proof was published by Gauss in 1801.

With the *Disquisitiones arithmeticae* (1801), the theory of binary quadratics crystallized into its classic shape. Systematizing and completing details in the work of his predecessors, Gauss also added many new ideas of his own. Among the innovations was one which was to prove most unfortunate: Gauss constructed his entire theory on forms $ax^2 + 2bxy + cy^2$ with a, b, c integers. The even middle coefficient $2b$ makes the accompanying algebra more elegant but needlessly complicates the arithmetic and leads to cumbersome refinements in classification. To an algebraist this may seem a trivial detail. But a moment's reflection will show that as the subject under investigation is rational arithmetic and not algebra, the insistence that the middle coefficient be even is likely to cause unavoidable complications.

Modern practice for binary quadratics (and to a lesser extent for ternaries), following Kronecker, has returned to the unrestricted integer coefficients of Lagrange. Consequently it is necessary to retain two vocabularies and to know which is being used in referring to papers on the subject.

The first man to master the synthetic presentation of Gauss was Dirichlet, who in 1863 summarized his personal studies and his recasting of the *Disquisitiones* in his *Zahlentheorie*. The successive editions (1871, 1879, 1893) of this text[3] and Dirichlet's earlier original contributions made the classical arithmetic of Gauss accessible to all without undue labor. A more significant advance of Dirichlet's in arithmetic generally will be noted later in connection with the analytic theory.

Up to 1847 the *arithmetical* theory of quadratic forms had been confined to binaries and ternaries. It might be thought that the extension to quadratics in 4, 5, 6, . . . indeterminates would be a matter of simple routine, like the passage from three dimensions to n in analytic geometry. Hard experience quickly corrects this misapprehension; the difficulties of a detailed investigation increase rapidly with the number of indeterminates and even necessitate the invention of new principles.

The first significant departure from the tradition of binaries and ternaries was F. M. G. Eisenstein's (1823–52, German) *arithmetical* determination in 1847 of the number of representations of an integer as a sum of six or eight squares. This was followed in 1847 and 1850 by an arithmetical determination of the number of representations of an integer without square factors as a sum of five or seven squares. In all cases only results were indicated, with no hint of the methods used. There is no doubt, however, that Eisenstein's procedure was purely arithmetical and not analytic. Although all his results have long been details in the general theory of n-ary quadratic forms, they are of more than casual interest historically, as it was partly owing to them that the arithmetical theory was created.

To trace the development, we must return to the determination of the number of representations of an integer as a sum of two squares by Legendre in 1798 and, more simply, by Gauss in 1801, and to Euler's unsuccessful struggle for forty years to prove that every positive integer is a sum of four integer squares. Euler's failure was the steppingstone to Lagrange's success in 1772, and thence to his own a year later. But neither obtained the number of representations. Quite unexpectedly the required number dropped out as an unsought by-product of an identity in elliptic theta constants, which Jacobi encountered in 1828 while developing the theory of elliptic functions. The like results for 2, 6, 8 squares are evident from other formulas in Jacobi's *Fundamenta nova* of 1829. Those for an odd number of squares, lying much deeper, do not follow from similar identities. In passing, the problem of three squares was a famous crux in the arithmetical theory of quadratic forms until Legendre in 1798 published the first proof that all positive integers except those of the form $4^h(8k + 7)$ are sums of three integer squares.

From all this it is clear that Eisenstein made a significant advance when he obtained his results for five and seven squares arithmetically. Possibly it was this work that moved Gauss to assert that "There have been only three epoch-making mathematicians, Archimedes, Newton, and Eisenstein." If Gauss ever did say this (it is merely attributed to him), it is the most astounding statement in the history of mathematics. But as he may have said it, and as anything Gauss said about mathematics is to be taken seriously, we may briefly examine its tenability

Like Abel and Galois, Eisenstein was an "inheritor of unfulfilled renown," also of poverty and ill-health, and there is no guessing what he might have accomplished had he lived. But he enjoyed (after a fashion) about two years more of life than Abel, and eight more than Galois. His principal achievement outside of arithmetic was in elliptic functions, where he partly anticipated certain details of the Weierstrassian theory. His own analysis halted under the fatal disability of conditional convergence.[4] On the other hand, his applications of elliptic functions to the laws of cubic and biquadratic reciprocity were among the least expected things in arithmetic. Against Gauss' assertion are the facts that for one reference in living mathematics to Eisenstein, there are hundreds to Abel and Galois, to say nothing of Riemann and Dedekind, who were pupils of Gauss, or Eisenstein's less famous contemporary Kummer. Even in the narrowly limited domain of arithmetic, Eisenstein's influence has been slight in comparison with that of Dedekind. In the older form, now obsolete, of the theory of higher reciprocity laws, Eisenstein's work of 1850 was vital; but here again the generative concept of prime ideal divisors was Kummer's. It seems unlikely, then, that posterity will revise the almost universal verdict of 1945, that up till then the three epoch-making mathematicians were Archimedes, Newton, and Gauss.

The principal steps toward a general theory of n-ary quadratics appear to have been as follows. Perfecting and greatly extending the theory for $n = 3$ as left by Gauss in the *Disquisitiones*, Eisenstein in 1847 introduced new principles for the classification of ternaries into orders and genera. Hermite in 1850 simplified the theory of reduction for ternaries, and in 1851 devised his general analytic method of continual reduction. The theory of ternaries was further developed by Eisenstein in 1851–2; by H. J. S. Smith (1826–1883, Irish) in 1867; by E. Selling (German) in 1874, and by many others in the 1850's–70's. In 1864 and again in 1867, Smith initiated one form of the general theory of n-ary quadratics from which Eisenstein's theorems on five and seven squares were easily obtainable. Owing partly to the conciseness of the exposition, these and other detailed consequences were overlooked, and the problem of five squares was proposed by the French Academy for its Grand Prix in 1882. Brevity in mathematics is sometimes the soul of obscurity. Smith elaborated the relevant parts of his general theory of 1864–7, and shortly after his death shared the prize with H.

Minkowski (1864–1909, Russian; Germany) then a student of eighteen at the beginning of his too brief career. Thus, after an unnecessary delay of over a quarter of a century, the general arithmetical theory of n-ary quadratics was launched with complete éclat.

Poincaré, Minkowski, and others further developed the theory in the two succeeding decades. With the exception to be noted immediately, little that could be considered basically new was done till C. L. Siegel (German) in 1935 gave a profound reworking of the entire theory.

The new acquisition, that of the geometry of numbers, was created almost entirely by Minkowski, although special instances[5] of it occur in the early (posthumously published) work of Gauss; in a project of Eisenstein's (1844); in Dirichlet's (1849) asymptotic evaluations of sums of arithmetical functions; in the work already cited on the reduction of ternary quadratics; and in the semi-geometrical presentation of the theory of elliptic modular functions by H. J. S. Smith in 1876, and the similar revision of binary quadratics by Poincaré in 1880.

One basic principle is so simple as to appear ridiculous: if $n + 1$ things are stored in n boxes, and no box is empty, exactly one of the boxes must contain two things. The solution of a trick problem popular some years ago follows from this principle of geometrized arithmetic: state necessary and sufficient conditions that there shall be at least two human beings in the world with the same number of hairs on their heads.

The first published results in the geometry of numbers appear to be Eisenstein's geometrical proof (1844) of the Gauss lemma for the proof of the law of quadratic reciprocity, and his formula (1844) for the number of solutions of $x^2 + y^2 \leqq n$ in integers x, y, where n is given. A lattice point being defined as a point whose coordinates are integers, the last is equivalent to finding the number of lattice points contained by a circle, including its circumference, with center at the origin and radius $\sqrt{n}$. Minkowski developed meager hints like these into a powerful method which was applied with conspicuous success by himself and many others to difficult questions in the theories of forms, especially linear forms with real coefficients, and algebraic numbers. It is not necessary, of course, to restate an arithmetical problem geometrically; but doing so suggests to those with spacial intuition in n dimensions analytic processes which they might not imagine otherwise. With or without appeal to spacial imagery, the type of problem suggested by the geometry of numbers inspired much work in the analytic arithmetic developed since about 1910 by the schools of E. Landau (1877–1938, German), G. H. Hardy (1877–, English), J. E. Littlewood (1885–, English), and S. Ramanujan (1887–1920, Indian) in England. Arithmetic thus repaid its heavy debt to analysis by showering some of the foremost classical analysts of the twentieth century with an abundance of difficult problems.

The arithmetical theory of forms of degree higher than the second was responsible for much less. Eisenstein initiated (1844) the theory of binary cubics, and in so doing came upon the first algebraic covariant in history. But he did not exploit his discovery, although he realized its suggestiveness. The arithmetic of binary cubics was reworked by the British mathematicians G. B. Mathews and W. E. H. Berwick in 1912.

Progress up to 1945 in the arithmetic of forms beyond this point, with three exceptions, was inconsiderable. The norm of an algebraic integer is the product of all its conjugates; the norm equated to unity defines the units of the field concerned. Dirichlet (1840) proved the basic theorems for such units, although there is yet no practicable way of obtaining them even in special cases for fields of degree higher than the third. This work generalized Lagrange's on Fermat's equation. Dirichlet's problem of units and its immediate extension to the representation of any number by a general norm are the most immediate generalization of the theory of binary quadratics. Today they are a topic in algebraic numbers. The origin of this farthest outpost in the systematized theory of forms was Lagrange's remark (1767) that the norm of a general algebraic number repeats under multiplication, and this in turn can be traced back to Fibonacci's identity. It may interest some to know that Dirichlet's inspiration came to him in church while he was listening to the music on an Easter Sunday.

The second exception to the general rule of sterility beyond real quadratics was Hermite's introduction in 1854 and 1857 of the forms since known by his name. In the binary case, a Hermitian form is of the type $axx' + bxy' + b'x'y + cyy'$, where a, c are real constants, b, b' conjugate imaginary constants, and the variables in the pairs x, x' and y, y' are conjugate imaginaries, so that the entire form is real and hence capable of representing real numbers. From Hermite's arithmetical theory of these forms in two or more variables evolved the extensive theory of

Hermitian forms and matrices, which after 1925 became familiar to physicists through the revised quantum theory. Hermite also (1849) initiated the closely related arithmetical theory of bilinear forms,[6] thus starting much algebra that is now standard in a college course, including parts of the theory of matrices and elementary divisors. The last originated explicitly in H. J. S. Smith's discussion of systems of linear diophantine equations and congruences (1861), and was developed independently by Weierstrass and by G. Frobenius (1849–1917, German) in the 1870's–80's. The point of historical interest here is that all these extremely useful techniques of modern algebra, which after 1925 became commonplaces in mathematical physics, evolved from quite useless problems in the theory of numbers.

The third and last exception to general sterility connects the arithmetic of forms with that other major outgrowth of ancient diophantine analysis, the Gaussian concept of congruence. Dickson in 1907 began the congruencial theory of forms, in which the coefficients of the forms are either natural integers reduced modulo p, p prime, or elements of a Galois field. The linear transformations in the theory, corresponding to those in the classical problem of equivalence, were similarly reduced, and hence modular invariants and covariants were definable. By 1923 the theory was practically worked out, except for two central difficulties, by Dickson and his pupils. Simplified derivations for some of the results were given (1926) by E. Noether by an application of her methods in abstract algebra.

Before passing on to congruences, we note an outstanding advance in the older tradition of diophantine analysis. If

$$f(z) \equiv a_n z^n + a_{n-1} z^{n-1} + \cdots + a_1 z + a_0$$

is an irreducible polynomial of degree $n \geqq 3$ with integer coefficients, and if

$$H(x, y) \equiv a_n x^n + a_{n-1} x^{n-1} y + \cdots + a_1 x y^{n-1} + a_0 y^n$$

is the corresponding homogeneous polynomial, then $H(x, y) = c$, where c is an integer, has either no solution or only a finite number of solutions in integers x, y. This is the capital theorem (1909) of A. Thue (Scandinavian). A generalization[7] was given by Thue himself, and another (1921) by Siegel. After all that has been said about the paucity of general methods and the plethora of fragmentary results in diophantine analysis, Thue's theorem speaks for itself. It was proved by elementary methods.

The theory of congruences

The subject of congruences in rational arithmetic is usually assigned to the multiplicative division, although it is concerned chiefly with the detailed investigation of one highly specialized type of diophantine equation, $a_nx^n + \cdots + a_1x + a_0 = my$, in which x, y are the indeterminates, and the coefficients $a_n, \ldots,$ a_1, a_0, m are given constant integers, with $a_n \neq 0$, $m \neq 0$. The essential point is that one of the indeterminates, y, occurs only to the first degree. The cases $m = \pm 1$, being of no interest, are excluded. The equation rewritten as a congruence is

$$a_nx^n + \cdots + a_1x + a_0 \equiv 0 \bmod m;$$

m is called the modulus; n is the degree; and solving the diophantine equation is equivalent to finding all integer values of x, called the roots of the congruence, that make the polynomial on the left a multiple of m. If $x = c$ is a solution, so also is $c + km$, where k is any integer. Since $c + km \equiv c \bmod m$, it suffices to find all solutions whose absolute values do not exceed $|m|/2$. These are said to be incongruent modulo m. The statement of the generalized problem for one congruence in several indeterminates, or for a simultaneous system of such, is immediate. We refer here to what has already been said about congruences in connection with algebraic structure.

Gauss does not disclose[8] what led him to this cardinal concept of modern arithmetic. But its systematic use in the earlier sections of the *Disquisitiones* enabled him to unify and extend important theorems of Fermat, Wilson, Euler, Lagrange, and Legendre on arithmetical divisibility and, in the famous seventh (concluding) section, to give a reasonably complete theory of binomial equations in algebra. A few of the older results on divisibility may be restated in the language of congruences to illustrate the general ideas.

If p is prime, $x^{p-1} - 1 \equiv 0 \bmod p$, has exactly $p - 1$ incongruent roots. This is Fermat's theorem, proved essentially by the implied method of congruences by Lagrange (1771), who showed also (1768) that a congruence of degree n has not more than n incongruent roots for a prime modulus. The theory of residues of powers, originating with Euler in 1769, is concerned with the general binomial congruence $x^n - a \equiv 0 \bmod m$. It has numerous applications in algebra, particularly in the theory of equations and in finite groups. One crucial unsolved problem may be noted. If p is prime, r any number not divisible by p, and if $p - 1$ is the least value of n for which $r^n - 1 \equiv 0 \mod p$, r is called a primitive root of p. A prime p always has exactly $\phi(p - 1)$ primitive roots, where $\phi(n)$ is Euler's function denoting the number of positive integers not greater than n and prime to n. The problem is to devise a practicable, non-tentative method for finding at least one primitive root of any given prime. Between Euler's initiation of the subject in 1769 and 1919—a century and a half—232 lengthy articles and short notes were published on binomial congruences. None made any substantial progress toward a solution of this crucial problem.

Another outgrowth of Euler's power-residues is the entire theory of reciprocity laws, already noted in connection with modern algebra. Yet another is the extensive theory of functions defined for integer values of their variables, all of which evolved from the theorem of Gauss that $\Sigma\phi(d) = n$, where the summation extends to all divisors d of the fixed integer n, and ϕ is Euler's function.

Practically every arithmetician of note and a host of humbler workers have contributed to the theory of congruences since Gauss started the subject in 1801. For all their efforts, two central problems of the theory defy solution: to assign criteria on the given coefficients of a system (one or more) of congruences to decide whether or not the system is solvable and, if it is, to find all its incongruent solutions non-tentatively. For a single congruence of the first degree in any number of unknowns, the problem is more completely solved than are most of the solved problems in arithmetic, and likewise for a simultaneous system of such congruences, the solution having been given (1861) by H. J. S. Smith. The higher reciprocity laws represent the farthest advance in the case of binomial congruences; their complexity may hint that the general problem is intractable by methods known up to 1945. It must be remembered that these problems and others like them in classical arithmetic have not been the easy sport of mediocre men; some of the most powerful mathematicians in history have wrestled with them.

Congruences were responsible for one theory of far more than merely arithmetical interest. The notation for a congruence suggests the introduction of appropriate 'imaginaries' to supply the congruence with roots equal in number to the degree of the congruence when there is a deficiency of real roots. As in the corresponding algebraic problem, it is not obvious that imaginaries can be introduced consistently. That they can, was first proved in 1830 by Galois, who invented the required 'numbers,' since called Galois imaginaries, for the solution of any irreducible congruence $F(x) \equiv 0 \bmod p$, where p is prime. He thus obtained a generalization of Fermat's theorem, and laid the foundation of the theory of finite fields. As remarked by Dickson,[9] "Galois's introduction of imaginary roots of congruences has not only led to an important extension of the theory of numbers, but has given rise to wide generalizations of theorems which had been obtained in subjects like linear congruence groups by applying the ordinary theory of numbers." Galois was eighteen when he invented his imaginaries.

We pass on to the third and last division of rational arithmetic which we shall consider. Here great progress has been made, most of it since 1895.

Applications of analysis

Since the time of Euler, analysis has been applied incidentally to rational arithmetic; but it was only in 1839, with Dirichlet's *Recherches sur diverses applications de l'analyse infinitésimale à la théorie des nombres*, that limiting processes entered organically into the theory of numbers. Before Dirichlet, such analysis as was used remained in the background,[10] arithmetical results being obtained by the device of comparing coefficients in two or more expansions of a given function by different algorithms. This technique originated with Euler in his work of 1748 in the theory of partitions, a subject which he initiated in 1741.

After Dirichlet, the next organic uses of analysis were Hermite's (1851) in his method of continual reduction, and Riemann's (1859) in the distribution of primes. But it was not until the twentieth century that modern analysis was systematically applied to additive arithmetic. Simultaneously there was an unprecedented advance in the application of analysis to the multiplicative division. We shall presently describe only enough to illustrate the radical difference between the old and the new. With the exception of a very few outstanding acquisitions like Thue's theorem, it seems probable that the early twentieth century will be remembered in the future history of rational arithmetic chiefly for its achievements in the analytic theory.

There remains, however, what some consider a desideratum: to obtain those results of the analytic theory which do not involve a limiting process in their statement without an appeal to continuity. Thus Dirichlet proved analytically that there are an infinity of primes of the form $an + b$, where a, b are constant relatively prime positive integers and n runs through all positive integers. Attempts by Emmy Noether and others in the 1930's to obtain this non-analytic theorem without the use of analysis failed. On the other hand, all of Euler's and Jacobi's theorems on partitions, and all of the non-analytic theorems on numbers of representations in certain quadratic forms, first obtained analytically by Jacobi and others, have been proved without analysis. The reasons for failure in one instance and success in another superficially indistinguishable from it are not understood. It will be convenient to call demonstrably avoidable analysis inessential; and to speak of analysis which has not yet been proved inessential, or which leads to final results implying a use of continuity, as essential. Kronecker would probably not have admitted essential analysis into arithmetic, and might even have declared that its products are as inexistent as irrationals.

A classic example of the use of analysis later seen to be inessential occurs in many of the applications of elliptic and modular functions to the Gaussian theory of binary quadratic forms. In the hands of Kronecker, Hermite, and a score of less notable mathematicians, a close connection between binary quadratics and the theory of complex multiplication of elliptic functions was developed after 1860 into an extensive department of arithmetic. One detail of all this intricate theory exemplifies the analytic peculiarities. From a passage in the *Disquisitiones* it appears that as early as 1801 Gauss had effected the difficult determination of the number of classes of binary quadratics having a given determinant. The first published determination was Dirichlet's of 1839, in which analysis was essential. (An outstanding desideratum is a 'finite' proof of Dirichlet's results.) For forms of a negative determinant, Kronecker in 1860 found several remarkable formulas whereby the number of classes can be calculated recursively without analysis. These formulas appeared as by-products of Kronecker's investigations in elliptic functions, and were the heralds of several hundred by later writers, many of whom used elliptic modular functions to obtain their results. We shall not pursue this matter further here, as a detailed account belongs to the specialized theory of numbers and we can attend only to matters of more than local significance.

The point of interest here is that whereas analysis was essential in Dirichlet's derivation of the class-number, and might therefore have been reasonably expected to appear essentially in the deduction of the recurrence relations, it actually proved to be inessential for the latter. Arithmeticians who insist (there are such) that a method containing analysis essentially[11] belongs to analysis and not to arithmetic would claim that Kronecker's formulas, not Dirichlet's, are the arithmetical solution of the class-number problem.

There is more than a pedantic difference between the two opinions, at least historically. Experience has shown that the search for proofs and theorems independent of essential analysis frequently turns up unexpected simplicities and reveals new arithmetical phenomena. Gauss emphasized the desirability of multiplying proofs in arithmetic with a view to making the abstruse clear. However, arithmetic is sufficiently broad and difficult to permit all types of workers to follow their own inclinations. Beginning about 1917, the general trend was toward essential analysis.

The theory of partitions illustrates the historical discontinuity between inessential and essential analysis. If $P(n)$ denotes the total number of ways the positive integer n is obtainable as a sum of positive integers, it is obvious, as noted by Euler in 1748, that $P(n)$ is the coefficient of x^n in the expansion of $\left[\prod_{r=1}^{\infty}(1 - x^r)\right]^{-1}$ into a power series in x. With unsurpassed manipulative skill, Euler derived numerous identities between this infinite product and others suggested by problems in partitions, thus anticipating many formulas in elliptic theta constants deducible from Jacobi's presentation (1828–9) of elliptic functions. An extensive literature sprang from these discoveries. Much of it is algebraic, and in none is analysis essential. Indeed, Sylvester, no analyst, desiring to understand the subject and being too impatient to master elliptic functions, developed a hint thrown out by N. M. Ferrers (1829–1908, English) in 1853 into a graphical theory in which some of the properties of partitions can be inferred from point lattices. But although the pictorial representation may have enabled Sylvester and some others to avoid inessential analysis in their thinking, it added nothing new to the theory of partitions.

Among other results of the Euler-Jacobi tradition were numerous elegant formulas whereby $P(n)$ and other partition functions could be calculated recurrently. Each afforded a complete arithmetical solution of the problem of computing the function of n concerned for any value of n. That is, all were useless as aids to practical computation for any but inconsiderably small numbers.

The break came in 1917, when G. H. Hardy (1877–, English) and S. Ramanujan (1887–1920, Indian) applied their new analytic methods to the derivation of an asymptotic formula for $P(n)$, which put the practical evaluation of $P(n)$ into touch with common sense. Before this, it had required a month's labor by a leading expert in the older methods to compute $P(200)$; only six terms of the asymptotic formula gave

$$P(200) = 3{,}972{,}999{,}029{,}388$$

with an error of .004. This detail typifies the computational superiority of formulas depending on essential analysis over the corresponding exact theorems preferred by pure arithmeticians. The analysis used is applicable to many other numerical functions appearing as coefficients in expansions of certain functions of a complex variable not continuable beyond the unit circle. This work of 1917, like most of the essential analysis in modern rational arithmetic, was of more than local interest in its own domain. It instigated a greatly increased activity in refined classical analysis and in the modern theory of inequalities.

One of the most famous problems in all arithmetic, that of the distribution of primes, yielded to analysis only in 1896, when J. Hadamard[12] (1865–, French) and C. J. de la Vallée-Poussin (Belgian) proved independently that the number $N(x)$ of primes $\leqq x$ is asymptotically equal to $x/\log x$, that is, the limit of $N(x) \cdot [x/\log x]^{-1}$ as x tends to infinity is 1. This is usually called the prime number theorem. The relevant history would fill a book.[13] Legendre, Gauss, and others had proposed formulas inferred from actual counts of primes; but it cannot be said that any of these tentative efforts materially furthered the final success. The great Russian mathematician, P. Tchebycheff (1821–1894) in 1850–1 made the first considerable advance since Euclid in the theory of primes; but much sharper analysis was needed than any available until the last decade of the nineteenth century. What appears likely to remain for some time the 'best' proof of the theorem is Landau's reformulation (1932) of that of N. Wiener (1894–, U.S.A.), who deduced the result almost as a corollary from his work on Tauberian theorems. The latter, so named by Hardy after the German analyst Tauber, evolved from the converse of Abel's theorem on convergent power series.

Even the briefest notice of the theory of primes must mention the famous conjecture known as Riemann's hypothesis, which is to classical analysis what Fermat's last theorem is to arithmetic. Euler (1737) noted the formula $\Sigma n^{-s} = \Pi(1 - p^{-s})^{-1}$, the sum extending to all positive integers n, and the product to all positive primes p. The necessary conditions of convergence hold for complex values of s with real part >1. Considering Σn^{-s} as a function $\zeta(s)$ of the complex variable s, Riemann (1859) proved that $\zeta(s)$ satisfies a functional equation involving $\zeta(s)$, $\zeta(1 - s)$, and the gamma function of s. He was thus led to the theorem that all the zeros of $\zeta(s)$, except those at $s = -2, -4, -6, \ldots$, lie in the strip of the s-plane (Argand diagram for s) for which $0 \leq \sigma \leq 1$, where σ is the real part of s. His theorems would be even more interesting than they are if all the zeros in the strip should lie on the line $\sigma = \frac{1}{2}$. Riemann conjectured[14] that this is so. Attempts to prove or disprove this conjecture have generated a vast and intricate department of analysis, especially since Hardy proved (1914) that $\zeta(s)$ has an infinity of zeros on $\sigma = \frac{1}{2}$. Although the question was still open in 1945, scores of profound papers bristling with thorny analysis had enriched the literature of analytic arithmetic for almost a third of a century, some of them based, however, on the supposition that Riemann's conjecture is true.

This bold technique of inference from doubtful conjectures was something of a new departure in arithmetic, where the tradition of Euclid, Lagrange, and Gauss had stickled for proof or nothing. The rationalized justification for the novel procedure, were any needed, was the unrealized hope that by transforming a dubious hypothesis into something new and strange, an accessible equivalent would sometime, somehow, drop out. Still on the assumption that Riemann's hypothesis and other unproved conjectures of a similar character are true, numerous profound theorems on the representation of numbers as sums of primes, or in other interesting forms, were skillfully deduced by some of the most refined analysis of the twentieth century. Should any of these boldly conceived but unborn theorems ever materialize, they will be among the most remarkable in arithmetic.

Adhering more strictly to the Euclidean tradition of proof before prophecy, the Russian mathematician I. M. Vinogradov, beginning about 1924, developed new methods in the analytic theory of numbers, and in 1937 apparently came within a reasonable distance of proving another famous guess concerning primes: every even number >2 is a sum of two primes. C. Goldbach (1690–1764, Russia) in 1742 confided this conjecture to Euler who, while believing it to be true, confessed his inability to prove it. Anyone who has inspected the analysis by which Vinogradov proved (1937) that every odd number beyond a certain point is a sum of three odd primes will sympathize with Euler. The best previous result was that of L. Schnirelmann (Russian), who proved (1931) that there is a constant n such that every integer >0 is a sum of n or fewer primes. But the method of proof, according to Landau, was incapable of further refinement. Vinogradov's theorem was conjecturally derived in 1923 by Hardy and Littlewood from an unproved mate of Riemann's hypothesis.

From 1896 till 1940 a major part of analytic arithmetic originated in the theory of rational primes. Some of this was extended to algebraic numbers, as when Landau (1903) obtained the prime ideal theorem corresponding to, and including, the prime number theorem. Here the necessary analysis proceeds from Dedekind's generalization (1877) of Riemann's $\zeta(s)$ to algebraic number fields. Another, less extensive, research generalized Dirichlet's work on the class-number of binary quadratics, being concerned with the number, proved to be finite, of distinct classes of (integral) ideals in an algebraic number field. The explicit determination of this number in an approachable form is one of the unresolved cruxes in arithmetic. All of this work belongs to the multiplicative division.

Equally prolific of new analysis and far-reaching theorems in arithmetic was the advance beginning in 1909 with Hilbert's solution of Waring's problem. The English algebraist E. Waring (1734–1798) emitted the conjecture (1770) that every integer $n > 0$ is the sum of a fixed least number $g(s)$ of sth powers of integers $\geqq 0$. For $s = 2$, this is the result proved by Lagrange and Euler that every positive integer >0 is a sum of four integer squares $\geqq 0$. Since no integer $4^h(8k + 1)$ is a sum of three squares, it follows that $g(2) = 4$; and it is known that $g(3) = 9$. Waring himself proved no single case of his problem; nor did he offer any suggestion for its solution. For all that he or anyone else in the eighteenth century knew, $g(s)$ might not exist.

It so happens, however, that Waring's guess was one of those few in the theory of numbers that have started epochs in arithmetic. Little of any significance issued from Waring's conjecture until about a century and a half after it had been made on only scanty numerical evidence. The theorem itself might be guessed after an hour's figuring. For example, $g(4)$ may be 19 as stated by Waring, a result which had not been proved as late as 1945.

It used to be imagined by romanticists that Waring and other rash guessers in arithmetic knew mysterious methods, now lost, of extraordinary power. There is no evidence that they did not. But professionals who appreciate the inherent difficulties of arithmetic believe that the lost methods, with the possible exception of Fermat's, are mythical. Gauss, for example, when urged in 1818 to compete for a prize offered by the French Academy for a proof or disproof of Fermat's last theorem, expressed himself quite forcibly on the undesirability of facile guessing in arithmetic. Including Fermat's theorem in his remarks, Gauss declared that he himself could manufacture any number of such conjectures which neither he nor anyone else could settle.

All questions of mathematical ethics aside, it is at least possible that stating difficult problems with no hint of a method for attacking them is more detrimental than advantageous to the progress of arithmetic. Unless we are eager to believe that certain individuals are divinely inspired and can foresee what course mathematics should follow to accord with the inscrutable verities of Plato's Eternal Geometer, we may suspect that baseless guessing is likely to deflect talented originality into artificial channels. Waring was an accomplished algebraist, but there is no evidence that he was inspired; and it seems like nothing but blind luck that his easy guess led to anything more profound than trivialities. That it did finally prove extremely stimulating may appear in the long run disastrously unfortunate. For there is little doubt that Waring's delayed success was largely responsible for the return about 1920 to the pre-Lagrangian tactic of deduction based on conjectures. Of course Lagrange and Gauss may have been mistaken or merely pedantic in their rejection of published guessing as a stimulus to progress, and the nineteenth-century caution may have been excessive. If so, the early twentieth century will doubtless be long remembered as the dawn of a new era in arithmetic.

Hilbert's proof (1909) of Waring's conjecture established the existence of $g(s)$ for every s, but did not determine its numerical value for any s. The curiously ingenious proof, shortly simplified by several mathematicians, depended on an identity in 25-fold multiple integrals, and like much of its author's mathematics aimed only at existence without construction. Its historical importance is less that it was the first solution of an outstanding problem than that it incited analysts to find at least a bound to the numerical value of $g(s)$ for any given s. It was the latter problem and the cognate one to be described presently that were largely responsible for the explosive outburst of analytic arithmetic in the 1920's-30's. As already implied, this work marks an epoch in the theory of numbers.

Hardy and Littlewood in 1920–8 invented the analytic method for Waring's problem which was to remain the standard till Vinogradov, having started in 1924 from methods similar to those of the English arithmeticians, developed his own more penetrating technique in the 1930's. The problem affiliated with $g(s)$ is that of finding $G(s)$, defined as the least integer n such that every positive integer beyond a certain finite value is the sum of n sth powers of integers $\geqq 0$. Thus the best value of $g(4)$ up to 1933 was $g(4) \leqq 35$, in contrast with Hardy and Littlewood's $G(4) \leqq 19$, while it was shown (1936) that $G(4)$ is either 16 or 17. For $s > 6$, Vinogradov's improved methods gave (1936) much smaller values of $G(s)$ than those obtained previously.

Although the pioneering methods were thus superseded, their influence on the development of asymptotic analysis remained incalculable. Utilizing results of Vinogradov's, Dickson and S. S. Pillai (1902–, Indian) in 1936 proved independently an explicit formula for $g(s)$, valid for all $s > 6$, except possibly for certain doubtful cases.[15] It is gratifying to report that since the first edition (1940) of this book, these doubtful cases have been disposed of (1943) by I. M. Niven (1915–, U.S.A.).

Thus, after 169 years Waring's guess was finally proved. In addition to instigating a vast amount of acute analysis, the problem had suggested numerous others solvable by similar methods, such as the representability of all, or 'almost all,' positive integers as sums of polynomials taking integer values for integer variables, or as sums of squares and primes. As this is written, there is no sign of an abatement in the output of analytic arithmetic. Two of the leading experts in the subject, Ramanujan and Landau, died before their time. The others more directly responsible for the creation of new methods were still active in 1945, and a crowd of younger men was coming on.

Another isolated result solidified by modern analytic methods may be cited for its curious implications regarding the suppressed work of the initiator of the modern era in arithmetic. C. L. Siegel (1896–, Germany, U.S.A.) in 1944 gave the first proof of a statement (1801) by Gauss concerning a certain asymptotic mean value in the theory of the binary quadratic class number. As the relevant formula could hardly have been inferred from numerical examples, it would be interesting to know how Gauss satisfied himself of its correctness. In any event, it is indicative of the lawless difficulties of the theory of numbers that a result stated by Gauss should have stood in the classic literature for 143 years without proof.

The 1920's–1930's witnessed the beginning of an era in arithmetic comparable to that inaugurated by Gauss in 1801. Analysis, the mathematics of continuity, had at last breached outstanding problems in the domain of the discrete. That explicit integer values for numerical functions like $g(s)$, $G(s)$ should be obtainable by analysis would have seemed miraculous to the arithmeticians of the nineteenth century. The like holds for the modern work in the theory of primes, and in other parts of multiplicative arithmetic. It is therefore not true, at least in the theory of numbers, that all the great mathematicians died before 1913.

In this chapter, as in others, literally hundreds of worthy names have been passed over without mention, and likewise for dozens of extensive developments to which scores of workers in the past two centuries have devoted their lives. The topics described are, however, a fair sample of some of the best that has been done in rational arithmetic since Fermat.

CHAPTER 15

Contributions from Geometry

With a literature much vaster than those of algebra and arithmetic combined, and at least as extensive as that of analysis, geometry is a richer treasure house of more interesting and half-forgotten things, which a hurried generation has no leisure to enjoy, than any other division of mathematics. Continually changing ideals and objectives in the development of geometry since the seventeenth century have made it impossible for students and working mathematicians to be aware of hundreds of theorems, and even extensive theories, that the geometers of the late nineteenth century prized as objects of rare beauty.

On a rather humble level, for example, it was held by competent geometers in 1940 to be a sheer waste of effort for a student contemplating a career in geometry, or in any division of living science or mathematics, even to glance at the so-called modern geometry of the triangle and the circle, created largely since 1870. Yet it has been said, no doubt justly, that almost any theorem of this intricate and minutely detailed subject would have delighted the ancient Greeks. And that, precisely, is the point. All the classic Greek geometers were buried or cremated two thousand years ago. Geometry in the meantime has advanced. By 1900 at the latest, special theorems in Euclidean geometry were no longer even a tertiary objective of creative geometers, no matter how beautiful or how interesting they might appear to their authors.

This does not imply that such theorems were valueless to those who could appreciate them; they preserved more than one isolated teacher from premature fossilization. They may also have irritated some who later became skilled geometers into finding out what modern geometry is about. On the other hand, many a working mathematician of the 1930's looked back with something akin to rage in his heart on the months or years squandered on this sort of geometry, or on the interminable properties of conics, at the very time of his life when his capacity for learning was greatest and when he might have been mastering some living mathematics.

In defense of this waste, if such it were, it was argued that English schoolboys still took a keen pleasure in these intriguing puzzles of their forefathers. No doubt they did. But the further claim that such a training made first-rate geometers is contradicted by the evidence. The attempted justification on the grounds of mental discipline may be left to the psychologists. In any event, it seemed slightly fatuous to impart discipline through outmoded fashions when so much of equal difficulty and vital necessity had to be mastered if one were to think geometrically in the manner demanded by a continually advancing science and mathematics. The foregoing opinions, it may be stated, were those of three of England's foremost mathematicians in the 1930's, all of whom had made high marks in this prehistoric sort of geometry in competitive examinations for English university scholarships.

However, even in the most elementary geometry an original and ingenious mind may occasionally think of something to do rather unlike what is already classical. The more orthodox Greek geometers, we saw, limited themselves to a straightedge and compass in their permissible constructions. Why not dispense with one or other of these traditional implements? It occurred to G. Mohr (Danish) in the seventeenth century to see what constructions could be performed with a compass alone, and L. Mascheroni, (1750–1800, Italian) actually wrote a book on the geometry of the compass. Napoleon Bonaparte is said to have been highly elated by his solution of a simple problem in Mascheroni's geometry. Others hobbled themselves by using a straightedge alone, or this with one given point in the plane, and so on. Finally it occurred to E. Lemoine (1840–1912, French), to attempt to assign a measure of the complicatedness of a geometrical construction. He presented an account of his proposals (1888–1889, 1892–1893) at the International Mathematical Congress held in connection with the Chicago World Fair of 1893. He succeeded in defining the simplicity of a construction in terms of five operations of elementary geometry, such as placing one point of a compass on a given point; the simplicity was the total number of times these operations were used. It was thus possible to assign marks to different constructions for the same figure, but a method for finding the construction with an irreducible number of marks seemed to be lacking. Once more it was demonstrated that the only royal road to elementary geometry is ingenuity. Another type of problem the beginner meets may have suggested the hotly controversial and partly discredited theory of enumerative geometry: how many lines, circles, etc., satisfy a prescribed set of conditions? Or, to take an instance from the 1940's, what practical use, if any, can be made of Pascal's theorem in conics? Almost exactly 300 years separate Pascal's discovery of the theorem and its application by aeronautical engineers to lofting; naval architects might well have used it earlier. At a slightly more advanced stage, almost anyone can invent his own peculiar system of coordinates and proceed to elaborate the geometry it suggests. Many have.

Rising to a considerably higher level of difficulty, we may instance what the physicist Maxwell called "Solomon's seal in space of three dimensions," the twenty-seven real or imaginary straight lines which lie wholly on the general cubic surface, and the forty-five triple tangent planes to the surface, all so curiously related to the twenty-eight bitangents of the general plane quartic curve. If ever there was a fascinating snarl of interlaced theories, Solomon's seal is one. Synthetic and analytic geometry, the Galois theory of equations, the trisection of hyperelliptic functions, the algebra of invariants and covariants, geometric-algebraic algorithms specially devised to render the tangled configurations of Solomon's seal more intuitive, the theory of finite groups—all were applied during the second half of the nineteenth century by scores of geometers who sought to break the seal.

Some of the most ingenious geometers and algebraists in history returned again and again to this highly special topic. The result of their labors is a theory even richer and more elaborately developed than Klein's (1884) of the icosahedron. Yet it was said by competent geometers in 1945 that a serious student need never have heard of the twenty-seven lines, the forty-five triple tangent planes, and the twenty-eight bitangents in order to be an accomplished and productive geometer; and it was a fact that few in the younger generation of creative geometers had more than a hazy notion that such a thing as the Solomon's seal of the nineteenth century ever existed.

Those who could recall from personal experience the last glow of living appreciation that lighted this obsolescent masterpiece of geometry and others in the same fading tradition looked back with regret on the dying past, and wished that mathematical progress were not always so ruthless as it is. They also sympathized with those who still found the modern geometry of the triangle and the circle worth cultivating. For the difference between the geometry of the twenty-seven lines and that of, say, Tucker, Lemoine, and Brocard circles, is one of degree, not of kind. The geometers of the twentieth century long since piously removed all these treasures to the museum of geometry, where the dust of history quickly dimmed their luster.

For those who may be interested in the unstable esthetics rather than the vitality of geometry, we cite a concise modern account[1] (exclusive of the connection with hyperelliptic functions) of Solomon's seal. The twenty-seven lines were discovered in 1849 by Cayley and G. Salmon[2] (1819–1904, Ireland); the application of transcendental methods originated in Jordan's work (1869–70) on groups and algebraic equations. Finally, in the 1870's L. Cremona (1830–1903), founder of the Italian school of geometers, observed a simple connection between the twenty-one distinct straight lines which lie on a cubic surface with a node and the 'cat's cradle' configuration of fifteen straight lines obtained by joining six points on a conic in all possible ways. The 'mystic hexagram' of Pascal and its dual (1806) in C. J. Brianchon's (1783–1864, French) theorem were thus related to Solomon's seal; and the seventeenth century met the nineteenth in the simple, uniform deduction of the geometry of the plane configuration from that of a corresponding configuration in space by the method of projection.

The technique here had an element of generality that was to prove extremely powerful in the discovery and proof of correlated theorems by projection from space of a given number of dimensions onto a space of lower dimensions. Before Cremona applied this technique to the complete Pascal hexagon, his countryman G. Veronese had investigated the Pascal configuration at great length by the methods of plane geometry, as had also several others, including Steiner, Cayley, Salmon, and Kirkman. All of these men were geometers of great talent;

Cremona's flash of intuition illuminated the massed details of all his predecessors and disclosed their simple connections.

That enthusiasm for this highly polished masterwork of classical geometry is by no means extinct is evident from the appearance as late as 1942 of an exhaustive monograph (xi + 180 pages) by B. Segre (Italian, England) on *The nonsingular cubic surface*. Solomon's seal is here displayed in all its "complicated and many-sided symmetry"—in Cayley's phrase—as never before. The exhaustive enumeration of special configurations provides an unsurpassed training ground or 'boot camp' for any who may wish to strengthen their intuition in space of three dimensions. The principle of continuity, ably seconded by the method of degeneration, consistently applied, unifies the multitude of details inherent in the twenty-seven lines, giving the luxuriant confusion an elusive coherence which was lacking in earlier attempts to "bind the sweet influences" of the thirty-six possible double sixes (or 'double sixers,' as they were once called) into five types of possible real cubic surfaces, containing respectively 27, 15, 7, 3, 3 real lines. A double six is two sextuples of skew lines such that each line of one is skew to precisely one corresponding line of the other. A more modern touch appears in the topology of these five species. Except for one of the three-line surfaces, all are closed, connected manifolds, while the other three-line is two connected pieces, of which only one is ovoid, and the real lines of the surface are on this second piece. The decompositions of the nonovoid piece into generalized polyhedra by the real lines of the surface are painstakingly classified with respect to their number of faces and other characteristics suggested by the lines. The nonovoid piece of one three-line surface is homeomorphic to the real projective plane, as also is the other three-line surface. The topological interlude gives way to a more classical theme in space of three dimensions, which analyzes the group in the complex domain of the twenty-seven lines geometrically, either through the intricacies of the thirty-six double sixes, or through the forty triads of complementary Steiner sets. A Steiner set of nine lines is three sets of three such that each line of one set is incident with precisely two lines of each other set. The geometrical significance of permutability of operations in the group is rather more complicated than its algebraic equivalent. The group is of order 51840. There is an involutorial transformation in the group for each double six; the transformation permutes corresponding lines of the complementary sets of six of the double six, and leaves each of the remaining fifteen lines invariant. If the double sixes corresponding to two such transformations have four common lines, the transformations are permutable. If the transformations are not permutable, the corresponding double sixes have six common lines, and the remaining twelve lines form a third double six. Although the geometry of the situation may be perspicuous to those gifted with visual imagination, others find the underlying algebraic identities, among even so impressive a number of group operations as 51840, somewhat easier to see through. But this difference is merely one of acquired taste or natural capacity, and there is no arguing about it. However, it may be remembered that some of this scintillating pure geometry was subsequent, not antecedent, to many a dreary page of laborious algebra. The group of the twenty-seven lines alone has a somewhat forbidding literature in the tradition of the late nineteenth and early twentieth centuries which but few longer read, much less appreciate. So long as geometry—of a rather antiquated kind, it may be—can clothe the outcome of intricate calculations in visualizable form, the Solomon's seal of the nineteenth century will attract its devotees, and so with other famous classics of the geometric imagination. But in the meantime, the continually advancing front of creative geometry will have moved on to unexplored territory of fresher and perhaps wider interest. The world sometimes has sufficient reason to be weary of the past in mathematics as in everything else.

What is geometry?

In the typical episode of the nineteenth century just recalled, we see once more the continual progression from the special to the general, in the emergence of widely applicable methods from laboriously acquired collections of individual theorems, that characterized mathematics since 1800. The methods generalized from Cremona's have retained their vitality and interest, although the particular theorems in which they originated may have lost their attractiveness for a generation trained in newer habits of thought for which those very theorems were partly responsible. So again, in seeking the things that have endured in mathematics, we are led to processes and ways of thinking rather than to their products in any one epoch. We shall see also that the conception of geometry itself changed with time, until what was called geometry in one stage of the development would hardly have been recognized as such at an earlier stage.

Attempts to obtain from professional geometers a statement of what geometry is are likely to be only nebulously successful. Beyond agreeing, more or less, on "geometry is the product of a particular way of thinking," few geometers will commit themselves to anything less hazy. Accepting this for the present —we shall return to it in other connections—we assume that it has a meaning which can be 'felt,' if not understood; and we shall describe some of the main contributions of geometric thought to mathematics as a whole.

Numerous representative selections might be made; the topics described here were chosen as an irreducible minimum on the advice of men actively engaged (1945) in developing the geometry of the twentieth century. The principal topics are: the vindication of Euclid's methodology in the creation of non-Euclidean geometries, and the outgrowths of this in the modern abstract or postulational method; differential geometry from Euler, Monge, and Gauss to Riemann and his successors, with its profound influence on the cosmology and mathematical physics of the twentieth century; the principle of duality in projective geometry, and its final elucidation in the inventions (1831) of J. Plücker (1801–1868, German), also this most original geometer's conception of the dimensionality of a space; Cayley's reduction (1859) of metric geometry to projective; algebraic geometry, particularly its connection with Cremona (1863) and birational transformations and the analysis of abelian functions; Klein's program (1872) for the unification of the diverse geometries existing in his day, and the supersession of this program after 1916; and finally, the abstract spaces and topology of the twentieth century, which some believe to be the beginning of a new type of mathematical thinking.

Naturally, only a bare hint of so vast a territory can be given in the space at our disposal; and here as elsewhere in this account we shall note only general trends. A history of any one of the topics would fill a book larger than this. The little described, however, may stimulate some to find out more about the subjects mentioned. Three of the topics are best considered under analysis, where they will be noted. Klein's program, its successors, and topology are described in the chapter on invariance; what little can be said about the connection between algebraic geometry and analysis is deferred to the theory of functions of a complex variable; and the rise of theories of abstract space is followed in a later chapter as a consequence of the trend toward general analysis, first plainly noticeable in 1906, for which mathematical physics was partly responsible. If, after all this, we are still unable to state what geometry is, we shall at least have caught a glimpse of the mathematics created by geometers in the worship of their inexplicit ideal.

Euclid cleared of all blemish

In 1733 the Jesuit logician and mathematician G. Saccheri (1667–1733) completed his involuntary masterpiece, *Euclides ab omni naevo vindicatus*, in which he undertook to prove that Euclid's system of geometry, with its postulate of parallels, is the only one possible in logic and experience. His brilliant failure is one of the most remarkable instances in the history of mathematical thought of the mental inertia induced by an education in obedience and orthodoxy, confirmed in mature life by an excessive reverence for the perishable works of the immortal dead. With two new geometries, each as valid as Euclid's, in his hand, Saccheri threw both away because he was willfully determined to continue in the obstinate worship of his idol despite the insistent promptings of his own sane reason.

To 'prove' Euclid's parallel postulate, Saccheri constructed a birectangular quadrilateral by drawing two equal perpendiculars AD, BC at the ends A, B of a straight-line segment AB, and on the same side of AB. Joining D, C, he proved easily that the angles ADC, BCD are equal. The parallel postulate is equivalent to the hypothesis that each of ADC, BCD is a right angle. To 'prove' the postulate, Saccheri attempted to show the absurdity of each alternative.

If each angle ADC, BCD is acute ('hypothesis of the acute angle'), it can be proved that the sum of the angles of any triangle is less than two right angles; if each angle is obtuse ('hypothesis of the obtuse angle'), the same sum is greater than two right angles; if each angle is a right angle, the same sum is equal to two right angles. Determined to establish the third possibility, Saccheri deduced numerous theorems from each of the first two hypotheses, hoping to reach a contradiction in each instance. He disposed of the hypothesis of the obtuse angle by tacitly assuming that a straight line is necessarily of infinite length. The hypothesis of the acute angle was rejected by an improper use of infinitesimals.

Cleansed by faulty reasoning of all blemish, Euclid's geometry shone forth to its worshiper as the absolute and eternal truth, the one possible mathematics of space. Saccheri died happy, unaware that he had proved several theorems in two new geometries, each as sound logically as Euclid's. The devout geometer had unwittingly demonstrated that his unique idol was but one-third of a trinity, coequal with the others but not coeternal; for no geometry is the everlasting truth that Saccheri thought he had proved Euclid's to be. It seems rather strange that the good geometer should have rejected the hypothesis of the obtuse angle so confidently; but here perhaps the fault was Euclid's with his meaningless definition of a straight line. With a precise definition[3] of a straight-line segment as the shortest[4] distance between two points, the concept of a geodesic on a surface is almost immediate. The geodesics on a sphere (which is a surface of constant positive curvature) are arcs of great circles, the analogues of the 'shortest distances' in a plane. It is just possible, however, that Saccheri's disciplined education required him to believe in a flat earth.

Any geometry constructed on postulates differing in any respect from those of Euclidean geometry is called non-Euclidean. Saccheri's two rejected specimens were the first non-Euclidean geometries in history. Ever since the time of Euclid, geometers had tried to deduce the parallel postulate from the others of Euclid's system. No useful purpose is served by cataloguing scores of failures to achieve the impossible, although several disclosed interesting equivalents of the doubtful postulate. A bibliography[5] of non-Euclidean geometry up to 1911 lists about 4,000 titles of books and papers by about 1,350 authors; and since 1911 the subject has expanded enormously. Much of the more recent work was directly inspired by physics, especially general relativity. Of the tentative steps toward a recognized, valid, non-Euclidean geometry between Saccheri's (1733) and Lobachewsky's (1826–9), we need recall only two.

In 1766, J. H. Lambert (1728–1777, German) noted that the hypothesis of the obtuse angle is realized on a sphere, and remarked that a novel kind of surface would be required to represent the plane geometry corresponding to the hypothesis of the acute angle. Nothing came of this suggestion till 1868, when E. Beltrami (1835–1900, Italian) showed that the surface vaguely conjectured by Lambert is the so-called pseudo-sphere. This is the surface of constant negative curvature generated by the revolution of a tractrix about its axis; it had been noted by Gauss, but without application to non-Euclidean geometry. But this belongs to the modern development, and we shall note its peculiar significance later.

The first indisputable anticipation of non-Euclidean geometry was by Gauss. As a boy of twelve, Gauss recognized that the parallel postulate presented a real and unsolved problem; but not till he was well past twenty did he begin to suspect that this postulate cannot be deduced from the others of Euclidean geometry. It is not definitely known when Gauss undertook the creation of a consistent geometry without Euclid's fifth postulate. It is certain, however, that he was in possession of the main results of hyperbolic geometry (Klein's designation for the system constructed on the hypothesis of the acute angle) before N. I. Lobachewsky (1793–1856, Russian) published his complete system (1829), and therefore also before J. Bolyai (1802–1860, Hungarian) permitted his to be printed (1833) as an appendix of twenty-six pages in a semiphilosophical two-volume elementary mathematical work (*Tentamen*, etc.) by his father.

It used to be claimed on the flimsiest circumstantial evidence that J. Bolyai had been influenced by Gauss. As it is now generally admitted that there are no grounds whatever for this hypothetical action at a very great distance, we shall pass it with the fact that J. Bolyai's father, W. Bolyai (1775–1856), was a close friend of Gauss during his student days at the university.

Gauss never made any public claim for himself as an inventor of non-Euclidean geometry. His anticipations of a part of what Lobachewsky and J. Bolyai accomplished, almost simultaneously and independently, were found in his papers after his death. Although he himself refrained from publishing the revolutionary geometry, Gauss encouraged others to proceed in their efforts to construct a consistent non-Euclidean system. Two of his correspondents made considerable progress: F. K. Schweikart (1780–1859) and F. A. Taurinus (1794–1874), both German. The latter in particular obtained correct and unexpected results (1825–6) in non-Euclidean trigonometry. The earlier date coincides with that at which J. Bolyai is supposed to have convinced himself that hyperbolic geometry is consistent; the later with Lobachewsky's first paper, unaccountably lost by the Kazan Physico-Mathematical Society, on the new (hyperbolic) geometry. For reasons that are not exactly clear, Taurinus destroyed all copies of his own work on which he could lay his hands.

In the bald historical statement that Lobachewsky in 1826–9 and J. Bolyai in 1833, almost simultaneously and entirely independently, published detailed developments of hyperbolic geometry, we have recalled one of the major revolutions in all thought. To exhibit another comparable to it in far-reaching significance, we have to go back to Copernicus; and even this comparison is inadequate in some respects. For non-Euclidean geometry and abstract algebra were to change the whole outlook on deductive reasoning, and not merely enlarge or modify particular divisions of science and mathematics. To the abstract algebra of the 1830's and the bold creations of Lobachewsky and Bolyai can be traced directly the current (1945) estimate of mathematics as an arbitrary creation of mathematicians. In precisely the same way that a novelist invents characters, dialogues, and situations of which he is both author and master, the mathematician devises at will the postulates upon which he bases his mathematical systems. Both the novelist and the mathematician may be conditioned by their environments in the choice and treatment of their material; but neither is compelled by any extrahuman, eternal necessity to create certain characters or to invent certain systems. Or, if either is so conditioned, it has not been demonstrated that he is; and to an adult twentieth-century intelligence the multiplication of superfluous and mystical hypotheses is a pursuit even more futile than it was in the days of Occam.

In reporting this estimate of mathematics by informed opinion in 1945, we must also state that it was by no means universal. Many of the older generation still adhered to the Platonic doctrine of mathematical truths. Nor is there any reason to suppose that Plato will not again reign supreme in the minds of mathematicians. Less rational mysticisms than Plato's have prevailed since the rediscovery of the virtues of blind irrationalism in 1914. But while the majority of mathematicians still believe they can see through an ancient fraud to the nonentity behind it, we shall record briefly how the humanization of mathematics came about. The deflation of older beliefs, however, comprises the main contribution of non-Euclidean geometry to mathematical thought as a whole, and also, perhaps, the principal contribution of mathematics to the progress of civilization. For it seems improbable that our credulous race is likely ever to get very far away from brutehood until it has the sense and the courage to discard its baseless superstitions, of which the absolute truth of mathematics was one.

To appreciate fully the next item of more than local significance, we must describe a geometrical detail in each of four geometries; and we shall select that one, the existence of parallels, which precipitated an apparently interminable deluge of non-Euclidean geometries after the hyperbolic geometry of Lobachewsky and Bolyai. In 1854, G. F. B. Riemann (1826–1866, German) invented a 'spherical' geometry, in which Saccheri's hypothesis of the obtuse angle is realized. The designations 'hyperbolic' and 'elliptic' refer to Cayley's 'Absolute,' to be noted later; Euclidean geometry is similarly called 'parabolic.'

If P is any point in the plane determined by P and a straight line L not passing through P, there is, in parabolic geometry, precisely one straight line, L', through P which does not intersect L; L' is the unique parallel to L through P. In hyperbolic geometry, there are two distinct straight lines L', L'' through P, neither of which intersects L; moreover, no straight line through P and lying in the angle between L' and L'' meets L. Lobachewsky took L', L'' as his parallels to L. In both parabolic and hyperbolic geometry two straight lines intersect in one point. Any two geodesics on a sphere (arcs of great circles) intersect in two points, and there are no parallels. In Riemann's so-called 'spherical' geometry, space is unbounded but finite; every 'straight line' (geodesic) is of finite length; and any two straight lines intersect twice, thus negating Euclid's postulate that two straight lines cannot enclose a space. Riemann's 'elliptic' geometry can be visualized on a hemisphere if, as in his spherical geometry, 'straight lines' are arcs of great circles of the whole sphere, and if the two extremities of an arc are regarded as the same point. Other 'realizations' are easily constructed for all of the four geometries mentioned; our interest here is in the significance, or lack of it, of all such realizations. This marks the discontinuity in geometry as conceived before and after 1899. The three non-Euclidean geometries mentioned above are usually called classical; 'Riemannian geometry,' as used since 1916, is distinct from all these geometries.

For about thirty years after the invention of hyperbolic geometry, few mathematicians paid any attention to it; and none, it appears now, foresaw what the non-Euclidean geometries were to imply for the whole of mathematics. There seems to have been a feeling of distrust, or incredulity that the new geometries were either 'true' in the same sense as Euclid's, whatever that might be, or that they could ever be of any scientific value. These doubts were dispelled by Beltrami's brilliant demonstration (1868) that plane hyperbolic geometry can be interpreted as that of the geodesics on a surface of constant negative curvature, and likewise for spherical geometry and a surface of constant positive curvature. Since pseudospheres and spheres are familiar surfaces in Euclidean space, it was felt that the consistency of the classical non-Euclidean geometries had been demonstrated. Euclid's geometry might still be the most useful, but the others were equally 'true,' because it had been shown that they could be realized in Euclidean space. If this is not a fair statement of what the majority of mathematicians in the 1870's elicited from Beltrami's Euclidean realizations of the classical non-Euclidean geometries, we apologize and pass on.

After our discussion of the similar situation for ordinary complex numbers, we need not labor the issue here. Beltrami's realizations demonstrated nothing more conclusive than the theorem that Euclid's geometry and the classical non-Euclidean geometries were either all logically admissible or all logically inadmissible together; the internal consistency of no one of them had been proved. This is obvious now to the point of platitude; it was not obvious in 1868. As in the case of complex numbers, the problem of internal consistency for any of the geometries was referred by Hilbert to the like for the real number system.

Following the successors of Beltrami, we now descend to a deeper level, and note the emergence of the modern postulational method in geometry. Ignoring masses of details and numerous contributors, we can observe the steady progression toward abstractness in the work of three men, each typical of the more original thought of his generation: M. Pasch (German), G. Peano (Italian), and Hilbert.

We recall that B. Peirce, in the introduction to his linear associative algebra (1870), asserted that "Mathematics is the science which draws necessary conclusions." It was not until 1882, in the work of Pasch, that this concept of the nature of mathematics began percolating into geometry. Space for Newton was the absolute ultimate in which motion of rigid bodies was possible and took place; Leibniz, in some of his philosophizing, imagined space as a matrix of possible relations. Pasch all but eliminated both these conceptions in his restatement of geometry as a hypothetico-deductive system in the tradition of Peirce. Instead of attempting to state definitions of points, lines, and planes, as Euclid had done, Pasch accepted these as the unreduced elements of 'space,' and from postulated but unanalyzed relations between these atomistic concepts of his system proceeded to deduction.

The relations postulated were abstracted from the accepted geometric notions of centuries of working with diagrams. For example, it was explicitly stated as a postulate that two distinct points determine precisely one straight line. Pasch was therefore closer to Leibniz than to Newton; but he seems still to have believed in the existence of a 'space' in Newton's sense. Whether or not he did, his was the first clear-cut presentation after Euclid's of geometry as an exercise in postulational technique. Pasch went beyond Euclid in that he realized what he was doing, and did it deliberately, whereas Euclid seems to have been guided by visual imagery, and in consequence to have overlooked hidden assumptions. In any event, Pasch profoundly influenced the geometric thought of his contemporaries and successors. His conception of geometry, now an accepted commonplace, met the usual opposition encountered by anything new and disturbing. Although neither the physical nor the philosophical meaning, if any, of 'space' was affected by the completely abstract reformulation of geometry, the novelty affected some—Veronese of Pascal's hexagram for one—like a shocking blasphemy suddenly shouted in church. They quickly stopped their ears. Even if this new thing were consistent, it was too arid and too barren to be profitable mathematics. Geometers were permitted to retain their feeling for the ineffabilities of the geometric mode of thought; but geometry itself was reduced to logical syntax. From a distance of fifty years it is difficult to see why anyone got excited. Intuition and strict logical analysis can live in the same science without killing each other; and what one cannot do, the other can.

The next man with courage enough to be unpopular was G. Peano. At first he escaped notice except in Italy. But when (1888) he began attempting to reduce all mathematics to a precise symbolism which left but few loopholes for vagueness, too-slippery intuition, and loose reasoning, he was regarded with suspicion. With the help of several Italian collaborators, Peano in 1891–5 recast a considerable part of mathematics in his new symbolism. From the rational integers to geometry, proofs were based upon explicitly stated sets of postulates, intended to be necessary and sufficient for the proofs.

To attain the desired precision, parts of mathematical logic itself were symbolized more minutely than Boole and his successors from 1847 on had done; and frequently occurring phrases of technical mathematics were also reduced to symbols. The result was a universal shorthand for practically the whole of mathematics as it existed in the 1890's. Peano's pasigraphy was a step in the direction of Leibniz' universal characteristic, and one of the most powerful stimuli to the mathematical logic of the twentieth century. It drew down on its unlucky inventor's head the ridicule and abuse of some of the most eminent mathematicians of the twentieth century, including Poincaré. Nevertheless, the logical analysis of geometry by Peano and his followers made it plain to all but the willfully blind that geometry is an abstract hypothetico-deductive system without intrinsic content[6] other than that implied by arbitrarily prescribed sets of postulates.

The final step was taken in 1899 by Hilbert in his *Grundlagen der Geometrie*. Hilbert was a mathematician and a Prussian who could not be ridiculed or abused with impunity, even by Poincaré. Hilbert's postulational discussion of geometry was still a classic forty years after it appeared. This must be the record, or close to it, for an elementary work on geometry written since the days of Euclid. Legendre's *Eléments de géométrie* (1794) passed through more editions than Hilbert's *Grundlagen;* but the two works are not comparable. Legendre's *Eléments* was addressed to schoolboys, and is of historical interest—in the derogatory sense—chiefly because of fallacious attempts to prove Euclid's fifth postulate. Hilbert's classic inaugurated the abstract mathematics of the twentieth century. Its latest edition might also be read with profit by intelligent schoolboys.

With a minimum of symbolism, Hilbert convinced geometers, as neither Pasch nor Peano had succeeded in doing, of the abstract, purely formal character of geometry; and his great authority firmly established the postulational method, not only in the geometry of the twentieth century, but also in nearly all mathematics since 1900. Again we emphasize that intuition was not ousted from mathematics by the abstract attack. Nor were the applications of postulational analysis more than a small fraction of twentieth-century mathematics. But they were a potent catalyst for that mathematics, and they attracted hundreds of prolific workers.

Among other subjects to profit by the revived and refined methodology of Euclid was non-Euclidean geometry. One of the most curious geometries invented through a deliberate application of the postulational method was M. W. Dehn's non-Archimedean system, which linked the similar triangles of the ancient Babylonians to the non-Euclidean geometry of the nineteenth century. Suppressing the axiom of Archimedes, Dehn constructed a geometry in which similar triangles exist, and in which the angle-sum for any triangle is two right angles. But parallels are not unique as in Euclid's geometry, an infinity being possible. Equally unforeseen consequences followed on applying the postulational method to projective geometry. The American geometers O. Veblen and W. H. Bussey constructed (1906) finite projective geometries in which a 'plane' contains only a finite number of 'points' and 'lines.' These finite geometries reduced the debates of the early nineteenth century on the 'space' of projective geometry to sequences of empty noises. Veblen and J. W. Young produced (1907) a set of completely independent postulates (in the sense of E. H. Moore (1862–1932)) for projective geometry which long remained a standard, and which must have convinced the most obstinate that geometry is a formal hypothetico-deductive exercise in logic. Americans probably did more to exploit the postulational technique after Hilbert's *Grundlagen* than their European colleagues, and their analysis was, on the whole, much sharper and clearer. The method was applied to algebra, geometry, arithmetic, topology, and other subjects by dozens of skilled mathematical logicians, among whom E. V. Huntington (1874–) may be specially mentioned for his exhaustive analyses of postulate systems in many fields.

After the formal character of mathematics had been admitted by many, intellectual inertia proceeded once more on its time-honored course. If mathematics, and in particular geometry, is an arbitrary creation of human beings, then surely the like is not so for traditional logic? In some extrahuman sense the logic which had lasted all of twenty-three centuries must be an absolute that not even mathematicians can defy. We shall see in the proper place that this absolute, too, was abolished, but not till 1920.

Returning to Saccheri (1733), we see now that after all he accomplished his purpose. In a sense that was to prove of incomparably greater significance for the future of mathematics than a proof of the parallel postulate could ever have been, Saccheri had cleared Euclid of all blemish. Although his work was ignored and forgotten for over a century after his death, Saccheri deserves as much credit as Lobachewsky and Bolyai for having taken the decisive step toward the abolition of mathematical absolutism.

The essential part of Euclid's doctrine, strict deduction from explicitly stated assumptions recognized as such, began to come into its own only with the unconscious creation by Saccheri of non-Euclidean geometry. After more than two thousand years of partial understanding, the creative power of Euclid's methodology was gradually appreciated; and Saccheri, had he lived till 1899, would have seen a profounder meaning than he intended in the title of his masterpiece, *Euclides ab omni naevo vindicatus.* The import of what evolved from non-Euclidean geometry transcends geometry in epistemological value. The extramathematical by-product seems to stand a better chance of enduring than the vast accumulations of technical theorems which delight geometers, and which tomorrow may join the twenty-seven lines on a general cubic surface in the museum of mathematical art.

A meaningless controversy

A skilled geometer might devote his entire working life to one species of curves, say hypocycloids or bicircular quartics, and find something new and interesting to himself every day. But he would scarcely come upon general principles. If history is any criterion, generalizations in geometry have not been reached by piling theorem on theorem, but by purposeful efforts to slash through jungles of special results, or by equally conscious attempts to find out why certain techniques furnish geometric theorems easily, while others demand more labor than their output justifies. A case in point is the prolonged struggle of the synthetic geometers during the first half of the nineteenth century to ascertain 'why' analytic methods were, apparently, so much more powerful than those of pure geometry. We shall trace the main lines of this fluctuating struggle in the following section. For the moment we consider the outcome, as this possibly is the item of greatest interest for mathematics as a whole.

To state the conclusion first, analytic methods are more powerful than those of pure geometry because the experience of more than a century has shown them to be so. No philosophy of 'space' and its 'geometry' has added anything of significance to this statement of brute fact; although the profane might suggest that as imaginary points, lines, etc., entered geometry through the formalisms of elementary algebra, and not through visual perceptions of diagrams, any attempt to disguise all algebraic concepts in an elaborate diagrammatic terminology could result only in unprofitable artificialities. This, at any rate, was the prevailing opinion among geometers in the 1930's. They still continued to exercise their esoteric geometric intuition with telling effect, but only a few persisted in trying to modulate every uncouth configuration through delicately cadenced involutions that would have ravished the pure geometric ear of Poncelet.

It is conceded by all that the technical vocabulary of imaginary points, etc., is of great utility, and serves much the same purpose as the terminology of analysis. The statement that $f(x)$ is continuous at $x = a$, for example, condenses several assertions into one that can be used as a unit in deduction. It is unnecessary to recur in each instance to the meaning of continuity, for its simpler implications have been worked out once for all, and can be applied without further thought. Similarly, the statement that a certain curve is a bicircular quartic, say, implies that each of the circular points at infinity is a node on the curve; and the standard elementary properties of nodes and the so-called circular points are applicable as units in investigating the curve.

But this admitted utility of a geometrized algebraic vocabulary is not what distinguishes the geometers of the nineteenth century from the majority of their successors. The leading synthetic geometers of the past attempted to find the circular points, etc., in the 'space' of common experience, because they confused 'physical' or 'real' or 'a priori' space—without attempting to explain what they understood by such space—with the abstractions of their algebra and the inadequacies of their diagrams. The conception of geometry as a hypothetico-deductive system was about sixty years in the future when (1822) J. V. Poncelet (1788–1867, French) published his *Traité des propriétés projectives des figures*. This classic of the synthetic method was largely responsible for one of the most fruitful and, as is now widely believed, least meaningful controversies in the history of mathematics. The point at issue was the 'reality' of 'space' as represented in geometry, and the crux of the dispute was the 'existence' in 'space' of imaginary elements.

The controversy left a substantial residue of powerful geometric methods; and these, or their direct outgrowths such as the algebra of birational transformations, proved most useful in other departments of mathematics, particularly analysis. But the main question in dispute among the geometers, it now appears to many, was a pseudo question without meaning. That so much of value issued from a vacuous logomachy may encourage us to hope that by the year 2000 the controversies of 1945 over the foundations of mathematics will have evaporated, leaving only a few powerful methods on which all can agree and which some can use.

It must be said in concluding these remarks that they present only one competent estimate of the controversy in question, and that opinion is not unanimous. Equally competent contrary estimates are readily available.

Contributions from projective geometry

The return to favor of projective geometry after its neglect during the eighteenth century is first noticeable in the *Géométrie de position* (1803) and the *Essai sur les transversales* (1806) of L. N. M. Carnot (1753–1823, French). This military genius, who in 1793 saved the French Revolution from the united reactionaries of Europe, set himself the task of showing that pure geometry is as powerful as the analytics of Descartes. Introducing negatives into synthetic geometry, and exploiting the invariance of the cross ratio of the four points in which a transversal cuts a pencil of four straight lines, Carnot derived many of the classical theorems of elementary projective geometry, including those associated with the complete quadrangle and the complete quadrilateral. Carnot might have concluded his work with the accurate prophecy of his royal enemy, "After me, the deluge." It arrived with Poncelet, one of the most enthusiastic lovers of pure geometry and most cordial haters of all forms of algebra and analysis in the history of mathematics.

Poncelet's ambition appears to have been to undo everything the successors of Descartes had done, or at least to do it better by strictly synthetic methods. As a prisoner of war from Napoleon's disastrous Russian campaign, Poncelet, the young officer of engineers, spent a cold but profitable winter (1813–14) at

Saratoff on the Volga, thinking out the pure geometry of his *Traité des propriétés projectives des figures*, which he published in 1822 after his return to France. He tells all, or nearly all, about it in his autobiographical introduction. The reservation, if justified, is hardly an argument in support of Poncelet's main thesis that analytic methods are inferior to synthetic. For G. Darboux (1842–1917), another great French geometer, let the analytic cat out of the synthetic bag before a large and appreciative audience at the St. Louis (U.S.A.) congress in 1904. According to Darboux, "We know, moreover, by the unfortunate publication of the Saratoff notes, that it was by the aid of Cartesian analysis that the principles which serve as the base of the *Traité des propriétés projectives* were first established." As this blunt statement of fact seems not to be generally known to geometers, it may be that others besides Darboux have considered it unfortunate. The proponents of analytic methods might consider this awkward disclosure extremely fortunate.

But, by whatever means Poncelet first established his general principles, he put new life into a branch of geometry that was almost dead. Observing that certain properties of a plane configuration, such as the collinearity of three points in Pascal's theorem, are unaltered by projection, Poncelet undertook a systematic investigation of such phenomena and defined the 'graphic' (our 'projective') properties of figures to be those which are independent of the magnitudes (measures) of distances and angles. As we shall see later, Cayley forced the metric properties banned by Poncelet into a more inclusive projective geometry, in which the imaginaries that plagued Poncelet are given a place of honor. Poncelet himself (1822), in accordance with his restatement of Kepler's principle of continuity, introduced the line at infinity, and demanded in his plane geometry that every circle cut this line in the same two imaginary points. All pairs of simultaneous equations representing circles were thus provided with the correct number of common solutions. G. Monge (1746–1818, French) had already used pairs of imaginaries to symbolize real spacial relations; but Poncelet was more concerned with attempting to find a 'real' justification for imaginaries in geometry.

Poncelet's principle of continuity amounts to the theorem of analysis, that if an analytic identity in any finite number of variables holds for all real values of the variables, it holds also by analytic continuation for all complex values. The attempt to 'realize,' or to disguise, this elementary fact of analysis in a supposedly existent space involved Poncelet in a furious controversy with Cauchy. The analyst insisted that the geometer's reasoning, if not entirely illegitimate, was a needlessly complicated restatement of simple algebra; the geometer justified his tactics by proving numerous theorems with much greater ease than seemed possible by analysis. The dispute ended in a draw. But, as will be noted in connection with topology, Poncelet's intuition outran his logic, and his attempted justification of his famous principle rested on a void. Nevertheless, he continued to derive correct geometrical theorems with astonishing facility.

The principle of continuity was generalized in 1874–9 by H. Schubert (1848–1911, German), who went far beyond Poncelet's boldest in his 'calculus of enumerative geometry.' Schubert's 'principle of the conservation of number,' as the generalization was called, likewise rested on nothing that could now be recognized as a foundation. It asserted the invariance of the number of solutions of any determinate algebraic problem in any given numbers of variables and parameters under variation of the parameters, or under substitution of special values for them, in such a manner that none become infinite, due account being taken of multiple solutions and solutions at infinity. This somewhat dangerous method was used with brilliant effect by several of the leading geometers of the nineteenth century, including M. Chasles (1793–1880, French), J. Steiner (1796–1863, Swiss), Cayley, and J. G. Zeuthen (1839–1920, Danish) the last of whom profited by the more glaring oversights of his predecessors. Opinion on these subtle questions was still divided in 1945, the geometers affirming that their reasoning was sufficiently rigorous, the algebraists dissenting.

All of the men just mentioned are conspicuous, and some famous, in the history of mathematics: Steiner, called by his contemporaries "the greatest [pure] geometer since Apollonius," who could not write till he was fourteen; Chasles, a creative geometer and the judicious historian of geometry, whose *Aperçu historique sur l'origine et le développement des méthodes en géométrie* (1837) is still a classic of mathematical historiography; Cayley, the all-round mathematician, whose development (1846–) of the theory of algebraic invariants provided geometers with a new outlook on algebraic curves and surfaces; and last, Zeuthen, remembered in geometry and as an historian of mathematics.

Leaving the discredited principle of continuity, we pass to

that of duality which, with its generalizations, left as substantial a residue of new and useful methods in geometry, algebra, and analysis as any mathematical invention of the nineteenth century. In its classic form, the principle seems to have been first clearly stated, but not fully understood, in 1825–7, by J. D. Gergonne (1771–1859, French). Gergonne noted that if in certain theorems of plane geometry the words 'point' and 'straight line' be interchanged, with corresponding changes for collinearities of points and intersections of lines, etc., independently provable 'dual' propositions result. He inferred that the original in all cases implies the dual, which therefore need not be proved independently. By this 'principle of duality,' Brianchon obtained his theorem as the dual of Pascal's. Gergonne also noted the corresponding principle in space of three dimensions, point and plane being duals, and the straight line self-dual. At one stroke an already vast empire of geometry was doubled in extent; and it was a self-denying geometer indeed who refrained from the practice, which quickly became epidemic, of publishing lengthy parallel columns of dual theorems.

Gergonne, like Poncelet, was a military man. Both laid claim to the principle of duality. Poncelet insisted that the principle was a consequence of the method of poles and polars, which he had used so brilliantly in his own geometry of conics; Gergonne contended that poles and polars were not the root of the principle. Gergonne was right, but a conclusive, simple demonstration of this fact was not forthcoming till Plücker attacked the question algebraically, and gave the usual explanation by means of point and line coordinates to be found in most textbooks on projective geometry. But although Gergonne was right, Poncelet was not entirely wrong. His contention was all but saved by the fortunate circumstance that the order and class of a conic each equal 2.

Failing to make Gergonne withdraw his claim to the principle by fair means, Poncelet resorted to foul, and succeeded in demonstrating that although a geometer may have been an officer, he is not necessarily a gentleman. The campaign of personal abuse and defamation of character which Poncelet waged against the comparatively obscure Gergonne makes the Newton-Leibniz controversy look like a love feast in Arcady. Today the law would interfere; but in the heroic age of projective geometry, mathematicians were free to persecute their defenceless enemies like the heroes some of them had been.

The next great contribution to all mathematics for which

Poncelet's methods were partly responsible is in Vieta's tradition of transformation and reduction. If a reversible transformation between the respective systems of coordinates in two spaces, or between different coordinate systems in the same space, can be established, theorems in either system can be translated immediately into theorems in the other, the correspondence providing the bilingual dictionary. If the restriction that the transformation be reversible is removed, the dictionary reads one way only, say like French into English, and configurations in the first space are mapped onto others in the second, but not vice versa. In either case there may be certain singular loci which must be excluded from the statements of theorems; an example will be given in connection with birational transformations. There is no limit other than practical utility to the transformations that might be constructed in accordance with these very general specifications. If easily manipulated transformations which alter the order of a curve (or surface, etc.) can be produced, the geometric gain is obvious.

Numerous useful transformations of the kinds described have been constructed since Poncelet, in his method of reciprocal polars, first made a geometric element of one kind (a point) correspond to one of another kind (a line). Possibly the most extensively investigated transformations have been the birational, in which the coordinates for either of two spaces are rationally expressible in terms of those of the other. This allusion must suffice here, as birational transformations are best described in connection with the theory of algebraic functions and Riemann surfaces, and must be deferred to a later chapter.

Continuing with our selection of great principles that originated in projective geometry, we note next one of the least expected and most surprisingly simple generalizations in the evolution of mathematics, Plücker's theory of spacial dimensionality. We have already seen that Cayley (1843) and Grassman (1844) independently arrived at the notion of n-dimensional space, and that the latter defined the n-dimensional manifolds which were to play a capital part in Riemann's geometry. Plücker in his *Analytische-geometrische Entwickelungen* (1831) observed not only the analytic meaning of duality, but also the germ of an idea which was to generalize the duality of Gergonne and Poncelet far beyond the obvious resources of pure geometry. He noted that the general equation of a straight line in plane Cartesian geometry contains two variables and two parameters, and that both the variables and the parameters enter the equation linearly. If the roles of the variables and parameters are interchanged, the equation becomes that of a point. It is to be noticed that the number of parameters is the same, two, in both cases: the plane is a two-dimensional space, or manifold of two dimensions in points and in lines; and we say that the plane contains ∞^2 points and ∞^2 lines. These simple observations were the origin of Plücker's vast generalization.

If a class of elements is such that a unique element of the class is specified when any particular numerical values are assigned to each of n numerical-valued parameters, the class is called a number-manifold, or a number-space, of n dimensions, and is said to contain ∞^n elements. The n parameters written in a prescribed order are called the coordinates of the general element of the class. For example, the general equation of a conic in the plane of Cartesian coordinates contains precisely five parameters; a particular set of values of the parameters specifies a unique conic; the plane is therefore a five-dimensional manifold, or space, when conics are taken as the basic space-elements, and it contains ∞^5 such elements. If this seems strange at first, it is no more so than the commonplace that the plane is a two-dimensional manifold of points. In Plücker's geometry, dimensionality is not an absolute attribute of space, but depends upon the basic elements constituting the space. A Cartesian plane, for example, is three-dimensional in circles. All of this was generalized early in the twentieth century to spaces in which the values of the parameters are not necessarily numbers; the resulting geometries are those of the various abstract spaces intensively studied since Fréchet's creation (1906) of the first. These will be noted in other connections.

Generalizing the classic duality for configurations of points and lines in plane geometry, Plücker stated a principle of duality for any two classes of configurations that have equal dimensionalities and are both linear in the respective coordinates, equal in number, determining the common dimensionality. Each of the classes is a number-space as already defined. The 'geometry' of each may be interpreted in many ways. For example, in the three-dimensional space constituted by all circles in a Cartesian plane, an equation between the three coordinates of a circle defines a family of ∞^2 circles. We might proceed, as many did, to investigate in detail the properties of such families, defined by equations of degrees 1, 2, ... Very simple plane representations, easily visualized, of the classical non-Euclidean geometries have been constructed from families of circles. Precisely as in the familiar Cartesian geometry in which points are the basic elements, in Plücker's geometry we translate the algebra of systems of equations into properties of families of straight lines, conics, cubics, and so on. The duality which Gergonne and others believed to be an absolute attribute of 'space' peculiar to the intuitive, diagramed 'space' of elementary projective geometry appeared in Plücker's geometry as a trivial consequence of an unnecessarily restricted way of choosing systems of coordinates.

Plücker's abandonment of the deceptions of visual intuition for the explicitness of algebra and analysis finished something the classical non-Euclidean geometries had only half done. The arbitrary freedom in the mathematical construction of 'spaces' and 'geometries' at last made it plain that Kant's a priori space and his whole conception of the nature of mathematics are erroneous. Yet, as late as 1945, students of philosophy were still faithfully mastering Kant's obsolete ideas under the delusion that they were gaining an insight into mathematics. As Kant appealed to his mathematical misconceptions in the elaboration of his system, it is just possible that some other parts of his philosophy are exactly as valid as his mathematics. Against this it has been contended that Kant's mathematics remains 'true' in a higher realm of 'truth' beyond the comprehension of professional mathematicians, whose grudging science leaves them but little time to explore the really important questions of their subject. The difference of opinion may well be left there. The most significant residue of Plücker's work was the demonstration once more that geometry as practiced by geometers is an abstract, formal discipline. It should not be necessary to repeat that any experiences which may have suggested a particular set of assumptions for a geometry are irrelevant to the mathematical development.

On the strictly technical side, Plücker elaborated (1865) in great detail the geometry of what we ordinarily picture as a Cartesian three-dimensional space, the point-space of elementary solid geometry and rigid bodies, but with straight lines instead of points as the basic elements. Since the equations of a general straight line in the Cartesian space involve exactly four independent parameters, Plücker's 'line geometry' is that of a four-dimensional space. Equations between the coordinates represent various families of straight lines; a family of ∞^1 straight lines is called a range, one of ∞^2, a congruence, and one of ∞^3, a complex; and these species are further classified according to the degrees of algebraic equations in the four line coordinates. The theory then proceeds partly by analogy with the familiar geometry based on point coordinates in point-spaces of two and three dimensions. For example, instead of the quadric surfaces defined by the general equation of the second degree in point coordinates (x, y, z), the geometric configurations defined by an equation of the second degree in line coordinates (p, q, r, s) are investigated and classified into types, analogously to the classification of quadrics into pairs of planes, cones, cylinders, ellipsoids, hyperboloids, etc. This particular detail is the geometry of the quadratic line complex; the problem of classification led to much interesting algebra of the type associated with the reduction of matrices to canonical form. In line geometry, a curve is visualized as an envelope of straight lines, not as a locus of points.

The inevitable question, 'What was the good of it all?,' is unanswerable. But for those who insist upon a scientific or industrial application for all mathematics, it may be recalled that Plücker's line geometry found an immediate interpretation in the dynamics of rigid bodies. A four-dimensional geometry of the late 1860's thus justified its creation and its existence to a generation that believed in machinery with all its heart, with all its mind, and with all the soul it had.

Synthesis versus analysis

Returning now to 1827, when A. F. Möbius (1790–1860, German) introduced homogeneous coordinates in his highly original work, *Der barycentrische Calcul*, we shall follow the struggle of synthetic methods against coordinates until both emerged victorious in the 1860's. Although no sharp line divided the contestants or separated the geometers of different nationalities, the prolific Italian school preferred synthesis after the 1860's, while the majority of French, German, and British geometers made greater use of analysis. At the beginning of the period, Steiner was the unapproachable champion of synthesis; Plücker, the unrivaled master of analysis.

Plücker is usually regarded as the true founder of the method of homogeneous coordinates, in his *Analytische-geometrische Entwickelungen* (1828, 1831), which also exploited abridged notation, usually attributed to E. Bobillier (1797–1832, French), and the simple analytic equivalent of Gergonne's principle of duality. This was followed in 1835 by the *System der analytische Geometrie*, in which, incidentally, curves of the third order were completely classified. Attempts by Cramer and others in the eighteenth century to discipline the swarms of curves of the fourth order had failed. In his *Theorie der algebraischen Curven* (1839), Plücker had a better success. This, however, was totally eclipsed by a discovery of a new kind and of the very first magnitude, 'Plücker's equations' connecting the order, the class, and the numbers of double points, double tangents, and points of inflection of algebraic plane curves. Cayley pronounced this discovery one of the greatest in the history of geometry. It became one of his life-interests to extend Plücker's equations to the singularities of skew curves (twisted curves, curves in space) and surfaces. G. Salmon, a fine geometer and algebraist who abandoned mathematics for theology, also made notable contributions to this elusive subject. It has extensive ramifications in modern algebra and analysis, the last through the theory of algebraic functions and their integrals.

It was in this prolific third decade of the nineteenth century that line coordinates were invented. It is generally agreed that Plücker and Cayley imagined these coordinates independently. Many special surfaces of some interest were exhaustively investigated by both point and line coordinates. Two may be recalled as representative of the more interesting specimens collected and anatomized in this detailed sort of work: Kummer's (1864) quartic surface, which is the so-called singular surface of the quadratic line complex, and which is represented parametrically, as Cayley was the first to discover (1877), by hyperelliptic theta functions; and the wave surface in optics, parametrized by elliptic functions. Through its generalizations to higher space, Kummer's surface by itself generated an extensive department of geometry. Numerous French, German, Italian, and British geometers elaborated—perhaps overelaborated—this intricate specialty well into the twentieth century. But the general quartic surface in 1945 still presented unsolved problems, and possibly was too complicated for profitable attack by the weapons available. Interest in such matters had declined rapidly in the twentieth century, especially after 1920, and they seemed to belong definitely to a glorious but buried past.

While Plücker, Cayley, and many others were creating modern analytic geometry with astonishing rapidity, the most ingenious pure geometer since Apollonius was engaged, with spectacular but severely restricted success, in attempting to forge synthetic geometry into an implement of what he hoped would be universal applicability. Steiner's *Systematische Entwickelung der Abhängigkeit geometrischer Gestalten von einander* (1832) unified the classical methods of pure projective geometry and applied them with amazing skill to numerous special problems. Incidentally, Steiner proposed several theorems which, presumably, he had discovered by pure geometry, as challenges to be proved by other geometers. The last detail in one of these defied proof by analytic methods till the early 1900's.

The powerful method of inversion is also attributed to Steiner (1824), although others also invented it. It was independently come upon (1845) through physical considerations by W. Thomson (Lord Kelvin). He and other physicists applied it effectively in its synthetic form to electrostatics, calling it the method of images. Conversely, problems in potential theory can be disguised as exercises in inversion. The trick works both ways because inversion is a conformal transformation.

Inversion was one of the first non-linear transformations to be studied deeply in geometry, although quadratic and cubic transformations had been familiar in algebra since the sixteenth century. Special birational quadratic transformations were used by Poncelet (1822), Plücker (1830), Steiner (1832), and systematically (1832) by L. I. Magnus (1790–1861, German). As a detail of historical interest, Magnus (1833) compounded two quadratic transformations to obtain a quartic transformation, by which straight lines correspond to quartic curves. He was thus enabled to read off theorems on quartic curves from their images in straight lines. In somewhat the same direction, E. de Jonquières established (1859) a special correspondence between straight lines and curves of order n with a prescribed multiple point of order $n - 1$. Geometers are interested in these historical minutiae because Cremona, who in 1863 set himself the problem of determining all birational transformations of order n between the points of two planes, apparently overlooked this earlier work, and as late as 1861 believed that if $n > 2$ no such transformations exist. Geometers point out that had Cremona been conversant with the algebraic notion of closure, as in a group, he would have drawn the correct inference immediately from what was already well known. However, when he realized his oversight, he made rapid progress. What amounts to a capital theorem in the particular birational transformations named after him will appear in another guise when we discuss algebraic functions. It suffices to note here that M. Noether, J. Rosanes (German) and W. K. Clifford (1845–1879, English) proved almost simultaneously (1870) that a Cremona transformation can be generated by compounding quadratic transformations. The prolific Italian school, from Cremona in the nineteenth century to Severi in the twentieth, developed the resulting algebraic geometry mainly by geometric methods; and indeed the corresponding algebra and analysis quickly become unmanageable. The permanent gain from all this somewhat confused development appears to be the methodology of establishing correspondences between classes of different types of geometric configurations.

Another extensive division of geometry that developed from the geometry and analysis of the 1820's–30's is concerned with the intersections of a variable curve with the curves of a linear series; yet another, with the geometric properties of the intersections of two plane curves; and still another, with geometry on curves and on surfaces; and last, the representation of one curve or surface upon another. Parts of these advanced theories belong to algebraic geometry, parts to analysis, the latter through the parametric representation of curves and surfaces by means of certain special functions intensively studied during the nineteenth century. Nothing further can be said here about these highly technical developments; a little more will appear by implication in connection with analysis. But in taking leave of them, we record that literally hundreds of men from the 1860's to the 1930's devoted the best years of their working lives to these kinds of geometry. Where so many did work of high quality, it would be invidious to single out individuals. One, however, was outstanding, the fertile and industrious R. F. A. Clebsch (1833–1872, German).

The decisive battle in the war between the purists and the analysts lasted twelve years, from 1847 to 1860. The first date marks the publication of K. C. G. von Staudt's (1798–1867, German) *Geometrie der Lage;* the second, the revised version of this 'geometry of position' in the same author's devastating masterpiece *Beiträge zur Geometrie der Lage* (1856, 1860).

It may be said at once that the uncompromising purist von Staudt drove the enemy from the field, but that the analytic geometers retired in good order with all their machinery intact. The victor was left to enjoy the fruits of his barren victory alone. In proving that geometry could, conceivably, get along without analysis, von Staudt simultaneously demonstrated the utter futility of such a parthenogenetic mode of propagation, should all geometers ever be singular enough to insist upon an exclusive indulgence in unnatural practices. This may not have been what von Staudt intended; it is merely what he accomplished. If the total exclusion of algebra and analysis from geometry must result in any game as complicated and as artificial as von Staudt's, then the game is not worth its candle, and geometric purity has cost more than a normal geometer should be willing to pay. None of this detracts from the merits of what von Staudt did. His purification of geometry remains one of the masterpieces of mathematical reasoning. Somebody, no doubt, had to do once for all what von Staudt did, whether it was worth doing or not. Its lasting contribution to mathematics is the unintended self-destruction of the ideal of total geometric purity.

Observing that cross ratio involves the concept of distance in the line segments from which the ratio is compounded, and remarking that projective geometry professes to be concerned with those geometric properties that are independent of distance and angle, von Staudt proposed to cut the vicious circle by eliminating measures, and therefore numbers, from geometry. The root of the trouble seemed to be that coordinates or their numerical equivalents, presumably extraneous to projective geometry, were subtly implicit in all the classical developments of the subject. The program of von Staudt would reduce number to form, the exact opposite of what Pythagoras proposed and what Kronecker believed he had accomplished. If both von Staudt and Kronecker achieved their aims, number and form may be one. But it seems more plausible that whatever identity, if any, underlies both is merely an irreducible abstract structure of mathematical logic on which both are based. Such speculations as these, however, were far in the future when von Staudt purified geometry. His theory of what he called 'throws' gives a purely projective algorithm for cross ratio and imaginaries. Most remarkably, the algorithm distinguishes between a complex number and its conjugate; conjugate imaginaries appear as the double points of an involution on a real straight line. It is interesting to note here a similarity between von Staudt's mathematical thought and Dedekind's: faced with a finite problem in arithmetic, Dedekind resorted to infinite classes in his solution; determined to expel imaginaries from geometry, von Staudt replaced them by infinities of real points.

It is sometimes asserted that von Staudt was not wholly successful in his attempt to geometrize real and complex numbers. The abstract geometries of the twentieth century would seem to support this contention. For although it may be possible to geometrize the numbers with which von Staudt was concerned, it seems unlikely that any algorithm whatever could reduce the elements of an abstract space to anything either more or less abstract than what they already are. The problem solved by von Staudt, if he did solve it, is of a kind that has been clearly formulated only by the modern postulational method, which was not in existence in the 1850's.

Cayley encountered a problem of the same genus as von Staudt's in his projective theory of metric geometry. This will be described shortly. Cayley's projective equivalent (1859) of metric distance is based on cross ratio, and therefore involves the very notion of distance which it was designed to eliminate. Cayley himself was aware of this, but he did not attempt to remove the vicious circle. It is probably correct to say that neither the nature of von Staudt's and Cayley's problems, nor the logical analysis necessary for satisfactory solutions, was understood before the twentieth century.

The struggle between the purists and the analysts, as typified in two of its heroes, illustrates certain general phenomena in the development of mathematical thought of more than geometric interest. Plücker's career might form the basis for a study of mental inertia. Steiner's contemporaries, as already noted, called him "the greatest geometer since Apollonius." Some even substituted Euclid for Apollonius in their meed of admiration for Steiner's synthetic genius. Plücker was not called anything much; he was rather ostentatiously ignored by nearly all the elite of geometry. Or at least he personally felt that his fellow geometers were smugly indifferent to his work; and he abandoned mathematics for physics, where he is still remembered. Toward the close of his life, Plücker emerged into the light again to compose his great treatise on line geometry, *Neue Geometrie des Raumes gegrundet auf die Betrachtung der geraden Linie als Raumelemente*, published posthumously (1868–9) under the sympathetic editorship of Klein.

Plücker's return to geometry was partly occasioned by the warm appreciation of Cayley for his work. Cayley appears to have been the one first-rate mathematician who had an adequate conception of what Plücker was doing for geometry. Steiner's dazzling brilliance blinded the majority to Plücker's incomparably more massive achievements. Plücker's geometry was neither pretty nor—vile but just word—elegant as Steiner's was. Steiner flaunted his incapacity for analysis, although some of his colleagues insinuated that "the old fox" knew a great deal more than he would admit and, like Poncelet in his fundamental work, occasionally concealed in synthesis what he had discovered by analysis. But even if this is no more than a malicious canard, Steiner was a contemporary of Apollonius in this thinking. Apollonius would have understood Steiner's geometry immediately and, with a few days' practice, might even have beaten his modern rival at the ancient game. But to understand and appreciate what Plücker was doing, Apollonius would have needed a new brain of a kind they did not produce in ancient Greece.

If anyone in the nineteenth century is to be dubbed the greatest geometer since Apollonius, Steiner now seems to be an unlikely candidate for the honor. Yet fashion turned its broad back on Plücker and favored Steiner with its sweetest, silliest smile. As has happened more than once in the history of mathematics, the man with new and fruitful ideas had to die before he might enjoy whatever satisfaction there may be in the esteem of one's fellow workers.

Projective metrics

As the last of the major contributions of projective geometry to mathematical thought which we shall describe, we select Cayley's reduction (1859) of metric geometry to projective. Cayley gave details only for plane geometry; but with suitable modifications his method can be extended to space of any finite number of dimensions in which a numerical 'distance function' is defined for any pair of elements in the space.

Abstracting the familiar intuitive properties of the distance between any two identical or distinct points in a plane, geometers lay down the following postulates for the distance, $D(p, q)$, between the elements p, q of any space whose elements are $p, q, r, \ldots$. (1) To any two elements p, q (identical or distinct), there corresponds a unique real number, their distance, $D(p, q)$. (2) $D(p, p) = 0$. (3) $D(p, q) \neq 0$, if p, q are distinct. (4) $D(p, q) = D(q, p)$. (5) $D(p, q) + D(q, r) \geq D(p, r)$. The last is called the triangle (or triangular) inequality; it has already been noted in another connection, and will occur again.

It was observed in effect by Cayley and E. Laguerre (1834–

1886, French) independently that these five postulates for distance have a solution $D(p, q)$ in plane geometry other than the usual one giving the distance between two points as a function of their coordinates by means of the Pythagorean theorem. With the new definition of distance, and a corresponding one for angle, Cayley converted metric geometry with its usual definitions of distance and angle into a species of projective geometry. In short, he showed that the metric properties of Euclidean space can be reinterpreted as projective properties. Although the details are too technical for brief description, a hint may be given of Cayley's approach. The quotations are from his sixth memoir on quantics (1859) and his own notes on it in his collected mathematical papers.

> . . . the theory in effect is, that the metrical properties of a figure are not the properties of the figure considered *per se* apart from anything else, but its properties when considered in connection with another figure, viz., the conic called the absolute." "Metrical geometry is thus a part of descriptive [projective] geometry, and descriptive geometry is *all* geometry, and reciprocally. . . .

Regarding Cayley's '*all*,' we must remember that he was writing in 1859. Cayley at first honored his 'absolute' with a capital 'A,' a deserved tribute to the magnitude of his invention. But on learning that the Absolute was commonly used by metaphysical theologians to designate a certain extraspacial, extratemporal Entity, Cayley, who was a devout Christian, hastily descended to lower-case 'a.' Cayley's absolute can be imaginary.

It may have been the additive property of collinear distances that suggested Cayley's projective distance and his absolute. For if p, q, r are collinear points, and if the straight-line segments pq, qr, pr are taken with their proper signs according to the usual rule, then $pq + qr = pr$. This resembles the theorem for the logarithm of a product. In any event, Cayley defined the distance $D(p, q)$ between two points p, q in terms of a logarithm, as follows. The join of p, q cuts a certain fixed conic, Cayley's 'absolute,' in two points p', q'; when p, q are any fixed points, the four collinear points p, q, p', q', taken in a certain order, determine a unique cross ratio; a constant, k, times the logarithm of this cross ratio is Cayley's definition of $D(p, q)$. It is easily seen that this $D(p, q)$ satisfies the stated postulates for a distance function.

Thirteen years after Cayley's reduction of metric properties to projective by means of his absolute, Klein (1871) noticed that the projective definitions of distance and angle provided a simple unification of Euclidean geometry and the classical non-Euclidean geometries. These geometries, Klein showed, differ basically only in their respective distance functions. In Cayley's definition, the constant k and the conic fixed as the absolute can be so chosen that the respective classical geometries of Lobachewsky and Bolyai, Riemann, and Euclid are completely specified according as the absolute is real, imaginary, or degenerate.

This striking result of Klein's was a fitting climax to half a century's striving for clarity in the projective geometry restored to life by Poncelet. Still greater things were to come a year later (1872) in Klein's famous *Erlanger Programm*. This will be noticed in connection with invariance. Klein's program dominated much of geometry for almost half a century. It was superseded by younger ideas that became popular only with general relativity after 1916, but which had their origin in Riemann's revolutionary work of 1854. We shall consider this next.

From cartography to cosmology

The problem of constructing flat maps of the earth's surface was one origin of differential geometry, which may be roughly described as the investigation of properties of curves and surfaces in the neighborhood of a point. Still roughly, it is required to specify the geometry of a sufficiently small neighborhood with sufficient accuracy, the specification to be valid for the neighborhood of any point on the curve or surface investigated. Another origin of this 'local' geometry was the study, in the seventeenth and eighteenth centuries, of tangents, normals, and curvature, the calculus having provided adequate means for a general attack. A third source is evident in the dynamics of the eighteenth century, particularly in constrained motion, as in the dynamics of a particle restricted to move on a prescribed surface. With problems of these general types occur also their obvious inverses. For example: given a particular formula for the geodesic distance between any two neighboring points, to determine the most general surface for which the formula holds; or to classify surfaces with respect to their lines of curvature. Many of these differential problems have immediate generalizations to space of any finite number of dimensions. The resulting theories, as might be anticipated even from these meager hints, are of vast extent and have close connections with differential equations and mathematical physics.

Attempting neither a history nor a catalogue of what has

been done since 1700 in differential geometry, we shall select a few typical incidents in the main line of progress, sufficient to connect the physical algebra already discussed with the analysis, differential equations, mechanics, mathematical physics, and the non-Riemannian geometries of the twentieth century to be described in subsequent chapters. The increasing attention paid to quadratic differential forms from Gauss (1827) to Riemann (1854), Christoffel (1869), and Lipschitz (1870), then from Ricci (1887) to Einstein and others (1916–), blazes an easily followed trail from the cartography of the earth's surface to the mapping of a large sector of cosmology on differential geometry. A map is not necessarily a picture on a sheet of paper. The maps of theoretical physics are mathematical descriptions of physical phenomena.

Like so much else in modern mathematics, differential geometry got its first real start in the analysis of "the myriad-eyed Euler," who overlooked nothing in the mathematics of his age, totally blind though he was for the last seventeen years of his life. In 1760 he investigated lines of curvature. This work inspired Monge to his own more systematic investigations (1781) in the same direction, and to his general theory of curvature, which he applied (1795) to the central quadrics. Equally significant for the future of mathematics was Monge's elucidation of the solutions of partial differential equations by means of his theory of surfaces. The geometric language in which partial differential equations are frequently discussed originated in this early work of Monge.

Another of Monge's inventions, his descriptive geometry, is of less mathematical interest than his analysis of differential equations, but possibly of greater technological importance. Without descriptive geometry of some sort, the engineering sciences of the nineteenth century would have developed much more slowly than they did. Monge's scheme for representing solids on one plane diagram by means of two projections, a 'plan' and an 'elevation' on two planes originally at right angles to each other before being laid flat, facilitated the visualization of spacial relations, and provided a uniform graphics for solving such problems as determining the curves in which two or more surfaces intersect. Cut-and-try methods might waste a great deal of metal in fitting two pipes of different dimensions at a given angle. This problem is solved with no waste as one of the earlier exercises in descriptive geometry. Practical mechanical drawing, without which the construction of modern machinery would hardly be feasible, evolved from Monge's simple scheme. It is seldom pleasant to give the devil his due; but history compels us to state that a problem in fortifications was the origin (1763) of descriptive geometry. The French militarists thought so highly of Monge's invention that they forbade him to publish it, and for about thirty years kept it a secret for their own use. Monge's account of the subject was first published in 1795–6.

Continuing with what since about 1920 has been called classical differential geometry, we note the *Applications de géométrie et de méchanique* (1822) of E. P. C. Dupin (1784–1873, French). Dupin's work was prophetic in several respects. Although the indicatrix was not invented by Dupin, he made more effective use than had his predecessors of this suggestive conic in which a plane parallel to, and 'infinitesimally near to,' the tangent plane at any point of a surface intersects the surface. Analytically, the indicatrix introduces a *quadratic* differential form into the geometry of certain curves (the asymptotic lines) on a surface.

This is analogous to a method of approximation in mathematical physics, where the state of a medium in the neighborhood of a point is obtained to a sufficient degree of approximation by neglecting infinitesimals of order higher than the first in the Taylor expansion of the function expressing the exact state of the medium at any point. This procedure is not universal; but where it is applicable, it is one source of linear differential equations in the physical sciences. Geometrically, the indicatrix is useful in the study of two of the most interesting families of curves on surfaces, the asymptotic lines and the lines of curvature. Dupin also investigated triply orthogonal families of surfaces, not as a barren exercise in the differential calculus, but because certain instances of such families are of the first importance in potential theory and other departments of mathematical physics. This aspect of differential geometry will be noted in another connection, when we follow the contributions of physics to mathematics, especially in Lamé's conception of coordinates. Another detail of Dupin's geometry was to assume an unforeseen significance in the 1890's, when Klein and M. Bôcher (1867–1918, U.S.A.) observed that the surfaces called cyclides, invented by Dupin, afford a unified geometric background for a wide class of differential equations of scientific importance. A cyclide is the envelope of a family of spheres tangent to three fixed spheres. Dupin's geometry was thus one source of much analysis of the nineteenth century. Triply orthogonal systems of surfaces, for instance, were the occasion for one of Darboux' more famous works, extending to 567 pages, which in turn partly inspired G. M. Green (1891–1919, U.S.A.) to a notable simplification (1913) of the general theory as an application of the so-called projective differential geometry of E. J. Wilczynski (1876–1932, U.S.A.). The last is based in part upon a pair of simultaneous partial differential equations of the second order. Green's 27 pages (1913) incidentally included the meat of Darboux' 567.

Projective differential geometry as practiced in the third decade of the twentieth century offered an interesting example of national preferences in mathematical technique. The two principal schools, the American and the Italian, sought essentially the same objectives, but by radically different methods. Each progressed far in its own direction; both were effectively halted, at least temporarily, by obstacles apparently inseparable from their respective methods. Theoretically adequate for any problem that might arise in the subject, the American method was retarded by wildernesses of unavoidable calculations. A less prosaic but equally discouraging difficulty, to be described presently, blocked the Italian approach to the generality of a projective differential geometry of higher space.

The American school followed the lead of Wilczynski, who presented his theory, with numerous applications to special problems, in a series of memoirs, beginning in 1901, and in a treatise (1906) on the general method. Wilczynski had been a pupil of L. Fuchs (1833–1902, German), under whom he acquired a mastery of the theory of differential equations as it was at the close of the nineteenth century. It was therefore but natural that he should base his geometry on a complete independent system of invariants and covariants of a system of one or more linear homogeneous differential equations. A fundamental set of solutions of the equations uniquely determines the several geometric objects investigated, up to a projective transformation. Under appropriate transformations of the dependent and independent variables in the differential equations and in the parametric equations of the accompanying geometric objects, the objects and the forms of the differential equations are invariant, although the coefficients of the equations will usually be changed. The covariants basic for the geometry are functions of the new coefficients, their derivatives, and the new dependent variables, which differ at most by a factor from the same functions of the original variables and coefficients; an invariant is a covariant not containing the dependent variables or their derivatives. The Lie theory of transformation groups (described here in the chapter on invariance) is the implement of calculation for obtaining the covariants and invariants as necessary preliminaries to the geometry. Probably almost anyone who has ever seriously attempted to solve differential equations by the Lie theory will appreciate the labor inherent in any such heroic project as Wilczynski's and agree with Galois that, whatever the nature of its unchallenged merits, the theory of groups does not afford a practicable method for solving equations. Galois of course was speaking of algebraic equations, but his opinion, in the judgment of experts in the Lie theory, carries over to differential equations. Beyond a not very advanced stage of complexity, the calculations become prohibitive to even the most persevering obstinacy. The Italian method circumvented the Lie theory.

About 1913 the Italian school headed by G. G. Fubini (1879–1943) approached projective differential geometry through differential forms, arriving at systems of differential equations of the type from which Wilczynski had started. By restricting the analysis to systems in which the coefficients are legitimately specialized, and thereby simplified, by permissible transformations, the basic covariants are reduced to fairly manageable shape. The method of calculation is the absolute differential calculus, or tensor analysis, of M. M. G. Ricci (1853–1925, Italian), which was noted earlier in connection with the general progress of recent mathematics toward structure. The Ricci calculus, however, originated in the algebra of quadratic differential forms. It was therefore inapplicable to the higher differential forms hinted at in passing by Riemann in his dissertation (1854) on the hypotheses which underlie geometry. But these forms are those appropriate for a projective differential geometry of higher space. The Italian method seemed definitely to be inextensible to a variety of m dimensions in a space of $n > 4$ dimensions, for $1 < m < n - 1$. Nor is there a covariant quadratic form for these cases. It is noted in another connection that the lack of an absolute calculus for differential forms in higher space may be supplied, if, for example, physical speculations should render a serious effort to develop such a calculus scientifically profitable. The Ricci calculus did not come into its own geometrically until it was publicized by the relativists, when the geometers adopted and further developed it. The projective differential geometries of the American and Italian schools do not seem to have attracted physicists.

These somewhat miscellaneous details have been recalled to underline the estimate of classical differential geometry which was that of a majority of professionals in the 1920's. Since its inception in the work of Euler and Monge, differential geometry had expanded somewhat lawlessly, until by 1900 it embraced a loosely coordinated collection of special problems and incomplete theories, thrown together with no detectable aim and without any clearly defined objective. Such, for example, was substantially the opinion of Hadamard. In contrast with this disorderly luxuriance, the differential geometry that became popular with the application of Riemannian geometry to physics and cosmology in general relativity was unified and given definite aims by the absolute differential calculus, or tensor analysis, of Ricci and Levi-Civita. When at last a uniformity in method was recognized, interest in classical differential geometry all but collapsed. Numerous special results obtained in the older tradition had long since passed into the general structure of infinitesimal geometry and analysis; but creative work in differential geometry took a new direction. Among other special developments which had seemed promising in the early 1900's, but which had lost much of their appeal by the 1920's, the projective differential geometry of one American school joined the classics which are respected but seldom cultivated. The line of descent from the old to the new, as already indicated, was from Gauss, through Riemann, to the tensor calculus.

Gauss (1827) made the first systematic study of quadratic differential forms in his *Disquisitiones generales circa superficies curvas*, in which the main theme is the curvature of surfaces. The forms investigated are in two variables only. With its relevance for the deformation of surfaces and the applicability of one surface on another, Gauss' theory is a direct descendant of cartography. This aspect, however, was not that which suggested the far-reaching generalization of differential geometry by Riemann. Geodesy also was one of Gauss' major interests (1843, 1847) in applied mathematics; and it too is partly a matter of quadratic differential forms, the line element on a spheroid being the square root of a quadratic differential form in two variables with variable coefficients. Taking the final step in this direction, Riemann, in one of the most prolific contributions ever made to geometry, passed immediately to the general quadratic differential form in n variables, with variable coefficients, in his vital classic on the foundations of geometry, *Über die Hypothesen welche der Geometrie zu Grunde liegen*, 1854.

Riemann's taste for speculative philosophy has made parts of his great essay needlessly difficult for mere mathematicians. Fortunately for geometry, Riemann's mysterious description of a manifold can be ignored; for when he proceeds to mathematics, he actually uses nothing more abstruse than an n-dimensional number-manifold. It would be interesting to know whether Riemann imagined himself the originator of this notion. But as he seldom mentions other mathematicians in any of his work, even where it is plain that he has profited by their ideas, it is impossible to say how much, if anything, he owed to others. The general manifolds which Riemann attempted to define, but which he did not use, might be interpreted as the abstract spaces of the twentieth century.

The mathematics of Riemann's geometry interweaves two fundamental themes: a generalization of the Pythagorean theorem to any space (number-manifold) of n dimensions; curvature in such spaces. If

$$(x_1, \ldots, x_n), (x_1 + dx_1, \ldots, x_n + dx_n)$$

are the coordinates of neighboring points in the space, and if ds is the infinitesimal distance between these points, it is postulated that $ds^2 = \Sigma\Sigma g_{ij}dx_idx_j$, in which the double summation refers to $i, j = 1, \ldots, n$; the g_{ij} are functions of $x_1, \ldots, x_n$; and $g_{ij} = g_{ji}$. In laying down this postulate, Riemann recognized that it gives a sufficient, but not necessary, specification of an elementary distance which is to retain the cardinal properties of a distance function; and he explicitly stated other possibilities. These had not been exploited (at least in print) as late as 1945, although as early as 1924 H. P. Robertson (1903–, U.S.A.) had investigated the analogue, for these possibilities, of the tensor calculus appropriate for the Riemannian geometry of general relativity. The metric geometry of a particular Riemannian space is determined by the g_{ij} occurring in the ds^2 for the space. Ignoring special cases, Riemann proceeded at once to his generalized curvature, guided partly by analogy with the Gaussian theory for a two-dimensional space. He then made the remarkable conjecture that his new metrics would reduce questions concerning the material universe and the "binding forces" holding it together to others in pure geometry.

Bolder even than Riemann, Clifford confessed his belief

(1870) that matter is only a manifestation of curvature in a space-time manifold. This embryonic divination has been acclaimed as an anticipation of Einstein's (1915–16) relativistic theory of the gravitational field. The actual theory, however, bears but slight resemblance to Clifford's rather detailed creed. As a rule, those mathematical prophets who never descend to particulars make the top scores. Almost anyone can hit the side of a barn at forty yards with a charge of buckshot.

The next long stride after Riemann's toward modern differential geometry was the determination by Christoffel (1869) of necessary and sufficient conditions that a quadratic differential form of the kind in Riemann's ds^2 be transformable into another by a general functional transformation on the variables. The same problem was also treated by Lipschitz (1870). Christoffel's solution proved the more useful. In the course of his analysis, Christoffel invented the process named covariant differentiation by Ricci (1887), and used it to derive a sequence of tensors from a given one. Beltrami and others, especially of the Italian school of geometers, used what are essentially tensors; but it remained for Ricci to isolate and perfect the tensor calculus as an independent algorithm.

The algebra of tensors as a generalization of vectors was mentioned in an earlier chapter. The further development of Riemannian geometry will be noted in connection with invariance. We may conclude this sketch with a summary indication of the mathematical reason for the scientific utility of tensors.

A functional transformation on the variables of a tensor transforms the tensor into another whose components are linear homogeneous functions of the components of the original tensor. A tensor, like an ordinary vector, vanishes if and only if each of its components vanishes. A transformation of the kind stated is, geometrically, a general transformation of coordinates, when the variables are interpreted as coordinates in a space of the appropriate number of dimensions. It follows that if a tensor vanishes in one system of coordinates, it vanishes in all; the homogeneity is the decisive factor. This is equivalent to saying that if a system of equations is expressible as the vanishing of a tensor, then the system will be invariant under all transformations of the variables in the system. But this is precisely the condition imposed by one of the postulates of general relativity on a system of equations, if the system is to be an admissible mathematical formulation of an observable sequence of events in physics or cosmology.

CHAPTER 16

The Impulse from Science

This chapter is introductory to the six following, in which we shall describe certain typical developments in the evolution of analysis from the seventeenth century to the twentieth. Analysis, perhaps more clearly than algebra or geometry, exhibits the constant influence of science on the general development of mathematics.

We saw that the calculus owed as much to kinematics as it did to geometry. From the death of Newton (1727) to the twentieth century, science continued to stimulate mathematical inventiveness. Of subsequent additions to mathematics originating at least partly in science, the most highly developed are the vast domain of differential equations, the analysis of many special functions arising in potential theory and elsewhere, potential theory itself, the calculus of variations, the theory of functions of a complex variable, integral equations and functional analysis, statistical analysis, and differential geometry. By 1800, the calculus of variations and differential equations had advanced sufficiently to be recognized as autonomous but interdependent departments of mathematics; the statistical method was still an embryonic possibility in the theory of probability; while the theory of functions of a complex variable had yet to wait a quarter of a century for systematic development by Cauchy, although some of the basic results were implicit in the applied mathematics of Lagrange and others in the eighteenth century.

In following the growth of rigor from 1700 to 1900, we noted a constantly sharpening precision of mathematical logic, and saw that attempts to provide a self-consistent foundation for analysis led in the early twentieth century to a period of confusion and a recognition of the necessity for ever more subtle reasoning. Leaving all doubts behind for the present, we now enter an untroubled region where the end justifies the means. The end is the increase of scientific knowledge, to which mathematics is but one of several means. As in previous accounts, we shall attend only to typical features illustrative of general trends. There are first two matters of possibly wider significance to be noticed: the influence of eighteenth-century mathematics on society; and the response of society, especially after the Napoleonic era, to mathematical research.

Mathematics in the Age of Reason

The most significant contribution of eighteenth-century mathematics to civilization was a rational outlook on the physical universe, for which dynamical astronomy and analytic mechanics were mainly responsible.

The eighteenth century has been called the Age of Reason, also an age of enlightenment, partly because the physical science of that century attained its freedom from theology. In the hundred years from the death of Newton in 1727 to that of Laplace in 1827, dogmatic authority suffered the most devastating of all defeats at the hands of scientific inquiry: indifference. It simply ceased to matter, so far as science was concerned, whether the assertions of the dogmatists were true or whether they were false. At the beginning of the period, it was customary to seek a teleological explanation for the principles of mechanics to accord with the orthodox theology of the time; when Laplace died, all such irrelevancies had been quietly ignored for forty years. Mechanics had at last come of age. Absolute truth, as revealed by science, fled to pure mathematics, where, according to some, it still resides.

The French Revolution, beginning in 1789, accompanied the change; and we might be tempted to ascribe the maturing of the exact sciences wholly to that very thoroughgoing upheaval. But the final liberation had occurred in the preceding year with the long-delayed publication of Lagrange's analytic mechanics. Here, for the first time, a masterpiece of mathematics and science of the very first rank stood erect on its own mathematical and scientific feet without external support. No mysterious spirit of nature was invoked; the work undertook to *describe*, not to *explain*, the mechanical behavior of material systems.

A free translation of a few sentences will indicate two of the respects in which Lagrange's mechanics differed radically from its predecessors in both science and mathematics. In his preface, Lagrange writes:

> I have set myself the problem of reducing this science [mechanics], and the art of solving the problems appertaining to it, to general formulas, whose simple development gives all the equations necessary for the solution of each problem. . . . No diagrams will be found in this work. The methods which I expound in it demand neither constructions nor geometrical or mechanical reasonings, but solely algebraic [analytic] operations subjected to a uniform and regular procedure. Those who like analysis will be pleased to see mechanics become a new branch of it, and will be obliged to me for having extended its domain.

From this it is clear that Lagrange fully realized the significance of what he had done. The following quotation, typifying the spirit of the entire work, indicates his grasp of the abstract nature of mathematical mechanics: "The second fundamental principle of statics is that of the composition of forces. It is founded on this supposition: . . . " Thus the *principles* of mechanics are founded on *suppositions*, that is, on postulates, and are not eternal truths revealed to a groping mankind by the grudging generosity of some supernatural intelligence. It is mathematical and scientific rationalism like this that validates the claim of the eighteenth century to be called an age of reason.

Such clarity of mind as Lagrange's, however, was the rare exception among mathematicians and scientists in his day and for over a century after his death in 1813. Lagrange's most prominent contemporary in the exact sciences, the self-confident Laplace, convinced himself and two generations of eager philosophers that the Newtonian mechanics of the heavens was absolutely and eternally true; and on this basis he sought to establish the everlasting stability of the solar system.

Almost aggressively hostile to the pretensions of the older absolutism, the would-be skeptic Laplace substituted one dogmatic creed for another. It was largely due to the successes of his own celestial mechanics and his widely appreciated popular exposition of the mathematical consequences of Newtonian gravitation that a crude mechanistic philosophy afflicted nearly all physical scientists and many philosophers of the nineteenth century.

Eighteenth-century mechanics was also partly responsible for the speed with which machinery overwhelmed civilization in the early nineteenth century. Instead of remaining the private servant of the intelligentsia, the mechanistic philosophy incontinently shared its inestimable benefits with the proletariat. Hundreds of thousands to whom Lagrange and Laplace might have lectured for years with no transfer of ideas were converted by the dumb, inerring accuracy of their monotonous machines.

That a fully developed abstract theory appeared first in modern mathematics from the applied side is less remarkable than it may seem at first sight. Mechanics and mathematical physics generally had no such crushing burden of tradition to throw off as had geometry. Mathematical mechanics was little more than a century old when Lagrange saw what it was. Only about sixty years before Lagrange published his mechanics, Saccheri's willful faith in the sanctity of Euclidean geometry had compelled him to ignore the promptings of his own acute reason. Possibly if Archimedes rather than Galileo and Newton had formulated the basic 'laws' of dynamics, Lagrange might have hesitated to deflate the foundations of his system to "suppositions." But the postulates of mechanics had not had time to fossilize into eternal truths, and Lagrange was not tempted to outrage his reason—at least in mechanics. But in his effort to rigorize the calculus by basing it on Taylor's expansion, he was as tradition-bound as Saccheri, and possibly for the same reason. The problems of continuity were as old as those of geometry, and the same almost superhuman intransigency was demanded to flout tradition and advance in a totally new direction.

The eighteenth century may indeed have been the golden age of reason in philosophy and human affairs generally that it is said to have been. The exact sciences, as we have just noted, also submitted to reason in that hard-headed century. But in pure mathematics, there was a marked decline from the standard which the ancient Greeks set themselves. The best that reason could do when confronted with a problem in continuity was Lagrange's curious attempt to rigorize the calculus. No classic Greek mathematician could have deluded himself so completely as did Lagrange, the greatest mathematician of his age and one of the greatest of all ages.

The feeling for sound reasoning in mathematics seemed to have been temporarily lost. Except only when logical rigor was almost unavoidable to even moderate competence, as in finite algorithms and combinatorial mathematics, the kind of reasoning that satisfied the leading mathematicians of the so-called age of reason would have shocked Eudoxus and Archimedes. Yet

Archimedes, no mere mathematician but a mechanist of the first rank, was Lagrange's idol. It may be significant that, of all his own great work, Lagrange prized least highly his contributions to the theory of numbers, where without rigid proof for even the seemingly most obvious theorems there is nothing. These had exacted his greatest efforts, and he doubted whether they had been worth their cost. There is no record of Archimedes' having esteemed his practical mechanics above his mensuration of the sphere.

Social stimuli since the death of Newton

The transition from supernaturalism to rationalism in the exact sciences did not take place in a social vacuum. Of several hypotheses to account for the worship of mathematicized reason in the post-Newtonian age, that of economic determinism is the most elastic. In brief, all the work in celestial and analytic mechanics was occasioned by the demands of navigation and ballistics. However, anyone interested may search the technical works of Laplace and Lagrange on mechanics and find no reference to sailing or gunnery. This does not disprove the thesis that the initial impulse for the mechanics of the eighteenth century may have been the mercantile desirability of a reliable nautical almanac and the military necessity for hitting whatever is aimed at. It merely illustrates the verifiable fact that once a mathematical theory has been initiated, whatever its origin, it proceeds by a sort of intellectual inertia to become abstract with no application in sight. We shall see many instances as we proceed. Potential theory, for example, in spite of its mechanical origin, ceased long ago in those of its divisions that interest professional mathematicians to have any discernible relevance for science or technology. These abstrusities may become practically useful tomorrow; but only should time be reversed and evolution unfold inward will any application of them become their origin.

The like holds for the calculus of variations, except that in this instance theology rather than science was the initial source in the eighteenth century. P. L. M. de Maupertuis (1698–1759, French) propounded (1747) a somewhat obscure form of the mechanical principle of least action because he credited his parsimonious deity with an aversion to avoidable effort or other waste. This might be construed as theologic, not economic, determinism unless, as some might insist, the theology was economically determined in the same way as the ballistics.

Asserting that "Nature always acts by the shortest path," Fermat also had gone behind observable phenomena in deriving his optical principle of least time. But Newton framed no superfluous hypotheses in determining (1687) the surface of revolution offering the least resistance to motion in the direction of the axis through a resisting medium. Newton's problem, were it proposed for the first time today, might be attributed to economic determinism, on account of its possible application to the marine torpedoes which had yet to be imagined in Newton's backward time. The problem, however, has actually been attributed to Newton's very early advice to a young friend that the latter study ballistics. From these and numerous other examples that might be cited, it seems clear that the truth of a sociological theory when stretched to include *all* mathematics may occasionally vary inversely as its degree of elasticity.

The fact seems to be that, if the mathematicians of the eighteenth century were motivated by anything less obvious than the desire to do mathematics and to earn their livings while following their inclinations, they were unaware of it. The conditions under which the great creators worked were basically different from those of the nineteenth and twentieth centuries. If an economic motivation is to be found, it might be profitably sought in the domestic and foreign policies of Frederick the Great of Prussia (1712–1786), Catherine the Great of Russia (1729–1796), the kings of Sardinia, Louis XVI of France (1754–1793), and Napoleon Bonaparte (1769–1821). The obvious demands of civil, naval, and military engineering made the development of mathematics imperative; and these rulers were clear sighted enough to see that the simplest way of getting mathematics out of a mathematician is to pay his living expenses. At various stages of his career, Euler was attached to the courts of Catherine and Frederick; Lagrange was similarly supported through the government-subsidized Turin Academy, a related military school, and later by Frederick, Louis, and Napoleon. Daniel Bernoulli, often called the founder of mathematical physics, was employed by Catherine. Monge and Laplace were employees of successive French governments in various capacities from military engineering and the training of government engineers to affairs of state, and likewise for Fourier and a number of less distinguished mathematicians. But before the Napoleonic era, once these men had advised their employers on the technical questions, usually simple, proposed to them, they were free to spend their working time as they chose. Consequently, an enormous amount of mathematics having no detectable application was created. That much of it proved of practical value years or decades later does not alter the fact that its motivation was not economic. All this was printed with the rest largely at public expense in the proceedings of government-subsidized academies. Up to the Napoleonic era, the learned societies were the most important agencies for the publication of research in mathematics.

With the eruption of French 'Liberty, Equality, Fraternity' in 1789, a rapid democratization of mathematical research began. Under Napoleon, the leading French mathematicians earned part of their keep by helping to train civil and military engineers at the Ecole Polytechnique. Others at the Ecole Normale Supérieure taught prospective teachers. Again the major part of the new mathematics produced by the leaders was of no immediate practical value, nor was it undertaken with a view to possible applications. Napoleon no doubt was partly responsible for this liberality. Provided the schools supplied him with a steady flow of competent civil servants and expert engineers to fill the rather frequent vacancies that might be anticipated in a militaristic regime, he was content. Some of the mathematicians compiled excellent texts for the students; others carefully prepared their few lectures a week; and nearly all did research in their ample spare time. The situation was not unlike that in a few of the more enlightened European and American universities of the twentieth century.

The next and last marked change in the social status of mathematics and mathematicians dates from the decade 1816–26 following the end of the Napoleonic era in 1815. The universities and technical schools increased rapidly in importance as centers of mathematical research; the learned societies were no longer financially able to cope with the torrent of new mathematics that gushed from a hundred sources; and, most important of all, what was everybody's business became nobody's business.

Whatever else may be said for democracy, it has consistently fostered the individual freedom of mathematicians. No mathematician in a democracy is constrained to create mathematics at public expense. All may earn their livings as they please and find what time they can to advance their hobby. Nor are mathematicians as a class debauched by having their researches printed at public expense, although such work can be neither patented nor copyrighted and may be used by anyone without payment of any kind. In the United States the more liberal universities subsidize the publications of their staffs.

It is admitted by the majority of educated persons that a technological and scientific civilization without mathematics is an impossibility. Apparently the most efficient way of getting the necessary job done is to leave it to the initiative of individuals on their own time after a more or less exhausting day's work. Perhaps rather unexpectedly, the result has been a vastly increased output of mathematics since the close of the Napoleonic era over all preceding history.

The strongest stimuli have been the constantly growing demand for scientific and technological instruction to keep pace with the rest of modern civilization, and the vast expansion of media for mathematical publication since the German engineer A. L. Crelle (1780–1855) in 1826 subsidized the first high-grade mathematical periodical. In 1940 there were about 280 such periodicals[1] devoted wholly or in part to the publication of research in mathematics, and the pressure on existing outlets was steadily increasing. Few if any of these publications would survive for two months if forced to pay their way in a competitive society. They are supported by the whole mathematical fraternity without distinction as to race, nationality, or creed. The average mathematician subscribes to as many of them as he can afford, even if but very few of the severely technical articles printed in a year are within his comprehension.

It might be difficult to account for this curious phenomenon on strictly economic grounds. A majority of the subscribers do no research themselves; so it cannot be polite hints from superiors that they shed luster or notoriety on their employers by getting their names into print that account for the altruistic subscriptions. Nor will a mathematically literate reader scanning the abstracts of current research agree that any considerable percentage of the articles printed by the hundreds every month were inspired by economic or other practical needs.

Several thousand[2] periodicals devoted to engineering and other exact sciences take care of the immediately practical applications of mathematics. But these are not the journals to which mathematicians subscribe. Many of these more practical journals pay their way in the competitive market. This, possibly, is the nub of the distinction between pure and applied mathematics.

In following the influence of science on mathematics, it should be remembered that mathematics and its applications are different things. A treatise or a monograph on mathematical physics, for example, may be a mass of formulas and equations from beginning to end, and yet make no contribution whatever to mathematics. If the general fact is not obvious, its extreme cases, as in bookkeeping or the calculation of characteristic functions in quantum mechanics, may illuminate the distinction. As a further aid to comprehension, it is a fair guess that out of a hundred thousand persons picked at random on the streets of New York, or Chicago, or London, or Paris, or Moscow, or Tokyo, not one would know the name of the man whom professional mathematicians almost unanimously considered was the foremost member of their guild since about 1912. He died in 1943, inactive; but his fame is secure even if the average man in the street (or in cultured society) is never likely to hear of him. Of the random hundred thousand, many would instantly name a theoretical physicist who deeply resents being called a mathematician.

CHAPTER 17

From Mechanics to Generalized Variables

Of all the exact sciences, mechanics, the simplest, has probably been the most influential in the development of modern mathematics. The amount of known mathematics applied to a science is no measure of the importance of that science in mathematical evolution; it is the new mathematics inspired by a particular science alone that weighs.

Thus in the first twenty years (1925–45) of its existence, quantum mechanics used an enormous amount of mathematics, from special functions to modern algebra, but did not suggest any essentially new mathematics.[1] General relativity, on the other hand, drawing less heavily on mathematics, was directly responsible for the direction taken by differential geometry about 1920. This newer geometry might have been developed almost forty years earlier. All the necessary technique was available; but it was not until the successes of relativity showed that Riemannian space and the tensor calculus were of more than mathematical interest that differential geometers noticed what they had been overlooking.

Before considering the mechanical origins of certain parts of analysis, we shall give brief summaries of the relevant progress in mechanics in the eighteenth and nineteenth centuries partly responsible for the mathematics.

The search for variational principles

One purpose of the eighteenth-century mechanists was the invention of principles from which the mechanics of Galileo and

Newton could be deduced, and the development of mathematics adequate for the deduction. The main mathematical outgrowths were the calculus of variations; a vast theory of differential equations; a heterogeneous collection of special functions; the beginnings of the theory of line, surface, and volume integrals; more than a hint of n-dimensional space;[2] the origins of potential theory; and certain basic results in what subsequently became the theory of functions of a complex variable.

The first comprehensive principle of post-Newtonian mechanics was D'Alembert's, published in his *Traité de dynamique* (1743): the internal actions and reactions of any system of rigid bodies in motion are in equilibrium. Or, as often expressed: in a dynamical system the reversed effective forces and the impressed forces are in equilibrium.

Supplementing Newton's principles of the conservation of momentum and of the center of mass, Euler and Daniel Bernoulli (1700–1782) independently stated (1746) the principle of conservation of areas. All of these foreshadowed the concept of invariance.

Euler's *Mechanica, sive motus scientia analytice exposita* (1736), was a halfway house between the purely geometrical and synthetic mechanics of Newton's *Principia* (1687) and Lagrange's *Méchanique analytique* (1788). Euler sought to replace synthetic methods by analysis, and was largely successful. Visual geometric intuition, however, was still used, as in the resolution into tangential and normal components in curvilinear motion. Much of the scrappy geometry of curves and surfaces embellishing antiquated texts on the calculus under the comprehensive rubric 'geometrical applications' originated in this way. Possibly mechanics is also partly responsible for classical differential geometry and the intrinsic geometry of curves.

Taking a considerable step toward a general method, C. Maclaurin (1698–1746, Scotch), in his *Complete system of fluxions*, 1742, advanced beyond Euler by using three fixed axes for the resolution of forces. The advantage of Maclaurin's procedure over Euler's is comparable to Descartes' use of one coordinate system to display any number of curves. But the mathematical formulation of each type of problem still required special devices.

Introducing his generalized coordinates, Lagrange in 1760 turned away from mere ingenuity, and started toward the general equations of motion on which he based his analytic mechanics of 1788. The equations of motion for a holonomic dynamical system were then obtained in a form adaptable to the special coordinates most convenient for particular problems. The distinction between holonomic and non-holonomic systems may serve to illustrate certain concepts of mechanics which were partly responsible for the calculus of variations in its earlier form.

To exhibit these, it will be necessary to use the "arbitrary infinitesimal displacements" in terms of which applied mathematicians frequently think. It is not easy to give a mathematically sound treatment of the related variational operator[3] δ and the infinitesimal displacements that lead to useful results before they finally disappear from the calculations; and one rather extreme school advocates abandoning all pretense of deriving the dynamical equations by such means. It would be closer to modern science to state the general equations of motion as postulates, the sole function of these equations being the mathematical statement of dynamical problems, for which the equations themselves are adequate. Their deduction by more or less mystical reasoning dating from ancient Greece and the Middle Ages is of purely historical interest, contributing nothing to understanding or utility. However, as these vestiges of an older mode of thought are still helpful to the majority of applied mathematicians, we shall follow tradition even where it is now asserted by rigorists to be unsound, and by modernists in theoretical physics to be meaningless.[4]

The configuration of a dynamical system, regarded as being composed of material particles subject to constraints (as that all the particles move only on given surfaces) and geometrical conditions (as that the distance between any two given points of a rigid body is constant), is specified at time t by n coordinates $q_1, q_2, \ldots, q_n$, where n is finite. If, for example, the Cartesian coordinates of the rth particle at time t are x_r, y_r, z_r, the system is specified by $3m$ equations

$$x_r = f_r(q_1, \ldots, q_n), \qquad y_r = g_r(q_1, \ldots, q_n),$$
$$z_r = h_r(q_1, \ldots, q_n); \qquad r = 1, \ldots, m.$$

Let each of the generalized coordinates $q_1, \ldots, q_n$ receive an arbitrary infinitesimal increment; say the increments are $\delta q_1, \ldots, \delta q_n$. There is not necessarily a physically possible displacement of the system corresponding to $\delta q_1, \ldots, \delta q_n$; if there is, the system is called holonomic; if not, non-holonomic.

A holonomic system specified by $q_1, \ldots, q_n$ is said to have n degrees of freedom.

Lagrange's equations for a system with n degrees of freedom, for which a potential function exists,[5] can now be stated.

The derivatives of $q_1, \ldots, q_n$ with respect to t being denoted by $\dot{q}_1, \ldots, \dot{q}_n$, and the difference $T - V$ between the kinetic energy T and the potential energy V of the system by L, the equations of motion are

$$\frac{d}{dt}\left(\frac{\partial L}{\partial \dot{q}_r}\right) - \frac{\partial L}{\partial q_r} = 0, \qquad r = 1, \ldots, n.$$

L is called the Lagrangian function, or the kinetic potential, of the system.

The point of historical interest here is in the 'small displacement' $(\delta q_1, \ldots, \delta q_n)$ of $(q_1, \ldots, q_n)$. By a route which need not be retraced, its interest being mechanical rather than mathematical, $(\delta q_1, \ldots, \delta q_n)$ descended from the virtual displacements used by Stevinus, Descartes, and others in statics. Virtual displacements appear fully matured in the principle of virtual work, which was one of the clues followed to an analytic mechanics by Euler and Lagrange. An extremely liberal interpretation of ancient and medieval mechanical speculations has enabled some scholars to detect elusive hints of virtual work all the way back to the Greek philosophers. Virtual displacements, virtual velocities, and virtual work are obviously in the general direction of a calculus of variations. The next major advance in analytic mechanics was in the same direction. It finally reduced the mathematics of statics and dynamics to a topic in the classical calculus of variations.

The statement of several mechanical theorems of the seventeenth and eighteenth centuries had suggested to Euler that all natural phenomena present extrema, and that physical principles, including those of mechanics, should be expressible in terms of maxima and minima. For example, Huygens had shown that Fermat's optical principle of least time holds for media whose index of refraction varies continuously from point to point; James and John Bernoulli had found the catenary as the arc of fixed length passing through two fixed points and having the lowest center of gravity; John Bernoulli's problem (1696) of finding the curve[6] of quickest descent under gravity from one fixed point to another in a vertical plane had been correctly solved by L'Hospital, Leibniz, Newton, James Bernoulli, and John himself; and last, Euler had sought a function whose variation equated to zero would yield the differential equations of dynamics. For a single particle, the ingenious Euler observed that if the velocity v is given as a function of the coordinates of the particle, the desired equations are obtained by minimizing $\int v\, ds$, where ds is an element of the path in which the particle moves. Otherwise expressed, the equations of motion are found on performing the variation $\delta \int v\, ds$ and equating the result to zero. In this way Euler was led to minimize definite integrals in which the integrand is of the form $f(x, y)\ (1 + y'^2)^{1/2}$, $y' \equiv dy/dx$, and the integration is with respect to x.

It is to be noted that Euler, guided by intuition, sought *minima* to express natural 'laws,' possibly because he was of the same pietistic cast of mind as Maupertuis. But Jacobi in his lectures on dynamics (edited, 1866), produced an almost trivial mechanical problem in which the action is a *maximum*. It is therefore customary to speak of *stationary* values of definite integrals for the expression of physical laws, rather than to prejudge the issue by expecting a *least*, a definite integral whose variation vanishes being said to represent a stationary value. The vanishing of the variation is insufficient to secure either a maximum or a minimum, although in many physical situations it is otherwise evident that a definite one of these must occur, and it is seldom necessary to proceed further. But modern science does occasionally demand more than shrewd guessing regarding extrema. Thus, in 1939, R. C. Tolman encountered a capital problem in astrophysics for which scientific intuition seemed insufficient, and for which the more refined techniques of the calculus of variations were at least helpful.

Euler's project was completed in 1834–5 by Hamilton, who showed that the dynamical equations are obtainable from a simple stationary principle, which, for a conservative system, is $\delta \int_{t_0}^{t_1} L\, dt = 0$, where t_0, t_1 are the initial and final times for the passage of a dynamical system with the kinetic potential L from one given configuration to another. A verbal equivalent is as follows. Of all possible motions by which a dynamical system may pass in a given time from one given configuration to another, the actual motion will be that for which the average value of the kinetic potential is stationary. The analytic equivalent of 'possible' is the process of variation, which evolved, at least in mechanics, from virtual displacements. Hamilton also gave a variational principle for non-conservative systems.

The variational principles of mechanics are far from exhausted by those noted. Thus Gauss (1829) reformulated and generalized D'Alembert's principle in his own of least restraint; and Hamilton's principle, also that of least action, were extended in the 1890's to non-holonomic systems. In his *Die Prinzipien der Mechanik* (1894), H. R. Hertz (1857–1894, German) reworked the subject in terms of geometrical imagery. With the development of metric differential geometry for space of *n* dimensions in the second half of the nineteenth century, it was apparently inevitable that the stationary principles of dynamics should be rephrased in the language of geodesics. But here, as elsewhere after Lagrange, whatever gain there may have been was scientific rather than mathematical, new techniques in geometry and analysis suggesting reformulations of mechanics. Lie's theory of contact transformations also was elaborated for its mathematical interest long before it was applied (1889) to unify the differential equations of dynamical systems, although the connection between dynamics and contact transformations was implicit in Hamilton's work of 1834–5. The like appears to be true of Poincaré's integral invariants (1890), also of the topological methods first applied to dynamics by him and since extensively developed by a prolific company of pure mathematicians.[7] In the last instances, however, outstanding problems of dynamical astronomy, such as that of three bodies mutually attracting one another according to Newtonian gravitation,[8] were the ultimate source of the mathematics.[9]

Enough has been given to suggest that mechanics was an important source of the calculus of variations. Before passing on to a summary account of the development of that branch of analysis, we may glance at the scientific significance of variational (stationary) principles in general.[9]

Competent opinion is sharply divided. In the tradition of Maupertuis, Euler, and their predecessors, one side professes to see cosmic profundities in the derivation of Lagrange's equations from a variational principle. The profundities are no longer theological, as in the eighteenth century, but concern unapprehended necessities of the physical universe. Consequently it is claimed that a real but not wholly understood scientific advance is made when the differential equations of a physical theory are shown to be obtainable from a variational principle, as was done, for example, quite early by Hilbert in general relativity.

The other side holds that variational principles are incapable of adding to science anything not already known in a form better adapted to calculation. For this side, any variational principle in science is at most only a concise restatement of more or less ancient history which might become useful, because it is easily remembered, should all working scientists suddenly forget the mathematics they actually use to obtain results that can be checked against observation. This side further asserts that the reformulation of a science in terms of a variational principle is what the average modern pure mathematician does when he attempts to contribute to science, a task for which he is not fitted.

Less immoderate observers occupy a middle position, pointing out that if physicists had scrutinized the duality in Hamilton's optics and dynamics, in which the principles of least time and least action were shown to be interrelated, they might have come upon de Broglie's waves and Schrödinger's wave mechanics about ninety years earlier than they did. But this belongs to the elusive metaphysics of might-have-been, and cannot be considered as a promising suggestion for the future.

There remain, then, only the two extremes with no tenable ground between them. Their respective creeds reflect those of the corresponding sides in mathematics. The disciples of Maupertuis would favor the vision of mathematics as eternally existing and necessary truth; their opponents would see mathematics only as a humanly created language adapted to definite ends prescribed by human beings. It is a matter of individual preference which of these, if either, is considered the worthier.

Functions as variables

Problems of maxima and minima in the differential calculus seek those values of the independent variables for which a given function of them assumes a greatest or a least value. The variables represent real numbers.

In the calculus of variations it is required to determine one or more unknown functions so that a given definite integral involving those functions shall assume greatest or least values. The variables here are functions. As the simplest example, it is required to find the shortest arc joining two fixed points (x_1, y_1), (x_2, y_2). All the infinity of arcs $y = f(x)$, $x_1 \leqq x \leqq x_2$, joining the two points satisfy the end conditions $y_1 = f(x_1)$, $y_2 = f(x_2)$. The shortest will be that one (or those, should there be several) of this infinity which makes $\int_{x_1}^{x_2} (1 + y'^2)^{\frac{1}{2}} dx$, where y' denotes dy/dx, a minimum.

The solution here is intuitively evident, and therefore open to suspicion. It is also 'obvious' that the plane closed arc of given length enclosing the maximum area is a circle; and likewise for a sphere as the surface of given area enclosing the maximum volume. But 'obvious' as the last two theorems are, the Greek geometers[10] attempted to prove them by elementary geometry. Discounting the legend of Queen Dido and her bull's hide,[11] we have in the first of these isoperimetric problems of the ancients the earliest concerning maxima and minima that have been rigorously solved *only* by the calculus of variations. Solutions were delayed till the second half of the nineteenth century.

The simplest mechanical problems of the seventeenth and eighteenth centuries involving minima, such as that of the brachistochrone,[12] transcend intuition but are still within range of ingenious geometry. John Bernoulli in 1697 solved the brachistochrone problem elegantly by special devices, using nothing more advanced than an integration. His brother James far surpassed him in the same year with an inelegant but more general method of solution applicable to a wide class of problems. James Bernoulli's[13] signal merit was his recognition that the problem of selecting from an infinity of curves one having a given maximum or minimum property was of a novel genus, not amenable to the differential calculus and demanding the invention of new methods. This was the mathematical origin of the calculus of variations.

The development of the subject is detailed and intricate, especially in the recent period; and we can give only the briefest summary sufficient to indicate the part played by the calculus of variations in the development of modern analysis. Intrinsically more difficult than some of the other major divisions of analysis, such as the classical theory of functions, the calculus of variations has attracted relatively fewer specialists. But those who have made it their chief concern seem to have been embarrassingly prolific.

For our purposes here, the points of greatest interest are the early emergence of the new calculus as an independent department of analysis concerned only incidentally with the mechanical and geometrical problems in which it originated, and the progression toward a theory of functions of a non-denumerable infinity of variables. The calculus of variations itself is not concerned with the last; but the minimizing arcs of the theory suggest an infinity of variables in two respects. An extremal (a minimizing or maximizing arc) subject to given end conditions is one of an infinity of variable arcs; the arc itself is an infinite set of points. These hints appear to have been partly responsible for the theory of functions of lines ('functionals') and the geometry of spaces of an infinity of dimensions.

There were roughly six stages in the development. The first extended from the last decade of the seventeenth century to about 1740, and is typified by the work of the Bernoullis. The second opened in 1736 with Euler's[14] differential equation giving a necessary condition for a minimizing curve. In 1744 Euler gave a systematic exposition of his method, with needed revisions.

Abandoning Euler's semi-geometrical attack, Lagrange (1762, 1770) passed to the third stage with an analytic method which furnished the differential equations of the minimizing curves. He introduced the variational operator δ and developed its algorithm, greatly simplifying and extending most of the work of his predecessors. With Lagrange, the calculus of variations become an autonomous division of analysis.

The fourth stage, 1786–1837, began with Legendre, who investigated the second variation of an integral to find criteria for distinguishing between maxima and minima. This was analogous to the use of the second derivative for the like purpose in problems of maxima and minima solvable by the differential calculus. Legendre's criteria were inconclusive; Jacobi (1837) gave a critical evaluation of Legendre's analysis, discussing when it would lead to the desired end and when it would not. Jacobi was thus led to his geometrical interpretation of his own criterion in terms of the conjugate point[15] which he defined.

For about forty years—a long time in modern mathematics—after Jacobi's advance, there was no significant progress. But analysis in the meantime was undergoing a basic revision. Weierstrass, "the father of modern analysis," was transforming the mathematics of continuity into a rigid logical system bearing but little resemblance to the intuitive analysis of most of his predecessors. His lectures[16] of 1879 at the University of Berlin on the calculus of variations mark the beginning of the fifth stage. With almost Gaussian indifference to fame, Weierstrass contented himself with lecturing on his revision of the theory; and although his work was not printed in his lifetime, it profoundly influenced the entire future development through the research and teaching of his students. Of the latter, one in particular may be mentioned here, O. Bolza (1857–1942, German) whose lectures over several years at the University of Chicago were responsible for the highly productive American school in the modern calculus of variations. Bolza attended Weierstrass' lectures of 1879.

In addition to rigorizing the entire subject as it existed in his time, Weierstrass made extensive additions of his own. To him are due a new sufficiency condition and the first acceptable sufficiency proofs, for which he invented his fields of extremals. For the geometrical interpretation of his analysis he used the parametric equations of curves, with a consequent gain in generality. This step may have been suggested by the like in differential geometry, which had been current since Gauss (1827) made extensive use of it in his study of surfaces. It goes back even farther, to Lagrange's generalized coordinates in dynamics as functions of the time; but Weierstrass was the first to apply it to the calculus of variations.

The Weierstrassian period lasted into the twentieth century. Its standards of rigor persisted as problems of increasing generality were attacked by modern analysis. There were notable applications to differential geometry in the 1890's, as in the work of G. Darboux on geodesics, later generalized by several other men. The sixth stage dates from 1899–1900, beginning with Hilbert's proof of his differentiability condition for a minimizing arc, assuring in many problems the existence of an extremal, and his exploitation of the invariant integral since named after him. Finally, in 1921–3, L. Tonelli, (1885–, Italian) opened a new chapter, proceeding from Hilbert's work to a revision, concerned principally with existence theorems, of the entire calculus of variations.

The individuals whose names have been mentioned are not, of course, the only men whose labors have created the calculus of variations; nor are the few advances noted an adequate measure of the rich complexity of this subtle division of analysis. Scores of men have contributed hundreds of theorems, until here as elsewhere in modern mathematics what was a narrow specialty in the early nineteenth century began in the twentieth to split into still narrower specialties, each with its assiduous corps of cultivators. Only an expert who has devoted his working life to the subject can take in the whole of it or estimate the vitality of its several subdivisions. The same is true for any major department of modern mathematics; and it may be taken for granted that any short report on a particular topic can indicate only a few of the salient characteristics.

The same general features as in the rest of recent mathematics stand out in the development of the calculus of variations, with one possible difference: some of the most difficult problems appeared early and were partly solved by ingenious men who could not possibly have realized how hard the problems were. Otherwise the progress from special problems to others more inclusive or less restricted followed the familiar pattern of generalization with increasing rigor. Instead of problems concerning arcs with fixed end-points, problems with variable end-points were considered, the earliest being James Bernoulli's (1697) of the curve of quickest descent under gravity from a fixed point to a fixed vertical straight line, a problem with one variable end-point. Generalization in another direction proceeded by modification of the integrand in the definite integral to be minimized. A third type of generalization combined the first two, superimposing generalized end-conditions on the function to be minimized. A far-reaching generalization of this kind was O. Bolza's of 1913, which included several famous problems as special cases, among them Lagrange's of 1770 and A. Mayer's of 1878. Since about 1920 the greatest activity in this direction has been in the United States; indeed, shortly after 1900 the calculus of variations became a favorite field of research with American mathematicians, of whom G. A. Bliss (1876–) and his numerous pupils, and M. Morse (1892–) were particularly active.

Although we cannot discuss special problems, one may be mentioned for its historic interest. J. Plateau's (1801–1883, Belgian) problem (1873), first proposed by Lagrange, to determine the surface of least area with a given boundary is solved physically by the soap film which spans a wire model of the boundary. A complete mathematical solution was given only in 1931 by J. Douglas (1897–, U.S.A.).

In the calculus of variations we have seen the first extensively developed department of analysis in which functions of variables other than those discussed in the ordinary calculus are considered. This long step forward was to prove of more than local significance. Much of the analysis of the twentieth century is concerned with functions of generalized variables, and with the corresponding abstract spaces created to provide the appropriate geometrical description of the analysis. Looking back on the analysis of the eighteenth and nineteenth centuries, we observe many trends toward what has been called general analysis. Enough of these generalizations to indicate the development of some kind of general analysis, and the need for it, will be described in later chapters.

CHAPTER 18

From Applications to Abstractions

In the progression toward general theories of analysis, the special functions devised in the eighteenth and early nineteenth centuries for the solution of problems in dynamical astronomy and mathematical physics played a dominant part in determining the course of modern analysis. From the historical record, it seems incredible that some of the special functions, for example those of Bessel and E. Mathieu (1835–1890, French), would ever have seriously engaged the attention of mathematicians had it not been for the initial impulse from science. But not all of the most extensively investigated functions can be credited exclusively to scientific necessity. Thus the multiply periodic functions developed inevitably from the straightforward evolution of the integral calculus. A few typical cases[1] will suffice to illustrate the general trends.

A central problem of applied mathematics

The hardest thing in any applied mathematics is to strip a scientific or technological problem of enough details, and no more, to bring it within the capabilities of skilled mathematicians and still preserve sufficient of the actual problem to make the solution not utterly irrelevant for practical applications. Observation presents us with no motion immune to friction, and no incompressible fluid; yet the classical hydrodynamics[2] of incompressible fluids without viscosity has had many applications. The all-important problem of deciding what concepts are to be made central in the mathematical description of natural phenomena is of a like character, and requires the same rare combination of scientific insight and mathematical tact for its successful solution. Velocity in kinematics, entropy in thermodynamics, also force, action, and energy in dynamics illustrate the point. A more recent instance is correlation in the statistical method.

The great mathematicians of the eighteenth century excelled in this most difficult field. The distinction between pure and applied mathematicians did not exist, nor was it necessary, when the Bernoullis, Euler, d'Alembert, Clairaut, Laplace, Legendre, and Monge were at their best. It was largely due to their colossal output of both pure and applied mathematics that it became humanly impossible by the middle of the nineteenth century for a man to attain the first rank as a scientist and as a mathematician.

As we look back on all this seething activity, we observe the hesitant beginnings of theories which were to occupy thousands of industrious mathematicians from the early nineteenth century to well within the twentieth. Following one of these along the clue provided by the Bessel functions, which are among the most useful functions in mathematical physics, we shall be led to a central problem of applied mathematics. This problem generated numerous special functions; and from these in turn some of the major divisions of modern mathematical analysis evolved.

Investigating the oscillations of heavy chains, Daniel Bernoulli[3] (1700–1782, Swiss) in 1732 encountered the function later called a Bessel coefficient of order zero. Bessel coefficients of order $\frac{1}{3}$ had appeared earlier in a problem of James Bernoulli's[4] (1654–1705). The vibrations of a stretched membrane led Euler[5] in 1764 to more general Bessel coefficients, and seven years later Lagrange encountered the same functions in elliptic motion. In 1824, the mathematical astronomer F. W. Bessel (1784–1846, German), needing these functions in his investigation of a perturbative function in dynamical astronomy, developed several of their more useful properties. Thereafter, the Bessel coefficients and their immediate extensions, the Bessel functions, appeared in physical science almost as frequently as the circular functions, and chiefly for the reasons indicated next. What follows is relevant for our entire subsequent discussion of the influence of the physical sciences on mathematics.

The advantages of special coordinate systems adapted to specific problems were familiar to geometers before a similar specialization in applied mathematics was recognized as an ultimate source of the indispensable special functions, such as Bessel's, of astronomy and physics. In discussing physical situations involving symmetry about a straight line, for example, it is convenient, indeed almost mandatory, to use cylindrical coordinates (r, ϕ, z), just as it is to use spherical coordinates (r, θ, φ) where there is symmetry about a point. When Laplace's equation[6] $\nabla^2 u = 0$ is transformed from rectangular to cylindrical coordinates the variables are separable, and Bessel's differential equation drops out as that which r must satisfy. The same equation appears similarly in the transformation to spherical coordinates of the equation[7] $\kappa\nabla^2 v = \partial v/\partial t$, to which Fourier was led in his analysis of heat conduction. A typical problem of great generality connected with this equation may serve to illustrate the central problem of applied mathematics which we have in view, that of boundary-values, in which special functions, such as Bessel's, are only details of calculation.

The typical problem is to find a solution of Fourier's equation subject to the following conditions. At each point (x, y, z) of the interior of a homogeneous isotropic solid, the temperature v (satisfying the equation) is to be a continuous function of x, y, z, t, having continuous first and second partial derivatives with respect to x, y, z, and having $\partial v/\partial t$ continuous. The temperature v throughout the body at the initial time $t = 0$ is to be given by $v = f(x, y, z)$, where f is an arbitrary continuous function; and the solution v, obtained as a function of x, y, z, t, must be such that its limit as t approaches zero is $f(x, y, z)$. It may be assumed that if two bodies of different conductivities are separated by a common boundary, the temperatures of the bodies at any point of the boundary are the same.[8] The problem is easily modified to take account of radiation into a surrounding atmosphere: the loss of heat per unit area of the boundary is to be proportional to the difference in temperature between the surface and the atmosphere, in accordance with an empirically established law of cooling. Finally, the temperature at any point (x, y, z) of the boundary at time t may be prescribed as a given continuous function $F(x, y, z, t)$. The solution v of Fourier's equation satisfying these conditions is unique. Special problems of this type leading to Bessel functions are the flow of heat in a circular cylinder or in a sphere whose surface is maintained at zero temperature.

This typical problem is a specimen of boundary-value problems, in which it is required to construct that solution of a given differential equation, ordinary or partial, that fits prescribed initial conditions. If properly posed, the problem has a unique solution; but, as will appear, it is not always obvious that all the conditions of a given situation have been included in the mathematical formulation, or that, if included, they are analytically compatible. The theory of such problems is coextensive with a vast tract of mathematical physics, and has given rise to equally extensive tracts of pure mathematics connected, if at all, only remotely with practical or scientific applications.

Many of the classical boundary-value problems in mathematical physics lead to analogues of Fourier's project of expanding an 'arbitrary' function $f(x)$ in a trigonometric series in the x-interval $-\pi$ to π, say $f(x) = \frac{1}{2}a_0 + \sum_{n=1}^{\infty} (a_n \cos nx + b_n \sin nx)$, where the coefficients a_0, a_n, b_n are to be determined. Under certain restrictions, the coefficients are given by

$$\pi a_m = \int_{-\pi}^{\pi} f(y) \cos my \, dy,$$
$$\pi b_m = \int_{-\pi}^{\pi} f(y) \sin my \, dy, \quad (m \geqq 0).$$

The point to be noted here is that $f(x)$ is expanded in terms of the solutions $\cos mx$, $\sin mx$, of the ordinary differential equation $d^2u/dx^2 + m^2u = 0$.

A central problem of mathematical physics is a generalization of this: it is required to expand a suitably restricted function $f(x)$ in a series of the form $c_0 + \sum_{n=1}^{\infty} c_n\phi_n(x)$, where the functions $\phi_n(x)$ are solutions of a given ordinary linear differential equation. The possibility of the expansion being assumed, the problem amounts to calculating the coefficients c_0, c_1, c_2, ... The conditions under which the series converges must then be determined, if the expansion is to be usable.

It seems conservative to say that the majority of those special functions which have been most exhaustively investigated since the early eighteenth century entered mathematics in this way through the differential equations of astronomy and physics. Although many of them, like the Bessel coefficients, appeared first in a rather haphazard manner in mechanical problems of the early eighteenth century, their wider significance began to emerge only with the problem of separation of variables in the partial differential equations of potential theory and other departments of mathematical physics. This led directly to the expansion problem just described, and to the modern theory of boundary-value problems which furnishes the desired coefficients and justifies the expansions. This phase of the general development of analysis will recur frequently as we proceed.

Once the special functions had fulfilled the more immediate scientific purposes for which they had been invented, they were exploited by numerous analysts whose interests were purely mathematical. Scientific applications were not even remotely envisaged in the continually refined generalizations[9] of the analysis that had sufficed for physical problems. From one point of view, this rapid transition from the immediately applicable to the abstract with no application in sight seems only natural and typical of the general progress of mathematics. Admitting that the development is typical, we may nevertheless question its curiously fortuitous character. The Bessel functions may serve once more as an illustration.

It has often been said by analysts with a taste for elegance that no mathematician left to his own devices would ever have dreamed of inventing anything so uncouth mathematically as the Bessel functions; or, if by chance he had imagined such things in a nightmare, he would have done his utmost to forget them on coming to his senses. Such elegancies as these functions may exhibit in the refinements of twentieth-century analysis, as in the theory of various transforms or in applications to the theory of numbers, were unimaginable to the eighteenth-century mathematicians, whose motives in investigating special cases of the functions were wholly scientific or practical. Whether defined by infinite series or by a differential equation, there was nothing about the Bessel coefficients as first presented to suggest that they and their generalizations might repay exhaustive investigation on their own account.

The like holds for many of the other special functions conceived in science and born into technology, for example the Mathieu[10] functions, introduced (1868) to analyze the vibrations of an elliptic membrane. All the intricate analysis that developed from these scientific origins seems strangely parasitic and accidental to those who believe that mathematics evolves in response to the dictates of an indwelling and eternal necessity.

To these, some of the most highly prized acquisitions of modern mathematics are mere by-products of chance. There is, they maintain, neither reason nor necessity in the selection of what particular things are to be developed; and almost any choice other than that actually made would produce results equally pleasing to a mathematician. So say the practical realists, who also occasionally take mathematicians to task for fleeing to abstractions when seemingly more fertile fields await cultivation.

Against this opinion, it is contended that mathematicians as a class prefer the problems of pure mathematics to those of applied because to do so is merely to follow the line of least resistance. Centuries of trial and error have shown in what directions advances may be anticipated for a moderate expenditure of thought; and the same process of elimination has simultaneously suggested the means of progress. This appears to be the basic reason for the phenomenal popularity of abstract algebra, abstract spaces, and general analysis in the twentieth century. Of the endless variations implicit in the syntax of mathematics at any stage of its development, those following most closely what has already been explicated are usually selected for further elaboration. We shall see a striking and historically important instance of this presently, when we consider elliptic functions.

Mathematics and scientific intuition

It is 'intuitively evident' that electricity applied to a bounded conductor will reach a definite and unique distribution when the conductor is fully charged and no more electricity flows onto it. But it is not evident mathematically. Intuition in mathematics frequently acts as a decoy to credulity. It was so in the evolution of the calculus, and it is so here. The physically evident assertion about the conductor may in fact conceal an ineradicable incompatibility. It may be too sweeping a generalization from crude observations.

The difficulties begin when intuitive notions of a boundary are made precise. Are Peano's area-filling curves, for example, to be admitted as boundaries? When these and similar conditions have been agreed upon, intuition has departed. Let intuition state offhand what will be the distribution of electricity on a one-sided conductor, or on a body like a cactus pad with spines tapering off exponentially to infinity. But, it may be legitimately objected, neither of these abnormalities ever appears in nature or technology. Granting this, we are left with the severely practical problems of deciding which conductors submit to mathematical analysis and of excluding from our calculations those that do not. Until these are solved with moderate completeness, our electrostatics will be applicable only 'in general.' That is, it will supply only dubious information.

A famous crux of mathematical physics shows just how deceptive intuition unrestrained by reason can be. As this—Dirichlet's principle—was of the first importance in the evolution of analysis in the nineteenth century, we shall describe it in some detail.

By a semiphysical argument, Gauss in 1840, and W. Thomson (Lord Kelvin, 1824–1907, Scotch) in 1847, using the calculus of variations, believed they had established the existence of a continuous solution V of Laplace's equation having assigned values on any given closed surface and minimizing the integral[11]

$$\iiint \left[\left(\frac{\partial V}{\partial x}\right)^2 + \left(\frac{\partial V}{\partial y}\right)^2 + \left(\frac{\partial V}{\partial z}\right)^2\right] dx\, dy\, dz,$$

the integration extending throughout the volume enclosed by the surface. It is intuitively evident from the physical situation of which this is the mathematical abstraction that the required V exists. Following Riemann (1851), we therefore assert that the mathematical existence of V is assured by that of the physical problem, and call this Dirichlet's principle, although Dirichlet himself was not so rash as to state it. Dirichlet did, however, follow (1856) Gauss and Thomson in assuming the existence of a minimizing V, a much milder assumption than Riemann's that, because a problem seems to make sensible physics, it must have a mathematical solution.

Unfortunately for intuition, the principle in either form is false. Weierstrass in 1870 proved that the required minimum value of V is not attainable within the domain of continuous functions. What seemed intuitively to be a meaningful problem was thus shown to be a disguised incompatibility. The like holds for the corresponding principle in two dimensions instead of three.

The *principle* being fallacious, what is called Dirichlet's *problem* supplanted it: to find a function $V(x, y, z)$ which, together with its first and second partial derivatives with respect to x, y, z, shall be uniform (single valued) and continuous throughout a given closed region R, and which shall take preassigned values on the boundary of R.

Dirichlet's principle was responsible for a vast amount of pure mathematics after Riemann's appeal to it in the two-dimensional case in his theory (1851) of functions of a complex variable (to be described in a later chapter). As this theory was one of the most extensively cultivated fields of analysis in the latter half of the nineteenth century, it became important to determine restrictions under which Dirichlet's problem is solvable. The outcome was a large division of the modern theory of the potential.

A list of the developers of this highly specialized topic reads like a directory of the leading analysts from Riemann (1826–1866) to Poincaré (1854–1912), and down to the present. For a critical account to 1929, we must refer elsewhere[12] because, after all, potential theory is but one department of dozens in modern analysis, and we can attend here only to general movements. Our present interest in the subject is incidental: it is a typical example of the physical origin of much pure analysis, and of the necessity for more than acute physical intuition in the correct formulation and solution of important problems in applied mathematics.

We note briefly the scientific and historical origins of the theory. Discussing Newtonian gravitation, Lagrange in 1773 (and 1777) observed that the components of attraction at a given point in space, due to a distribution of mass-particles, are obtainable as the space-derivatives of a certain function of the positions of the particles. Thus Lagrange invented what is now called the potential V for a Newtonian gravitational field due to a discrete distribution of mass-particles. Laplace (1782) showed that for a point in empty space the potential V due to a continuous distribution of matter satisfies $\nabla^2 V = 0$; and S. D. Poisson (1781–1840, French) derived (1813) the corresponding equation $\nabla^2 V = -4\pi\rho$ for points within the attracting mass, the density ρ at an interior point being given as a function of the coordinates.

The next long step forward was taken in 1828 by G. Green (1793–1841, English), in his fundamental *Essay on the application of mathematical analysis to the theories of electricity and magnetism*. This contained the extremely useful result known as Green's theorem for the reduction of certain volume integrals to surface integrals.

It may be noted in passing that Stokes' (actually Kelvin's) companion "platitude of mathematical analysis"[13] for the reduction of certain surface integrals to line integrals, which also is of constant use in mathematical physics, made its first public appearance as a problem in a Cambridge examination paper of 1854. Whether any of the examinees solved the problem appears not to be known. But it seems likely that if anyone did turn in a solution acceptable to Stokes, he could not satisfy a modern examiner[14] with the same solution. Like Dirichlet's problem, Stokes' theorem, its proof, and its generalizations have developed into a thriving industry of modern analysis. A concise report of what has been done on this detail alone would occupy a chapter.

Enough has been said to indicate the strictly physical origin of potential theory, in which Dirichlet's problem is only an incident, although one of the first importance. It may be remarked, however, that in the interests of historical justice Laplace's prolific equation should be renamed after Lagrange, who used the equation as early as 1760 in his work on hydrodynamics.

The partial rehabilitation of Dirichlet's discredited principle dates from 1899, when Hilbert proved that under suitable restrictions on the region in which V is defined, on V itself, and on the values assumed by V at the boundary of the region, Dirichlet's problem is rigorously solvable. But it is no longer intuitive in any sense. The unique historical importance of Dirichlet's problem is that it was the first in potential theory to raise the question of existence.

Double periodicity

We pass on to the origins of one of the most extensive departments of nineteenth century analysis, in which practical utility quickly gave way to purely mathematical interest. The history of multiple periodicity is a perfect foil to that of the extremely useful Bessel functions.

The knowledge that many natural phenomena are periodic in time, or approximately so, is probably as old as the emergence of the human race from brutehood. Day and night, the recurrence of the seasons, the waxing and waning of the moon, the physiology of the human body, and many other unescapable facts of daily life must sooner or later have forced the existence of natural periodicity on even the most rudimentary intelligence.

Philosophical extrapolations of single periodicity preceded mathematical formulations by thousands of years. Long before Greece was civilized, the sublimely imbecilic vision of Plato's Great Year, revived in Friedrich Nietzsche's (1844–1900, German) insane dream of an Eternal Recurrence, had evolved from such banal phenomena as the periodicity of the seasons. Fortunately for the sanity of mankind, poetic philosophers have yet to hear of elliptic functions, whose double periodicity leads at once to a two-dimensional Time. In this infinitely ampler time, with its ∞^2 eternity, history repeats itself indefinitely in the parallelograms of a skewed chessboard extending to Infinity in all directions. But the ratio of two sides of any lozenge is real if, and only if, the sides are parallel, when the ratio is Unity.[15]

The mathematics (as opposed to the mysticism) of periodicity originated in 1748 with Euler's completely correct determination of the values of the circular functions when the argument is increased by integer multiples of a half-period. Euler, incidentally, was the first to emancipate the circular functions from slavery to diagrams, and to consider them as numerical-valued functions of a numerical variable. The hyperbolic functions, with one pure-imaginary period, followed immediately as obvious consequences of Euler's exponential forms of the circular functions. They are usually ascribed to V. Riccati[16] (1707–1775, Italian), about 1757; their simple theory was developed in detail by J. H. Lambert (1728–1777, German).

None of this indispensable work suggested that more general functions having two distinct periods, and including both the circular and the hyperbolic functions as degenerate cases, might exist. Abel's discovery in 1825 of these doubly periodic, or elliptic, functions, as they are called, is one of the outstanding landmarks in the history of analysis. The elliptic functions are of the first importance historically, not so much on their own account as for what they instigated. Their singularly rich and symmetrical theory became an invaluable testing ground for the vastly more inclusive theory of functions of a complex variable and for its prolific offshoot, the theory of algebraic functions. These will be considered in a later chapter; for the present we are interested in the genesis of elliptic functions.

The unfortunate term 'elliptic integral,' for historical reasons only, designates any integral of the form $\int \frac{F(z)dz}{\sqrt{R(z)}}$, in which $R(z)$ is a polynomial of the third or fourth degree in z, and $F(z)$ is rational in z. The rectification of the arc of an ellipse leads to a special integral of this type; hence the name. Of mechanical problems leading to elliptic integrals, the most elementary is that of finding the duration of one complete oscillation of a simple pendulum.

The early work on elliptic integrals has long been of only antiquarian interest. A small sample will suffice to indicate its quality. Being unable to evaluate a special elliptic integral appearing in a problem of elasticity (noted in a later chapter), James Bernoulli in 1694 expressed his conviction that the integration was impossible by means of elementary functions. He was right; but a proof of the impossibility lay far beyond his resources. Maclaurin (1724) translated Bernoulli's problem into a geometrical construction, which would have been an advance had he shown what means were necessary and sufficient to carry it out.

The first work to transcend the obvious was that of the Conti di Fagnano (1682–1766, Italian), who in 1716 proved that two arcs of any given ellipse may be determined in an infinity of ways so that their difference is a segment of a straight line. The significance of this is that Fagnano's methods are suggestive of those used by Euler[17] in his proof (1761) of the addition theorem for elliptic integrals. But Fagnano's most remarkable achievement was his discovery that a quadrant of a lemniscate can be divided into n equal parts by a Euclidean construction, where n is an integer of the form $2^m h$, $h = 2, 3, 5$.

As thus stated, the last may give Fagnano slightly more than his due; but he had the substance of it. The next published hint of a general theory behind such constructions occurs in the *Disquisitiones arithmeticae* (1801, p. 593, Art. 335), where Gauss remarks that his theory of cyclotomy "can be applied to many other transcendental functions [beyond the circular], for example to those which depend on the integral $\int \frac{dx}{\sqrt{1 - x^4}}$." This particular elliptic integral was one of those discussed by Fagnano in the work just cited; its inversion leads to the special case of elliptic functions sometimes called lemniscatic functions. It would be interesting to know whether Gauss was inspired by Fagnano's work; Euler frequently expressed his admiration for what his most sagacious predecessor in elliptic integrals had done.

Another early hint of greater things to come appeared[18] in 1771, in the discovery with which J. Landen (1719–1790, English) succeeded in astonishing himself: "Thus beyond my expectation, I find that the hyperbola may in general be rectified by means of two ellipses." Landen's ingenious analytic reformulation (1775) of his geometrical theorem is recast today in the transformation of the second order (more generally, of order 2^n, n an integer) in elliptic functions.

But all of this early work, including much by d'Alembert, was haphazard in comparison with Euler's systematic attack on elliptic integrals and their geometrical applications. Embedded in an enormous mass of hideous formulas and intricate calculations, two items in Euler's contribution outrank all the rest in historical significance. The first was the addition theorem (1761) for elliptic integrals, rated by Euler's contemporaries and immediate successors as the most amazing tour de force of manipulative skill in eighteenth-century analysis.

Euler's second major contribution was of far greater importance both historically and mathematically, as by an almost ludicrous mischance of fate it misdirected progress for all of forty years after his death in 1783. In the introduction to a memoir[19] of 1764, Euler advocated the incorporation of elliptic *arcs* into analysis on a parity with logarithms and circular *arcs*. (Note the italicized word.) Abandoning the fruitless efforts of his predecessors and contemporaries to integrate elliptic differentials in finite terms by means of functions then known, Euler boldly proposed that elliptic *integrals* be recognized as *primitive* new transcendents to be investigated on their own merits. If this is not what he meant, he proceeded in all of his own analysis as if it were. So great was the momentum of Euler's algoristic ingenuity that before he could realize his initial mistake he was carried completely out of sight of the right turning which he had missed. That he, of all mathematicians, should have gone astray in this particular matter is one of those mysteries in the evolution of mathematics that pass all understanding. The master who had initiated the modern theory of the circular *functions* failed to observe the greater opportunity which his Providence kept crowding on him and which, had he given it even a casual glance, must have appeared to a mathematician of his particular quality as the most natural thing in the world. Instead of considering elliptic *arcs* as the basic new transcendents, and thereby endowing an already overburdened integral calculus with a new wealth of uncouth formulas, Euler might easily have followed the simple lead of trigonometry. His oversight in adopting the *elliptic integrals* instead of their corresponding *inverse functions* as the data of his problem led him into a morass of tangled algebra, precisely as if he had attempted to develop trigonometry by an exclusive use of the inverse circular functions—his 'circular arcs'—$\sin^{-1} x$, $\cos^{-1} x$, $\tan^{-1} x$, etc. The far greater complexity of the theory of elliptic over circular arcs bogged him deeper at each step.

Realizing that Euler's heroic explorations in the wilderness of elliptic integrals had not got the undauntable pioneer very far in spite of many treasures found along the way, A. M. Legendre (1752–1833, French), in 1786 set out on his own explorations. For nearly forty years he followed Euler's trail, systematizing and civilizing as he went. It is at least conceivable that uncritical reverence for the works of his great predecessor was partly responsible for Legendre's personal misadventure.

More systematic than Euler, and taking more time to his work, Legendre reduced his chaos of refractory material to as coherent a whole as seems to be possible. To him are due the three standard forms of elliptic integrals to which any elliptic integral is reducible. Legendre's integrals are of course not the only canonical forms possible, and many others have been proposed; but Legendre's retain their usefulness. Forty years of unremitting labor by a master could not fail to produce much of value, if only for its suggestiveness. In particular, Legendre's work on the algebraic transformation of elliptic integrals directly inspired Jacobi's first notable success.

Legendre presented systematic accounts of his theory in 1811–17, in his *Exercices de calcul intégral sur divers ordres de transcendentes et sur les quadratures*, amplified in 1825–32 in the three volumes (with supplements), *Traité des fonctions elliptiques et des intégrales eulériennes*. The title of the second is responsible for a prevalent confusion in some historical accounts: Legendre's personal work is concerned with elliptic *integrals*, not with elliptic *functions*. The distinction, which became of epochal significance in 1827 with the publication of Abel's inversion of elliptic integrals, is comparable to that between night and day. Before Abel, nothing was publicly known of elliptic functions as they did not exist outside the private papers of Gauss until Abel invented them.

In addition to providing invaluable hints to Abel and Jacobi for the theory of elliptic functions, Legendre's treatises furnished Cauchy and others with numerous definite integrals, explicitly evaluated, on which to test the efficiency of integration by Cauchy's method of residues. The like holds for Legendre's systematization of the beta and gamma functions as they existed in his day. But here again it must be remembered that Legendre's analysis of 1827 became hopelessly archaic with the creation of modern methods by the great analysts of the nineteenth century, beginning with Cauchy in 1825. The contrast between the old and the new is strikingly evident on comparing Legendre's discussion of the gamma function with that of Weierstrass in 1856, only twenty-three years after Legendre's death.

In taking leave of this fine mathematician of the eighteenth century, we may remember him as a man of the highest character, whose only ambition was the advancement of mathematics. If Legendre was so far outdistanced in his own lifetime by younger men—Gauss in arithmetic and the method of least squares, Abel and Jacobi in elliptic functions—it was partly because his own labors had laid the necessary steppingstones. And although Legendre misjudged Gauss and hated him with a venomous hatred, he was the first to welcome and publicize the works of Abel and Jacobi which rendered obsolete his own efforts of forty years. The veteran of seventy-odd not only showed himself incapable of jealousy for his vigorous young rivals in their early twenties, but took pains to understand their work and to expound it in an amplified edition of his own. Such liberality of spirit is no commoner in mathematics than it is elsewhere.

Abel revolutionized the subject, and at the same time opened the floodgates of nineteenth-century analysis, in 1827 with a simple remark, "I propose to consider the inverse functions." Instead of regarding the elliptic integral

$$\alpha = \int^{x} \frac{dx}{\sqrt{(1 - c^2x^2)(1 + e^2x^2)}}$$

as the primary object of investigation in which α is considered as a function $\alpha(x)$ of x, Abel reversed the problem and regarded x as a function, which he denoted by $\phi(\alpha)$, of α. This *inversion* of the integral was the essential first step which Abel's predecessors had overlooked. Its 'naturalness' after it had been taken was obvious from the analogy with (x properly restricted)

$$\beta \equiv \int_0^x \frac{dx}{\sqrt{1 - x^2}} = \sin^{-1} x, \qquad \sin \beta = x.$$

Abel's first capital discovery[20] concerning the new functions was their double periodicity: $\phi(x + p_1) = \phi(x)$, $\phi(x + p_2) = \phi(x)$, where p_1, p_2 are constants whose ratio is not a real number. Thus the *elliptic function* $\phi(x)$ *is doubly periodic.*

Impressed by the great wealth of new ideas that entered mathematics as a direct consequence of Abel's simple remark, Jacobi some years after Abel's death characterized inversion as the secret of progress in mathematics: "You must always invert." If science or mathematics presents us with an awkward situation in which y is given as a function of x, say $y = f(x)$, we should examine the inverse situation, $x = f^{-1}(y)$, as Abel did when he inverted elliptic integrals and discovered that the inverse functions—the elliptic—are doubly periodic. Jacobi, who balanced his enthusiasm for mathematics with a sense of the ridiculous and who kept his tongue in his cheek when he pontificated, did not intend his prescription to be gulped down as a panacea. He was one of the least professorial of professors who ever lectured to an advanced class.

Jacobi's classic *Fundamenta nova theoriae functionum ellipticarum*, published in 1829, the year of Abel's death, exploited the consequences of inversion and double periodicity, and made the new functions easily accessible to the mathematical public. Even if, as is now generally conceded, Abel's was the priority in the two basic discoveries, Jacobi made the theory his own and contributed enough to entitle him to rank with Abel as one of its creators.

In awarding priority to Abel at the expense of Gauss, we have followed the modern custom of dating ownership from first publication. But with the printing of Gauss' posthumous papers and the scientific diary which he kept as a young man, it is known that Gauss was in possession of the double periodicity of the lemniscatic function in 1797. Early in 1800 he had discovered the general doubly periodic functions, anticipating Abel by a quarter of a century. His posthumous papers also contain numerous formulas relating to the elliptic theta constants, rediscovered and brilliantly applied by Jacobi. But Gauss, possibly for lack of an opportunity to develop and systematize his discoveries, published nothing on elliptic functions. Nor did he make any public claim[21] to have anticipated Abel and Jacobi. In estimating the place of Gauss in mathematics, it is customary to credit him with what he actually did. Thus non-Euclidean geometry and elliptic functions are two of the items which have counted in ranking Gauss with Archimedes and Newton, although he published nothing on either.

Elliptic functions have been given more space than their position relative to modern analysis might justify in a general account, because they clearly mark the beginning of a prolific epoch and were responsible for several major activities in the algebra, arithmetic, and analysis of the nineteenth century. Double periodicity not only opened up boundless new territories; it also marked the definite end of a road which had been followed since Euler's creation of analytic trigonometry. Jacobi proved (1834) that, if a single-valued function of one variable is doubly periodic, the ratio of the periods cannot be a real number; and that single-valued functions of one variable having more than two periods are impossible.[22]

Further technicalities would take us too far off the main road. However, three details of Abel's and Jacobi's early work were to prove so prolific of new mathematics all through the nineteenth century that they must be mentioned in passing. The first is Abel's discovery of complex multiplication, most conveniently described in terms of the Weierstrassian elliptic function $\mathrm{p}(u)$, $\equiv \mathrm{p}(u|\omega_1, \omega_2)$, with periods $2\omega_1$, $2\omega_2$, arising from the inversion of a certain standard elliptic integral involving the square root of a polynomial of degree three. The choice of $\mathrm{p}(u)$ implies no restriction. If n is a rational integer, $\mathrm{p}(nu)$ is expressible as a rational function of $\mathrm{p}(u)$. Seeking all other n's for which a similar theorem holds, Abel found the following unexpected[23] result. If c is a complex number such that $\mathrm{p}(cu|\omega_1, \omega_2)$ is rationally expressible in terms of $\mathrm{p}(u|\omega_1, \omega_2)$, $\mathrm{p}(u)$ is said to admit a complex multiplication by c. In order that such a c may exist, it is necessary and sufficient that ω_1/ω_2 be a root of an irreducible algebraic equation of the second degree with rational integer coefficients. This should be enough to suggest that the theory of complex multiplication is intimately connected with the arithmetic of binary quadratic forms.[24] The development of this hint occupied scores of algebraists, beginning with Kronecker (1857) and Hermite (1859).

The second item is Jacobi's representation of his doubly periodic functions as quotients of what are now called elliptic theta functions.[25] The thetas are not doubly periodic; one of Jacobi's four is $\vartheta_3(x|\tau) \equiv \sum_{n=-\infty}^{\infty} q^{n^2} \cos 2n\pi x$, where $q \equiv e^{i\pi\tau}$, $\tau \equiv \omega_2/\omega_1$, $|q| < 1$. The others are obtainable from this by simple linear transformations on x, for example,

$$\vartheta_3(x + \tfrac{1}{2}|\tau) \equiv \vartheta_4(x|\tau).$$

As the values of x for which the thetas vanish are readily determined, the analytic character[26] of the elliptic functions is put in evidence, and from this the Fourier expansions are obtained. The corresponding theta constants ($x = 0$) had been investigated by Euler in the 1750's and by Gauss[27] about 1800. But it was only when Jacobi discovered their connection with elliptic functions that their symmetrical theory emerged. Apart from their own extensive theory, the elliptic theta functions proved of great importance as clues to more general theta functions. These will be noted when we come to functions of a complex variable.

The third advance that opened up another vast expanse of nineteenth-century analysis also originated with Abel and Jacobi. It is required to exhibit the connections (algebraic relations) between elliptic functions, or between theta functions, whose respective pairs of periods are obtained from one another by linear homogeneous transformations with rational integer coefficients and non-vanishing determinant. This, the transformation theory, includes as special cases the problem of real multiplication, as for $\wp(nu)$ noted above, and that of division of the periods by a rational integer. As will appear later, a single detail of this theory, that of the elliptic modular functions, expanded in the late nineteenth century into an independent branch of mathematics. Its connection with the general quintic was noted in an earlier chapter. As might be anticipated from its formulation, the general problem draws heavily for its modern solution on the theory of linear groups.

Even these meager hints, displayed against a background of unprecedented activity in all departments of mathematics, should suffice to suggest the extent and intricacy of the theories that evolved from Abel's discovery of double periodicity. Each of half a dozen or more leaders elaborated the entire theory[28] or some favored subdivision according to his personal conception of symmetry and grace. Eighty years of this rugged aestheticism endowed analysis with a welter of conflicting notations and trivial distinctions without much of a difference, through which even an expert picks his way with exasperation. Almost in spite of themselves, the leaders rapidly acquired hosts of partisan followers. Mathematicians of all capacities began swarming into the new territory within a decade of its discovery. Several were pupils of Jacobi, but others quickly found leaders with different ideals.

The causes of this mass migration are not far to seek. Unlike the special functions devised primarily for the solution of physical problems, such as the Bessel coefficients, the elliptic functions seemed to have been created to round out and extend the integral calculus as it had evolved since the days of Newton and Leibniz. With the applications to the theory of numbers by Kronecker and Hermite in the late 1850's, it seemed also as if Gauss must have elaborated his arithmetical theory of binary quadratic forms especially for these unforeseen consequences of Abel's and Jacobi's early discoveries. Comprehensive syntheses to correlate these unexpected coincidences were sought as the century aged, and were found in the Galois theory of equations, the algebraic theory of fields, and the arithmetic of quadratic number fields.

Algebraic curves and surfaces[29] also absorbed enormous quantities of elliptic functions. Applications to classical applied mathematics were made simultaneously, especially to rigid dynamics and problems in potential theory. But it must be admitted that most of these practical applications have remained of greater interest to pure mathematicians than to working scientists. The rotation of a rigid body,[30] for example, yields numerous elegant exercises in the elliptic theta functions; but few engineers who must busy themselves with rotation have time for elegant analysis. The like holds for the occurrence of elliptic functions in practical applications of conformal mapping. When faced with one of these enticing horrors, the experienced designer turns to his drafting board. Contrasted with the Bessel functions, the elliptic functions are incomparably more beautiful and less useful. Yet—or possibly on that account—they were preferred many to one by the leading mathematicians of the nineteenth century because, in a sense that any mathematician will understand, they were closer to the 'natural' development of mathematics.

CHAPTER 19

Differential and Difference Equations

Continuing with the mathematics directly inspired by science, we shall indicate next four of the principal stages, not discussed in other connections, by which differential equations became a major discipline of modern pure mathematics. In its later development, this great episode is complementary to the evolution of the Galois theory of algebraic equations and the emergence of algebraic structure. Once more we shall see mere ingenuity being gradually displaced by coordinated attacks, and again we shall note the distinction between mathematics as practiced in the recent period and nearly all that preceded the nineteenth century.

Discounters of ingenuity do not mean to disparage intuition and insight in any assault on basically new problems. They merely emphasize that the characteristic strategy of modern mathematics favors the mass attack, where feasible, rather than any number of brilliantly executed raids. General methods, not individual gains, are the order of the modern day in mathematics. Ingenuity still has its function, even in a general offensive; but it is of a more comprehensive kind than any that sufficed in the past. The problems of modern mathematics are not isolated, and to overcome them coordinated efforts on a wide front are increasingly necessary.

Five stages

The first stage in differential equations, opening with Leibniz in the 1690's, closed about seventy years later. Roughly, what was accomplished in this period amounts to the first eight weeks' work in the usual introductory college course. Remembering that mechanics, dynamical astronomy, and mathematical physics were intensively cultivated all through and after this period and that numerous problems of analysis originated thus, we must also bear in mind that there was no adequate discussion of differential equations before Cauchy in the 1820's obtained the first existence theorems. This inaugurated the second stage. The third opened in the 1870's–80's with the application by M. S. Lie (1842–1899, Norwegian) of his theory of continuous groups to differential equations, particularly those of Hamilton-Jacobi dynamics. The fourth stage, beginning in the 1880's with the work of E. Picard (1856–1941, French) developed naturally from the third. Here the aim was to construct for linear differential equations an analogue of the Galois theory of algebraic equations.

Each stage after the first marked a definite and abrupt advance. The second paralleled the rigorizing of the calculus by Cauchy, and might have been anticipated from the general trend in analysis. The third, Lie's, even in retrospect, appears to have been unpredictable. Each of the periods has left a substantial residue in living mathematics; the last three posed many problems which still engage scores of specialists. What may be the beginning of a fifth stage opened in the 1930's, paralleling the modern development of abstract algebra. We pass on to a brief indication of some of the outstanding acquisitions in each of these stages and the accompanying developments in finite differences. Before the last three stages can be discussed, the concept of invariance must be described. This will be done in the next chapter.

The reign of formalism

Both Newton and Leibniz in the seventeenth century solved simple ordinary differential equations of the first order. It seems to have been believed in this earliest stage that the functions then known would suffice for the solution of the differential equations arising from problems of geometry and mechanics; and the aim was to find such explicit solutions, or to reduce the solution to a finite number of quadratures. Even when a solution was exhibited as a quadrature, it does not seem to have been suspected that the required integration might necessitate the invention of new transcendents. In fact, it was not until the 1880's—a stretch of two centuries—that definite knowledge concerning the extreme rarity of differential equations integrable in this rudimentary sense was obtained. Very roughly, if a differential equation is written down at random, the odds against its being solvable in terms of known functions or their integrals are infinite.

The first faint hint of generality was Newton's (1671) classification of ordinary differential equations of the first order into three types, and his method of solution by infinite series.[1] The coefficients of the assumed power-series solution were found as usual. There was no discussion of convergence. Without explicit statement of the assumption, it was assumed that the existence of a physical problem guarantees the existence of a solution of the equivalent differential equation. This seemingly reasonable supposition remained unquestioned in applied mathematics from Newton to Riemann. Its viciousness was first unmasked, as we saw in connection with Dirichlet's principle, only in 1870.

Another early forward step was taken by Leibniz, who stumbled on the technique of separating variables. Nearly two centuries were to elapse before Lie's theory showed when and why this familiar device should succeed. Among other early advances, the homogeneous linear differential equation of the first order was reduced to quadratures by Leibniz (1692); and James Bernoulli (1690) solved the equation of the tautochrone by separation of variables. His brother John (1694) circumvented dx/x, not well understood at the time, by first applying an integrating factor. Incidentally, the discovery of integrating factors proved almost as troublesome as solving an equation. Another hint of more general tactics appeared in Leibniz' (1696) change of the dependent variable. John Bernoulli also used this device. By the end of the seventeenth century all the usual elementary and inadequate tricks for first-order equations were known.

In addition to problems of the differential calculus on tangents, normals, and curvature of the types common as exercises in textbooks, the calculus of variations also had stimulated ingenuity in solving differential equations. Thus James Bernoulli's (1696) isoperimetric problem (noted in another connection) led to a differential equation of the third order which John reduced to one of the second.

Before 1700, John also attacked the general linear homogeneous differential equation with constant coefficients. To dispose of this detail here, a complete discussion of such equations was given in 1743 by Euler, who also (1741) devised the classical method for non-homogeneous linear equations.

The name of Count Riccati[2] (1676–1754, Italian) is familiar to every student in a first course. What is usually called Riccati's equation (1723) persistently defied solution in finite form. In accordance with the taste of the age, the 'real' problem was to impose sufficient restrictions on the variables to render the transformed equation finitely solvable by separation of variables. The Bernoullis claimed to have at least partial solutions; and in 1725 Daniel noted that if m is of the form $-4n/(2n \pm 1)$, where n is a positive integer, $dy/dx + ay^2 = bx^m$ is solvable in finite terms. By 1723 at latest, then, it was recognized that even an ordinary differential equation of the first order does not necessarily have a solution finitely expressible in terms of elementary functions. But anything approaching a proof of the impossibility, in general, of such a solution lay far in the future.

Singular solutions were noted unexpectedly early, the first[3] instance being due to Taylor (of Taylor's series) in 1715. Clairaut, whose name decorates a special type of equation in a first course, followed in 1734 with a fuller discussion of singular solutions, while Lagrange (1774) unsuccessfully attempted a general theory. Other noteworthy early contributions were Euler's (1756, 1768), and Legendre's (1790), neither of conspicuous merit. This detail continued to interest analysts throughout the nineteenth century, a satisfactory treatment gradually emerging through the labors of Boole (1859, 1865), G. Darboux (1870, French), A. Cayley (1873, English), G. Chrystal (1896, Scotch), and many others. Perhaps the most important contribution of singular solutions to the general progress was their intimation that mathematicians had not grasped in the first stage what was to be understood by 'solution' of an ordinary differential equation.

The unsystematic beginnings of several other extensive divisions of the modern theory can be traced to eighteenth-century attacks on problems in elementary infinitesimal geometry, dynamics, mathematical physics, and celestial mechanics. The most influential for the future of analysis entered through partial differential equations. As these will be considered in other connections, we continue with ordinary equations, noting only three of the more outstanding items as fair samples of the most significant.

Prophetic of a deluge that was to come, Euler's (1762) solution by infinite series of a second-order equation substantially the same as Bessel's almost reconciled the old masters to the introduction of new transcendents. Haphazard elementary devices began to give way to more systematic treatment in Euler's invention (1739) of the method of variation of parameters,[4] elevated into a general procedure by Lagrange in 1774. The germ of the extensive modern theory of linear differential systems appears in d'Alembert's work of 1748; while Lagrange (1762) introduced the fertile concept of the adjoint equation.

The leaders in this period were Euler, Lagrange, and Laplace, of whom Euler was the most inventive and Lagrange the most mathematical, with Laplace occupying an intermediate position, adapting the analysis of his contemporaries to his immediate needs in celestial mechanics, mathematical physics, and probability. Laplace's individual contributions were considerable, particularly to the development of special functions arising in the boundary-value problems of Newtonian potential theory. But the frequent awkwardness of his analysis exemplifies the reverse of Einstein's dictum that "elegance is for tailors." Solvers of the universe, or even of the relatively trivial solar system that absorbed the major part of Laplace's efforts, can get on well enough, it seems, without elegance or even common orderliness in their calculations. Contrasted with Lagrange's uniformly concise, general expositions, the analysis of his contemporaries appears haphazard and opportunistic. It may be for this reason that the analysis of Lagrange's analytic mechanics is more vital in applied mathematics today than is much of what his contemporaries created. It has been said that style is the preservative of literature; and it is conceivable that a slight regard for the elementary decencies of pure mathematics might go a long way toward retarding obsolescence in applicable mathematics. Lagrange's consistent pursuit of generality is the clue to his mathematical survival.

Difference equations

The persistent scientific importance of differential equations may be credited to the mistaken assumption that "nature does not make jumps." On this obsolete hypothesis, continuity is the only permissible medium for constructing mathematical models of the physical universe. All change is assumed to be continuous; 'natural laws' are expressible as equations between rates of continuous change. Thus differential equations are the proper mathematics of physical science. But with the advent (1900) of quanta in radiation, and of genetics in biology, it was seen in the first decade of the twentieth century that not all natural phenomena are conveniently described in terms of continuity. The old Greek conflict between the continuous and the discrete broke out anew, the extremists of one faction contending that continuity in nature is an illusion, and that differential equations were about to be wiped from the slate to make way for equations in finite differences. Should an advancing science ever urgently demand an analysis of difference equations, mathematicians will have their work laid out for a hundred years or more. Such equations so far appear to be intrinsically less tractable than differential equations.

The differential and integral calculuses,[5] with their corresponding differential, integral, integro-differential, and functional equations, have provided science with usable methods of calculation because they are based on the facile technique of limits. The resulting continuous analysis still furnishes the easiest means of approximating to exact numerical measures, even when the phenomena under investigation are known to vary discontinuously. This is the case, for instance, in statistical mechanics where, if rigorously exact numerical results were sought, they would have to be calculated by strict combinatorial analysis in which continuity is meaningless. Such calculations are at present wholly beyond human capacity when the numbers involved are of the order of magnitude believed to obtain in science. It is therefore necessary to compromise by adapting the analysis of finite differences—discontinuous, stepwise change—to that of continuity.

In particular, solutions of difference equations are obtained by means of the theory of analytic functions of a complex variable. The outcome is a region of analysis which attracts a few hardy compromisers, but which the majority of mathematicians seem to find singularly repulsive. The functions produced exist and do indeed satisfy the equations. But if this awkward forcing of the round pegs of classical continuous analysis into the square holes of essentially discrete patterns is the best mathematics can do, then something less cumbersome and closer to an intelligible language than mathematics in this instance has proved itself will likely be sought by our successors. There may be a limit to the scientific utility of mathematical analysis. If there is, it seems to have been attained in some of the analytic solutions of linear difference equations hopefully offered to practical computers, who must produce reasonably accurate numerical results or lose their jobs. These defects of the now classic analytic theory of linear difference equations do not, of course, detract from its purely mathematical interest. Nor are they a reflection on the skill and sagacity of the creators of the modern theory, among whom are counted some of the foremost analysts of the nineteenth and twentieth centuries. They are but one more hint that the neglected mathematics of the discrete is still looking for its Newton and its Leibniz. That neither has yet appeared may be because the discrete is mathematical bedrock. But whether or not it is, we may anticipate that significant progress will demand genius at least equal to that responsible for the mathematics of continuity.

As noted in an earlier chapter, the calculus of finite differences is usually said to have emerged as a distinct division of mathematics with Taylor's *Methodus* of 1715–17. According to professional opinion, there are but two periods in the entire evolution of the subject. The first, covering almost exactly two centuries, extends from Newton in the 1680's to Poincaré in the 1880's; the second, from Poincaré on. There was no peer of Cauchy in the early 1800's to breathe new life into the discrete as there was into the continuous. The first period falls into two well-marked stages, the earlier ending in the 1820's. From then to the 1880's, finite differences stagnated, except for a riotous interlude of symbolic algorithms reaching its climax with the publication of Boole's treatise in 1860. These methods, as presented in the mid-nineteenth century, have long been inadequate for an exact discussion in either the practical or the theoretical sense.

The early development was intertwined with that of interpolation, beginning with Wallis' quadrature of the circle in 1655, continued some time before 1676 by Newton in his derivation of the binomial series, and passing thence to the construction of usable interpolation formulas. Sometimes called the difference and sum calculus, the basic operations of the classical theory of finite differences are Δ and its inverse,[6] Σ. These correspond respectively to differentiation and integration in the continuous calculus. The difference symbol Δ was used in 1706 by John Bernoulli with its customary meaning; Σ was used, if not first introduced, by Euler in 1755. If the 'step' is h, $\Delta u(x)$ denotes

$u(x + h) - u(x)$, and there is usually no loss in generality in modifying the notation so that $h = 1$.

The history of interpolation formulas[7] is complicated and controversial. We are interested here only in the following items: such formulas were a potent stimulus in the seventeenth and eighteenth centuries to the independent evolution of the Δ, Σ calculus; they were developed principally to facilitate numerical computation in astronomy, the construction of tables, and mechanical quadrature. Although dates and priorities may be controverted, those following give a sufficiently accurate scale for the general progress: Newton's formula, 1687; Euler's, usually called Lagrange's, 1775, also attributed to Waring, 1779; Gauss', 1812.

The other principal source of the Δ, Σ calculus was combinatorial analysis and the theory of probability as developed in the eighteenth century, especially by Laplace. Many problems in probability beyond the types offered in elementary texts are best attacked by preliminary translation into difference equations, which, however, may prove intractable. One detail originating in such work has attracted hundreds since James Bernoulli[8] in 1713 first defined the numbers and polynomials named after him. Euler followed in 1739 with his numbers; thenceforth the Bernoulli and Euler numbers and polynomials occupied a central position in the Δ, Σ calculus, especially in connection with the extremely useful Euler (1730) and Maclaurin (1742) sum formula. The history of these numbers and polynomials with their numerous generalizations would fill a sizable book. Yet they are but one incident of many in the general development. We need note here only that the first serious attempt to investigate the remainder in the Euler-Maclaurin formula appears to have been Poisson's of 1823. This disregard of what now seem the most rudimentary precautions demanded by scientific sense is typical of analysis before the Cauchy reformation of 1821.

Little that would now be considered either practically adequate or mathematically rigorous has survived of the eighteenth-century work on difference equations. Suggestive analogies with differential equations were pursued, without much success. The outstanding exception is Lagrange's complete algebraic theory (1759, 1775, 1792) of recurring series. Earlier discussions, such as de Moivre's (1722, 1730, 1738), lacked generality. Laplace's method of 'generating functions' (1766, 1773–4–7, 1809), invented primarily for use in mathematical probability, while yielding results of great utility, suffered from an almost total absence of anything that would have been called proof after Cauchy's and Abel's discussion of convergence in the 1820's. Some might insist that all this dead analysis was necessary fertilizer for a healthy crop, and they may be right; there is no way of telling. However, when at last a sound treatment of linear difference equations appeared, it sprang from a basically different soil that did not exist before 1825. Cauchy's creation of the theory of functions of a complex variable in that year made possible a satisfactory analysis of such equations some sixty years later.

The modern theory opened with Poincaré's memoir of 1885. The general linear difference equation is

$$\sum_{i=0}^{n} p_i(x)u(x+i) = \phi(x),$$

in which the $p_i(x)$ and $\phi(x)$ are given functions and $u(x)$ is to be found. If $\phi(x) = 0$, the equation is called homogeneous. The case of constant coefficients is the problem of recurring series solved by Lagrange. The very special equation

$$u(x+1) - u(x) = \phi(x),$$

in which $\phi(x)$ is an analytic function of x, already presents a serious problem in analysis. The special case of this in which $\phi(x)$ is a polynomial had been completely discussed by eighteenth-century methods when C. Guichard (1861–1924, French) in 1887 considered the equation for any analytic $\phi(x)$, thereby giving the first complete theory of a difference equation involving an analytic function. Poincaré's revolutionary contribution (1885) initiated the theory of asymptotic solutions, in analogy with the like for differential equations; and in 1909–13, H. Galbrun (French) continued in this direction with asymptotic solutions for large values of the variable. Such work is of practical utility.

A later phase dates from 1911, when G. D. Birkhoff (1884–1944, U.S.A.) and R. D. Carmichael (1879–, U.S.A.) almost simultaneously published their investigations, the former in systems of homogeneous linear difference equations, the latter in analytic solutions of linear difference equations. As in his work (noted later) on systems of linear differential equations, Birkhoff made most effective use of matric algebra. This work was definitive. Both continued their researches intermittently for about a decade. Carmichael's branched out to generalizations of a type of factorial analysis that had occupied scores of mathematicians from Newton to Weierstrass.

Factorial series and expansions in terms of factorials, appearing initially in the problem of interpolation, raise more difficult questions of convergence than those presented by power series. One of Weierstrass' earliest projects (1842, 1856) dealt with factorials; the first thoroughly satisfactory discussion is said to have been J. L. W. V. Jensen's (1859–1925, Scandinavian) of 1884. Factorials have a high sentimental value in the history of mathematics, as it was through them that Crelle got his first glimpse of Abel's genius. Without Crelle's generous support, it is probable that Abel would have gone to his early grave with his greatest work unachieved. Of moderns who made notable contributions to this branch of analysis, S. Pincherle (1853–1936, Italian) was the most prolific. Much was done in the modern theory by Scandinavian analysts, of whom N. E. Norlund (1885–, Scandinavian) may be specially mentioned.

The calculus of finite differences, with its unfashionable theory of difference equations, is seldom offered in American universities beyond the rudiments sufficient for an elementary course in statistical methods. This seems rather unfortunate; for if the remarks at the beginning of this section are even slightly justified, much remains to be done, and the changing scientific scene of the twentieth century may necessitate the creation of a usable mathematics of the discrete. If nothing better than what now exists is possible, then that should be demonstrated. Mere talent is not likely to get very far in such a program.

We return to the easier but still sufficiently difficult field of differential equations.

Existence and special problems

Although we cannot discuss special problems in any detail, one, Pfaff's,[9] may be noted here for three reasons: it is typical of the enormous expansion of mathematics by specialization throughout the nineteenth century; Pfaff's problem played a central part in Lie's theory of contact transformations, a theory which revolutionized the treatment of the differential equations of dynamics and altered the aspect of geometry; it was first successfully investigated (1814) by J. F. Pfaff (1765–1825,

German), the only noteworthy mathematician in the Germany of the late eighteenth century. Pfaff befriended young Gauss and recognized his ability, and Gauss reciprocated by appreciating Pfaff's high qualities of character and intellect. When Gauss' work began to become known throughout all Europe, Laplace was asked, "Who is the foremost mathematician in Germany?" —"Pfaff." "But what of Gauss?" "Oh," Laplace replied, "Gauss is the first mathematician in Europe."

Pfaff's problem[10] (1814) is concerned with the solution of a total differential equation in $2n$ or $2n - 1$ variables for which not all the integrability conditions are satisfied. Euler (1770) found the necessary integrability condition for

$$P\,dx + Q\,dy + R\,dz = 0,$$

(P, Q, R functions of x, y, z), and believed that unless this condition obtains there is no solution, because the equation then cannot be derived from a single primitive. G. Monge (1746–1818, French), exhibited (1784) two equations as the equivalent of the single total equation. Pfaff observed that in general the total equation, under the conditions stated, is equivalent to a system of n or fewer equations; his problem is to determine these equations; the modern treatment discusses also the integration of this equivalent system.

Inspection of the works[9,10] cited will show that in little more than a century this isolated detail of analysis as it was in the early nineteenth century had swelled prodigiously and intricately. Numerous other details expanded even more impressively, until by the end of the century a mastery of all mathematical analysis was claimed by nobody, and seemed to be permanently beyond human powers. Differential equations, contributing their huge share to this superabundance of wealth, added to the rich confusion. Too often for comfort, mathematics in the nineteenth century followed the same formula of glut without digestion as the rest of civilization in that heroic age of expansion at any cost. But, according to the abstractionists of 1940, the discarnate spirit of simplicity was then about to descend and bless all mathematics, and the more rococo masterpieces of the nineteenth century were to be preserved only in museums frequented exclusively by historians. Sufficient provocation for this wishful prophecy will appear as we proceed, if indeed it is not already evident. Seeing only the obstructive masses of details, many of them already in an advanced state of obsolescence, the prophets of an abstract millennium have occasionally overlooked the greater syntheses, such as Lie's theory of continuous groups, by which the nineteenth-century mathematicians themselves coordinated innumerable special results of their predecessors.

We pass on to the fundamental question of existence. The first existence theorems for differential equations were developed by Cauchy in his lectures[11] of 1820–30. It is proved that if $f, \equiv f(x, y)$, and $\partial f/\partial y$ are real, uniform, continuous functions for all real x, y such that $|x - x_0| \leqq a$, $|y - y_0| \leqq b$, and if $|f|$ is bounded in this rectangle, then the real equation

$$dy/dx = f(x, y)$$

has a unique solution $y = F(x)$ satisfying a given boundary condition $y_0 = F(x_0)$. The proof exhibits a theoretical means for constructing the solution to any desired degree of accuracy. Cauchy extended his process to n such first-order equations in n independent variables, equivalent to one equation of the nth order; and R. Lipschitz (1832–1903, German) improved the method (1876), possibly with a view to making it less unpractical. Probably Cauchy himself, and certainly many following his lead, extended the method, with suitable modifications, to the complex domain. E. Picard (1856–1941, French) in 1893 perfected an existence theory based on the different method of successive approximations, which is considered more practical than the Cauchy-Lipschitz. Modern machines in 1945 were said to offer some hope of a greater efficiency than any of the purely mathematical methods[12] for obtaining numerical solutions.

Numerical solutions received a strong impetus in the world war of 1914–18, when modernized German artillery, having pounded the Belgian forts to bits, rendered the older ballistic tables partly obsolete. It was remarked at the time that the improved methods for the solution of differential equations in the mechanics of the heavens effectively did their bit in creating a terrestrial hell. This, and equally curious arguments, were strongly urged in the United States of 1917–19 as sufficient grounds for government subsidy of pure mathematics. With the alleged return to normalcy, ballistics receded from the foreground of academic propaganda, and the applications of mathematics to genetics, agriculture, intelligence tests, usury, and other peaceful pursuits were vociferously advertised. Though it is only remotely connected[13] with differential equations, the use of Latin squares[14] in the design of experiments[15] in crop improvement and elsewhere may be cited as a striking instance of the seemingly most useless mathematics harboring unsuspected germs of profit. But even such appealing applications as this were no more successful than differential equations in coaxing bounty from the politicians, and mathematics continued on its beggarly way, ostentatiously proud of its incorruptible purity.

Thus far only linear equations have been noted. With nonlinear equations of order higher than the first, new phenomena appear: solutions may have movable singularities dependent on the initial values. The study of such equations by M. Hamburger (German) in 1877, L. Fuchs in 1884, and others was continued by Picard in the 1880's and by P. Painlevé[16] (1863–1933, French) in the 1890's. To the latter are due (1893) certain remarkable equations not integrable in terms of elliptic functions or any of their degenerate cases. These equations, of which the simplest is $d^2w/dz^2 = 6w^2 + z$, define new transcendents. A somewhat similar type of analysis appears in the scientifically more useful linear homogeneous differential equations whose coefficients are functions of the independent variable with known singularities. The classic instances are the hypergeometric equation, first competently investigated by Gauss in 1812, the more general equation (1857) for Riemann's P-function, and the extensive class of equations called Fuchsian in honor of L. Fuchs (1833–1902, German). These developments had a twofold significance in the growth of analysis: many of the important ordinary differential equations of mathematical physics are special cases of those mentioned, and the properties of their solutions were unified in the general theory; the researches of Fuchs and his school aided in preparing the way for Poincaré's creation of automorphic functions. This allusion must suffice here; the relevant details will be noted in later chapters.

The useful method of solution due to G. Frobenius (1849–1917, German), frequently explained in texts, belongs to this period (1873). In this work, as in his contributions to the theory of groups, hypercomplex number systems, and other parts of algebra, Frobenius proved himself one of the master technicians of the nineteenth century. The quality of his mathematics, also the detailed type of problem that interested him, afford an illuminating contrast to the more philosophical and abstract approach of his contemporaries Kronecker and Dedekind. Anyone wishing to form a firsthand estimate of the distinction between the high-average mathematics of the late nineteenth century and that of the twentieth might compare certain of Frobenius' classic memoirs on given topics with corresponding work dated after 1920 by competent mathematicians of the younger generation.

Symbolic comedy in three acts

Accompanying the theory of existence, the somewhat shady episode of symbolic methods attracted considerable attention, especially in England. Despite their utility, the earlier applications of these devices were scarcely reputable mathematics, because no explicit formulation of the conditions under which they give correct results accompanied their use. They have a long and rather listless history, beginning with Leibniz and ending in the twentieth century with satisfactory theories of linear, distributive operators. A few highlights must suffice here. The point of general interest to be noted is this: if an empirical formalism produces almost invariably consistent and useful results, some underlying mathematical justification is indicated; and instead of censuring the toilers of formalism for living and laboring in a mathematical slum, purists might more profitably expend their talents in making the slum sanitary. Whatever these unassuming workers may lack, it certainly is not imagination. The use of divergent series in astronomy will furnish a striking example when we come to real variables; O. Heaviside's (1850–1925, English) extremely useful operational calculus, noted presently, is a classic instance of the final triumph of creative insight over academic pedantry.

Leibniz (1695) noted that the symbol d of differentiation, now D in the symbolic method, has some of the properties of ordinary algebraic 'quantities,' for example, the rules for positive integer exponents. The total separation of operative symbols from their operands was accomplished in the early 1800's. All but achieved by L. F. A. Arbogast (1759–1803, Alsatian) in 1800, it was consummated by B. Brisson (French) in 1808, M. J. F. Français (French) in 1811, and F. J. Servois (French) in 1812. From 1836 to 1860 the British school exploited the symbolic method in the differential and integral calculuses, the calculus of finite differences, and both differential and difference equations. Some of this work survives in elementary texts.

Among the most energetic proponents of the method, Boole[17] and D. F. Gregory (1813–1844, Scotch) made outstanding contributions. A little-known application was Boole's proof (1857) of Abel's prolific theorem (1826, described in a later chapter) in the integral calculus.

The last notable advance in the symbolic method was Heaviside's operational analysis (1887–98) of certain differential equations in electromagnetics, important also in electrical engineering and elsewhere.[18]

Heaviside's struggle for recognition from the official censors of mathematical morals illustrates once more the adage that the way of the transgressor is hard. The congenitally obedient will see in Heaviside's defeat a merited rebuke to contumacious intelligence; born rebels against the sort of respectability that protects itself behind an authoritative front may read his transgression as a challenge to revolt. Provided the rebel can consolidate his revolt, he may achieve the same sort of retarded tolerance that failed to sweeten the bitterness of Heaviside's wretched old age. Otherwise, he may be remembered only with the circle squarers.

Heaviside's assault on the bastions of orthodoxy was repulsed with heavy losses. Outraged by the innovator's lack of mathematical rigor, the Royal Society of London refused to publish Heaviside's papers, although he was a member of the Society. The 1890's and 1900's enjoyed quite an acrimonious debate on the merits of the new operational calculus. The official verdict pronounced the novelty too disreputable for notice by self-respecting mathematicians. Handsome amends were made, however, five years after Heaviside's death, when it had been discovered that the new calculus was quite moral after all. The following estimate (1930) by a recognized British analyst[19] may perhaps err on the side of generosity, a pardonable fault under the circumstances.

> Looking back on the controversy after thirty years, we should now place the Operational Calculus with Poincaré's discovery of automorphic functions and Ricci's discovery of the Tensor Calculus as the three most important mathematical advances of the last quarter of the nineteenth century. Applications, extensions and justifications of it constitute a considerable part of the mathematical activity today [1930]—I am thinking of the recent work of H. Jeffreys, Norbert Wiener, Bromwich, Carson, Murnaghan, March and others.

The cream of the jest, however, did not rise till 1937. This may be particularly commended to inexperienced students looking forward to a mathematical career, these innocents often believing that the pursuit of mathematics elevates the human mind to cold, glittering heights far above the foibles of mankind. Following the trite pattern, the Heaviside tragi-comedy degenerated in three acts into broad farce: the Heaviside method was utter nonsense; it was right, and could be readily justified; everybody had known all about it long before Heaviside used it, and it was in fact almost a trivial commonplace of classical analysis. The supple yet somewhat hoity-toity acrobats[20] in the third act have been too busy balancing their dignity against the slip in the first act to notice what their gymnastics prove.

American engineers began using Heaviside's methods about 1910, and the demonstrated utility of the operational calculus at length attracted professional analysts. By 1926 a belated baptism of engineering's dubious offspring had conferred on it all the rights and privileges of mathematical legitimacy. Among Heaviside's unsolicited mathematical rehabilitators, T. J. I'A. Bromwich (1875–1929, English) in 1921, P. Levy (1886–, French) in 1926, and H. Jeffreys (1891–, English) were the most active. If there is any moral implicit in this episode, it may be that suggested by Heaviside's satirical gibe at the purists: "This series is divergent; therefore we may be able to do something useful with it." However, lest Heaviside's belated triumph be taken as an incitement to unintelligent rebellion for rebellion's sake, it may be recorded that prior to his revolt, the pioneers in integral equations[21] were already developing the analysis which was to justify Heaviside's calculus while definitely prescribing its inherent limitations. A curious application of this calculus is that (1938) by B. van der Pol (Dutch) to the theory of primes, illustrating at least the heuristic value of the method. By 1930, Heaviside's calculus was part of the routine training of electrical engineers in the more progressive technical schools.

Systems; Cauchy's problem

Systems of ordinary differential equations appeared early in dynamical astronomy, specifically in the Newtonian problem of three bodies. Attempts throughout the eighteenth and nineteenth centuries to find solutions in special cases, also to discover further integrals (beyond those given by the fundamental principles of dynamics) in the general case, seem to have been largely responsible for the elaborate theory of simultaneous systems. The analysis of n simultaneous linear differential equations of the first order, with coefficients analytic in a parameter, was notably advanced by Poincaré in his new methods (1892–9) of celestial mechanics. Euler in his development of the lunar theory, about 1750, had obtained solutions of such systems as power series in the parameter; but it remained for Cauchy to justify the analysis. With the first satisfactory marine chronometers (J. Harrison's, 1765), the moon was no longer the navigator's one reliable clock, and the importance of the lunar theory in determining longitude at sea diminished almost to the vanishing point. But in its purely mathematical aspects the problem of three bodies continued to dominate one major department of differential equations. The theory of perturbations, with all its cumbersome analysis, can be traced back to attempts to describe the motion of the moon.

Of much notable modern work on simultaneous systems of the first order, that of G. D. Birkhoff (1884–1944, U.S.A.) instigated further progress. An interesting detail of the twentieth-century developments of this theory is the incidental use of the algebra of matrices. In the early 1900's, the connection with integral equations and the related boundary problems was developed by Hilbert, Birkhoff, M. Mason (1877–, U.S.A.), and many others. The scientific importance of this connection will appear in a later chapter. Parallel with these developments, another rapidly forged into prominence. Beginning in 1881, Poincaré initiated a radically new philosophy in both the theory and practice of differential equations. The primary objective here was a qualitative or topologic study of all the solutions of a system and the relations between them before the consideration of quantitative aspects. From this the vast theory of recurrent motions, periodic orbits, and dynamical stability developed. Once more the origin is in the dynamical astronomy of the eighteenth century, particularly in Laplace's attempt to establish the stability of the solar system. Among other causes which rendered Laplace's efforts obsolete were the unprecedented increase of accurate astronomical knowledge all through the nineteenth century, and relativity. But although the oversimplified Laplacian synthesis has been modified out of recognition, the problems which it posed, many times refined and generalized, remain as suggestive as they were in 1827 when Laplace died. The twentieth-century revision of classical celestial mechanics, as in the qualitative dynamics of Poincaré,

Birkhoff, and many others, induced an expansion of topology, which in turn reacted on qualitative dynamics. It is of interest to recall that Poincaré began one of his notable advances (1892) by proceeding from the methods invented (1877) by G. W. Hill for analyzing the motion of the moon's perigee, in which infinite determinants appeared for the first time. To Hill's work also can be traced the modern theory of differential equations with periodic coefficients, to which many Americans, including F. R. Moulton (1872–) and W. D. MacMillan (1871–), made noteworthy contributions. Both singly and doubly periodic coefficients were admitted in this work of the twentieth century.

Partial differential equations will be noted when necessary in relevant contexts. However, two episodes of more than incidental significance may be remarked here, as neither enters conveniently in subsequent discussions. The first was the attempt to define the 'general' and 'complete' solutions of a partial differential equation of order higher than the first. Jacobi in the 1830's (published, 1862) had given a satisfactory method of formal integration for systems of partial differential equations of the first order in one dependent variable. But for equations of the second order new principles are demanded, and there are several possibilities.[22] Laplace in 1777, A. M. Ampère (1775–1836, French) in 1815–20, Boole in 1863, and Darboux in 1870 proposed methods for defining and obtaining integrals, none of which as late as 1908 was regarded as wholly satisfactory. Such detailed progress as has been made relates almost entirely to single equations of the second order in not more than three independent variables. The most important equations of classical mathematical physics fall in this category, which accounts for the intricate theory that has grown up round it.

The second item to be noted here, Cauchy's problem of 1842, may be regarded as a natural outgrowth of boundary problems in mathematical physics as it was in the early nineteenth century. It is closely related to the first item, but with a distinction of cardinal significance. Although it sounds reasonable, the problem of finding the general solution of a partial differential equation of the second order may be improperly posed. If several distinct solutions, each with some desirable qualities but none with all, are feasible, the seemingly reasonable problem may be too general to be useful. Accordingly, it becomes a meaningful project to isolate from all the possibilities one which shall, in general, eventuate in a unique solution, and to investigate the exceptional cases where there is not a unique solution. This is the problem of interest in mathematical physics. Cauchy's problem is of this type; its special cases include many that appear in classical physics. The general problem concerns a second-order linear partial differential equation in one dependent and n independent variables; and it is required to determine a solution such that it, and its first partial derivative with respect to one of the independent variables, evaluated for the value zero of that variable, shall be given functions of the remaining $n - 1$ variables. The exceptional cases where the problem has not a unique solution are also to be determined and analyzed. Among those who discussed this problem after Cauchy, Sonja Kowalewski (1850–1891, Russian) in 1874, and Darboux (1875) may be mentioned from the nineteenth century; in the twentieth, Hadamard and others of the French analytic school developed the general theory. By consulting the work cited,[23] some conception may be gained of the extent to which this specialty had expanded in eighty years.

One famous name has not been explicitly mentioned in connection with differential equations, because it has been taken for granted. The specific contributions of Weierstrass were less significant, possibly, than the overwhelming influence of his fundamental reformation of analysis, including differential equations.

Toward systematization

Only a handful of specimens from a multitude have been exhibited in the preceding sample. Items were selected with two ends in view: to suggest that by the 1870's the chaotic masses of analysis miscalled a theory of differential equations were ripe for a systematic attempt at coordination; to offer some moral justification, if nothing more substantial, for the feeling of oppression which drove many of the younger analysts after 1900 to seek relief in general analysis and complete abstraction. The second of these motivations will be amplified as we proceed. For the moment we note a significant step toward a general analysis: equations in a denumerable infinity of variables were considered as early as 1908 by E. H. Moore. Thereafter several mathematicians of the American school[24] discussed differential equations of infinite order. With the accompanying generalizations of the concept of a variable in the calculus of variations and the theory of functionals, the progression toward some type of general analysis seems to have been inevitable. Concurrently with this movement another, beginning in the 1880's, reached its objective earlier, in the first structural theory for differential equations. This joint evolution is naturally quite complicated, involving developments in algebra, geometry, and analysis. For reasons to be noted presently, we shall follow its main lines along the paths indicated by applied mathematics.

The impulse from science is plain in differential equations, and needs no further emphasis. As we proceed, the influence of applied mathematics on the evolution of theories of functions will become increasingly obvious. With algebra, however, the orderly progression from applications to abstractions was reversed: abstractions preceded applications in the most important contribution yet made to science by algebra, the concept of invariance.

In geometry, the constant interplay between pure and applied mathematics continued all through the nineteenth century. Cartography and geodesy, for example, are assignable to either division; their outgrowths in differential geometry also are either pure or applied mathematics according as the emphasis is on geometry itself or on its interpretation in cosmology and the physical sciences. But it is immaterial by what label a particular advance be designated; the point of interest is the continual interaction between the geometry created for practical or scientific ends and that developed for its own sake.

It is not mandatory, of course, to follow applied mathematics as a clue through the intricate maze. Several other approaches will occur to anyone who has done a little exploring. Geometry, for instance, also threads the labyrinth. But what is geometry? Attempts to obtain a usable description were partly responsible for the choice here of applied mathematics. The best definition so far offered is due to a distinguished American geometer:[25] "Geometry is what geometers do." If it be asked what geometers do, the reply is equally satisfying: "Geometers do geometry." These definitions were not intended facetiously. They were a frank acknowledgment that to a majority of those who call themselves geometers, all things are geometry. But with applications there is no difficulty. Anyone believes he understands what is meant by applied mathematics, especially if, as must be asserted by consistent Marxians, no good mathematics is pure.

CHAPTER 20

Invariance

First seriously considered in 1841, invariance had become one of the dominating concepts of mathematical thought only thirty years later. A short prospectus of certain features of the general development, before closer description of some, may serve to indicate the basic importance of invariance in algebra, geometry, analysis, and theoretical physics. Topologic invariance, not mentioned in this prospectus, is described later by itself.

General features

A comprehensive formal definition of invariance might be difficult to fabricate and unilluminating once it was constructed.[1] The following informal description[2] gives the gist of the matter more intelligibly. "Invariance is changelessness in the midst of change, permanence in a world of flux, the persistence of configurations that remain the same despite the swirl and stress of countless hosts of curious transformations."

An immediate example is the conservation of energy as conceived in nineteenth-century physics; another, the constancy of the cross ratio of four collinear points under projection; another, the permanence of the order of the points on any line of a coordinate mesh on an elastic map as the map is stretched in any manner without tearing. An instance from elementary algebra goes back to 1773 and to the unsuspected germ of the entire theory of invariance with all its ramifications throughout modern mathematics. In the course of his arithmetical investigation of binary quadratic forms, Lagrange noted incidentally that if x be replaced by $x + \lambda y$, and y by y, in $ax^2 + 2bxy + cy^2$, so that the form becomes $Ax^2 + 2Bxy + Cy^2$ where $A \equiv a$, $B \equiv a\lambda + b$, $C \equiv a\lambda^2 + 2b\lambda + c$, the discriminants $b^2 - ac$, $B^2 - AC$ are equal.

Lagrange's discovery is a special case of the (historically) first general theorem in algebraic invariants: a linear homogeneous transformation of the variables in a quadratic form converts the form into another, whose discriminant is equal to that of the original form multiplied by a factor[3] depending only on the coefficients of the transformation. This theorem, due to Boole in 1841, is usually considered the origin of the theory of algebraic invariants. Gauss (1801, *Disquisitiones arithmeticae*) had used the special cases for quadratic forms in two and three variables; but neither he nor Lagrange noticed the suggestive clue to an undiscovered continent of algebra.

A form in algebra is as already defined in the arithmetical theory, except that the coefficients are now not restricted to integer values, but are arbitrary constants. Boole (1842) followed up his success with the discovery of certain functions of both the coefficients and the variables of forms having the same property of invariance with respect to linear transformation. To distinguish these from the first species involving only the coefficients of the forms, they were later called covariants.[4]

When it is remembered that Boole was one of the principal founders of symbolic logic, he has a strong claim to the sour honor of having been the most greatly underrated mathematician of the nineteenth century. Self-taught, Boole lacked the alleged advantages of an orthodox training in the mathematics of his time. His originality was so aggressive that it might have overcome even the thorough education which he regretted having missed through poverty. In any event, he became a most unacademic mathematician, to the lasting benefit of mathematics and philosophy.

Cayley immediately recognized the significance of Boole's discoveries, and in 1845 began the systematic development of a theory of algebraic forms and their covariants and invariants with respect to linear homogeneous transformations, briefly designated as the theory of algebraic invariants, or of algebraic forms, or the algebra of quantics.[5] For about half a century after Boole's initial discoveries (1841), 'modern higher algebra' was practically synonymous with the theory of algebraic forms. Simultaneously, projective geometry was used as a convenient mould in a praiseworthy but futile attempt to squeeze shapeless masses of the new algebra into a semblance of intuitiveness. Increasingly complicated formulas were tortured into ever more intricate theorems, until the one conceivable justification for such exercises evaporated. It was impossible, by geometry or any other human mathematics, to make formulas of inordinate length interesting. The golden age of aimless calculation in algebra quietly expired in the early 1890's under the accumulations of its own gold. Its legacy to the modern abstract algebra of the twentieth century was Hilbert's (1890) basis theorem: a polynomial ideal has a finite basis.

While the algebraists were reveling in their interminable calculations, the physicists pursued a more elusive manifestation of invariance through the hypotheses of conservation of mass and energy. Like the formal algebraists, they deluded themselves into believing that they had unveiled imperishable absolute truths for the guidance of all future generations. It was not until 1915–16 that the deeper significance of physical invariance was explicitly formulated in Einstein's principle of covariance.[6] What had been mistaken for eternal necessity was seen to be a platitude of mathematical language: one coordinate system is as good as another. If the differential equations of mathematical physics are to reflect this indifference to any particular coordinate frame of reference, they must be invariant under all transformations of the space-time coordinates. Before the principle of covariance in mathematical physics was enunciated, critical philosophers had objected to the restricted relativity of 1905 because it still favored one very special system[7] of space-time coordinates. But to remedy the defect, more than acute criticism was necessary. A complete mastery of a somewhat neglected field of differential geometry was demanded.

In addition to algebra and physics, a third activity of nineteenth-century mathematics influenced the development of invariance profoundly. Riemann's investigation (1854) of the hypotheses underlying geometry and his analysis (1861) of a problem in heat conduction suggested a new type of invariance, in which general analytic transformations and (quadratic) differential forms replace the linear transformations and algebraic forms of algebraic invariance. From this, modern differential geometry and the geometrized physics of relativity developed. The Riemann-Christoffel tensor, basic in relativistic cosmology and gravitation, made its first appearance in Riemann's[8] posthumously published notes on heat conduction.

The 1870's contributed the final major impulse to the evolution of invariance in the extraordinarily rapid development of Lie's theory of transformation groups. In passing, it may be noted that Lie was one of the extremely rare great mathematicians who showed neither interest nor talent in mathematics in their youth. He was impressed, but not roused to activity, at the age of twenty-one by Sylow's lectures on the Galois theory of equations. When Lie finally succumbed to geometry at twenty-six, he accomplished in weeks the normal work of years. Association (1869) with the enthusiastic Klein for some months, about a year after he had found himself, converted Lie to groups and invariance. The history here is enlivened by more than a touch of human interest. The early cordial friendship between these two outstanding nineteenth-century mathematicians cooled as Lie gradually became morose and suspicious in his later years. While Klein was doing his energetic best to sell Lie's ideas to the mathematical public, Lie imagined Klein was trying to appropriate some of the retarded kudos that accrued from the invention of transformation groups. The stage was set for another bitter wrangle over priority, when Klein adroitly withdrew behind the curtain, leaving Lie to face the claque alone.

Academic squabbles between sensitive mathematicians seldom end so tamely. With eminent men not averse to dueling[9] in public, it is scarcely to be wondered at that humbler mathematicians dissipate much of their time and energy in backbiting and futile envy of their intellectual betters. This very human trait was characterized by R. Baire (1874–1932, French), himself a subtle analyst and one of the more neurotic mathematicians of the twentieth century, as "the agglomeration of mediocrity." These observations are not offered in any spirit of derogation, but merely as a plain statement of academic fact, for the consideration of young idealists about to seek refuge from humanity in a life dedicated to mathematical research.

Lie's theories and methods had a far-reaching influence on mathematics during the last third of the nineteenth century. Thus Klein's famous program[10] of 1872 for a unification of the principal geometries then existing was partly suggested by Lie's theory of transformation groups. Invariance certainly is one of the dominant ideas in much of Lie's work. And as will appear later, the concepts of group and invariance are complementary in Klein's synthesis. After relativity, the group concept in geometry retained its suggestiveness, but sank to a minor position with the invention of numerous geometries beyond those unified by Klein's program. Again, the classical theory of algebraic (projective) invariants found its final resting place as a subordinate detail in Lie's theory, which also contributed directly to mathematical physics in its applications to differential equations. The method of contact transformations, for example, rendered much of the classical Hamilton-Jacobi treatment of the dynamical equations of mere historical interest. In pure mathematics, Lie's methods reacted on the theories of differential and integral invariants.[11] The latter, in Poincaré's hands, (1891–), became a most effective tool in celestial mechanics. Integral invariants passed thence in the 1920's into the type of mathematical physics associated with general relativity and conservation laws. Both differential and integral invariants played dominant parts in the revitalized geometry of the twentieth century which, according to some critics,[12] advanced far beyond the bounds set by Klein's program.

By 1915, Lagrange's 'mustard seed' of 1773 had grown into a forest overshadowing vast regions of mathematics, and it was still sending up new shoots. When general relativity emerged, invariance was firmly established as a seemingly permanent addition to mathematical thought. With the popularization of relativity, invariance attracted the attention of philosophers. Lagrange's unconsidered trifle thereupon assumed an importance in the history of philosophic thought that would have astonished that mild skeptic of all philosophies, including his own.

If history were capable of performing any useful function, it might be expected to induce the present to learn something from the past, instead of merely remembering it in mutually contradictory epics garbled according to taste and prejudice. In the present connection, we recall that others besides Lagrange failed to recognize the germ of something great when they saw and handled it in passing. Gauss missed it completely; Boole half recognized it; Eisenstein let it slip through his fingers; and it remained for Cayley to seize and develop the opportunity thrust at all algebraists of his generation. Although the algebraic theory initiated by Cayley in the 1840's had sunk to a minor status by the 1880's, the emergence of invariance as a new unifying principle, to be deliberately sought and followed in mathematics, dates from the era of excessive and somewhat futile calculation in the algebra of forms.

The wider scientific significance of invariance was overlooked by one who, in retrospect, might have been expected to recognize it instantly. Riemann's cosmic imagination foresaw the partial geometrization of mathematical physics as early as 1854, only to miss the deeper generalization hinted in his own work on heat conduction seven years later. The full mathematical, scientific, and philosophical import of invariance was perceived only when general relativity displayed it for all to see, an historical fact that must be recorded despite the hostility of absolutist ideologies to any form of relativism. Looking back on what now appears as an unnecessary delay in the emergence of this dominating concept of modern scientific thought, we may wonder whether busy young mathematicians of the twentieth century are not overlooking tremendous trifles like Lagrange's which, a century hence, will be cultivated to the neglect of what now seems to overshadow the mathematical heavens.

Algebraic invariance[13]

Algebraic invariants were first called hyperdeterminants by Cayley in his initial paper of 1845, as he then regarded the phenomenon of algebraic invariance as a generalization of the rule for multiplying determinants. It is remarkable that this early work introduced and used the symbolic method, already implicit in Grassmann's algebras of 1844, which was later simplified, expressed in a more suggestive notation, and employed with amazing skill and effectiveness by the German algebraists. We may recall here in passing the principal stages in the evolution of determinants.

Possibly as early as 1100 B.C., the Chinese[14] solved two linear equations in two unknowns by a rule equivalent to the usual method by determinants; and there is a legend that Seki Kowa (1642–1708, Japanese), Newton's equal if not superior,[15] did about the same in 1683. Seki, by the way, is almost credited with the invention of what is alleged to be a form of the integral calculus, "although we have no positive knowledge that he ever wrote on the subject."[16] As for the invention (the yenri) itself, only the most generous of historians might detect in it anything less crude than the earliest Greek steps toward the method of exhaustion. Shortly after Seki's hypothetical anticipation of determinants, Leibniz (1693) gave a rule for simultaneous linear equations equivalent to that of the Chinese. This rule was amplified (1750) by G. Cramer (1704–1752, Swiss) and simplified

(1764) by E. Bézout (1730–1783, French). But in spite of their antiquarian appeal, it is difficult to see that any of these extremely interesting developments had anything whatever to do with determinants.

The next contributions have a more valid claim. A. T. Vandermonde (1735–1796, France) improved the notation, isolated (1771) as objects for independent study what were subsequently seen to be determinants, and gave a systematic account of the little then known. Lagrange (1773) discovered useful identities, much later recognized as very special cases of the characteristic property of reciprocal determinants; and Laplace (1772) gave a clumsy statement of his rule for the expansion of a determinant. The German combinatorialists of the late eighteenth century easily smothered several useful results under a thick notation; and Wronski discussed (1811) several special forms, but still in a singularly uncouth disguise. Gauss (1801) made incidental use of identities of the same type as Lagrange's.

A long step forward was taken (1812) by J. P. M. Binet (1786–1856, French) in the rule for multiplication which, under suitable hypotheses, suffices to define determinants. The same year, the subject was finally launched by Cauchy, who used the $S\pm$ notation and gave general proofs of the basic theorems. Thereafter determinants were part of the equipment of all working mathematicians, but still a perspicuous notation was lacking. The year 1841 is a landmark in the subject: summing up his researches of several years, Jacobi gave a masterly presentation of the fundamentals, including his own functional determinants (Jacobians); Cayley invented the alluring notation of a square array between vertical bars, and used it effectively.

Cayley's notation, one of the happiest in nineteenth-century mathematics, was to prove almost too suggestive. It conjured up swarms of special determinants, a few useful, the majority merely curious. By 1920, T. Muir (1844–1934, Scotch) had found it necessary to use five volumes totalling approximately 2,500 pages for a mere *history* of determinants. In the same year the subject was absorbed as a minor detail and greatly simplified in tensor algebra, a reduction due directly to the popularity at the time of relativity. The more useful identities, such as Kronecker's, were then obtained with ease, and likewise for the expansion theorems. The classical algebra of quantics can be similarly absorbed. Determinants as cultivated in the nineteenth century will doubtless long remain the monumental example of elaboration for no apparent purpose. When mathematics descends to the level of nineteen-twentieths of the work in determinants it is no longer even a reputable game.

Cayley soon dropped the designation 'hyperdeterminant' in favor of 'algebra of quantics,' and set himself the following comprehensive problem, which he pursued in all its ramifications through numerous papers and a famous series (1854–78) of ten memoirs on quantics. An m-ary n-ic, once more, is a homogeneous polynomial, with arbitrary constant coefficients, of degree n in m independent variables; such polynomials Cayley called quantics. The variables in any given quantic are subjected to a nonsingular linear homogeneous transformation, and the transformed quantic is then reduced to a quantic in the new variables. If the original is an m-ary n-ic, so is the transformed quantic. It is required to find all those polynomials (rational integral algebraic invariants and covariants) in the coefficients alone, or in the coefficients and the variables, of the original quantic that differ only by a factor,[17] which is a power of the determinant of the transformation, from the similarly constructed polynomials in the coefficients alone, or in the coefficients and the variables, of the transformed quantic.

Early experience soon suggested two further problems, each of cardinal significance for the subsequent development of modern algebra and its numerous applications, for example to algebraic geometry. The first concerns the existence of a so-called fundamental system: to determine whether there exists a set of invariants for a given quantic such that any invariant of the quantic is expressible as a polynomial in members of the set; the like for covariants, the quantic itself being counted among its covariants. This problem is immediately generalized to the corresponding one for any finite set of quantics, of any degrees in any numbers of variables. The second problem, that of syzygies in Sylvester's planetary language, is to find all independent irreducible algebraic relations among the invariants of any finite set of quantics, and likewise for covariants. These descriptions of the two capital problems of the algebraic theory of forms must suffice here; for detailed statements we refer as usual to any standard text.

Not to prejudge the issue, a crucial possibility was ignored in the description of each problem: does there exist a *finite* fundamental system; is there a *finite* number of independent syzygies from which all are deducible by rational operations? By one of the most fortunate slips in the history of mathematics, Cayley (1856) convinced himself, in the special case of binary (two-variable) quantics of degree higher than the sixth,[18] that the answer to both questions is negative: there is not a fundamental system consisting of only a finite number of members; nor are there only a finite number of syzygies from which all are rationally derivable. Cayley was deceived, the correct answer in both cases being affirmative. Falling foul of arithmetic, he had assumed a certain system of linear diophantine equations to be independent when they were not.

Cayley's numerous successes, quickly followed by those of the prolific Sylvester, unleashed one of the most ruthless campaigns of totalitarian calculation in mathematical history. From 1846 to 1867 the British school of Cayley, G. Salmon (1819–1904, Ireland), and J. J. Sylvester (1814–1897, English), with Hermite cooperating in France, led an enthusiastic army of algebraists and projective geometers through the fertile territory of binary quantics of all degrees, quadratic and bilinear forms in any number of variables, and ternary cubics. Proceeding entirely on foot, all shunned the powerful symbolic mechanism perfected from Cayley's hyperdeterminants by the German divisions of S. H. Aronhold (1819–1884), who founded the modern symbolic method about 1850[19], P. Gordan, its complete master, and R. F. A. Clebsch (1833–1872), who systematized (1871) and vigorously applied the German symbolism to geometry in classic treatises[20] today but seldom opened. Headed by the resolute F. Brioschi (1824–1897), a corps of Italian specialists joined the triumphal march, calculating furiously as they stormed into the lush meadows. The backward U.S.A. maintained its neutrality till 1876.

The surfeit of easy victories might have bogged the march in 1868; but it surged far beyond, into the 1890's, by its sheer momentum.[21] Sylvester in 1876 hastened from his native London to the laggard United States, where he served ably as mathematical ambassador till he was recalled at his own request in 1883. His unflagging zeal swelled the army of algebraists with a band of raw recruits who mistook the united kingdom of binary quantics, symmetric functions, and eliminants for the wider democracy of mathematics. The *American Journal of Mathematics*, founded by Sylvester in 1878, began storing up sheaves of calculations against an imminent famine that has yet to arrive.

Such misdirected foresight was not peculiar to the algebra of quantics in mathematics since 1850. In the accompanying theory of groups, for example, especially permutation groups, there was a similar panic.[22] Once the means for raising unlimited supplies of a certain crop are available, it would seem to be an excess of caution to keep on producing it till the storehouses burst, unless, of course, the crop is to be consumed by somebody. There have been but few consumers for the calculations mentioned, and none for any but the most easily digested.

Nevertheless, the campaign of calculation for the sake, apparently, of mere calculation did at least hint at undiscovered provinces in algebra, geometry, and analysis that were to retain their freshness for decades after the modern higher algebra of the 1870's had been relegated to the dustier classics. Only four need be noted here.

Profiting by Cayley's oversight of 1856, Gordan in 1868 proved the existence of finite fundamental systems of invariants and covariants for any binary quantic, and in 1870 did the same for systems of such forms. Gordan's proof, using the symbolic calculus of Aronhold and Clebsch, was constructive: by a sufficient expenditure of brute force the fundamental systems could actually be dragged out and exhibited. For this memorable victory over the barbaric hordes of algebraic formulas, Gordan was crowned "King of Invariants" by his romantic German admirers. He occupied the throne exactly twenty-two years, until Hilbert, a mere stripling of twenty-eight, in 1890 snatched the crown from Gordan's ageing head and rammed it firmly down on his own.

Reducing calculation to a minimum, Hilbert proved the finiteness theorems for the fundamental system and the syzygies of any set of quantics. Great as this achievement was, it was completely overshadowed by the principal implement of proof. Hilbert's basis theorem has passed into the very foundations of algebra. It marked an advance of the first order, not only in algebra but also in modern arithmetic and algebraic geometry. It is fundamental in Kronecker's modular systems and the classic theory of algebraic fields, two of the three main sources of modern abstract algebra, the third being the Galois theory of equations. It was Hilbert's reliance on general reasoning and existence proofs rather than calculation in this matter that exasperated Gordan to cry out, "This is not mathematics; it is theology!"

Gordan's distress was prophetic. Theologians are not noted for their tolerance of one another's creeds, as was demonstrated once more in the half-century of mathematics following Hilbert's proof of his basis theorem. The two most aggressive factions of mathematical theologians—in Gordan's sense—of the 1930's, the abstract-algebraists and the topologists, found much to dispute. According to an expert observer[23] bulletining from the front in 1939, "In these days the angel of topology and the devil of abstract algebra fight for the soul of each individual mathematical domain." Overlooking the possible transposition in the mythology of this amazing brawl over something that may not exist, we merely note that it was precipitated in the first instance by Lagrange's legacy to Boole and Cayley of a single mustard seed. May the better angel win, if anything is to be won.

The second great domain to profit more or less indirectly by the labors of the algebraists was geometry. There can be but little doubt that the rehabilitation of analytic projective geometry by Cayley, Möbius, Plücker, Clebsch, L. O. Hesse (1811–1874, German), and many others, as a partial concomitant of the work in algebraic invariants, was the determining factor in Klein's synthesis of 1872. The algebraists and projective geometers missed the kernel of the matter. Klein[24] saw it when he recognized that certain manifestations of invariance are accompanied by an appropriate group: the operations of the relevant group leave unchanged the invariants concerned; conversely, all the operations leaving certain objects unchanged form a group. (The foregoing is only a very rough description, subject to limitations and exceptions.[25]) Here at last was a comprehensive outlook on the welter of special theories constituting geometry as it was in the early 1870's, when Lie's transformation groups provided the means for unifying all under Klein's program. This will be our concern in a later section. We have already noted that other epochal advance due directly to the algebra of quantics, Cayley's reduction (1859) of metric geometry to projective by the adjunction of suitable invariant loci.

The third province to benefit by the devices of calculation introduced in the algebra of quantics was the theory of linear differential operators, especially as developed in Lie's groups. Here the suggestion offered by algebra to analysis was not very explicit, and may have had no influence at all. But the reaction in the opposite direction of Lie's analysis on algebraic invariants was decisive, in that it forced the algebraic theory into a subordinate place as a detail in the vastly more comprehensive theory of invariance under transformation groups. The link here, connecting Boole and Lie, spans nearly forty years. Boole used linear differential operators as a means for the successive generation of invariants and covariants; Cayley (1846) and Sylvester (1851–2) did likewise more systematically, while Aronhold (1851, in lectures) and Cayley characterized covariants and invariants of binary quantics as the polynomial solutions of certain linear partial differential equations. The last, in Sylvester's warlike phrase, is equivalent to annihilating the polynomials concerned by the corresponding differential operators, called annihilators. Sylvester (1878) generalized the method of differential operators to any quantic; E. Study (1862–1922, German) and A. Hurwitz (1859–1919, Swiss) in 1889 and 1894 respectively extended and simplified the method. The final outcome of this operational approach to the algebraic theory of forms showed that invariants and covariants are particular functions invariant under the infinitesimal transformations of a certain Lie group. Thus, as noted by Lie in the grand summary (1893) of his lifework,[26] the nineteenth century's greatest effort in formal algebra—as distinguished from the more abstract, structural algebra originating with Galois—was absorbed in analysis. We shall return to this presently.

The fourth and last outcome of the calculations in quantics to be noted here, the benzene comedy, is not yet (1945) comparable in significance to any of the other three, although it may outlast them all in historic interest should atomic physics turn away from classical analysis to imitate modern algebra more closely. The entire episode is curiously like some in medicine, where old wives' brews and folk-messes such as a few of the Chinese prescriptions have been justified on modern scientific grounds, although the primitive science accompanying the remedies no longer makes sense. So it was with the "chemico-algebraical theory" (1878), the offspring of Sylvester's fecund imagination as, according to his own account, he lay tossing about unable to sleep. "That sublime invention of Kekulé," the theory of chemical valence, accompanied by the benzene ring, burst upon the insomniac's vision in the revelation that chemical combination "is tantamount to the construction of an invariant (or in Professor Gordan's language, the final 'Ueberschiebung') of a quantic, or of the derivee of a quantic, or set of quantics, with itself." Clifford saw that it follows almost immediately that to the simplest of all organic compounds, marsh gas or methane, CH_4, "corresponds" an invariant of a binary quantic and four linear forms.

As late as 1900, Gordan was taking seriously Sylvester's superficial analogies between graphical formulas in chemistry and the algebra of quantics, and was writing papers on the subject. But it had yet to be shown that chemistry or any other science will yield up all its secrets to middle-aged algebraists, who have seldom been inside a laboratory, tossing about in bed. Sylvester was not the only pure mathematician who has attempted to solve the universe in his head; nor was his the only vision that facts have quickly dissipated.

The classical algebra of quantics found an interpretation[27] in atomic physics in 1930–1, but for no reason even remotely resembling the speculations of 1878–1900. The algebra of quantics appeared in the modern quantum theory as an epiphenomenon of the calculations. It remains to be seen whether a majority of physicists were right in their estimate of this curious quantum-quantics as nothing more profound than a mathematical pun.

The classical theory of quantics was reworked incidentally (1939) in its broad outlines by H. Weyl (1885–, Germany, U.S.A.) in *The classical groups, their invariants and representations*. The groups considered are on n variables, and are the general linear group and its subgroup of all non-singular or all unimodular linear transformations, the orthogonal group, and the groups having as invariant a non-singular antisymmetric bilinear form. The first two are those of interest for the classical theory. The field of the coefficients and coordinates is an abstract infinite field, as in the theory of linear groups elaborated by L. E. Dickson (1874–, U.S.A.) in the early 1900's. The renovation consists principally in fitting the theory into the general scheme of modern abstract algebra, including the theories of semi-simple algebras and group representations. The last was inaugurated in 1896–1903 by G. Frobenius (1849–1917, German) for finite groups, and developed thereafter by him and many others. By means of the commutator algebra of a completely reducible matric algebra, the representation theory of certain finite groups is applicable to the infinite groups analyzed by Weyl. For the analysis of tensors by means of symmetry operators, permutation groups serve to separate the tensor representations of the general linear groups into their irreducible components. A generalization of the older methods makes the invariants depend on quantities transforming under an arbitrary irreducible representation. The 'quantities' are vectors appertaining to a specific representation of the group. The commutator algebra (in Weyl's terminology) of a given algebra is the algebra of all matrices commuting with every matrix of the given algebra; it is useful in the representation theory of semi-simple algebras.

In the calculation of group characters, analytical considerations offer an alternative to the purely algebraic methods, the finite summations of the latter being replaced by integrations over the group manifold, the invariant element of volume of a compact group having been suitably defined. It will be seen shortly that F. Klein (1849–1925, German), in his comprehensive 'Erlanger Programm' of 1872 for the codification of geometry as it existed at the time, essentially reduced geometry to the study of equivalence under certain groups of transformations. In the modernized presentation of Klein's project, groups appear as groups of automorphisms and of preferred coordinate systems of the base space. The automorphisms leave invariant the characteristic relations of the particular geometry concerned. These automorphisms induce transformations on the vector subspaces of the base space. As in the development of Klein's geometry as an application of groups, the group here has a dual aspect. Weyl's synthesis also includes parts of the topology of the groups discussed. In its application of a wide variety of methods to a given situation, whether of algebra or geometry, it is reminiscent of Klein's universality in both of these fields as they were in his day.

The synthesis by transformation groups

In the following rough description of transformation groups, it is assumed that all conditions necessary and sufficient to insure the existence of all the transformations described are satisfied.[28] The functions $f_1, \dots, f_n$ are such that the n equations

(1) $x_i' = f_i(x_1, \dots, x_n; a_1, \dots, a_r), i = 1, \dots, n$

are solvable for the n independent variables $x_1, \dots, x_n$ as functions of $x_1', \dots, x_n'$; and $a_1, \dots, a_r$ are r parameters; both the variables and the parameters may vary continuously.

The n equations (1) define a transformation of the variables $x_1, \ldots, x_n$ to new variables $x'_1, \ldots, x'_n$; solution of the equations gives the inverse transformation defining the variables $x_1, \ldots, x_n$ as functions of $x'_1, \ldots, x'_n$. All this is for a specified choice of parameters $a_1, \ldots, a_r$. If for another choice, $b_1, \ldots, b_r$, the $x'_1, \ldots, x'_n$ are transformed to new variables $x''_1, \ldots, x''_n$ by a transformation of the same sort (that is, by the same functions $f_1, \ldots, f_n$) as before, so that

$$(2) \quad x''_i = f_i(x'_1, \ldots, x'_n; b_1, \ldots, b_r),\ i = 1, \ldots, n,$$

the variables $x'_1, \ldots, x'_n$ may be eliminated from (2) by substitution of their values from (1). It will not usually happen that the outcome of the elimination is expressible in the form (1), say as

$$(3) \quad x''_i = f_i(x_1, \ldots, x_n; c_1, \ldots, c_r),\ i = 1, \ldots, n,$$

where $c_1, \ldots, c_r$ are functions of $a_1, \ldots, a_r, b_1, \ldots, b_r$ alone. But should such be the case, the transformations (1) then have one of the cardinal properties of a group:[29] the set of all such transformations is closed under successive performance of transformations, as in passing from (1) through (2) to (3). The theory of transformation groups is concerned solely with sets of transformations having this property of closure. But more is now customarily demanded, as will be noted presently. When sufficient further requirements to make the closed set a group are imposed, the resulting group is called an r-parameter transformation group.[30] Because $f_1, \ldots, f_r$ are continuous functions of the variables and parameters, the group is said to be continuous; and if the number r of parameters is finite, the group is called a finite continuous r-parameter group. There are also infinite continuous groups, in which the number of parameters is infinite. These have been less extensively investigated, and will not be further mentioned here. Discontinuous r-parameter groups are similarly defined when the parameters are restricted to range over a set of discrete numbers. The rest of the description, given only for one-parameter groups, is readily extended to r-parameter groups; it illustrates the essential points. As customary in the classic treatment, we shall use the ancient language[31] of infinitesimals, δa, δx_1, δx_2, whose squares are to be discarded; all this can be readily rephrased[32] in terms of neighborhoods of points, etc. If desired, it can also be abstracted and generalized up to a certain stage. The one-parameter set of all transformations

$$x_1' = f_1(x_1, x_2; a), \qquad x_2' = f_2(x_1, x_2; a)$$

is assumed to have the closure property as described above for r-parameter sets.

It is now assumed[33] that for some value a_0 of the parameter a the resulting transformation is the identity, so that

$$(4) \qquad f_1(x_1, x_2; a_0) = x_1, \qquad f_2(x_1, x_2; a_0) = x_2;$$

that is, for $a = a_0$ the transformation is $x_1' = x_1$, $x_2' = x_2$. Thus the set, by assumption, contains the identical transformation. With the same a_0, and δa an infinitesimal increment of a, the transformation

$$x_1' = f_1(x_1, x_2; a_0 + \delta a), \qquad x_2' = f_2(x_1, x_2; a_0 + \delta a),$$

on expansion by Taylor's theorem, gives

$$(5) \qquad x_1' = f_1(x_1, x_2; a_0) + \left(\frac{\partial f_1}{\partial a}\right)_{a=a_0} \delta a,$$

higher powers of δa being neglected, with a similar expression for x_2' having f_2 in place of f_1. The partial derivative in the above, evaluated at $a = a_0$, contains as variables only x_1, x_2, since a_0 is fixed; say this evaluated derivative is $\xi(x_1, x_2)$. By the assumed continuity of f_1, f_2 (4), (5) give

$$\delta x_1 \equiv x_1' - x_1 = \xi(x_1, x_2)\delta a;$$

and in the same way,

$$\delta x_2 \equiv x_2' - x_2 = \eta(x_1, x_2)\delta a,$$

in which $\eta(x_1, x_2) \equiv \left(\frac{\partial f_2}{\partial a}\right)_{a=a_0}$. Unless ξ, η vanish identically, or unless either becomes infinite, the above define an infinitesimal transformation. The possible exceptions may be ignored, since Lie proved that a one-parameter group always contains an infinitesimal transformation and indeed essentially only one. For simplicity, it is assumed that every transformation in the set has a unique inverse. Conversely to the above, Lie showed that every infinitesimal transformation determines a one-parameter group. He also proved that any such group has an invariant, say $I(x_1, x_2)$, such that, for all transformations of the group, $I(x_1', x_2') = I(x_1, x_2)$. Such an invariant is said to admit the one-parameter group of transformations. In particular, the invariant admits the infinitesimal transformations of the group.

We indicate first the connection with differential equations. In the sense just described, a differential equation may admit infinitesimal transformations. For the historically important case when the variables are separable, and hence when the equation is solvable by quadratures, Lie gave a method for constructing the infinitesimal transformations leaving the equation invariant; the construction indicates the steps to be taken to reach the solution. The advantage of Lie's method over its predecessors is the obvious one of uniformity and generality over haphazard ingenuity. It is applicable both to ordinary differential equations of any order and to partial differential equations.

It is interesting to recall that Galois, who was Lie's idol, indirectly inspired the application of continuous groups to differential equations. In a letter of 1874 to A. Mayer, Lie[34] observes that "In the theory of algebraic equations before Galois only these questions were proposed: Is an equation solvable by radicals, and how is it to be solved? Since Galois, among other questions proposed is this: How is an equation to be solved by radicals in the *simplest* way possible? I believe the time is come to make a similar progress in differential equations."

Although Lie himself did not construct an analogue of the Galois theory for differential equations, his methods were the point of departure for Picard's initial success (1883, 1887), from which E. Vessiot proceeded to his own complete structural theory (1892) of linear differential equations. Vessiot (1892) stated his fundamental theorem as follows: To every linear differential equation of order n corresponds a finite continuous group of linear homogeneous transformations on n variables, which has properties similar to those of the group of permutations for an algebraic equation. It is shown, among other things, that the integration of a given equation by means of auxiliary equations is thus related to the progressive reduction of the group of transformations for the equation. Further, if the auxiliary equations are restricted to certain types suggested by about two centuries of experience, the method of solution indicated by the group is the only one possible.

One objective of any structural theory is to determine what operations are necessary and sufficient, also what mathematical objects must be invented, to provide a solution for a problem of a prescribed kind. For example, in the Galois theory of algebraic equations, the intrinsic structure of the group of permutations of the roots of an equation, as determined by the composition series of the group concerned, fixes the nature of the algebraic irrationalities to be adjoined to the field of the coefficients so that the equation may have roots equal in number to its degree. In the analogous structural theory for differential equations, it is required to determine the nature of the operations necessary to furnish a general integral for a system of differential equations.

Just as the algebraic theory characterizes the nature of the irrationalities required for the solution of a given algebraic equation, so does the structural theory of differential equations characterize and classify the functions defined by a system of differential equations. As noted, the initial impulse for a structural theory came from Lie's transformation groups. In spite of Picard's[35] deprecatory estimate of his own contribution, which, historically, inaugurated the project, as "only a very natural extension to an analytic problem of the extremely fruitful ideas introduced into algebra by Galois," the problem of devising a structural theory for differential equations was no facile exercise in principles already classic.

Departing from Lie's philosophy of groups, but first using his methods, J. Drach (1871–, French) in 1893 took another direction.[36] This led in 1898 and later to a structural theory of a different type, in which the concepts of reducibility[37] and group of rationality for any system of ordinary differential equations are defined and elaborated. The resulting theory might be termed algebraic, in that it proceeds directly from the differential equations and can be developed entirely independently of Lie's theory. Conversely, the theory of Lie groups is recoverable from Drach's quasi-algebraic structural theory. Among many applications to the differential equations of geometry and mechanics, one in particular harmonized exquisitely with the spirit of 1945: all those laws of air-resistance which render the equations of exterior ballistics integrable by quadratures are determined.

Although all these structural theories of a major division of analysis originated in the late nineteenth century, they are more in the spirit of the general analysis of the twentieth. Their primary objectives are to discover what can be done rather than to do it, and to give criteria for what cannot be done. For example, Picard showed that the general linear differential equation of order n is not integrable by quadratures. It is therefore futile to seek such a solution. This might have been guessed; but, as in Abel's proof that the general quintic is not solvable by radicals, a demonstration of impossibility definitely disposed of what might seem a reasonable problem. Once more the methodology of Abel and Galois made an outstanding contribution to the development of mathematics. In this connection it is interesting to recall Lie's opinion that the pattern of nineteenth-century mathematics was laid out by four men, Gauss, Cauchy, Abel, and Galois.

It was remarked in an earlier chapter that the 1930's saw the beginning of what may be a distinct phase in the theory of differential equations, in which the concepts of modern algebra (since 1910) are carried over, to a certain extent, to the theory of systems of differential equations algebraic in the variables and their derivatives. The direction here, not determined by the theory of groups, diverges widely from previous structural theories of differential equations. The type of result obtained is correspondingly different; thus there is an analogue, for an infinite system of differential forms, of Hilbert's basis theorem for algebraic forms. For further details we refer to the monograph[38] by J. F. Ritt (1893–, U.S.A.), who has been chiefly responsible for this theory. Another,[39] with still a different aim, is that of J. M. Thomas (1898-, U.S.A.).

Continuing with the direct outgrowths of Lie's theory of transformation groups, we note its immediate application to algebraic invariants. We cannot give details; but the following is a bare hint of the strategy. The variables $x_1, \ldots, x_n$ in a given system of forms are transformed by a group of linear homogeneous transformations having determinant 1. This group may be generated by its infinitesimal transformations. Any one of these, applied to the given system, induces an infinitesimal transformation in the coefficients of the forms. Hence also there is induced an increment in any function of these coefficients, in particular, in any invariant of the system. By the definition of an invariant, this increment must vanish. This condition yields a partial differential equation which the invariant must satisfy. Similarly, each of the infinitesimal transformations of the group leads to a partial differential equation satisfied by the invariant. The resulting simultaneous equations can be solved systematically to produce all the invariants of the given system. The method amounts to Sylvester's (1852, 1878); Lie systematized and fitted the differential operators of the algebraists into the more comprehensive frame of transformation groups, thereby revealing their necessity and sufficiency. The actual calculations for forms of higher degree or in many variables may be very prolix. But we have seen that such calculations serve no rational purpose until they are needed for something other than filler for mathematical periodicals.

For a comprehensive abstract of what Lie's synthesis accomplished, we must refer to the account[26] (about 12,000 words) by Lie himself. Writing in 1893, Lie gave reasons for claiming that his "theory of invariants of all continuous groups embraces all theories of invariants hitherto noted." Among numerous other items of the highest interest, we shall cite only three. "The principles of [Newtonian] mechanics," as Lie discovered by an application of his theory to the differential equations of that science, "originate in the theory of [continuous] groups. . . . Kinematics and its theorems are partly subsumed as very special cases in my general theorems. My researches on geodesics and the general problem of equivalence in differential equations, show by what principles mechanics can be successfully handled." Part of this work has passed into the standard modern treatment of analytic mechanics. The appropriate analysis is that of contact transformations, so called because if two curves or surfaces touch before transformation, they also touch after.

The application to kinematics was closely related to a great project in the foundations of geometry. In Euclidean space, a rigid body retains its shape and size[40] as it is moved about; all possible motions of the body generate a continuous group. Lie determined (c. 1870) all subgroups of the groups of Euclidean and non-Euclidean motions in space of three dimensions, thereby characterizing the corresponding geometries. This classic contribution to the foundations of geometry clarified and made exact the ingenious but somewhat vague attempt (1868) of L. von Helmholtz (1821–1894, German) to describe Euclidean space kinematically.

The third item is that of differential invariants, basic in infinitesimal geometry and in geometrized physics. Sylvester, in his theory of reciprocants (1885), had rediscovered several such invariants by his habitually ingenious methods; but he had been anticipated in the more systematic attack of G. H. Halphen (1844–1889, French). A close enough description of the subject is Poincaré's remark that "the theory of differential invariants bears the same relation to the theory of curvature that projective geometry bears to elementary geometry." Of Lie's predecessors, we need recall only Gauss, Riemann, Beltrami, Laguerre, and Halphen, all of whom, either in their work on infinitesimal geometry or in that on differential equations, encountered differential invariants. Lie justly claimed that his earlier general theory included and extended all those of his predecessors and contemporaries. Seventeen years after Lie's death, differential invariants assumed an importance in geometry and physics that he could hardly have foreseen.

One analytic device is perhaps the secret of the utility of Lie's theory and its applicability to classical physical science: reduction to linearity. Within the very mild restrictions of being what some analysts called civilized functions, the transforming-functions from which the theory proceeds are arbitrary. But iterations of even so elementary functions as polynomials of degree higher than the first, as would be demanded if the transforming-functions $f_1, \ldots, f_n$ were such, lead to prohibitive complexities in both algebra and analysis. Operating in the continuous domain, Lie saw that the proper mathematical tools are the corresponding infinitesimal transformations, which are linear in the infinitesimal increments $\delta a_1, \ldots, \delta a_r$ of the parameters $a_1, \ldots, a_r$ in an r-parameter group. Thus linearity, the basic simplifying assumption in much of both pure and applied mathematics, was preserved.

Dying in 1899, Lie had lived long enough to see his most prolific successor, the French geometer E. Cartan, well started on a profitable road which returned many times to continuous groups. Cartan's Paris thesis (1894) was so exhaustive in its disposal of certain outstanding problems that it practically abolished interest in finite continuous groups for many years. There remained the continuous groups with an infinity of parameters. Interest in Lie's theories and methods revived about 1920, after general relativity had given differential geometry a new direction.[41] In line with verbal geometrizations of analysis, the newer work generated geometries of group-manifolds and the like with astounding abandon. These severely technical developments are beyond description here; we note only that their very abundance was but one more contributing cause of the flight to complete abstraction.

Returning to Lie himself for a moment, we observe in his

highly individualistic career a distinction between general working conditions before and after the 1830's that must be taken into account in any appraisal of mathematical development during the recent period. In his later years, Lie acquired enthusiastic disciples and extremely able collaborators, notably F. Engel (1861–) and G. W. Sheffers (1866–), both German. But like some other masterpieces of nineteenth-century mathematics, Lie's theory was essentially a solo performance. It may be too early to judge whether the twentieth-century mania for cooperation in everything, with its monotonous mass production of a merely high-grade output, is deadening enough to blight a talent of the first order. But it is quite generally agreed that, of the most active performers in the 1930's, there was no soloist in Lie's class.[42]

Three explanations to account for the lack of outstanding innovators and great synthesizers have been proposed. Some think the innovators are still with us but too close for recognition. Staunch believers in the virtues of organization and the omnipotence of administrative machinery fervently hope that the age of unpredictable originality died forever with the nineteenth century; fifty competent laborers regimented in a common task can accomplish more, and do it quicker, than the most brilliant improviser. Sharing the optimism of the first apologists, while admitting that the second may be right, the third see the dearth of first-rank mathematicians[42] as the natural outcome of a movement started in the 1830's by Jacobi. That irrepressible innovator believed the infallible method to advance mathematics was for domineering professors in the leading universities to drill their own ideas, and as little else as possible, into as many advanced students as could be induced to scribble lecture notes. In short, Jacobi anticipated the Führer Prinzip. The success of Jacobi's personal venture incited other would-be leaders to acquire their own droves of obedient followers. When the leader is a Jacobi, the desired end may be attained. But when the leader himself is a man of second-rate intellect, with no very clear notions of where he is going or how to get there, the faithful procession may find itself following its guide into the barren wildernesses of a dead past.

Lie's opulent originality flourished a decade or more too late to be left in unorganized peace. The massive, slow-moving Norwegian Lie had nothing systematic about him; the friend of his early manhood, the mercurial Prussian Klein, was the dynamic personification of system. From 1869 to 1884, Lie developed his theory in his own disorderly way. Anything so novel and so extensive was naturally not appreciated at its full value immediately. Convinced that his untidy friend lacked systematization as well as recognition, Klein decided to organize him, detailing the meticulous Engel to supervise the gigantic task.

For five laborious years (1888–93), Lie degenerated into a monotonously productive factory. Three colossal volumes, *Theorie der Transformationsgruppen*, stand as an imperishable monument to disciplined labor. Had this universally acclaimed but seldom read masterpiece been Lie's sole legacy to posterity, his fame might well have been retarded for decades after his death. Even in 1893, when the final volume appeared, life was too short for the thorough assimilation at one sitting of 2,000 large pages of pure mathematics. The undidactic but clear and suggestive presentations in many of Lie's original papers were now smothered under systematized masses of details designed to make transformation groups palatable to weanlings and easily digestible by surfeited mathematicians. But the age of three-decker masterpieces was long past by the 1890's. In the yet faster tempo of the twentieth century, a mathematical paper over fifty pages in length must be of extraordinary quality if it is ever to be read seriously by anyone but its author and the unfortunate referee who must certify its correctness.

Engel's triumph in organizing Lie seems to have been a mistake. But it was not fatal. Not three hundred leaden volumes could have crushed and buried a talent like Lie's. His ideas had done their enduring work long before all the freshness and ferment were systematized out of them. As for Klein's part in this comedy of well-intentioned errors, it is only fair to record that he was actuated by the least inexcusable of all excuses for meddling with another man's life: the sincere and unselfish desire to make a friend do something distasteful for his own supposed good.

Codification of geometry by invariance

Anything in recent mathematics that retains its freshness for half a century is an extreme rarity. A single decade since 1850 may span profounder changes in the general outlook on a particular division of mathematics than did the entire eighteenth century. As the year 1900 is passed, the like holds for even briefer periods and any preceding fifty years. Perhaps the most conspicuous illustration of these facts is Klein's 'Erlanger Programm' (1872) for the codification of geometry. For about fifty years it persisted unmodified; about two years of the new geometries that followed general relativity (1916) rendered Klein's synthesis inadequate.

The best account of the Erlanger Programm is still Klein's own. As this is readily accessible in both German and English, we shall merely state its central concept as rephrased in 1918 by an American geometer.[42a] In addition to the technicalities the date of this quotation is to be noted, as another from the same expert exactly ten years later illustrates the rapidity with which modern mathematics progresses.

A *geometry* is defined as the system of definitions and theorems invariant under a given group of transformations. . . . Two groups of transformations are to be considered in connection with any geometrical relation: a group by means of which the relation may be defined; a group under which the relation is left invariant. The more restricted the group, the more figures will be distinct relatively to it, and the more theorems will appear in the geometry. The extreme case is the group corresponding to the identity [the transformation leaving everything considered invariant], the geometry of which is too large to be of consequence.

The second quotation from the same geometer[12] ten years later (1928) summarizes the radical change in outlook that had occurred in less than a decade:

This [Klein's] point of view was the dominant one for the first half century after it was enunciated. . . . It was a helpful guide in actual study and research. Geometers felt that it was a correct general formulation of what they were trying to do. For they were all thinking of space as a locus in which figures were moved about and compared [as was implied earlier in connection with Lie's revision of Helmholtz' kinematic geometry]. The nature of this mobility was what distinguished between geometries.

With the advent of Relativity we became conscious that space need not be looked at only as a "locus in which," but that it may have a structure, a field-theory, of its own. This brought to attention precisely those Riemannian geometries about which the Erlanger Programm said nothing, namely those whose group is the identity. In such spaces there is essentially only one figure, namely the space structure as a whole. It became clear that in some respects the point of view of Riemann [1854] was more fundamental than that of Klein.

Before describing the basic distinction between the two points of view, we shall quote another informed opinion[43] on the chameleon-like change that overtook geometry when exposed to general relativity in the decade following 1916. In the two quotations above, it is interesting to compare the status of "the group corresponding to the identity" in Klein's program, with that of "geometries . . . whose group is the identity" in the revised conception of geometry. It may be a misreading of the evidence, but it would seem that the trivial stone rejected by the master builder in 1872 was recovered and given a place of honor by his successors in the 1920's. However that may be, it is reassuring to learn that the new cornerstone was still supporting a massive buttress of modern geometry as late as 1945. Again we quote an expert.[43]

"What may be called the Euclid-Klein concept of geometry, that is the study of equivalence under a given group of transformations, was replaced by the idea that space has an intrinsic structure, consisting of a set of relations which may be, but in general are not, defined in terms of a transformation group."

The basic difference between the two conceptions of geometry exemplifies a logical platitude. If at least one postulate of a mathematical (hypothetico-deductive) system be suppressed, the system developed from the modified set of postulates is less restricted than the original. In this sense, the modified system is more fundamental and more general than the original system. Specifically, if one or more of the postulates of a group of transformations be dropped, and if invariance with respect to the transformations defined by the curtailed set of postulates be taken as the criterion for a "geometric object,"[44] the new geometry will be more general than that developed for invariance with respect to a transformation group. Thus in the new geometries the concept of invariance will survive, while that of a group may, but in general will not. This is the essential difference between the new geometry and the old; it may be outmoded by the time this is printed.

Klein's synthesis of geometry was an outstanding landmark in the mathematics of the nineteenth century. As in numerous other instances where an historical monument has been far surpassed, it does not follow that Klein's synthesis was abolished. It retained its usefulness in the limited domains it was designed to unify; and its technical methods, although superseded, undoubtedly suggested many things to be done in the revised geometry. In fact, so close is the formal resemblance between the old and certain earlier parts of the new that hostile critics characterized the latter as conspicuous examples of the facile sort of generalization which is beneath the notice of reputable geometers. But even if this had been justified, there remained a substantial residue of new geometry, some of it useful in the physical sciences, which Klein's methodology from its very nature could never have produced. For these and many other reasons, the revised program in geometry has a chance of being remembered as one of the capital achievements of mathematics in the twentieth century. Some supplementary remarks are therefore in order.

Returning to the Erlanger Programm itself, we note some of the things it accomplished or inspired. It unified, from the standpoint of groups, Euclidean geometry, the classic non-Euclidean geometries, projective geometry, and the conformal geometry growing out of Steiner's inversive transformation.[45] Riemannian geometry was included only trivially. By seeking the geometries corresponding to subgroups of the groups defining these respective geometries, Klein's approach revealed the interconnections between various geometries and suggested many interesting things to be done in both geometry and the theory of groups.

Regarding the last, it can hardly be claimed that the outcome was wholly fortunate for mathematics. The great vogue of groups that persisted well into the twentieth century was due partly to the popularity of the Erlanger Programm. The abler specialists in groups, men like Sylow, Jordan, Frobenius, Hölder, W. Burnside, E. H. Moore, Dickson, and Klein himself, were expert in a great deal of mathematics besides the theory of groups or one of its narrow subdivisions, such as permutation groups. Their intuition for what constitutes mathematics as opposed to mere calculation, which no amount of technical difficulty can elevate above triviality, is evident in the generality of classic theorems associated with their names. But it was possible for industrious laborers little advanced beyond mathematical illiteracy to obtain almost any finite number of permutation groups by obstinate grubbing, and to find by the same dull means all the finite abstract groups in certain narrow categories which they themselves had defined, apparently with the express purpose of dignifying their calculations with an air of pseudo generality. In permutation groups, for example, the first week of school algebra will give the prospective calculator all the manipulative skill he needs. But it will not give him the sense to distinguish between what is living and what is dead in mathematics. The intellectual maturity which is frequently cited as a prerequisite to a successful career in finite groups can be ignored by the mere collector of specimens. It is rather humiliating to American mathematicians to have to record that the greatest excesses in this troglodytic sort of activity occurred in the United States, beginning in the 1890's. Fortunately for the progress of mathematics in America, nearly all those who continued active abandoned groups, or at least gave up calculation, after they had satisfied the requirements for a Ph.D.

The success of the Erlanger Programm was also partly responsible for another tendency that did mathematics no particular good. When it was shown that a certain theory satisfied the postulates of group, it seems to have been assumed as a matter of course that the theory was thereby significantly advanced. To cite a trivial instance, when it is gravely announced that all the rational integers form a group with respect to addition, common sense will not stand open mouthed in dumb admiration, but will demand, 'What of it?'

Groups, like Klein's program, have not disappeared from mathematics; nor is there any likelihood that they will in the immediate future. But with the rapid development of geometry and algebra in the twentieth century, groups took their place as but one of several unifying concepts of recent mathematics. One of these, the pseudo-group, is crucial here. It occupies the same position with reference to the geometry that surpassed Klein's program that the group concept occupied with respect to that classic landmark. A pseudo-group[46] is a set of transformations such that, *if* the result of performing any two transformations of the set successively exists, or is defined, it is in the set, and the set contains the inverse of each transformation in the set. An invariant with respect to a pseudo-group is called a geometric object, and geometry is then identified with the study of pseudo-groups and their invariants. Klein's synthesis identified geometry with the study of transformation groups and their invariants.

It was remarked in one of the quotations above that 'space' used to be regarded as a 'locus in which.' In relativity (1916), space-time itself became the object of investigation, for the revolutionary reason that matter, of which material bodies are composed, appeared as an aspect of space-time itself. Instead of asserting that matter is present in a given region of space-time, it proved more fruitful to say that the region has a certain curvature, the measure of curvature varying from point to point in a manner corresponding—in the discarded terminology—to the amount of matter present. In the absence of matter, space-time is flat. This apparently mysterious way of restating and generalizing familiar hypotheses of the physical sciences was not done to bewilder common sense, but to furnish a workable mathematical imagery for certain parts of physics, notably the observable phenomena of gravitation. It justified itself in the prediction of new phenomena, later observed, and in predicting correct numerical measures where the Newtonian theory of gravitation predicted measures that disagreed with observation. There is no need to repeat the familiar story here; our present interest is in seeing that physical science gave the new geometries of the 1920's–30's their strongest impetus. There is an obverse to this: however abstract and artificial the new geometries may appear at first sight, they cannot be wholly useless scientifically, for they include as a very special instance the geometry of space-time. The Erlanger Programm included Euclid's geometry and Newton's space, both of which retained their utility in the science for which they were invented. But geometry did not end with Euclid, nor physical science with Newton, and the older frame proved too narrow for the expanded world picture of mathematics since 1826 and physical science since 1916.

As we saw in an earlier chapter that geometry descended in the late nineteenth century from its celestial status of eternal, necessary truth, it will be interesting to see what became of space in the ascent of geometry to ever rarer abstractions. Neither the philosophical nor the scientific answers to 'What is space?' are relevant here. We are concerned only with what space became for geometers. First, there is not one space, but any number. There is nothing startlingly novel in this; it was already coming into view in 1826 when Lobachewsky constructed the first non-Euclidean geometry. Second, the progressive abstraction of mathematics by the postulationalists, from the algebraists of the 1830's to the geometers of the 1890's, gradually convinced mathematicians that their space, too, was a by-product of their own mathematics. By the 1920's, the conception of space as an arbitrary creation of geometers had become a commonplace: "a space is a set of objects with a definite system of properties, called the structure of the space."[46] An example, to which we shall recur, is a flock of browsing sheep completely surrounding a few alert goats. Another of a somewhat similar kind is any finite number of points enclosed by a sphere of constant radius. Neither of these spaces need be metric, although a distance function for either is easily definable in numerous ways.

With this highly abstract definition of a space in mind, we return once more to Klein's program and its successors. It is interesting to observe the abstract identity between the following description of spacial structure and structure as described in connection with modern algebra, and further to note once more that the basic concepts originated with Galois. Two spaces are called equivalent or (simply) isomorphic if there is a one-one correspondence between the objects in the spaces which establishes a one-one correspondence between all the properties constituting the structures of the respective spaces. When this is applied to two spaces which are the same, there is thus defined what is called an automorphism of the space. It follows easily from these definitions that all the automorphisms of a given space form a group. When this group is the identity, Klein's program says nothing regarding the geometry of the corresponding space. But the identity is as valid a group of automorphisms as any other; and once more the necessity for a revised conception of geometry as a mathematical system is evident. For further particulars, with references to notable attempts in the 1920's to salvage the Erlanger Programm, we refer to technical expositions.[47]

The part played by physical science in this explosive outburst of geometry has been remarked several times already, but the record is not yet complete in its major aspects. The geometric framework of general relativity is Riemannian geometry; the need to include this geometry in any comprehensive concept of geometry necessitated the advance beyond Klein's synthesis. By 1918, Riemannian geometry in its turn appeared to be inadequate for a capital project in physics. The two main field theories, Einstein's (1915–16) of the gravitational field, and Maxwell's (1859–60) of the electromagnetic field, posed the obvious problem of constructing a field theory which would yield both on suitable specialization. Weyl (1918) constructed the first non-Riemannian geometry in an effort to produce the required unified field theory. Several later attempts by Einstein and others were no more successful than Weyl's in satisfying all the scientific demands. Although there was no accepted unified field theory as late as 1945, Weyl's attempt is of the first importance historically for the new direction it gave mathematics in the construction and investigation of non-Riemannian geometries. It was the first outgrowth of relativity that contributed anything basically new to mathematics. All the elaborate mathematical machinery of general relativity was in existence for about twenty to fifty years before relativity was invented. Einstein's prologomenon (1944) to a unified field theory of gravitation and electromagnetism introduced the new device of bivectors.

Physics, however, cannot be credited with everything that happened in geometry after relativity. One of the most fruitful ideas introduced into geometry since Lobachewsky was Levi-Civita's (1917) parallel displacement, a generalization of the notion of parallelism. This in turn was generalized by Weyl and others. The matter is too technical for description here, but it may be mentioned that the parallel displacement of a vector, say in Riemannian space, has some of the cardinal properties of parallelism in elementary geometry. For example, as a relation with respect to displacement along a fixed curve, this parallelism is symmetric and transitive. But a vector displaced parallel to itself round a closed curve will not in general return to its initial value, as is the case for vectors in Euclidean space. This anomaly may seem absurd, but a familiar fact of geography offers a very simple example.[48] If a ship sails a triangular course along arcs of great circles, so that the ship remains parallel to its initial direction as long as it sails on the same arc, the sum of all the angles measuring the changes in course is the spherical excess of the triangle, and hence is zero only if the earth is flat.

In conformity with the spirit of twentieth-century mathematics, it was inevitable that Levi-Civita's parallel displacement should be generalized. The language, at least, of geometry was preserved in the generalization to the displacement of a vector in the tangent space of the original Riemannian manifold. There seems to have been some as yet unjustified optimism that this and further generalizations of Riemannian geometry in the 1920's would provide a geometric framework for the unified field theory (to include the phenomena of gravitation and electromagnetism) desired by the theoretical physicists of the time. None did, and Einstein's geometry of bivectors (1944) was still on probation in 1945. General relativity had included some physical principles, in particular the principle of equivalence, in addition to its geometry. This, according to the physicists, was the source of its successes. According to the same authorities, the failure up to 1945 of the attempts at a unified theory might have been anticipated: each attempt was geometry and nothing more. The truism that, to get something empirically verifiable out of mathematics, something empirically known must be put into mathematics, appeared to have been overlooked. It must be mentioned, however, that not all theoretical physicists of the 1940's accepted this truism. To some it was not even true; it was definitely false. Eddington, for example, until almost the week of his death in 1944, continued to defend his capital creed that "all the laws of nature that are usually classed as fundamental can be foreseen wholly from epistemological considerations." Although the epistemology generated by Einstein's general relativity (1916) and Weyl's gauge-invariance theory of electromagnetism (1918) failed to predict any fundamental law of nature or to provide a satisfactory unified field theory, the geometry originally inspired by both persisted through the 1920's–1940's as the prolific source of one generalization after another of 'space,' 'displacement,' and 'geodesic.'

The illustration of the ship sailing along the arc of a great circle, so that it remains parallel to its initial direction, suggests a generalization of parallel displacement extensively developed in the 1920's–1930's by the American school of differential geometers. This 'geometry of paths,' begun in 1922 by O. Veblen (1880–) and L. P. Eisenhart (1876–), is concerned with 'paths,' or lines of constant direction, and is a generalization of the theory of geodesics in Riemannian geometry. Analytically, the geometry is an image of the properties of the integral curves of a special system of n linear homogeneous ordinary differential equations of the second order. In a suitably restricted region the system has a solution: through every point of the region, and in each of the directions at the point, an integral curve passes; there is one integral curve through two points. Hence, the solution defines a system of ∞^{2n-2} geodesics; these are the paths. Briefly, the theory is concerned largely with projective transformations of displacements and the accompanying projective invariants. Conformal transformations and invariants also are discussed for Riemannian manifolds. The geometry of paths was extensively developed by Veblen, Eisenhart, and T. Y. Thomas (1899–, U.S.A.) in the 1920's–1930's. Almost simultaneously a further generalization of displacement, introduced in 1922 by E. Cartan (1869–, French), mapped the space at a point of an n-dimensional manifold on a point in its neighborhood (in the usual sense of analysis).

These and other successive generalizations may have been suggested by geometric intuition, or by the analysis and algebra of transformation groups, or by the efforts of the physicists to approach their problem of the unified field by means of distant parallelism. Whatever their ultimate source, generalizations continued, and a significant break with linearity of connections at last seemed imminent. If the laws of physics are basically non-linear, as Einstein has asserted they must be, the geometry of non-linear connections may prove as useful in physics as physics has proved in geometry—although some of the geometers might be reluctant to acknowledge any obligation. The last generalization which need be mentioned in this small sample from the 1920's–1940's, Veblen's geometry of spinors, seems to have issued directly from Dirac's algebra of the quantum theory.

As we are interested in the broader aspects of mathematics in relation to civilization as a whole, we note in passing what part European culture played in these epochal advances in geometry. The Germany where Einstein and Weyl did much of their finest work expressed its appreciation of their efforts by robbing one of his personal property and driving both into exile. The United States received them. Returning to the intellectual ideals of Julius Caesar, the virile descendants of the ancient Romans rewarded Levi-Civita by depriving him of his professorship in his declining years before so effectively persecuting him that he suffered a collapse. These are but three of the more conspicuous cases among many of the same kind. Some see in this reversion to the Dark Ages only the transient shadow of a few survivors from the glamorous past as they strutted or stumbled their way across the stage of human stupidity. Others, looking more critically, observe that the farce would close in a week were it not supported by the will of the majority.[49] And while almost any civilized human being will sympathize with the victims, it is a fact that neither mathematicians nor scientists are a particularly homogeneous set. It will be remembered that mathematics has often been likened to music; but there, usually, the flattering comparison stops. It is closer to the facts to compare mathematicians with musicians, who, of all artists, are probably the most deeply imbued with the venemous hatreds distilled from professional jealousy. Anyone with an extensive acquaintance among mathematicians, unless he is so fortunate as to be a naive simpleton, must admit that only the clubbed fist of the law prevents some rival mathematicians from doing to one another privately what the state has done publicly to those whom the authorities dislike. This condition of course is by no means universal. But then, neither was the official persecution of mathematicians.

Before taking leave of geometers and their unstable conceptions of space and geometry, we note what at least some of them imagined the nature of their activities to be in the 1930's. Precise definitions of geometry serve no useful purpose, and attempts to formulate them savor too strongly of scholastic pedantry to suit the taste of the twentieth century. If we ask one who calls himself a geometer what geometry is, we may be fobbed off with the bland assurance that an objective definition would probably include the whole of mathematics.[50] The possibility of including metamathematics in geometry appeared to have been overlooked as late as 1945; but possibly the omission was intentional, as the inclusion might have caused acute internal discomfort. Except in the classic routines of postulate systems, geometry had been but little concerned up to 1945 with any of the sharper logical analysis that rendered the trite routines partly obsolete. In fact, anything approaching a searching analysis of the foundations of geometry, including topology, in a post-1930 spirit seemed to be nonexistent. But the almost unanimous professional opinion regarding the problematical outcome of such an analysis, should it ever be attempted, was that geometry would survive the ordeal practically unmaimed. Here again we have an instance of the essential immortality of mathematics, as in Euclid's I, 47. Something recognizably like the mathematically or scientifically useful theorems will stand, although their successive proofs may be shot from under them. And so, following some of the geometers of the twentieth century, we find ourselves once more in the mystic presence of Plato's Eternal Geometer. But no edict compels us to remove either our hats or our shoes.

Being denied an objective description of geometry, we either abandon the quest for enlightenment or accept the subjective alternative,[51] that "what a sufficient number of competent people have thought fit, on traditional or emotional grounds," to call a geometry is indeed a geometry. To those who lack either competence or the orthodox emotions, it is always like witnessing a miracle to see a geometer decide without a moment's hesitation that a certain piece of mathematics, which the uninitiated might mistake for analysis, is in fact geometry. If only all the competent people who call themselves geometers reached a unanimous decision in each instance, the miracle would carry conviction.

Finally, looking back on the progress of geometry from 600 B.C. to 1945, we note that large flocks of competent people all down mathematical history have been stampeded on several occasions by unorthodox goats who refused to be circumscribed by any number of competent people. There is no reason to suppose that the like will not happen again, not once but many times. Some Descartes or Lobachewsky or Riemann or Lie of the future will disrupt the contented circle and escape from academic respectability; the rest will follow. In fact, something not altogether orthodox began happening about 1912 in the very center of the flock. By 1939, the disturbance seemed on the point of resolving itself into the battle observed by Weyl.[23] With locked horns, the devil of abstract algebra and the angel of topology were doing their utmost to butt each other out of the circle. Which, if either, geometry is to follow, was not clear in 1945. But there was no doubt that topology then was one of the most aggressive of all the geometric or mathematical sciences. Accordingly, we note next some of the principal stages by which topology attained its maturity.

Intrinsic spacial invariance

With the extraordinarily rapid development of topology, or analysis situs, in the twentieth century, the concept of invariance penetrated more deeply than ever before into both geometry and analysis. Experts in topology claim that their methods render complicated situations in analysis spacially intuitive; and to judge by the results they obtain with apparent ease in some of the most intricate problems of modern dynamics, they must be right. But to less gifted mathematicians, by nature condemned to trudge through every step of a logical argument in order to credit its conclusion, topology must remain an untraveled highway, even though it may be the royal road to mathematics that Menaechmus assured Alexander the Great did not exist. One thing seems certain: to think topologically, the thinker must begin young. The cradle with its enchained teething rings may be a little too early; but the education of a prospective topologist should not in any case be deferred beyond the third year. Chinese and Japanese puzzles of the more exasperating kind, also the most devilish meshes of intertwisted wires to be taken apart without a single false move, should be the only toys allowed after the young topologist has learned to walk. With this apology for what follows, we proceed to one of several definitions[52] of topology before noting some of the problems to which it refers. The central idea is invariance with respect to a biuniform, bicontinuous transformation between two spaces. The following description is intended to convey only a rough idea of what the subject is about; as always we refer to readily accessible technical expositions for an exact account.

Topology is concerned with certain qualitative properties of a space as already defined, namely as a set of objects with a definite system of properties. The objects are usually called points; but the crude intuitive notion of a point is not the only one to which the topology of a general or abstract space is applicable. Of the properties assignable to a space, that of a neighborhood is central here. Neighborhood, like point, or object, is undefined except for the few seemingly trivial requirements to be stated presently. An instance of neighborhoods is the traditional one as in analysis and geometry, where the neighborhood of a point is specified by means of a distance relation or, intuitively, by a sufficiently small circle surrounding the point. Similarly for the notions of region and boundary; but once more the situations included in the abstract definitions are not restricted to these familiar examples. However, having them in mind may suggest one origin of the general formulation and make it seem less mysterious. We shall first state a modern definition of topology and then break it down into its constituents in terms of objects, sets (classes), and neighborhoods.

Topology is the study of those properties of spaces that are invariant under homeomorphic transformations.[52]

First, as to spaces. A space here is a set of objects together with an aggregate (set) of subsets, called neighborhoods, such that to each object correspond one or more neighborhoods, and every neighborhood of an object contains that object. Let S denote such a space, and let A be any subset of S. An object is called an inner object of A whenever there is a neighborhood of the object that is contained in A. The set remaining when all the objects of A are removed from S is denoted by $S - A$. An object is called a boundary object of A if every neighborhood of the object includes objects of both A and $S - A$; the boundary of A is the set of the boundary objects of A; and A is said to be open when its boundary is contained in $S - A$, closed when its boundary is contained in A.

Second, as to transformations. A transformation of one space S into another S' is the assignment of a correspondence between the objects in S, S' such that to every object in S there corresponds at least one object in S' (as in mapping, say); the transformation of a subset A of S is the set of all correspondents, under the transformation, of all objects in A. The transformation is said to be single valued, or uniform, whenever it assigns a unique correspondent to every object in S; it is said to be one-one whenever it also assigns a unique correspondent to every object in S'; it is said to be continuous whenever the transform of every open set A of S is an open set of S'; and finally the transformation is said to be homeomorphic when it is one-one and continuous both ways. Combining these definitions, we have the definition of topology given above.[52]

As remarked, the intuitive origins of topology are evident from familiar geometry and analysis. Roughly, topology is concerned with those intrinsic qualitative properties of spacial configurations which are independent of size, location, and shape. Thus, space being as described in the preceding section, and returning to the Chinese puzzles of our childhood and the more complicated wire meshes of our adolescence, we note that although the former may be squashed out of recognition they are still the same puzzles, and that the latter are still interlinked in the same way after crumpling, stretching, and other transformations that break no wire and introduce no new linkings. Similarly for knots; what Tait called the knottiness, beknottedness, and knotfulness of a knot are qualitatively unaffected by infinities of transformations that alter the appearance of the knot as a body in common solid geometry. In fact one of the classical and more difficult if as yet less useful problems of topology is that of characterizing and enumerating all possible knots. Gauss[53] considered (1794, 1823–7, 1844, 1849) the problem of knots but did not publish his notes; J. B. Listing (1808–1882, German) discussed the subject in his *Vortsudien zur Topologie* (1847); Tait (1876–7) attacked it with his customary vigor in the ardent faith that the quickly abandoned theory of vortex atoms of W. Thomson (Lord Kelvin) was about to unravel all the spectra of all the chemical elements. T. P. Kirkman (1806–1895, English) applied (1884) his rare combinatorial talent to the problem of enumeration; M. Dehn (1910) defined the group of a knot, an approach successfully followed by F. Reidemeister in 1926; and finally, J. W. Alexander (1888–, U.S.A.), reformulating the problem in modern language, made remarkable progress (1927–8) in defining definite, calculable invariants for distinguishing one type of knot from another. All this is for knots in space of three dimensions. Klein showed that knots are impossible in a space of four dimensions (and in $2n$ dimensions, $n > 2$).

Lest the whole project of analyzing knots seem trivial to the uninitiated, we recall that Gauss[54] (1833) was attracted to such topics partly by an important problem in electrostatics, whose solution is obtainable only by topological considerations of the kind demanded in the exact characterization of both unlinked and interlinked networks, a problem to which the German physicist G. R. Kirchhoff (1824–1887) made outstanding contributions (1847) of enduring practical importance. Chemists and physicists who groan at the mathematics required in modern spectroscopy should rather "thank with brief thanksgiving whatever gods may be" that the abandonment of vortex atoms relieved them of the necessity of mastering topology. But since numerous useful problems in electromagnetism lead to many-valued functions, topology in some guise is essential for an intuitive understanding of the accompanying analysis situs of Riemann surfaces.

There is, however, one problem of some chemical interest which demands topological considerations of another kind for its systematic solution: given any number of atoms with assigned valences, how many distinct compounds can be made from them, on the assumption that any graphical formula satisfying the valence conditions is admitted? The combinatorial machinery[55] for obtaining the required chemical graphs was devised by Cayley (1875) in what he called the theory of trees; it was rescued from oblivion and amplified in the 1930's by chemists when the question again became of scientific interest. Closely related to this problem is another: given any configuration consisting of line segments joining points, to determine whether all the lines of the configuration can be traced successively without traversing any line more than once. One of the historically first problems in all topology was of this type; Euler (1736) showed that such a traverse was impossible for the seven bridges of Königsberg. He then generalized the problem and also discussed the knight's path in chess. Another early theorem gave the numerical invariant $N_0 - N_1 + N_2, = 2$ of Descartes[56] and Euler (1752) connecting the numbers N_0, N_1, N_2 of vertices, edges, faces respectively of a simply connected polyhedron. Thus $N_0 - N_1 + N_2$ is an example of a topological invariant in space of three dimensions. Euler's formula was generalized (1852) by L. Schläfli (Swiss) to space of n dimensions. This brings us to a brief description[57] of the combinatorial approach to topology.

The basic ideas are those of simplexes, complexes (in a technical sense differing from Plücker's in line geometry, etc.) boundary, cell, and, as before, continuous one-one transformation. A k-simplex is the k-dimensional analogue of a tetrahedral region; a 0-simplex is a point, a 1-simplex a line segment, a 2-simplex a triangular plane region, and so on. The boundary of a k-simplex is composed of simplexes of dimensionalities $0, 1, \ldots, k - 1$, and consists of $k + 1$ 0-faces (called vertices), $\frac{1}{2}k(k + 1)$ 1-faces (called edges), . . . , and

$$\frac{(k + 1)!}{(i + 1)!(k - i)!}$$

i-faces, . . . , the numbers $k + 1, \ldots$ being the binomial coefficients in $(1 + x)^{k+1}$. A k-simplex is completely determined by its vertices, say $V_1, V_1, \ldots, V_k$, and this simplex itself may be denoted by $|V_0V_1 \ldots V_k|$, called the symbol of the simplex.

A complex is any finite set of simplexes such that no two simplexes of the set have a common point, and every face of a simplex of the set is itself a simplex of the set. The k-simplexes $(k = 0, 1, \ldots)$ of a complex are called the cells of the complex; 0-cells and 1-cells are called vertices and edges respectively. A complex is completely determined by those of its cells not on the boundaries of other cells.

If the complex consists entirely of n-cells and the cells on their boundaries, the complex is called an n-complex. A complex is representable schematically by the combined symbols of its determining cells, and hence ultimately in terms of the symbols of its vertices. In the work cited, Alexander showed that all operations on complexes may be arranged so that the operations are finally expressible as combinatorial operations on the symbols associated with the vertices of the given complexes and of complexes obtained from these by subdivision. Last, two complexes are said as before to be homeomorphic provided there is a one-one continuous correspondence between the points of one and those of the other.

For obvious reasons, the topology developed from these concepts is called combinatorial. One famous unresolved crux that has been attacked[58] by the methods of combinatorial topology is the four-color problem, said by De Morgan to have been first noticed[59] by practical map makers: four colors suffice for a map any two of whose regions with a common boundary are to be colored differently, provided a boundary does not consist of isolated points but is a segment of a line. It is easy to construct a map for which four colors are necessary.

A few further examples of topological theorems and definitions may suggest the two main directions, the combinatorial and the Cantorian, or analytic, which topology followed in the twentieth century. The famous theorem stated next suggests Cantor's point sets rather than the cells described above: a simple plane closed curve (topologically equivalent to a circle) separates the plane into exactly two regions, and the curve is their common boundary. If it seems childishly obvious that it is impossible to pass from the inside of a circle to the outside by a path in the plane of the circle without crossing the circumference, and similarly for a cube and its surface in space of three dimensions, the reader may state the corresponding theorem for a sphere in space of four dimensions. These things may be obvious, but they require proof. It is also evident but difficult to prove that two coordinate spaces of different dimensionalities, say the Cartesian spaces of two and three dimensions, cannot be mapped on each other by a one-one transformation continuous both ways. As a last example, we recall the ancient Egyptian rule for the volume of a truncated square pyramid, whose proof depends ultimately on the theorem that triangular pyramids with equal altitudes and bases equal in area have equal volumes. This is not provable without continuity.

In the definition of topology given earlier, a space S was described in terms of sets and neighborhoods. To what was stated there, we now add three equally 'natural' postulates concerning neighborhoods; the system so defined is the topologic space of F. Hausdorff (1914). First, two neighborhoods of the same object have a common subset which is a neighborhood of that object. Second, if an object y belongs to a neighborhood of an object x, then there is a neighborhood of y which is a subset of the neighborhood of x. Third, for any two different objects there are two neighborhoods having no point in common.

Dots surrounded by expanding and contracting circles make all this immediately reasonable; and it seems like a miracle that anything but trivialities could issue from such a set of postulates. But it will be noticed that these postulates for a topological space clarify our almost subconscious intuitions concerning the 'space' of everyday life, and that little more if anything can be said about the intuited space of banal experience that would command as ready assent. So it may not be miraculous after all that the mathematical system developed from these platitudinous postulates goes to the very heart of qualitative spacial relations. An example of a topologic situation far from intuitive is the theorem conjectured (1912) by Poincaré and proved (1913) by G. D. Birkhoff. It is given that a continuous one-one transformation takes the ring bounded by two concentric circles into itself in such a way as to advance the points of the outer circle positively and those of the inner negatively, and at the same time to preserve areas. It is required to prove that there are at least two points invariant under this transformation. Again it may seem to the casual onlooker inspecting this curious theorem that mathematicians have a positive genius for wasting their time on puzzles of no conceivable utility. But such is not the case, at least in this instance. Poincaré encountered the puzzle in his work on the restricted problem of three bodies, and was unable to proceed with his dynamics because it baffled him completely. Its solution was one origin of the twentieth century qualitative dynamics, in which the problems attacked and solved are unapproachable by the classical methods of Lagrange and Hamilton. The problem has been generalized in an extensive literature, much of it American.

We conclude this sketch with a bare indication of the principal epochs in the development of the topology which has been described, and a very tentative forecast of its future. Experts agree that the year 1895, when Poincaré published his *Analysis situs*, marks the end of the dark ages in the subject and the dawn of topologic enlightenment. Poincaré followed his first effort (123 pages) with supplements in 1899, 1900, and 1904, and with applications to algebraic geometry. His greatest achievement, possibly, was the inauguration of a rigorous combinatorial topology for space of any finite number of dimensions. In dismissing the contributions of Poincaré's predecessors as relics of a benighted past, topologists do not deny the interest and importance of earlier work, such as Euler's on polyhedra; they merely insist that before 1895 there was no systematic attempt to develop topology as an independent mathematical discipline on its own merits. Proofs of such results as were obtained in the dark ages were seldom if ever sound, and basic revisions of principle and method were necessary before much resembling mathematics was possible. But the methods and inadequately proved theorems of many before Poincaré appear to have determined the future course of the subject. In addition to those already mentioned, we note the three following.

A complete characterization of closed surfaces by means of topologic invariants was achieved by 1890 through the labors of several writers (Möbius, 1863; Jordan, 1866; Schläfli, 1872; Dyck, 1888); the invariants were orientation and non-orientation, and the genus corresponding to Euler's number

$$N_0 - N_1 + N_2$$

for polyhedra. The next two items are of far greater significance, both historically and mathematically.

When we trace the development of the theory of functions of a complex variable, we shall see that the researches of Lagrange, Laplace, Clairaut, and others in the eighteenth century on the Newtonian potential suggested to Riemann (1851) his personal version of the theory. So whatever issued from Riemann's semi-intuitive approach to functions of a complex variable and its outgrowths in the theory of algebraic functions can be credited partly to physics. Riemann thought deliberately in physical imagery. For his investigations (1857) in algebraic functions, Riemann made brilliant use of the many-sheeted surfaces he had invented, shown by Clifford (1877) to be topologically equivalent to a box with p holes, where p is the genus of the surface concerned. The box in its turn is equivalent, as shown by Klein, to a sphere with p handles or, as Hadamard once expressed it, a pot with p pierced ears. More will be said about these matters when we come to Riemann's theory; it suffices here to note that as early as 1857 topological considerations were being freely used in analysis.

One of the earliest definite analytic applications of topology (specifically, linked and nonlinked circuits) was the work of Gauss (1833) on electromagnetism already noted. Riemann was a pupil of Gauss, and may have had his attention directed to the possibilities of topology by his master. Gauss predicted that topology would become one of the most influential methods of mathematics, as it did about sixty years after his death. But it was the intensive study of Riemann surfaces that made scores of analysts and geometers familiar with the basic significance of topology in analysis. A few intuitive theorems concerning the topology of a Riemann surface devised to exhibit the properties of certain functions are more easily apprehended than their equivalents in pages of detailed analysis.

The second major impulse to topology of the period before Poincaré's work of 1895 was Cantor's theory (1879–84) of point sets (Mengenlehre). With the encyclopaedic report (1908) of A. Schoenflies (1853–1928, German) on Cantor's theory and its developments, point sets were firmly established as the groundwork of analysis. By 1920 Mengenlehre had become so popular, especially with the Polish school, that a bulky new periodical, *Fundamenta mathematicae*, was founded in Warsaw to celebrate the recently acquired freedom of Poland—lost nineteen years later to the anti-Cantorians—and to develop Cantor's theory in all of its ramifications. These included much topology, mostly from the analytic[60] and abstract points of view.

One of Cantor's most suggestive innovations was his conception of any geometric configuration as a set of points in Euclidean space. From this, especially as generalized to abstract spaces by Fréchet and his prolific school beginning in 1906, the analytic method in topology developed with disconcerting rapidity. We shall return to this when we consider theories of functions. But it may be remarked here that the general definition of topology given earlier is relevant for Fréchet's theory, in which neighborhoods, continuous mapping, distance, and convergence are defined for any set of objects by abstracting the familiar properties of these concepts as they occur in classical analysis; neighboring elements (or objects) are mapped (or transformed) into elements which, in the map, are neighboring. With this work of Fréchet's and that of Hausdorff already mentioned, topology passed in the second decade of the twentieth century into the postulational or completely abstract phase. But earlier approaches were by no means abandoned.

Still in the analytic tradition of point sets, L. E. J. Brouwer (1882–, Dutch) in 1911 opened what some experts consider a new era in topology, with his proof that the dimensionality of a Cartesian space (n-dimensional number-manifold) is a topologic invariant. With Cantor and Poincaré, Brouwer is regarded as one of the founders of modern topology. His work in point sets also is considered by many the most penetrating since Cantor's. It was Brouwer whom we had in mind while observing the struggle of geometry to escape from the circle of competence; possibly the escape may be effected by a topologic transformation in higher space rendering penetration of the circumference unnecessary.

After a short, tentative start in the analytic direction, Poincaré followed the combinatorial method already described, and set himself the capital problem of determining whether a given property of a complex is a topologic invariant (unchanged under homeomorphic transformations). He also rescued the integer characteristics known as Betti's numbers (E. Betti, 1823–1892, Italian) from temporary oblivion and showed their basic importance in topology. While we have no intention of giving a directory of the hundred-odd ingenious workers who in less than forty years elevated topology from little more than a bag of tricks to a major division of mathematics, we may state that, in the opinion of others than Americans, the American mathematicians J. W. Alexander, G. D. Birkhoff, S. Lefschetz, R. L. Moore, and O. Veblen were responsible for much of the finest work. They and their numerous students, some of whom in 1945 stood a fair chance of equaling their masters, made one of the notable American contributions to twentieth-century mathematics. Perhaps the other most active school of the 1920's–30's was the Polish; the Russians also excelled.

It remains to exhibit a small sample of the specific things topology has done for the rest of mathematics. The impact of topology on analysis has been remarked, and will be observed implicitly in connection with theories of functions. Kronecker's (1869, 1873, 1878) theorems on the zeros common to a system of real functions, when rephrased in the geometry of space of n dimensions, are seen to be 'intuitive' theorems in topology. From the work of Brouwer, the modern theory of dimensionality (to be noted in connection with analysis) developed; homeomorphic mapping in Euclidean space furnished information, not easily obtainable otherwise, regarding the solutions of differential and other functional equations; and a similar technique inaugurated a new era in dynamics, beginning with Poincaré's (1893–) methods in celestial mechanics. In classical geometry, Poncelet's principle of continuity, which boldly included imaginary elements as if they were real in questions of enumeration, also the "similar recklessness"[61] of H. Schubert (1848–1911, German) in his so-called enumerative geometry (1879) for determining the number of those points, lines, planes, etc., of a given configuration which satisfy certain conditions, were stripped (1929) of vagueness and mystery and decently clothed in mathematics. Finally, Lie's groups were generalized to a new species, topologic groups, themselves an incident in a rapidly expanding algebra of topologic fields, rings, etc. This encourages us to hope that our successors may yet witness a lasting peace by mutual absorption between the devil of topology and the angel of abstract algebra.[23]

On what in the middle 1940's was rather optimistically described as the undergraduate level of mathematics, topology made notable contributions to precision and clarity. Not that the contributions in question were beyond the understanding of any normal intelligence enterprising enough to master a few rudimentary abstractions, such as those described above, and sufficiently cautious to ignore seductive and misleading intuitions in favor of precisely defined concepts. Far otherwise: stubborn objection to exact definitions of curves, surfaces, lengths, areas, and volumes in the 1940's was more frequent from jaded geometers of the older tradition than it was from unsophisticated beginners. A frequent and natural question from intelligent neophytes in the calculus was "How do you know that this integral gives you the length of an arc of a continuous curve?" On being told that it is all a matter of definition, the enquirer may have been skeptical though still courteous. In this he differed from some of his elders, who contumaciously propagandized their obsolete creed that curves, surfaces, lengths, areas, and volumes are not defined by mortal mathematicians, but either are given 'naturally' to man or are merely imperfectly perceived intimations of an Eternal Curve, an Eternal Surface, an Eternal Length, an Eternal Area, and an Eternal Volume, respectively, laid up once for all in the Heaven of Eternal Ideas, and therefore forever inaccessible and invisible to a purblind humanity. Not even the most etherealized of abstractionists claimed that his conceptions of neighborhood, transformation, and the like were generated wholly without benefit of gross sensory experience; but this was not the same as insisting that only the crude pleasures of such experience, which may have thrilled the great mathematicians in Euclid's Greece or in the ever-glorious eighteenth century of Euler and Lagrange, should continue to be the sole source of satisfactions in the mathematics of the twentieth century. Something of the luxurious freedom and naive license of the heroic past may have been curtailed by the close analysis of thoughtlessly inherited intuitions of space and number. The critical inventiveness of twentieth-century geometry and analysis was content with less than its predecessors demanded, insisting on durability rather than magnificent display. Possibly it was harder, although to those who had never tried it seemed easier, to discover reasonably sound theorems by painstaking elaboration of a set of postulates than it had been to infer undoubtedly more spectacular consequences, true 'in general,' from hastily sketched diagrams and the hidden connotations of their inexplicit assumptions. In bringing these assumptions to the surface, where they could be seen by anybody, the abstractionists and the topologists did not uproot intuition, to let it wither like a noxious weed on the rubbish heap of the past, but nourished and strengthened it—'in general.' If curves and surfaces became somewhat more complicated than the geometers and analysts of the nineteenth century had imagined them, the new definitions did not preclude what was consistent in the old; they merely attempted to clarify it and make it explicit. The revised definitions of curves and surfaces exposed much that was unsuspected but perhaps partly implicit in the intuitive concepts abstracted originally from sensory experience. In short, a deeper intuition broadened and deepened a shallower. It justified itself as mathematics which its practitioners and others found even more interesting than some of the classical geometry and analysis of the past.

A revised definition of a curve involves several of the concepts already noted. One in particular is of historical interest: the separation of a set of elements into classes by means of an equivalence relation. We have seen that this device first appeared (1801) in the Gaussian theory of congruences, the special equivalence relation of congruence with respect to an integer modulus being used to separate the countably infinite totality of rational integers into a finite number of mutually exclusive classes. The totalities of interest in geometry and analysis are usually uncountably infinite. Only an indication of the manner in which equivalence entered the revised definitions of curves and surfaces need be given here. For further information the reader may consult the paper *Curves and surfaces*, 1944, by J. W. T. Youngs (1910–, U.S.A.), on which the following account is based.

If A is a metric space, $f(a) = b$ denotes a continuous transformation from A into another metric space. If B is the image of A under f, we say that f is a map from A onto B, and write this $f(A) = B$. A transformation $T(A_1) = A_2$ is topological if it is bicontinuous and biunique, these terms being as previously defined. A specific equivalence relation, $\sim$, is defined as follows: $f_1(A_1) \sim f_2(A_2)$ if and only if there is a topological transformation $T(A_1) = A_2$ such that $f_1(a_1) = f_2(T(a_1))$ for every a_1 in A_1. The notation for the equivalence class generated by f_1 is $[f_1]$. A definition of equivalence suited to a generalization of 'distance,' is Fréchet's (1924): $f_1(A_1) \sim f_2(A_2)$ if and only if, for every $\epsilon > 0$, there is a topological transformation $T_\epsilon(A_1) = A_2$ such that $\rho\{f_1(a), f_2(T_\epsilon(a_1))\} < \epsilon$ for every a_1 in A_1, where ρ denotes the distance from a to b. From this definition it follows that if the maps $f_1(A_1)$, $f_2(A_2)$ are in the same Fréchet equivalence class $[f]$, then the base spaces are topologically the same, so that equivalence classes may be classified in terms of the topological character of the base space. A class $[f]$ is called a curve or a surface depending on the base space A. If A, topologically, is a closed segment, $[f]$ is called a curve of the type of a closed segment. If A is the circumference of a circle, $[f]$ is called a curve of the type of a circle. If, topologically, A is a closed square, $[f]$ is called a surface of the type of a closed square; if A is a sphere, $[f]$ is called a surface of the type of a sphere.

The classical problems of defining length and area have been reformulated for curves and surfaces as just defined. This is the analytic problem; the topological problem is concerned with the structure of Fréchet equivalence classes of maps. Both have been quite extensively developed, but we shall not attempt to follow the theory further. A surprising feature is that the curve and surface types of equivalence classes, whose definitions differ so slightly, should lead to widely divergent results. The definition of distance $\rho\{[f_1], [f_2]\}$ for equivalence classes is important in the analytic development, and as some of the concepts involved in this definition have already been noticed in other connections, we may conclude with it.

For any given a in A there is the Euclidean distance $|f_1(a) - f_2(a)|$ from $f_1(a)$ to $f_2(a)$, where $f_1(A)$, $f_2(A)$ are maps. The deviation $d(f_1, f_2)$ of f_1 from f_2 over A is defined as max $|f_1(a) - f_2(a)|$, the maximum being taken over all a in A. This $d(f_1, f_2)$ is a non-negative, symmetric function of two maps which vanishes only if the maps are the same; it also satisfies the triangle inequality. It therefore satisfies the postulates for a distance function. Each member of $[f_1]$ has a deviation from each member of $[f_2]$; $\rho\{[f_1], [f_2]\}$ is defined to be the infimum of $d(f'_1, f'_2)$ taken over all f'_1 in $[f_1]$ and all f'_2 in $[f_2]$. There is thus defined a metric over the set of equivalence classes $[f]$.

This small sample of unintuitive geometry and analysis may suffice to suggest that some things which formerly seemed obvious are no longer so, or not even true, and that other things, unsuspected in the older development, are true, sometimes quite disconcertingly. To maintain an equable balance between the intuitive and the unintuitive, we quote the opinion of a proponent of the former on the latter. The American geometer J. L. Coolidge (1873–), writing in 1940, compared the two approaches:

> It is evident that in all of this [the geometry and analysis of abstract space] we are skating around the outskirts of abstract geometry, abstract algebra, point-set theory, large disciplines that have been pushed far in the second and succeeding decades of the twentieth century. We drift in this fashion far indeed from anything suggestive of space intuition. Moreover, a certain class of mathematicians . . . finds it hard to maintain interest in any mathematical theory which does not seem to connect with any concrete problem, or anything related to our world of sensible objects. If lines in S_4 [space of 4 dimensions] mean circles in our space, if points of a V_4^2 mean our lines, that conveys something natural to the mind. But a theory of abstract spaces that covers 300 pages! The dimensions that really count can be bought cheaper.

So much for the maintenance of balance.

Topology has been given what may seem more than its just allotment of space in a general account for a reason of the very first importance. Many feel that topology has added a new dimension to mathematical thought. The classical quantitative, metric, and arithmetic approaches appear to be incapable of leading to anything either valuable or comprehensible in the qualitative aspects of mathematics. Certainly topologic thinking is unlike any other in mathematics, and it requires either a type of mind distinct from the ordinary good mathematical intelligence, or a training radically different from that fostered by conservatism, if it is to be applied effectively. The kind of thinking demanded in modern abstract algebra may appear somewhat similar; but in its closer adherence to precise logical analysis and its almost total disregard of spacial intuition, it is basically of another character. Galois, for instance, had nothing of the topologist about him; Riemann, in no sense an algebraist, was a born topologist. Of scientists, M. Faraday (1791–1867, English) was a pure topologist, and a great one, in his visual thinking about electromagnetism. Maxwell also had a marked tendency to topologize.

It seems not altogether fantastic to imagine that a few centuries hence the qualitative habit of thought will have superseded the quantitative in the growing parts of mathematics. Certain indications in science,[62] and many in mathematics, point to the analysis of structure as the mathematics of the future. Stated roughly, it is not things that matter, but the relations between them; and if topology with its spacial visualizations of intricate relations between abstract 'objects' has made possible a rudimentary but still difficult analysis of relations, it may be the germ of the mathematics of the future. In any event, topology is so unlike previous mathematics that it hints of a major transition.

About a century and a half elapsed between Lagrange's disregarded discovery of a simple algebraic invariant and the emergence of invariance as a dominating concept in topology. At the beginning of the period, topology was little more than a heterogeneous collection of a few amusing puzzles. In the 1820's, the first adumbrations of qualitative spacial thinking appeared in the revived projective geometry of Poncelet and others; but it was not till the 1850's, when Riemann demonstrated the power of topologic thinking in analysis, that topology began developing as a new mode of thought applicable to far more than geometry. Riemann himself noted topologic invariance in his (unpublished) theorems on the genus of his surfaces and his anticipation of the Betti numbers. Thus once more we observe the enduring quality of invariance as a mode of thought, and note that a single bold innovator may do more than hundreds of the merely competent to deflect mathematics into a new channel.

Of all the more daring originators who created modern topology, one is of peculiar significance for the novel quality of his thought. Brouwer's critical revaluations of all mathematical reasoning appear to be new in kind. Should his criticisms be sustained, he may be remembered as the Zeno of the twentieth century. As will be seen in another connection, much of Brouwer's criticism is destructive. Curiously enough, this as yet (1945) unanswered criticism and its outgrowths destroy most of Cantorism and, with it, a large part of classical analysis and analytic topology. One task of the future will be to provide analytic topology with a consistent foundation. Brouwer and others have constructed postulational bases for point sets which some find satisfactory but which others eye with suspicion. In this connection we refer again to what was said in the Prospectus regarding proofs of Euclid's I, 47.

Another cloud, at present no bigger than a giant's hand, may presage a storm that will overwhelm topology and its new mode of thought before its creators have ceased thinking topologically. The first volume (1935) of a masterly treatise[63] on topology, which when completed will extend to three volumes.

contains over 600 pages of close reasoning; the total presumably will be about 2,000 pages, the same bulk that came within an ace of crushing Lie and smothering his theory. In the meantime, topology keeps on ramifying, and already there are as many different methods for developing the subject as there were in the classic, all but stationary theory of algebraic functions and their integrals. As in so much of recent mathematics, vital things may be buried and forgotten under mountainous deposits of esoteric details comprehensible only to the narrowest specialists.

But who is to judge what should be salvaged and what allowed to perish? It is a fair guess that any competent jury trying to settle this question for differential geometry in the 1890's would have voted to discard Ricci's tensor calculus. In fact, more than one competent mathematician stigmatized the Ricci calculus as a sterile formalism. Yet it helped in the 1920's to revolutionize differential geometry and to render much that was prized in the 1890's of only historical interest. Possibly such problems solve themselves as in the case of algebraic invariants, by evaporating when decades of grueling labor suddenly terminate in some simple generalization that abolishes the problem.

CHAPTER 21

Certain Major Theories of Functions

Over a thousand[1] special functions have been deemed[2] of sufficient interest to merit more or less detailed investigation since the beginning of the eighteenth century. Many are all but forgotten; some, such as the multiply periodic functions and those invented for use in mathematical physics, have generated vast theories, each cultivated with ardor at various times by its band of devotees. In several instances, the mere history of one kind of functions would fill a large book. No one theory is sufficiently comprehensive to embrace all the special functions. The need for general methods to derive properties common to functions in each of several classes should have been obvious by 1800; but it does not seem to have been appreciated until about 1825, when Cauchy began his systematic creation of the theory of functions of one complex variable.

To be useful, a theory of functions must not be so general that it yields only trivialities common to all the functions it includes. Here, as elsewhere in modern mathematics, the maxim of logic that the greater the extension the less the intension has been observed, probably subconsciously, by the creators of the three major theories. In the historical order of development these are the theory of functions of one complex variable, the theory of functions of real variables, and general (or abstract) analysis. There is also a much less highly developed theory of functions of several complex variables, of use in the theory of algebraic functions. All evolved with apparent inevitability from the existing analysis of their respective epochs; none was an arbitrary creation out of nothing.

Mathematically, the theory of functions of real variables is prior to the others; but historically it emerged as an independent discipline only in the last third of the nineteenth century, when the theory of functions of a complex variable was far advanced. As we have seen, the need for such a theory to justify the calculus was acute even in the seventeenth century. In the eighteenth century, the encounter with trigonometric series in the problem of the vibrating string pointed in the same direction; and after 1822, when Fourier's physical applications of such series became widely known, the necessity for a rigorous theory of functions of real variables was generally admitted. It began to arrive in its valid form only in the 1870's with the revision of the real number system.

Meanwhile, the complex theory had become one of the major activities of the nineteenth century. The official birthday of this theory is usually fixed in 1825, with Cauchy's memoir on complex integration. But it seems reasonable to regard the theory of functions of one complex variable as a natural outgrowth of Lagrange's introduction[3] in 1773 of the potential in Newtonian gravitation. The theories of Cauchy and Riemann, as will be seen, arise naturally in this connection.

The final stage in the development of theories of functions is the counterpart in its own domain of the general arithmetics, abstract algebras, and abstract spaces of the twentieth century. Here again we observe the continual progression toward a more abstract mathematics, and once more for the same reason: the practical necessity for unifying by means of underlying general principles those aspects of numerous theories that promise to be of more than transitory interest. In the next chapter we shall see a clear origin of this modern analysis in the work of V. Volterra (1860–1940, Italian), who in 1887 began his extensive researches on what were subsequently called functionals. General analysis finally emerged as a distinct division of mathematics in the first decade of the twentieth century, with the work of M. Fréchet (1878–) in France and that of E. H. Moore (1862–1932) in the United States. We shall retrace some of the principal steps by which analysis kept abreast of the rest of modern mathematics.

Real variables

The theory of functions of one or more real variables has been less concerned than the complex theory with the investigation of functions principally useful in applications. Nevertheless, this theory is readily traced to the dynamical astronomy of the seventeenth and eighteenth centuries, and to the mathematical physics of the eighteenth and nineteenth centuries. The initial impulse was approximative calculation of numerical results from solutions of astronomical and physical problems expressed as infinite series or other non-terminating algorithms, such as infinite products and infinite continued fractions. Infinite series have greatly outweighed all other infinite processes in both practical and mathematical importance.

Although the real theory may have been less useful than the complex in obtaining properties of special functions, its significance for the development of mathematics as a whole has been incomparably greater. It was in the real variable that the necessity for a rigorous theory of the number system of analysis was first recognized. As we have already remarked, the reconstruction of the real number system by Weierstrass in the 1860's and by Dedekind and Cantor in the 1870's led in the last three decades of the nineteenth century to a revaluation of all analysis and thence, in the twentieth century, to a profound reconsideration of the nature of all mathematical reasoning. This in turn initiated some of the most searching examinations of all deductive reasoning since the days of Aristotle. Thus the theory of functions of a real variable since the 1870's has increasingly acquired more than a merely local interest: its problems, solved and unsolved, are significant in fields far distant from technical mathematics.

Regarding the strictly mathematical theory, expert opinion is practically unanimous that before the reconstruction of the real number system there was nothing that could be called a theory of functions of a real variable. Thus it is asserted in a popular text[4] on infinite series that "No lecture or treatise dealing with the fundamental parts of higher analysis can claim validity unless it takes the refined concept of the real number as its starting point. . . . For a theory of infinite series . . . would be up in the clouds throughout, if it were not firmly based on the system of real numbers, the only possible foundation." If this is correct, to ignore the modern theory of the real number system in a course on analysis is to enter upon a carefree excursion to the nebulosities of Cloudcuckooland.

The theory of functions of a real variable is of such vast extent that only some of the main landmarks can be noted here,

There appear to have been three principal stages in the evolution, the first two of which were preparatory. The earliest extends from the seventeenth century to 1821; the next from there to the 1870's; and the last from the 1880's to the 1940's. The earliest period, that of Newton, Leibniz, the Bernoullis, d'Alembert, Euler, and Lagrange, set the problem. Lagrange's abortive attempt to rigorize the calculus has already been noted. The great master in this period was Euler, whose unsurpassed inventiveness endowed analysis with a wealth of algorithms which fixed the formal pattern of analysis for well over a century. Many of his techniques in infinite processes have remained as vital as they ever were.

It is seldom possible in the history of mathematics to fix the end of an epoch with the precision implied in the date 1821, marking Cauchy's lectures[5] at the Ecole Polytechnique on what he called algebraic analysis—infinite series, limits, continuity, the calculus. The chasm between the old and the new that opened up in 1821 is plainly visible on comparing two classics; Cauchy's just cited, and the third edition, 1819, of Lacroix' *Traité du calcul différentiel et du calcul intégral.* If the later work was mathematics, the earlier was not.

An equally abrupt transition marks the passage from analysis as it was before the reconstruction of the real number system to what it became afterward. It need hardly be emphasized that in neither transition were the gains of the preceding periods discarded as worthless. By far the greater part of what had been found to give consistent results was retained. But most of it was radically transformed. The like holds for the third period, beginning in 1886 with the reconsideration by Poincaré and T. J. Stieltjes (1856–1894, Dutch) of divergent series.

Unlike the earlier transitions, the passage to the third period was less a revision of basic concepts than an extension of established mathematics. As such, it was partly responsible for the general analysis of the twentieth century. The late 1890's and early 1900's saw the creation of a usable theory of divergent series; and in 1902, H. Lebesgue (1875–1941, French) revolutionized the theory of integration. It is generally agreed that the work of four French analysts of this period, Poincaré, R. Baire (1871–1932), E. Borel (1871–), and Lebesgue, opened a new era in real analysis comparable to that inaugurated by Cauchy and Abel about sixty years earlier. Attending only to certain critical points in the course of these developments, we shall consider first the transition from empiricism to mathematics in the use of infinite series.

Although the distinction between convergence and divergence was recognized[6] by Leibniz as early as 1673, it does not seem to have occurred to the analysts of the late seventeenth and early eighteenth centuries that an investigation of infinite series on their own merits might be of value even in numerical computation. In their quadrature of the hyperbola in 1668, revised by Newton the following year, N. Mercator and W. Brouncker had encountered the special logarithmic series $\sum_{1}^{\infty} (-1)^{n-1}/n$. Leibniz observed that the companion harmonic series $\Sigma 1/n$ is divergent; and in 1705 he stated a sufficient condition for the convergence of an alternating series. The Bernoullis, James (1654–1705) and John (1667–1748) (Swiss), recognized the divergence of the harmonic series in 1689; and James (1696) noted "the not-inelegant paradox" that $1/(1-x)$ when expanded as $1 + x + x^2 + \cdots$ gives $1 - 1 + 1 - \cdots = \frac{1}{2}$ for $x = -1$, a revelation which Leibniz and others attempted to justify by the theory of probability, but which the devout Euler humbly accepted. Less eminent mathematicians applied similar considerations to theology. Thus, by rearranging the terms of a divergent series, it was readily proved that $0 = 1$, and hence that God had created the universe out of nothing. Even Leibniz, in a philosophical mood, drew similar conclusions from the binary scale of notation.

More remarkable still, some of the series in dynamical astronomy were divergent and yet furnished correct numerical results on taking only the first few terms. Stirling's divergent series (1730) for $n!$ similarly gave correct approximations for large values of the positive integer n; and Euler's summation formula (published 1738, rediscovered by Maclaurin, 1742) also proved extremely useful without benefit of convergence. These dubious successes failed to convince d'Alembert, who declared (1768) that all divergent series are open to suspicion. But as nothing was understood of convergence at the time, these most astonishing miracles were generally accepted as natural phenomena. In a sense they were, as appeared when Poincaré in 1886 dispelled their mystery in his theory of asymptotic expansions.

All this raised the question of what might be meant by the sum of an infinite series. Euler almost attained the accepted philosophy in this matter and in fact in all infinite processes. There is no compulsion to define the 'sum' in any particular way; the definition is prescribed at will and is conditioned only by the uses, if any, to be made of infinite series and their 'sums.' It is universally demanded, however, that the definition shall not invariably lead to nonsense.

The customary desiderata for sums have been stated by Borel[7] substantially as follows: To establish a correspondence between series and numbers such that when the number corresponding to a particular series is substituted for the series in the usual calculations where the series occurs, correct results shall always, or nearly always, be obtained. The 'nearly always' is to be made precise by specifying those cases in which the correspondence is invalid. Usually practicing this, Euler all but stated an equivalent. But neither he nor anyone else before Cauchy (1821) and Abel (1826) made any systematic application *with valid general proofs*. The pioneering work of Gauss (1812) on the hypergeometric series has already been mentioned; but it is commonly agreed that Cauchy and Abel founded the theory of convergence, an essential step toward a theory of functions of real variables.

Divergent series were not abandoned without regret and misgivings. Apologizing for what he was about to do, Cauchy said, "I have been forced to admit certain propositions which may appear rather drastic; for example, a divergent series has no sum." Abel (1828) stigmatized divergent series as works of the devil, but was irresistibly attracted by them. Confessing his bewilderment that such series frequently give correct results, he declared his intention of finding the reason. But he died (1829) before he had time to carry out this part of his program, and the first success in this direction was Cauchy's. Seeking the rationale of Stirling's approximation to $n!$, Cauchy in 1843 created a satisfactory theory applicable to a wide class of divergent series. This fine work was forgotten. Klein has remarked that Cauchy's prodigious output resembles a jungle. But even jungles occasionally repay exploration; and another buried treasure in Cauchy's wilderness was the 'root test' (1821) rediscovered in 1892 by Hadamard.

Among the obvious demands for a useful theory of infinite series are these: If s, s_1, s_2 are the respective sums of the series S, S_1, S_2, and c, c_1, c_2 are constants, then to cS, $c_1S_1 \pm c_2S_2$, cS_1S_2 shall correspond respectively cs, $c_1s_1 \pm c_2s_2$, cs_1s_2, 'nearly always.'

Without explicitly formulating the problem thus, Cauchy and Abel solved it in its fundamentals. The outcome was the classical theory of convergence as usually presented in elementary texts. All that was lacking in this first systematic attack on infinite processes to make it thoroughly rigorous in the modern sense was a consistent theory of the real number system. With this supplied, the work of Cauchy and Abel has passed into the standard expositions of analysis practically unchanged.

One classic in particular of this period may be mentioned, Abel's investigation (1826) of the general binomial series. The importance of this series in the evolution of the Newtonian calculus was noted in an earlier chapter. Newton, it appears, inferred the formal expansion in 1676 or earlier from suggestions in Wallis' work on interpolation. A proof of the binomial theorem for positive integral exponents was given by James Bernoulli[8] in 1713; and in 1774 Euler devised an ingenious but incomplete proof for any rational exponent. Abel's was the first general proof that has stood. Thus, a century and a half after its first use in the calculus to obtain the derivative of x^n for any real n, the binomial theorem was finally proved.

This earliest work on the theory of convergence generated a major industry of nineteenth-century analysis. Once the importance of convergence was recognized, scores of analysts sought criteria for the convergence of both real and complex series. Of the tests usually appearing in textbooks on series are that of Gauss (1812), Cauchy's ratio test (1821), his integral test[9] (1837), and the logarithmic tests of Abel (1827), Kummer (1835), and De Morgan (1842). In 1873, du Bois-Reymond attempted a general theory of convergence tests, but was only moderately successful. With the completer and in part illusory synthesis of A. Pringsheim (1850–1941 German) in the 1890's, interest in such questions dwindled for a season, until it suddenly flared up again in the first decade of the twentieth century. The British, German, and Polish schools, partly impelled by difficult problems in the analytic theory of numbers, began manufacturing delicate tests for convergence, until by 1945 the theory of series had gone the way of all recent mathematics by developing into an independent field with its own corps of expert cultivators. The theory of Dirichlet series[10] alone has become a separate division of analysis. Modern work in this field dates from J. L. W. V. Jensen's (1859–1925, Scandinavian) proof (1884) that such a series has an abscissa of convergence.

We digress here to note briefly the parallel development of two other infinite processes, products and continued fractions.[11] The theory of infinite products is so intimately connected with that of infinite series that we shall pass it. Infinite continued fractions, on the other hand, afford a basically different mode of approximative calculation, and lead to more difficult questions of convergence. They are of special historical interest in the theory of functions of a real variable, as it was while investigating them that Stieltjes in the 1890's invented the species of integral since named after him.

The first systematic discussion of continued fractions was Euler's of 1737. Apart from sporadic appearances in the arithmetic of the Greeks, the Hindus, and the Moslems that can *now* be interpreted as results in continued fractions, the general algorithm was not isolated till P. A. Cataldi (1548–1626, Italian) discussed it in 1613. Prior to Cataldi's use of continued fractions, R. Bombelli in 1572 had approximated to square roots by means of them; and after Cataldi, Brouncker (1658) expanded $4/\pi$ in an infinite continued fraction. Lagrange in 1767 applied continued fractions to the numerical solution of algebraic equations, inventing a method of successive approximations which, he justly claimed, left nothing to be desired from the standpoint of completeness and rigor. But as it left nearly everything to be desired in the way of practicability, it rapidly lapsed into desuetude. Lagrange's experience with continued fractions is typical. Of high theoretical interest, as in proving the irrationality of certain numbers, continued fractions are too cumbersome an algorithm to be of much practical use.[12] Generalizations of continued fractions through simultaneous difference equations were investigated by Jacobi in 1850(?), and by E. Furstenau (German) in 1872. These are of considerable arithmetical importance, but their theory is still in embryo.

Common continued fractions began to become significant for the theory of functions when E. Laguerre (1834–1886, French) in 1879 converted a divergent power series into a convergent continued fraction. The modern analysis of continued fractions originated in the subsequent work of Stieltjes and in that of H. Padé (1863–, French) who showed (1892) among other things that Euler's problem of expanding a function in a continued fraction is improperly posed. Different expansions are obtainable, conditioned by the choice, from an available infinity of approximating rational fractions, of a particular sequence suitable for the successive convergents to the continued fraction. Padé's problem was to reduce this apparent chaos to a semblance of order, in which he had a partial success. In Laguerre's direction, Stieltjes (1894) established a correspondence between divergent series and convergent continued fractions, by which he was enabled to define integration for the series. For analysis as a whole, the significant gains partly due to continued fractions were Stieltjes' integrals and his version of the theory of asymptotic series, quickly superseded by Poincaré's.

Continuing with series, we shall note only certain details of the general development which stimulated the theory of functions of real variables. The commonplace of modern analysis that processes valid for finite algorithms may cease to have a meaning on passing to the infinite was long in being generally appreciated. In the eighteenth century, Stirling (1730) and Euler (1755) freely transformed series by rearrangement of terms without a qualm; and it was only in 1833 that Cauchy showed by an example that a conditionally convergent series by rearrangement of its terms yields different sums. Dirichlet (1837) proved the basic theorem that rearrangement of the terms in an absolutely convergent series does not affect the sum.

Differentiation and integration of infinite series deceived some of the greatest mathematicians in history, as also did the interchange of limiting processes. Cauchy's lapse in the matter of uniform convergence was noted in an earlier chapter. Even Gauss (1812) gave an incorrect proof of the theorem which he stated and used, now universally called Abel's, that if

$$f(x) = \sum_{n=0}^{\infty} a_n x^n$$

has radius of convergence 1, and if $\Sigma a_n = s$, then $\lim_{x \to 1-0} f(x) = s$. Abel proved this extremely useful result; Gauss[13] interchanged two limiting processes without, apparently, suspecting that justification was necessary. Again, as will be seen in a later chapter, Fourier series were freely used for many years before Dirichlet in 1829 first validated their use under certain restrictions. Even the keenest intuition was misled by the vagaries of such series.

We have already emphasized the central part played by trigonometric series in the evolution of rigor. To what has been said, we may add a few typical specimens of the unforeseen, to indicate once more the necessity for a clearer understanding of functions, limits, and continuity than was possible before the modern theory of the real number system. The Fourier series generated by a given function $f(x)$ may not converge in the whole interval for which it is obtained, or even at any point of the interval; and if the series does converge, its sum need not be $f(x)$. Again, du Bois-Reymond (1873) showed by an example that a continuous function is not necessarily representable by its Fourier series; while Weierstrass (1875) proved that continuity and differentiability together are more than is necessary.[14] All this and much more of a similarly unexpected kind stimulated Cantor to create his theory of point sets.

To conclude this sketch of series, we return to Cauchy's and Abel's expressions of regret at abandoning divergent series. As Borel has observed, the neglect of such series was unfortunate in that it cramped the freedom of inquiry and invention that had generated so much of lasting interest in thc eighteenth century. The microscope supplanted the telescope, and although some were enabled to describe the minute anatomy of analysis with accuracy, they missed larger things of possibly greater scientific value.

A new era of free invention opened in 1886 with Poincaré's method of asymptotic expansion. Divergent series were at last made fairly respectable, and it was seen why some of them, such as Stirling's and those of astronomy, could be used with complete safety to calculate numerical approximations with any desired degree of accuracy.

Shortly before this notable advance, O. Hölder in 1882 inaugurated the modern theory of divergent series with his method of summation by arithmetic means. E. Cesàro (1859–1906, Italian) followed in 1890 with similar processes for the summation of divergent series; Borel in 1895 proposed another usable method; LeRoy in 1900 gave another, and M. Riesz in 1909 yet another. In 1911, the English analysts Hardy and S. Chapman made a comparative study of the various methods of summation devised up till then; and in 1913 Hardy and J. E. Littlewood followed this with a detailed discussion of the relations between the methods of Borel and Cesàro. What appears to be the permanent residue from all this explosive outburst in the 1890's and early 1900's has long since passed into the standard texts on infinite series.

There is no obstacle to the invention of further methods, except possibly common sense. The consensus of informed opinion is that unless a new method of summation has some definite objective, there is no point in developing it. The usual desiderata for the definition of 'sum,' quoted earlier, have restrained invention in a reasonable way. But there are no grounds for anticipating that they or any other ban will continue to operate.

The foregoing sketch of infinite processes has merely suggested a few of the outstanding points of interest. A more detailed discussion would have revealed unmistakably a main trend of the general development which has been only hinted: the continually accelerating expansion of mathematics, and in particular of analysis, since the first rigorous work of Cauchy and Abel in the 1820's. In little more than a century, each of several minor subjects of the 1820's had swelled to enormous bulk, until by 1945 any one of them was sufficiently vast to absorb the energies of highly specialized experts. Such, for example, was the case with trigonometric series, or with the determination of the properties of a function whose Taylor series is given, a problem originating only in 1878 with Darboux' approximative study of functions of large numbers. The theory of real variables proper is yet more extensive. Continuity alone, for example, has so often been refined that its modern applications demand a sizable literature of their own.

All this is doubtless a cause for jubilation. But whether or not it is, the very fecundity of analysis drove some in the early twentieth century to the ultimate of complete abstraction. Unless the ever-swelling mass could somehow be controlled, unified, and directed, it seemed to stand a fair chance of expiring under its own weight. For mathematics unfortunately never completely dies to any considerable extent, although large areas of its ancient body may be temporarily forgotten in favor of fresher attractions.

The central theme in the theory of functions of real variables is continuity, with its implications for the classification of functions and limiting processes, especially differentiation and integration. By far the greater part of the theory is concerned with functions of a single real variable. The theory for several variables is not obtainable as an immediate generalization from that for one, essentially new phenomena entering with an increase in the number of variables, as for example in the order of performing the individual operations in multiple-limit processes. Accordingly, unless the contrary is indicated, real function shall mean a function of one real variable. The evolution of the theory will be sufficiently suggested by following the development of integration.

As we shall frequently mention the integrations of Cauchy (1823) and Riemann (1854) for real functions $f(x)$, we recall their definitions here. In Cauchy's definition, $f(x)$ is continuous in the interval $[x_0, x_n]$, and his integral is the limit of the finite sum

$$\sum_{i=0}^{n-1}(x_{i+1} - x_i)f(x_i), \qquad (x_0 < x_1 < \cdots < x_n),$$

as the length of the longest of the intervals $[x_i, x_{i+1}]$ tends to zero when n tends to infinity. In Riemann's definition, $f(x)$ is bounded in the interval $[x_0, x_n]$; U_i, L_i are the upper and lower bounds of $f(x)$ in $[x_i, x_{i+1}]$. We write

$$U \equiv \sum_{i=0}^{n-1} (x_{i+1} - x_i)U_i, \qquad L \equiv \sum_{i=0}^{n-1} (x_{i+1} - x_i)L_i.$$

If U, L have a common limit as n tends to infinity in such a way that the longest of the intervals $[x_i, x_{i+1}]$ tends to zero, this common limit is the Riemann integral of $f(x)$. Riemann's integration is obviously of wider scope than Cauchy's. By an obvious amplification, Cauchy's definition is extended to take care of a finite number of discontinuities in $[x_0, x_n]$.

Integration of course is as old as Newton and Leibniz, or even Archimedes; but the *theory* of integration was a creation of the twentieth century. Before 1902 there were innumerable studies of integration, many of which contributed to the development of the first general theory, Lebesgue's of that year. It may seem ungracious to insist that before 1902 nobody really understood what integration is all about, in spite of the vast mass of correct and valuable results obtained by integration since the invention of the calculus. But such appears to be the opinion of those who should know.[15]

To avert a possible misapprehension, we hasten to add that the integral calculus did not come to a sudden and glorious end in 1902. Quite the contrary: integration just began to thrive. Lebesgue's generalization of Riemann integration was but a beginning. Others quickly generalized Lebesgue, until methods of integration began multiplying on themselves as if they were alive. All this furious activity, according to the abstractionists, was but one more argument for a general analysis. And here again it must be remembered that to generalize is not to obliterate. All the valid gains of the past were conserved in the successive generalizations, and in many instances much that older methods were incapable of manipulating was handled with ease. Moreover, an enormous amount of new knowledge was acquired.

Integration from Newton and Leibniz to Riemann followed two complementary courses. Newton regarded integration as the operation inverse to differentiation; Leibniz considered integration as a limiting summation. Both were familiar with the connection between the two concepts. Indeed, without the fundamental theorem of the calculus, first proved by Cauchy,[16] neither Newton nor Leibniz could have made much use of his analysis. The historical development favored Newton until Riemann (1854) defined his integral, when it was soon discovered [17] that non-differentiable bounded functions having Riemann integrals exist. A further discordance entered in 1881 with Volterra's bounded differentiable functions not integrable in Riemann's sense. These two discords together abolished the mutual harmony between differentiation and integration that had persisted from Newton and Leibniz to Cauchy. But Riemann's integration was applicable in numerous important instances where Cauchy's was not. The problem of refining Riemann's definition so as to retain its desirable qualities and at least soften the discordances was open.

As the Leibnizian and Newtonian concepts of integration had originated in intuitive notions of velocities, accelerations, lengths, areas, volumes, and tangent lines, we can see now where the root of the difficulty lay. Until these vague notions, particularly length, area, and volume, were made precise, only inappreciable progress was possible. Precision arrived with the arithmetization of analysis by Weierstrass, Dedekind, and Cantor. Details would take us too far afield here, and we can give only a suggestion of what was done, referring to any standard text on the real variable for an adequate account.

The intuitive length, or measure, of a straight-line segment, say [0, 1], is deceptively clear so long as we imagine the segment to contain all the points corresponding to the real numbers x such that $0 \leqq x \leqq 1$. But if we imagine all the points for which x is a rational number removed from the segment, the intuitive concept of the 'measure' corresponding to the remaining class, or set, of points vanishes. The first desideratum then was a usable definition of 'measure' for a point set. Technically, a theory of the metric properties of point sets was needed. Several partly successful attempts were made by the German mathematicians in the 1880's, notably by H. Hankel (1882), O. Stolz (1884), Cantor (1884), and Harnack (1885). All, with further promising essays by Peano and Jordan, were superseded in 1894 by Borel's theory. This was developed by Lebesgue[18] and applied by him (1902) to his new integration in the classic memoir, *Intégrale, longueur, aire.*

Lebesgue's theory of integration is one of the outstanding landmarks in the evolution of analysis. Proceeding from his definition of measure for point sets, Lebesgue passed to integration, dividing the domain of integration into measurable sets, and taking the limit of a certain sum for all the sets as their number is indefinitely increased.

This process already contains the germ of the final species of generalization, in which metric properties are defined for classes (sets) of unspecified 'elements' or 'objects.' The 'elements' in pre-abstract analysis are specified as numbers. But Lebesgue's landmark was not washed away by the deluge of modifications which shortly descended upon it.

Before the passage to complete abstraction, more orthodox generalizations of Lebesgue integration supplemented and perfected its desirable qualities. For while it preserved the more useful aspects of both Newton's and Leibniz' conceptions, and effected the integration of wide classes of functions for which Riemann integration was only artificially applicable,[19] Lebesgue's generalization was not a complete synthesis of integration as summation and anti-derivation. A more inclusive union came with the further generalized integrations of A. Denjoy (French) in 1912 and O. Perron (German) in 1914. In the meantime, crossbreeds between various types of integrals came into being, for example the Lebesgue-Stieltjes integrals. Simultaneously, the theory of point sets suffered an enormous expansion to accomomdate all the new phenomena of integrability, limits, continuity, differentiability, and the numerous subspecies into which these had split.

From all this it might be suspected that by 1945 the theory of integration had begun to go the way of all mathematics of the recent period. It appears to be the fact that this one branch alone of the theory of functions of a real variable, after less than fifty years of existence, was proliferating at incredible speed like a tropical forest in a nightmare. For a time it was thought that such ventures as that of J. Radon, who in 1913 extended Lebesgue integration to abstract spaces for which a measure is definable, might regulate the growth. Twenty years later promising indications for a unified theory appeared in S. Ulam's theory of measure (1930–2) in general set theory, and in A. Haar's (1933) for certain metric spaces with its applications to the theory of continuous groups.

These more abstract methods developed from Fréchet's theories of abstract space, originating in 1906 (to be noted in the following chapter). It came as somewhat of a shock, therefore, and as an occasion for joy or despair according to temperament, to find a leading young abstractionist exulting in 1938 that his favored abstract space is superior to Fréchet's because it readily splits into eight others. By operating in one or more of these abstract spaces simultaneously, it is thus possible to generalize a given situation in 255 distinct ways. Fortunately some are vacuous. Whether we should laugh or weep is not exactly clear, but one thing seems certain: we can do nothing about the situation. Ever since the beginning of the nineteenth century, mathematics has developed by continually expanding and subdividing. But from a distance of two or three decades after any exceptionally active period, only those novelties which have proved their value in either mathematics or science remain visible. On the whole, then, it seems that we should rejoice in the irrepressible fecundity of mathematics.

The foregoing sketch has barely outlined only one aspect of the theory of functions of real variables. To leave it without mentioning the other side of the entire theory might seem partisan to those who believe there is another side. According to these heretics, the wealth of intricate theorems which make this theory one of the richest in mathematics is largely rainbow gold. Their heresy is based on the undisputed fact, already sufficiently described, that the reasoning in point sets is still (1945) deeply infected with paradox. Passing to the limit of skepticism, the doubters maintain that non-constructive existence theorems, as in the more abstract theories developed since 1905, establish nothing. Extremists even assert that such modes of reasoning as are employed in the accepted theory of functions of real variables are no longer of more than historical interest.

Three specimens of the kind to which objection is taken will suffice.

In historical order, the first is the famous Heine-Borel theorem[20] (1872, 1895) without which many accepted proofs would collapse. A generalized form of the theorem was stated by W. H. Young[21] (English) in 1902 as follows. "Given any closed set of points on a straight line and a set of intervals so that every point of the closed set of points is an internal point of at least one of the intervals, then there exists a finite number of the given intervals having the same property." The second is of a similar character: It is E. Zermelo's notorious axiom[22] (1904), which thus far has proved most useful, if not indeed indispensable, in much of analysis: "Given any class of mutually exclusive classes, of which none is empty, there exists at least one class which contains exactly one thing in common with each of the classes." The third and last specimen is one of the most highly prized in all analysis, Baire's classification (1899) of functions. Roughly it is as follows. Functions which are continuous in a given domain constitute the lowest class, say C_0; any function which is the limit of a sequence of functions belonging to C_0, and is not in C_0, is assigned to the next class, say C_1. Similarly for functions of the next class C_2, defined as limits of sequences of functions belonging to C_0 or to C_1, provided they are in neither C_0 nor C_1. And so on, it is said, indefinitely.

Borel's objections to Zermelo's use of his so-called multiplicative axiom touched off a spirited controversy[23] (1905) enlivened by several sharp differences of opinion. Admitting that Zermelo described no method for making the infinity of choices implied in his axiom, Hadamard asserted that the existence of a mathematical 'entity' is provable without explicitly defining the entity. Only by this type of validation, according to Hadamard, can the transfinite numbers of Cantor enter mathematics. Such existence proofs, of course, have been a commonplace in other departments of human aspiration since the logicians of the Middle Ages first employed then with conspicuous success to provide a firm foundation for theology. There was no apparent reason, therefore, why they should not be applicable with equal success in mathematical analysis.

But, like the medieval theologians, the leading analysts themselves disagreed over the exact measure of intuition admissible in their type of reasoning. Baire, for instance, would not grant that the subclasses of a given class are given; while

Lebesgue opined that Zermelo's somewhat mystical 'existence' was equivalent, when not clear, to 'mere' freedom from contradiction. This historic controversy of 1905 ended inconclusively, with Borel maintaining that while Cantor's theory of transfinite cardinals might be a valuable heuristic guide to valid proofs, it could not of itself furnish them. The submathematical analysis of the Middle Ages appeared to be on the point of rising from the dead. In 1912 it rose, in the intuitionism of the Dutch mathematician L. E. J. Brouwer (1882–), of whom more considerably later.

Whatever else such disputes may have instigated, they were partly responsible for the modern critical period. A remarkable and possibly encouraging feature of these doubts is that those expressing them were not deterred from creating new mathematics by the very means to which they objected. By a sort of Hegelian synthesis, destructive criticism and constructive action were reconciled in each of several individual minds. But a majority experienced no uneasiness, firm in their faith that their heavy investments in analysis were in no danger of being hypothecated to the usurious demands of modern logic.

Functions of a complex variable

In the classic differential and integral calculus, an independent variable ranges over real numbers. The class of all complex numbers being the largest closed under the four rational operations, it might be anticipated that anomalies in the behavior of functions of a real variable, especially in the integration of irrational functions, would be removable only by passing to functions of a complex variable. This was one of the impulses behind the creation of the theory of complex functions.

To substantiate later a possible line of descent from the eighteenth-century mechanics, we shall need a rather detailed description of what is sometimes called the fundamental theorem of analysis, Cauchy's[24] of 1825. A uniform function w of the complex variable $z(\equiv x + iy,\ x,\ y$ real variables, $i \equiv \sqrt{-1})$ is expressed in the form $u + iv$, where u, v are real functions of x, y, and are uniform. Conversely, if u, v are any real, uniform functions of x, y, $u + iv$ might be considered as a uniform complex function. But this is not the route taken in the *classical* theory[25] of functions of a complex variable; and the origin of the following drastic restrictions on u, v may have been in hydrodynamics and the geometry of conservative fields of force.

Analytic functions and their quotients are a primary concern of the classical theory: w (uniform) is an analytic[26] function of z if u, v, and their partial derivatives with respect to x, y are continuous, and moreover satisfy the so-called Cauchy-Riemann partial differential equations

$$\frac{\partial u}{\partial x} = \frac{\partial v}{\partial y}, \qquad \frac{\partial u}{\partial y} = -\frac{\partial v}{\partial x}.$$

It follows that u, v are solutions of Laplace's equations in two dimensions,

$$\frac{\partial^2 \phi}{\partial x^2} + \frac{\partial^2 \phi}{\partial y^2} = 0.$$

For brevity of statement only, the values of the complex variable z are represented, as in an Argand diagram, by points (x, y) in a plane of complex numbers, 'the z-plane'; those of w, $= u + iv$, as points (u, v) on another Argand diagram, 'the w-plane.' As a point in the z-plane traverses a continuous curve C, u, v assume different values and (u, v) traces a corresponding curve C' in the w-plane. A point z (value of z) in the z-plane at which w is not analytic is called a singularity of w. Provided C passes through no singularity of w, C' will be a continuous curve.

We shall consider in a moment the line integral[27] $\int_C w\, dz$ taken along any simple, rectifiable arc C lying wholly within a region R of the z-plane such that w is analytic for all points in R. It will be sufficient for our purposes to describe the meanings of 'simple' and 'rectifiable' without attempting rigorous definitions. An arc is simple if it does not cut itself, and rectifiable if its length, determined as the limit of a finite sum as in ordinary integral calculus, is finite. The 'boundary' of a region in what follows presently is either one simple, closed, rectifiable curve in the z-plane, or is such a curve enclosing others of the same kind. The positive sense of describing a boundary is as in mechanics: each of the bounding-curves is to be circuited as if a man walking round each in turn would always have the bounded region on his left. If the region has only one boundary (like a plate without holes) it is simply connected;[28] if it has more than one boundary (like a plate with holes in it), it is multiply connected. By the device of cross cuts from points on the outer boundary to each of the inner boundaries, and cuts between the latter, a multiply connected plane region is made simply connected; and it usually suffices to state the cardinal theorems of analytic functions for simply connected regions. To avoid repetition, a simple, rectifiable arc in the z-plane will be called a contour. A contour is not assumed to be closed unless so stated.

Let z_1, z_2 be any fixed points in a region R of the z-plane within which the function w, $\equiv f(z)$, is analytic; and let C denote any contour starting from z_1 and ending in z_2. Then it is proved that $\int_C f(z)dz$ is independent of the particular contour C. That is, the value of this integral depends only upon the end-points z_1, z_2. Otherwise stated, as z_1, z_2 are held fixed, and any contour C joining them is deformed continuously, without at any stage of the deformation passing outside of R, the value of the line integral $\int_C f(z)dz$ is unchanged. Since the sign of a line integral is reversed with the direction of integration, it follows that if K is any *closed* contour lying wholly within R, then $\int_K f(z)dz = 0$. If R is multiply connected, its boundary K is to be described in the positive sense.

This is substantially Cauchy's integral theorem:[29] if $f(z)$ is analytic at all points on, and at all points within, a closed contour K, then $\int_K f(z)dz = 0$, the integration being in the positive sense. The proof is immediate on writing $f(z)$ in the form $u + iv$, and $dz = dx + i\,dy$. It is then to be seen, on equating reals and imaginaries separately to zero, that

$$\int_K (u\,dx - v\,dy) = 0, \qquad \int_K (v\,dx + u\,dy) = 0.$$

But these follow at once on transforming the line integrals along K into surface integrals over the region bounded by K, by Green's theorem (1828) for two dimensions, and applying the Cauchy-Riemann equations to the results; each integrand then vanishes identically. Goursat's refinement (1900) of the theorem is harder to prove; but this need not concern us here.[30]

Although it is a detail, we may quote Cauchy's integral formula, one of the most useful results in analysis. If $f(z)$ is analytic at all points on, and inside, a closed contour C; and if a is any point inside C, then $2\pi i f(a) = \int_C \frac{f(z)dz}{z - a}$. Equally remarkable are Cauchy's formulas for the successive derivatives $f^{(n)}(a)$, $n = 1, 2, \ldots,$ of $f(z)$ evaluated at $z = a$:

$$2\pi i f^{(n)}(a) = n! \int_C \frac{f(z)dz}{(z - a)^{n+1}}.$$

Cauchy used these in his proof of Taylor's theorem for analytic functions.

Before proceeding to the mechanical analogy, we may note some of the principal developments of Cauchy's method for functions of a complex variable. Cauchy himself elaborated his calculus of residues as a means for evaluating real definite integrals. While there is no royal road to integration, indeed hardly more than a cowpath, Cauchy's calculus proved most effective. It was based on the conception of those special singularities, called poles, of certain functions. Several equivalent definitions may be given; the following is as simple as any.[31] Let $f(z)$ be analytic at all points except a, inside C, of a given region R bounded by a closed contour C. Then if $f(z)$ is expressible in the form $F(z) + \sum_{s=1}^{n} A_s(z - a)^{-s}$, where $A_1, A_2, \ldots, A_n$ are constants, $A_n \neq 0$, and $F(z)$ is analytic at all points of R, the function $f(z)$ is said to have a pole of order n at $z = a$, and A_1 is called the residue of $f(z)$ at the pole a. The following theorem[32] is the implement of power in applications to integration. If C is a closed contour within which the only singularities, if any, of $f(z)$ are poles, $\int_C f(z)dz$ is equal to $2\pi i$ times the sum of the residues of $f(z)$ at the respective poles within C.

Cauchy and his followers reveled in the useful or merely curious evaluations of definite integrals by the calculus of residues. With the growing demands of technology in the twentieth century, this calculus and other parts of the theory of functions of a complex variable, such as conformal mapping, were incorporated into the routine undergraduate course in the better technical schools training modern engineers. At the beginning of the century, the theory was reserved in the United States largely for somewhat dreary lectures to graduate students, who learned numerous existence theorems but who occasionally were embarrassed by quite elementary problems. Existence theorems seldom disturb an engineer who has learned to do really difficult things with the complex variable. But, save for such theorems, his dams would go out and his airplanes crack up even more frequently than in the romantic past. The designer is insured against disaster partly by his own practical sixth sense, partly by the seventh sense of pure mathematicians who provide him with reasonably safe devices of calculation. The nineteenth-century glamour of the theory of functions of a complex variable has no doubt dimmed as the theory has been increasingly applied to practical affairs. This, at least, is what a purist might lament. But the progressive world of swifter and shinier clipper transports, noisier bombers, and huger debts hails these modern applications of "the science venerable" as the dawn of the true romance of mathematics. Our successors will finally judge the issue, provided they shun our example and do not fall victims to their own anthropoid ingenuity.

One source of the practical utility of the theory is the fact that if $u + iv$ is an analytic function of $x + iy$, then u, v satisfy Laplace's equation in two dimensions. Solutions of this equation having continuous first and second partial derivatives are called harmonic functions, and similarly for Laplace's equation $\nabla^2\phi = 0$ in three dimensions. Here we are concerned only with the two-dimensional case. One of its simplest applications is to the plane irrotational flow of an incompressible, frictionless fluid. Physically and roughly, 'irrotational' means absence of vortices. Mathematically, the flow is irrotational if there exists a uniform function u, $\equiv u(x, y)$, defined for all points (x, y) in the plane fluid sheet, and having continuous first and second partial derivatives, such that the components X, Y of the velocity of flow at the point (x, y), parallel respectively to the x-axis and the y-axis, are given by $X = \partial u/\partial x$, $Y = \partial u/\partial y$. The function u is called a velocity potential. The historically relevant detail here is the last: the two components are obtained by taking the partial derivatives in the x, y directions. Incidentally, the practically important problem of turbulent flow escapes this mathematics of classical hydrodynamics.

It is readily seen that if the fluid is incompressible, then must $\partial X/\partial x + \partial Y/\partial y = 0$. The connection with analytic functions of a complex variable is evident on glancing at the Cauchy-Riemann equations. For if $w(= u + iv)$ is any analytic function of $z(= x + iy)$, then u is a suitable choice for a velocity potential, as also is v. If u be chosen, the curves $u =$ constant are the equipotential lines of the flow and the curves $v =$ constant the stream lines; and vice versa with u, v interchanged. From the geometry of the situation, the equipotential lines cut the stream lines orthogonally, checking the intuitive physics.

Numerous other physical applications are abstractly identical with that to hydrodynamics. In all, it is the existence of a potential that makes analytic functions applicable. In the flow of heat, temperature replaces the velocity potential in hydrodynamics; in the flow of electricity, it is the electric potential; in Newtonian gravitation it is the gravitational potential, the historical origin of all potential[33] theory.

Continuing with the sources of practical utility, we note conformal mapping. Without details, the source is as follows. In the notation $z = x + iy$, $w = u + iv$ already explained, the analytic function w defines a particular mapping of the z-plane on the w-plane. A contour traversed by the point (x, y) in the z-plane maps into a contour traversed by (u, v) in the w-plane; and if the two contours in the z-plane intersect at a certain angle, the corresponding contours in the w-plane intersect at the same angle. Thus, in this kind of mapping, angles are preserved; distances, except in trivial cases, are distorted. Gauss[34] in 1825 called such mapping conformal. Its utility is a commonplace to anyone who has ever done any serious work in classical mathematical physics, or who has taken a course in airplane performance and design. The more interesting features of such mapping appear when the map is investigated in the neighborhood corresponding to a singularity of the mapping function. One of the most frequently used conformalities is that (1869) of H. A. Schwarz (1843–1921, German), mapping any simply connected region bounded by an n-sided rectilinear polygon upon the positive half-plane. This is a special but practically important case of the general problem of conformal mapping: to find the function mapping one given region upon another. Conformal mapping is almost a branch of modern mathematics in itself; it offers an alternative approach to the theory of functions of a complex variable. In particular, the mapping of polygons, first discussed in 1867 by E. B. Christoffel (1829–1900, German), stimulated many others besides Schwarz to develop the entire subject, until now it has a vast literature of its own.

Although it belongs more properly to the development of differential geometry, we may note in passing the obvious origin of conformal representation in the practical demands of mapping the earth's surface. The first great classics in the subject were Lagrange's memoirs[35] of 1779, in which he described the contributions of his principal forerunners, J. H. Lambert (1728–1777, German) and Euler, and obtained a solution for any surface of revolution, which he applied to the sphere and the spheroid, the cases of first importance in geodesy. Gauss in 1825 attacked the general problem of conformally mapping either of two given surfaces upon the other, and in 1847 applied[36] his solution to geodesy. He does not mention Lagrange's work. With Riemann's version of the theory of functions of a complex variable (1851), conformal representation became an integral part of modern analysis, and its practical applications were no longer restricted to mapping and geodesy.

Closely related to the mapping problem is another of great historical and living interest, Dirichlet's, which we restate here: to demonstrate the existence of a function, harmonic within a closed region of space and having preassigned continuous values on the boundary of the region. To a physicist it may seem obvious that a hot solid in the steady state will have a definite and unique distribution of temperature at every interior point once the temperature over the surface is known; mathematicians have learned by experience that physical intuition is insufficient to solve Dirichlet's problem. Hilbert solved[37] it satisfactorily in 1901 under suitable restrictions adequate for almost any conceivable physical application. In two dimensions the problem reduces to that already described in connection with the Cauchy-Riemann equations and plane flow. Its historical interest for the theory of functions is that Riemann, as will be seen, proceeded from two-dimensional physical intuition to his theory (1851) of analytic functions.

We pass on to the influence of Cauchy's theory on analysis. What follows applies equally to the alternative theories, to be described presently, of Riemann and Weierstrass. In brief, the theory met the prime requisite of any professedly general doctrine by unifying, simplifying, and extending the special theories it was designed to include, and at the same time providing stricter and clearer proofs for known results in the included theories. Only the general trends can be noted here.

First, functions were classified with respect to broad principles suggested by experience. Two peculiarities stood out as having special significance. It was found that many of the useful and mathematically interesting properties of a function are intimately connected with its singularities and zeros, a zero being a value of the variable for which the function vanishes. Poles have been defined; other singularities are called essential; and a value z_0 of z at which $f(z)$ is analytic is an ordinary point. Another type of singularity, a branch point,[38] is relevant only for multiform (non-uniform, many-valued) functions. For at least one value of z, a multiform function of z has more than one value; for example, w defined by $w^2 = z$ has two values, $+z^{\frac{1}{2}}$, $-z^{\frac{1}{2}}$, for a given z. As z varies, each of the values varies, generating what is called a branch of the function. A value b of z for which two or more branches coincide, and at which the branches are interchanged as z describes a circle in the z-plane with center at b containing no other singularity, is called a branch point. In the example, zero is a branch point. This classification of singularities into poles, branch points, and essential proved adequate for most purposes.[39]

The simplest functions are polynomials and rational functions constructed as quotients of polynomials. Both were taken as the points of departure for generalization and classification of less elementary functions, especially by Weierstrass. As his theory blends here with Cauchy's, we shall briefly describe it. The sketch of Cauchy's theorem given earlier may seem to rely on geometric intuition. But this is an illusion; all appeal to the Argand diagram, contours, and the rest of the spacial imagery can easily be obviated, and this is done in an exact presentation.

Distrusting geometric intuition, Weierstrass discarded it entirely and constructed his strictly arithmetical theory of functions. Although he was slow in publication, his influence on nineteenth-century analysis was profound and probably greater than that of any other individual except Cauchy.[40] Intermittently from 1857 to his partial retirement in 1890, Weierstrass instructed his pupils at the University of Berlin in the new analysis he had created, or was creating; their lecture notes or personal researches gradually spread the gospel of Weierstrassian rigor. It was not always glad tidings for some whose supposed proofs of crucial theorems were seen to be incomplete or fallacious.

With his reconstruction of the number system as a basis, Weierstrass made power series the cornerstone of his theory. He introduced the fundamental concepts of radius and circle of convergence,[41] giving arithmetical definitions of both, and proceeded to transcendental generalizations of polynomials and rational functions. One of his aims was to characterize wide classes of functions in terms of their singularities. Thus a polynomial has no singularity for values of the variable in the finite part of the plane, but has a pole at infinity. Weierstrass proceeded from this to the class of functions called integral (or entire) transcendental, defined by power series of the form $\sum_{n=0}^{\infty} a_n z^n$ with infinite radius of convergence and an essential singularity at infinity. Simple specimens are the series for $\sin z$, $\cos z$, e^z.

Another basic concept of the Weierstrassian theory is analytic continuation. As this is the simplest means of establishing the abstract identity of the theories of Cauchy and Weierstrass, we shall describe it briefly. It is to be borne in mind that the following is only a loose sketch.

From Cauchy's theory, the representation of a function $f(z)$, analytic in the neighborhood of a point c, is readily obtained as a power series of the form $\sum_{n=0}^{\infty} a_n(z - c)^n$. This was in fact done by Cauchy himself in his proof (1831) of Taylor's theorem for analytic functions. The circle of convergence is centered at c and extends to the singularity of $f(z)$ nearest to c. For points inside the circle the series converges absolutely; for exterior points the function is undefined; and there may or may not be points on the circumference for which the function exists. Unless the singularities are everywhere dense[42] on the circumference, the region in which the function is defined by power series may be continuable. For if a point other than the center be taken within the circle, and a new circle with this as center and radius to the now nearest singularity be constructed, the function may be expanded as before for points inside the new circle. The process may be repeated for the new circle, and so on. The totality of power series expansions thus obtained by 'analytic continuation' defines the function at all points of a region, any point of which can be reached by an arc not passing through any singularity of the function. Weierstrass defined an analytic function as one power series together with all those obtainable from it by analytic continuation as just described. From this it follows without difficulty that analytic functions as defined by Cauchy and by Weierstrass are the same. It is therefore immaterial which terminology is used, although for integral properties of functions Cauchy's is held by the French analysts to be superior.

The transcendental entire functions defined by convergent series of ascending powers were partly suggested by polynomials. Rational functions similarly suggested transcendental functions having only pole singularities in the finite part of the plane. These are called rational transcendental, or meromorphic.[43] The rapid progress in elliptic functions in the early nineteenth century was partly due to Jacobi's discovery that these functions are meromorphic. In this connection, an episode of special significance may be noted; it afforded the first completely successful application of the theory of functions of a complex variable to the systematic deduction of a particular theory. Liouville in his lectures (1847, published 1851) at the Collège de France developed the theory of elliptic functions by means of Cauchy's analysis and a famous theorem of his own: If $f(z)$ is analytic for all values of z, and $|f(z)|$ is bounded as $|z|$ tends to infinity, then $f(z)$ is a constant.

Three further theorems of wide range may be cited. These again illustrate the passage from the finite case of polynomials and rational functions to the corresponding infinite case of transcendental functions. Weierstrass proved (1876) that a uniform function whose only singularities in the finite part of the plane are poles, with an essential singularity at infinity, is a quotient of two entire functions. He also obtained an infinite-product representation of a function which is analytic in the finite part of the plane, and which has an infinity of zeros so distributed that any circle of finite radius centered at the origin includes only a finite number of them. This recalls the fundamental theorem of algebra and the familiar decomposition of $\sin z$ into an infinite product. A third general theorem is reminiscent of the decomposition of a rational fraction into partial fractions. G. Mittag-Leffler (1846–1927, Swedish) in 1882 gave a method for constructing a uniform function having an infinity of assigned poles or essential singularities, with only a finite number of singularities in the finite part of the plane.

By 1850, enough analytical apparatus was available for an ordered attack on many special functions. A Taylor series is not adapted to some of the classical functions, such as Bessel's. A necessary implement was supplied[44] to fill the lack in 1843 by H. Laurent (1841–1908, French), whose expansion gives the coefficients a_n, b_n in the development of a function $f(z)$ in the form $\sum_{n=0}^{\infty} a_n(z-c)^n + \sum_{n=1}^{\infty} b_n(z-c)^{-n}$, valid for all points z

within the ring bounded by two concentric circles centered at c, provided $f(z)$ is analytic for all points within the ring and on its boundaries.

The gradual departure from the classic Weierstrassian theory, beginning in the 1890's, affords an interesting analogy with the earlier development of field theories in physics. Analysis after Weierstrass turned increasingly to integral properties of functions, and was less concerned with their behavior in the immediate neighborhood of a given point, for which expansions in power series had proved adequate. Analytic continuation had then served to define a function *in toto*. Theoretically perfect, this method had serious practical limitations. Weierstrass himself transcended the strictly local theory in his method of approximation (1885) to continuous functions by means of polynomials, a departure which opened an extensive tract of analysis concerned with the representation of functions of one or more variables by infinite series of continuous functions. From these and other modes of representation, for example by (summable) divergent series, the domains in which functions might be investigated were extended, and the neighborhood of a point surrendered some of its exclusive privileges. A step toward an analysis of functions independent of any particular algorithm was taken by E. Borel (1871–, French) in his theory of growth (croissance). Power series meanwhile were not neglected; in fact Hadamard's Paris thesis of 1892 on the Taylor series and its analytic continuation, already cited in connection with the distribution of primes, opened a new era in the study of functions by means of their singularities. Among other things almost entirely neglected in the older theory, the refined analysis of the twentieth century has included the study of functions, given by power series, for points on the circle of convergence. This hint of the wider scope of modern methods beyond those of the nineteenth century must suffice to indicate that analysis continued to advance after the old generals had retired from the field.

As the methods of Cauchy, Riemann, and Weierstrass shared equally in the enormous development of the theory of functions from about 1860 to the first decade of the twentieth century, we shall indicate Riemann's approach before noting one of the major gains due in part to the entire theory. Riemann was as much of a mathematical physicist at heart as he was a mathematician. His approach might have been that of Lagrange, had the latter pursued the plain hints of a general theory implicit in his own discoveries in hydrodynamics and harmonic functions.

Riemann's theory was outlined in his doctoral dissertation (1851), a mighty effort which Gauss praised with restrained enthusiasm. Gauss, according to Riemann, had long projected a work of his own along somewhat similar lines. That he never carried out his project may have been due to a belated appreciation of certain profound difficulties, such as Dirichlet's problem, in potential theory, which Riemann either overlooked or left for future investigation. For while the fruitful suggestiveness of Riemann's creations has been appreciated by all who have developed them, it seems to be generally agreed that Riemann proposed more problems than he solved. Although Riemann himself was not unduly concerned with rigor, either in his dissertation or in his subsequent work, his analysis has probably inspired as much sound mathematics as have the deliberately rigorous methods of Weierstrass. Riemann's theory appeals particularly to mathematical physicists and to analysts gifted with spacial intuition, that is, to mathematicians of the same type as himself. His geometric methods have been more popular with the German school than the French, and but few English-speaking mathematicians have made much use of them.

A possible mechanical origin[45] of Riemann's definition of w as a function of the complex variable z appears in the contrast between the derivatives dw/dz and dy/dx, where y is a real function of the real variable x. Omitting all refinements, we recall that dy/dx is the limit, as Δx tends to zero, of the difference quotient $\Delta y/\Delta x$. But Δx *must* tend to zero through real values; that is, Δx is constrained to vary on the axis of reals. For z complex, Δz may tend to zero by any one of an infinity of paths in the z-plane; there is no distinguished path as for Δx, and the limiting value of $\Delta w/\Delta z$ as Δz tends to zero is not necessarily independent of the path. Riemann *defined* w to be a *function of* z if and only if the limiting value dw/dz of $\Delta w/\Delta z$ *is* independent[46] of the differential dz. This gave him the Cauchy-Riemann partial differential equations already described, and hence also the connection with Laplace's equation in two dimensions.

A mechanical analogue of Riemann's definition is obvious. A conservative field of force is characterized by the existence of a potential V, defined at every point (x, y, z) of the field, whose gradient $(\partial V/\partial x, \partial V/\partial y, \partial V/\partial z)$ gives the components[47] (X, Y, Z) of the force at (x, y, z). In a conservative field, the work done by the forces of the field on a system moving from one configuration to another is independent of the path, depending only on the values of V in the two configurations. If the path is a closed contour, the total work is zero. Further, in empty space, V satisfies Laplace's equation $\nabla^2 V = 0$. The relevance of the two-dimensional case for analytic functions has been noted. Without laboring the point, it is plain that the analysis mimics the physics. The latter had long been classical when Riemann defined his dw/dz through independence of the path by which Δz tends to zero, and even when Cauchy proved his integral theorem.

It remains to indicate a possible impulse toward a theory of functions of a complex variable in the mechanics of the eighteenth century. The history of this detail is complicated and not yet fully worked out. We shall note only a few of the more significant incidents.[48] The Cauchy-Riemann equations appeared as early as 1746, and possibly even in 1743. For it was known to d'Alembert (1746), Euler (1749), and Lagrange (1762–65) that when $f(x + iy) = u + iv$, then $f(x - iy) = u - iv$, $\frac{\partial u}{\partial x} = \frac{\partial v}{\partial y}, \frac{\partial u}{\partial y} = -\frac{\partial v}{\partial x}$. Lagrange encountered these equations[49] in a hydrodynamical problem, and was led to their general solution in the form

$$u = iF(x + iy) - iG(x - iy), \qquad v = F(x + iy) + G(x - iy).$$

He then sought what is now called the stream function. This leads naturally to the integration of the functions F, G for complex values of the variables. But Lagrange carefully avoided the integral sign, carrying through the analysis by means of infinite series. The value of a function for complex values of the variable was in fact obtained by Taylor's expansion, and Lagrange showed that he was aware of the limitations of the analysis. That $\int(u\,dx + v\,dy)$ vanishes in general when the integration is round a closed contour and $\frac{\partial v}{\partial x} = \frac{\partial u}{\partial y}$, had already been noted (1743) by A. C. Clairaut (1713–1765, French) in his work on the figure of the earth, and d'Alembert had observed the exceptional case $u \equiv y/(x^2 + y^2)$, $v \equiv -x/(x^2 + y^2)$.

The last may have some bearing on a much-quoted remark by Gauss[50] in his letter of December 18, 1811, to F. W. Bessel (1784–1846, Prussian), as d'Alembert was interested in complex functions $f(x + iy)$ when he observed his exceptional case. Gauss stated that he had a proof of the fundamental theorem that "the integral of a function round a closed contour is zero when the integrand does not become infinite within the region bounded by the contour." This was fourteen years before Cauchy published his memoir on complex integration. There has been much inconclusive speculation why Gauss kept a theorem of this magnitude to himself. Perhaps he knew that he was merely restating what was already known. Or he may have discovered later that he had been substantially anticipated. He seems to have dropped the matter, and there is no evidence that he ever proved the theorem. Had he known G. Green's (1793–1841, English) theorem (1828) for two dimensions in 1811, he could have proved the integral theorem immediately. A proof without Green's theorem is more difficult; and it does not seem very probable that Gauss, who went astray the following year (1812) in the matter of Abel's theorem on power series, also in a double-limit process, could have invented a deeper proof for the integral theorem. The historical question (if it is of any importance) seems to be, *when* did Gauss become acquainted with Green's theorem?

Regarding complex integration itself, Euler was probably the first to obtain the value of a definite integral by replacing a real variable by a complex. Laplace (1782, 1810) investigated the validity of the process. It is asserted by P. Stäckel that S. D. Poisson (1781–1840, French) was the first really to use a line integral in the complex plane.

Finally, functions of a complex variable were freely used during the eighteenth century by J. H. Lambert (1728–1777, German), Euler, and Lagrange. But none of these hints bears much resemblance to the theory of functions of a complex variable as imagined by Cauchy in 1825. If the mechanical suggestions in the work of Clairaut and Lagrange be admitted as possible stimuli for the theory, this will not in any way detract from the merits of those who developed the theory into a major department of modern analysis.

Algebraic and automorphic functions

The general theory of functions found its most extensive applications during the nineteenth century in algebraic functions, their integrals, and the multiply periodic functions arising from Jacobi's inversion problem for such integrals. This whole field is sometimes (improperly) called the theory of abelian functions, although Abel himself was unacquainted with the $2p$-fold periodic functions of p variables, named after him, which arise from the inversion of abelian integrals. Abel, Jacobi, A. Göpel (1812–1847, German), J. G. Rosenhain (1816–1887, German), Weierstrass, Riemann, Hermite, H. Weber (1842–1913, German), and scores of less well-known mathematicians devoted some of their strongest efforts to developing the theory, which responded by suggesting new methods and novel problems in algebraic geometry, the general theory of functions, topology, algebraic numbers, and the number system of analysis.

Weierstrass, we recall, made the systematic and rigorous investigation of abelian functions his lifework; and it was while seeking satisfactory foundations for the necessary analysis that he was compelled to rigorize the real number system. Riemann's most individual invention, that of the many-sheeted surfaces known by his name, was devised to render the theory more intuitive. Some of Hermite's earliest and most suggestive work in algebra, analysis, and arithmetic originated in a continuation of Jacobi's contributions to abelian functions. R. F. A. Clebsch (1833–1872, German), Riemann's successor at Göttingen, started a new department of geometry in his application of abelian functions to algebraic geometry. The topological problems presented by Riemann surfaces were among the first to lift analysis situs from the domain of trivial amusements to that of serious mathematics. In brief, it would be difficult to overestimate the influence of algebraic functions on the pure mathematics of the nineteenth century. Accordingly, we shall give the theory of algebraic functions a more extended notice than might be justified by its position relative to the mathematics of the twentieth century.

The theory was developed after its origin in 1826 by so many specialists using different methods, that the finished product is beyond adequate description in less than a voluminous monograph. We shall merely recall the different approaches and enough of the main underlying concepts to suggest that analysis, algebraic geometry, and the modern higher arithmetic all contributed to the complete theory; and we shall begin with the many-sheeted surfaces of Riemann. These are useful for the representation of any multiform functions. A detailed description of these surfaces is out of the question here, and we shall indicate only what their inventor sought to accomplish—the reduction of multiformity to uniformity.

If $f(z)$ is an n-valued ($n \geqq 1$) function of z, its n values, or branches, say $f_1(z), \ldots, f_n(z)$, are permuted ($n > 1$) as z circuits a closed contour K with one branch point and no other singularity of $f(z)$ within or on K. Both Cauchy and Riemann sought a geometrical representation of the branches of a function. One object was to separate the branches, thus securing for multiformity ($n > 1$) the simplicities of uniformity ($n = 1$); another was to picture the mutual relations of the n branches with respect to the branch points. Cauchy proposed the device of loops issuing from a non-singular point in the z-plane, a loop being a closed contour passing through no singularity and enclosing only one singularity. The branch points being thus looped, the branches of $f(z)$ are permuted as z circuits the loops. The permutations form a group, whose multiplication table summarizes the interchanges of the branches for all possible circuits.

Riemann's radically different method exhibits $f(z)$ as a uniform function on an n-sheeted surface. At first glance the problem of reducing multiformity to pictorial uniformity seems as hopeless as squaring the circle. Riemann solved it in four steps. Instead of representing z in one plane, n planes, or sheets, $P_1, \ldots, P_n$, corresponding respectively to the n branches ($n > 1$) $f_1(z), \ldots, f_n(z)$ of $f(z)$, are assigned to z. As z varies in a particular sheet, say P_j, $f(z)$ takes only the values for the corresponding branch, $f_j(z)$. The n sheets are now superimposed on the Argand diagram; n points, say $z_1, \ldots, z_n$, lie one above the other on $P_1, \ldots, P_n$; to these points correspond the values of $f_1(z), \ldots, f_n(z)$ for one and the same z in the Argand diagram. The third step takes account of the branch points. At a particular branch point z_0 in the Argand diagram, two or more of $f_1(z_0), \ldots, f_n(z_0)$ are equal, and as z circuits z_0, the branches concerned are permuted. The sheets corresponding to these branches are imagined as "hanging together"—connected—at the branch point z_0; and similarly for each of the branch points of $f(z)$. Finally, passage from one sheet to another must be provided to picture the actual permutations of $f_1(z), \ldots, f_n(z)$ as z in the Argand diagram circuits the branch points in any manner, or describes any contour. This is accomplished by lines, which need not be straight but no one of which cuts itself, called branch cuts, each joining a pair of branch points (infinity may be a branch point), or passing from a branch point to infinity. Along each branch cut, bridges or membranes are imagined connecting the sheets in such a manner that the interchange of branches is imaged in the passage of z from sheet to sheet of the n-fold surface. If z in the Argand diagram by circuiting a particular branch point once changes the branch $f_r(z)$ to the branch $f_s(z)$, z in the Riemann surface passes by a bridge from the sheet P_r to the sheet P_s. There are usually several ways of making the connections. All, however, are equivalent, in the sense that each exhibits the n-valued $f(z)$ as a uniform function on an n-sheeted Riemann surface, in which the sheets are so connected that the permutations of the branches are consistently represented.

About thirty years before Riemann imagined his intuitive approach to algebraic functions, Abel (1826) gave the entire theory its first impulse in his masterpiece, specifically in the great theorem of the integral calculus which Jacobi called Abel's theorem in honor of its discoverer, and which Legendre, in the boastful words of Horace, described as *monumentum aere perennius*. It is customary now to phrase Abel's theorem in the language of algebraic geometry.[51] But Abel himself presented it purely analytically, as a generalization of the addition theorems for elliptic integrals, the first of which was given in 1761 by Euler. Abel's proof has been called "only a wonderful exercise in the integral calculus." As the usual geometrical statement of Abel's theorem demands extensive preliminaries, an analytic form[52] is preferable here.

Let $f(x, y)$ be a polynomial of the form

$$X_0y^n + X_1y^{n-1} + \cdots + X_n,$$

in which $X_0, \neq 0, X_1 \ldots, X_n$ are polynomials in x, having no common factor involving x. Further, let $f(x, y)$, considered as a polynomial in y, have a discriminant not identically zero. Then y defined by $f(x, y) = 0$ is called an algebraic function of x. In the following statement of Abel's theorem, $X_0 \equiv 1$, and $R(x, y)$ is any rational function of x, y. Abel's theorem is

$$\sum_{i=1}^{m} \int^{(x_i,y_i)} R(x, y)dx$$

$$= F(x_1, y_1; \ldots ; x_m, y_m) - \sum_{j=1}^{k} \int^{(u_j,v_j)} R(x, y)dx$$

in which: m is any positive integer; the lower limits of the integrals on the left are arbitrary; F consists of rational functions of $x_1, y_1, \ldots, x_m, y_m$ and logarithms of such functions; $u_1, \ldots, u_k$ are values of x which can be determined as the roots of an algebraic equation whose coefficients are rational functions of $x_1, y_1, \ldots, x_m, y_m$; and $v_1, \ldots, v_k$ are the corresponding values of y, any one of which, say v_j, can be determined as a rational function of u_j and $x_1, y_1, \ldots, x_m, y_m$.

In this, the relations determining $(u_1, v_1), \ldots, (u_k, v_k)$ from $(x_1, y_1), \ldots, (x_m, y_m)$ hold throughout the integration; these relations determine the lower limits of the k integrals on the right from the arbitrary lower limits of the m integrals on the left, k depends neither upon m nor upon the form of $R(x, y)$, and in general k does not depend upon the values of $(x_1, y_1), \ldots, (x_m, y_m)$, but only upon the equation $f(x, y) = 0$, which determines y as an algebraic function of x.

Such is a bald summary of Abel's theorem in something like its original form. A fuller statement shows how the equations determining the u_j, v_j from the x_i, y_i are constructed. We can only note the hints in this enunciation for the trifid development of algebraic functions and their integrals from Abel's initiation of the theory to the twentieth century.

Integrals of the kind occurring in Abel's theorem are called abelian integrals associated with $f(x, y) = 0$. In the transcendental method for developing the theory of algebraic functions, abelian integrals are basic. After Abel, this method was next most extensively cultivated (1859) by Riemann, whose theory has points of contact with the second, or algebraic-geometric theory. That such contacts might be expected is suggested by the fundamental equation $f(x, y) = 0$, which defines an algebraic curve in the x, y plane, and which also, when written in the form $X_0y^n + X_1y^{n-1} + \cdots + X_n = 0$, exhibits y as an n-valued function of x. Thus it might be anticipated that the associated Riemann surface and the algebraic curve would each somehow reflect the other. Before indicating roughly the basis for the correspondence between the curve and the surface, we note the third of the classic methods in the theory of algebraic functions, again suggested by the fundamental algebraic equation $f(x, y) = 0$. The totality of rational functions of x, y, such as $R(x, y)$ occurring in the statement of Abel's theorem, constitute a field in the technical sense of abstract algebra, and this special field is defined by $f(x, y) = 0$. It is thus suggested that the methods of algebraic fields and the techniques of the arithmetic theory of ideals can be extended to the integrands of abelian integrals. When developed, this generates the third type of theory, the arithmetic, of algebraic functions. This many-sided theory turned to the analysis, the algebraic geometry, and the modern higher arithmetic of the nineteenth and twentieth centuries for suggestions, and reciprocated by proposing numerous difficult problems in each of its subsidiaries.

Finally, originating with Abel in 1826 as a generalization of the addition theorem for elliptic integrals, the theory of abelian integrals suggested to Jacobi a generalized problem of inversion, similar to that which led from elliptic integrals to elliptic functions. It was no easy extension. Jacobi's (1832–4) inspired solution—he practically guessed it after an unsuccessful attempt to find the right way—for the simplest case is usually considered one of the most brilliant achievements of nineteenth-century analysis. As the nature of this can be briefly indicated, we describe it next.

The sums of the special abelian integrals

$$\int^{x_1} \frac{dx}{\sqrt{R(x)}} + \int^{x_2} \frac{dx}{\sqrt{R(x)}}, \int^{x_1} \frac{x\,dx}{\sqrt{R(x)}} + \int^{x_2} \frac{x\,dx}{\sqrt{R(x)}},$$

in which $R(x)$ is a polynomial of degree six, are functions, say s_1, s_2, of the upper limits x_1, x_2. Jacobi found that the symmetric functions $x_1 + x_2$ and x_1x_2 are uniform quadruply periodic functions of s_1, s_2. The partial analogy with the doubly periodic functions, arising from the inversion of an elliptic integral, suggested to Jacobi that his new functions might be expressible in terms of an appropriate generalization of the elliptic theta functions. He divined that these would be of the form $\Sigma\Sigma e^{am^2+2bmn+cn^2+2mu+2nv}$, where the summations $\Sigma\Sigma$ refer to all integers m, $n \gtreqless 0$; a, b, c depend on the periods; and u, v are linear functions of s_1, s_2. Jacobi's conjecture was proved correct (1851) by J. G. Rosenhain (1816–1887, German). With this and an earlier (1847) investigation of A. Göpel (1812–1847, German), the vast theory of the $2p$-fold periodic[53] functions in p variables was begun. After the pioneering work for $p = 2$, Riemann (1857) introduced the corresponding functions for any p, but his theta functions were not the most general possible. The development of a general theory of the thetas was part of Weierstrass' life-program; it was carried out principally by him [54] and his pupils in the second half of the nineteenth century.

It remains to indicate the source of the connection, through the fundamental equation $f(x, y) = 0$ and its associated Riemann surface, between algebraic functions and geometry. The underlying concept, that of birational transformations, generated one of the most extensive divisions of geometry of the nineteenth and twentieth centuries. Hinted at originally in a project of Apollonius (second century B.C.), birational transformations lay dormant till 1824, when Steiner and, later, others applied the special type called inversion to synthetic geometry.[55] But nothing approaching a general theory appeared until 1863, when the Italian geometer L. Cremona (1830–1903) began a systematic investigation of the particular kind of birational transformations since named after him. These are defined as follows.

The (x, y)-plane can be transformed into itself: to each point (x, y) corresponds a point (x', y'), in which x', y' are rational functions of x, y, and inversely; all this with the possible exception of certain singular loci. For example, the transformation $x' = x$, $y' = x/y$ and its inverse $x = x'$, $y = x'/y'$ establish a one-one correspondence between the (x, y) and the (x', y') except for points on the axes. Such birational transformations of the entire plane, with the possible exception noted, are called Cremona transformations. An algebraic curve whose equation is in x, y coordinates will be denoted by C, and similarly for x', y' and C'.

Clearly, C, C' correspond birationally pointwise under a Cremona transformation. But C, C' may correspond birationally pointwise by a birational transformation which does not extend to the entire plane. In such cases we say that the transformation is curve-to-curve, and we shall write $C \sim C'$. Evidently this 'equivalence,' $\sim$, is an instance of the abstract equivalence relation described in connection with algebra and arithmetic. Hence, with respect to $\sim$, algebraic plane curves fall into classes. All curves in a particular class are said to have the same genus. The genus is a non-negative integer p, to be described presently.

Before proceeding with these considerations, we note why their relevance for the theory of algebraic functions might be anticipated, and observe one of the reasons why some have preferred the arithmetic approach to the algebraic-geometric.[55]

The theory of algebraic functions is centrally occupied with an irrationality defined by the equation of an algebraic curve C. In particular, it was stated in Abel's theorem that the number

k of abelian integrals in a certain sum does not depend upon the particular form of the rational function $R(x, y)$ of x, y occurring as the integrand, where $f(x, y) = 0$ is the equation of C. This suggests that the theory will be essentially unaltered if C be replaced by C', where $C \sim C'$. We should therefore choose from all curves having the same genus one of the geometrically 'simplest.'

In an obvious sense, an algebraic curve whose only singularities are nodes is simpler than one with higher singularities. But an irreducible algebraic curve C is birationally transformable into another, C', having no singularities other than double points with distinct tangents, a basic theorem discovered, apparently, by Kronecker[56] in 1858. Proofs of this theorem were still exercising algebraic geometers in the 1940's. M. Noether's companion theorem (1871) states that any irreducible algebraic curve is transformable by a Cremona transformation into another whose only singularities are multiple points with distinct tangents. The arithmetic method in algebraic functions can be carried through without this somewhat troublesome reduction of singularities. But geometers naturally prefer to exercise their spacial intuition.[55] The mere vocabulary of geometry also suggests many promising lines of attack.

Only a summary indication of the relation between the algebraic curve $f(x, y) = 0$ and the corresponding Riemann surface can be given here. The relation in question concerns the genus p of the curve and the connectivity of the surface.

Simply (or singly) connected surfaces have already been described.[28] A surface is called doubly connected if by one cross cut it can be made singly connected. For example, the plane surface bounded by two concentric circles is doubly connected. If one cross cut renders a surface doubly connected, the surface is called triply connected, and so on. A surface without a rim may be provided with one by punching a small hole in it; the first cross cut starts from the hole. Thus a punctured sphere is simply connected; a punctured anchor ring is triply connected. It will be assumed that a Riemann surface has been suitably punctured when necessary.

By $2p$ appropriately defined cross cuts the Riemann surface corresponding to a curve C of genus p can be made singly connected. The genus of C, as in the theory of algebraic plane curves, is the integer by which the actual number of double points of C falls short of the possible maximum as given by Plücker's equations.

If $p = 1$, the corresponding abelian integrals $\int R(x, y)dx$ are elliptic. If $p = 2$, the integrals are called hyperelliptic; they lead to the quadruply periodic functions of two variables. There is an enormous literature devoted to this special case alone. Generally, to each algebraic curve corresponds a Riemann surface appropriate for all the curves having a given genus; conversely, to a Riemann surface there corresponds a class of curves, all of which have the same genus. A surface of genus p depends upon $3p - 3$ constants; to each set of values of these constants corresponds a class, or an infinity of classes, of algebraic curves.

Enough has been described to exhibit the theory of algebraic functions as an outstanding example of the spirit which animated mathematics in the nineteenth century. The theory presented many typical examples of generalization. For instance, the circular and hyperbolic functions are degenerate cases of the elliptic, themselves a mere incident in the theory of algebraic functions and their integrals. There were corresponding generalizations in the theories of algebraic curves, surfaces, and manifolds of n dimensions, one of which may be noted here.

The problem of uniformizing a curve (one-dimensional manifold) is that of finding the appropriate uniform functions $g(t)$, $h(t)$ of the parameter t such that the coordinates (x, y) of any point on the curve are given by $x = g(t)$, $y = h(t)$. The circle $x^2 + y^2 = 1$ is uniformized by $x = \sin t$, $y = \cos t$; the cubic without a double point is uniformized by elliptic functions; the quartic surface (two-dimensional manifold) with sixteen nodes (the maximum number possible) is uniformized by hyperelliptic theta functions. Uniformization thus appears as a natural outgrowth of the theory of algebraic functions. One of the most striking advances of nineteenth-century analysis was Poincaré's uniformization of any algebraic curve $f(x, y) = 0$ by means of the automorphic functions which he constructed in 1881–4.

Automorphic functions[57] rapidly expanded in the 1880's and 1890's into an independent department of analysis, to which the theories of groups, functions of a complex variable, differential equations, quadratic forms, and non-Euclidean geometry contributed at various stages. After Poincaré, one of the most active exploiters of the new field was Klein, whose wide knowledge of the several related subjects enabled him to apply his Riemannian intuition to the creation of a unified theory, expounded in a total of 1,296 octavo pages,[58] in which groups play the dominant part. It was syntheses such as this that induced some enthusiasts in the nineteenth century to prophesy that all mathematics worth remembering would ultimately be comprised in the theory of groups. They appear to have been badly mistaken.

The characteristic property of automorphic functions generalizes periodicity. Several such generalizations are conceivable; that relevant here, when represented on the plane of complex numbers z, is the simplest extension of periodicity as it appears in the circular (singly periodic) functions. For example, $\sin (z + 2n\pi) = \sin z$, for all integers n. Again, if $E(z)$ is an elliptic function with the periods p_1, p_2,

$$E(z + np_1) = E(z + np_2) = E(z),$$

which may be expressed thus: the value of $E(z)$ is invariant under the group of translations $z \to z + n_1p_1 + n_2p_2$, where n_1, n_2 run through all integers. This group is generated by iterations of $z \to z + p_1$, $z \to z + p_2$ and their inverses, $z \to z - p_1$, $z \to z - p_2$. But these are almost trivial instances of the transformation $z \to (az + b)/(cz + d)$, in which $ad - bc \neq 0$. There are various categories of linear groups generated by such transformations. With certain of these[59] are associated specific kinds of automorphic functions $f(z)$, so named because the value of $f(z)$ is unaltered when z is replaced by any $(az + b)/(cz + d)$ in the group concerned. To cite only one theorem on automorphic functions, any two of a very general kind belonging to the same group are connected by an algebraic equation.

The earliest (1858) instances of functions having the last mentioned property are the elliptic modular functions of Hermite. These are defined as uniform functions $F(z)$ such that $F(z)$ and $F\left(\frac{rz + s}{tz + u}\right)$, where r, s, t, u are integers and $ru - ts = 1$, are connected by an algebraic equation. These modular functions also generated a vast theory, developed by Hermite, Dedekind, H. Weber, and Klein, among many others. By 1890 this specialty had grown so enormously that even Klein required 1,488 large pages to expound its main features as it then was. Its applications include some of highly specialized interest to arithmetic. But neither the modular nor the automorphic functions found applications to science worth mentioning. The like appears to be true of the multiply periodic and theta functions in more than one variable. With the exception of some rather academic applications of the functions of two variables to hydrodynamics, there seem to be none. However, this dearth of applications may be only a reflection of the fact that few working scientists can afford to spend months digesting massive tomes on analysis which even professional mathematicians find somewhat cloying.

In compensation, the modular functions were initially responsible (1879) for Picard's theorem,[60] usually rated one of the finest in its qualitative department of analysis: in the neighborhood of an isolated essential singularity a uniform function takes every value, with one possible exception, infinitely often. In the same order of ideas, Landau, generalizing another theorem of Picard's, proved (1904) that if a_0 is any number, and a_1 any number different from zero, then there is a number $N = N(a_0, a_1)$, depending only on a_0, a_1, such that the function

$$f(z) = a_0 + a_1 z + a_1 z^2 + \cdots,$$

regular in the circle $|z| \leqq N$, takes in this circle one of the values 0,1. For many further 'qualitative' theorems concerning the general behavior of functions of a complex variable since the Weierstrassian era, we must refer to Landau's tract[61] of 1929. Himself one of the most skillful analysts of the twentieth century, Landau was forced to admit that "The literature on such questions is large; the individual memoirs are sometimes so lengthy that one has difficulty in selecting the most beautiful result and in reassembling the relevant proof." This frank admission from a master might well stand as a description of all recent mathematics.

The automorphic functions also contributed to the progress of pure mathematics, in the theory of linear homogeneous differential equations whose coefficients are algebraic functions of the independent variable. Obviously, the theory of algebraic functions and their integrals is but a special case of this. Riemann initiated this theory in his investigation of the differential equations suggested by that for the hypergeometric series. It was elaborated by L. Fuchs (1833–1902, German) and his pupils; but the full significance of this earlier work appeared only with Poincaré's invention of automorphic functions. Poincaré showed that, just as the elliptic and abelian functions suffice to integrate algebraic differentials, so do the automorphic functions serve to integrate linear differential equations with algebraic coefficients.

We conclude this sketch with a brief mention of another modification of periodicity. H. Bohr[62] (Danish) proceeded (1923) in a direction radically different from any thus far described, in his theory of almost-periodic functions $f(x)$ of the real variable x. Such a function is defined by the property that for every $\epsilon > 0$ there is a length, $L \equiv L(\epsilon)$, such that every interval $(\alpha, \alpha + L)$ contains at least one number $\tau, \equiv \tau(f, \epsilon)$, satisfying the inequality $|f(x + \tau) - f(x)| < \epsilon$ for all x. The resulting theory is a generalization of Fourier analysis. It has applications to Dirichlet series, and shows some promise of being scientifically useful. In the first nine years of its existence, the subject had swelled to such proportions that it demanded a separate treatise[63] for adequate exposition.

The pursuit of unity

Numerous amplifications and modifications would be required to make the foregoing sketch of algebraic and automorphic functions either exact or adequate. But inadequate as it is, it suggests the bewildering fertility of the leading nineteenth-century mathematicians. There seems to have been no bound to their inventiveness. The followers ably seconded the leaders, until by 1900 *a hundred and twenty* had made contributions of sufficient significance to be included in a standard treatise on abelian functions, and these were by no means all who had written on the subject. Probably about the same number of algebraic geometers had more or less directly furthered the theory of algebraic functions up to 1900.

In addition to those already mentioned, some of the most active contributors were Clebsch, P. Gordan, and G. Roch in the 1860's; A. Brill and M. Noether in the 1870's; and Klein, beginning in the 1880's. All were German, and none followed the arithmetic method. Roch's memoir (1865) on the number of arbitrary constants in algebraic functions has associated his name with Riemann's in a famous theorem, one of the basic results of the entire subject. Another noteworthy contributor was J. Lüroth (1844–1910, German), remembered for his topologic investigations (1871) of Riemann surfaces. The 1880's saw the entry of two leading French analysts of the time, Poincaré and E. Picard (1856–1941), into a field that seems capable of absorbing any amount of talent. The Italians also, continuing the tradition of Cremona, began to make algebraic geometry almost a national pastime. The magnitude of their effort can be judged by consulting the treatises (1921, 1926) of F. Severi (1879–1940, Italian) who, as a very young man in the early 1900's, put new life into algebraic geometry. Nor did the arithmetic method suffer from a lack of the highest talent—Kronecker in 1862, Weierstrass in 1875–6 with his application of analytic functions, Dedekind and H. Weber in 1882, K. Hensel and G. Landsberg (1865–1912, German) in 1902, J. C. Fields (1863–1936, Canadian) in 1906, Emmy Noether in 1919, most of whom, or whose followers, continued to develop the method from the 1860's to the 1930's.

The illustrious names in the foregoing sample from a dis-hearteningly extensive directory are sufficient to suggest that by 1926 the nexus of subjects that evolved from Abel's initial discovery of a hundred years before had begun to resemble a mathematical neophyte's nightmare. If all this intricate mass of detailed analysis had to be mastered before a beginner might hope to find something new and vital, the outlook for getting one's name into a revised directory was dark indeed. Leading experts in each of the several methods were still pouring out new theorems, presumably of deep significance, which other leading experts might understand only after much hard labor.

To interrupt one's own researches in order to follow those of another is a scientific pleasure which most experts delegate to their assistants. Consequently, the confusion of tongues increases as the square of the number of talkers, until only ever more select coteries of narrow specialists really understand the refinements of their esoteric vocabularies.

A practical way out of such an impossible muddle is simply to ignore it and go on to something else. This seems to be what happened shortly after 1900 in the theory of algebraic functions and their integrals. A majority of the more inventive young analysts made fresh starts in new and possibly less thorny fields. Not that algebraic functions were neglected after 1900; for a great deal was done in the subject, particularly after about 1920, when the impact of twentieth-century algebra and arithmetic began to be felt. But the character of the newer work[64] was noticeably different from that of the classics of the preceding century. For one thing, it exemplified a fresher rigor; for another, it was more general.

What happened in algebraic functions was not exceptional; and we have not intended to pillory them as a uniquely terrifying example of mathematical fecundity. They have been exhibited merely as a typical product of mathematical development in the nineteenth century. Specialty after specialty kept splitting into new and narrower specialties, which in turn swelled and subdivided, and so on, apparently without any inhibiting principle to regulate their growth.

The need for some control was recognized by a few of the leading mathematicians themselves, of whom Klein was the most energetic. In the last two decades of the nineteenth century, Klein's was a name to conjure with, especially in Germany and the United States.[65] Some of the most influential chairs of mathematics in American universities were occupied by men who had gone abroad in the 1880's and 1890's to study under Klein. And in his own country, a whole generation of German theoretical engineers and industrial scientists remembers Klein with gratitude for his efforts to make higher mathematics useful to them.

In pure mathematics, Klein's enthusiasm and great personal charm made lifelong converts of practically all his pupils. The exact boundary between masterly instruction and propaganda has yet to be determined. One of the most convincing expounders of mathematics that ever lived, Klein persuaded his followers as effortlessly as breathing that his own recipe was about to bring order out of chaos and at last to unify all mathematics—or nearly all. His remarkable lectures before the Chicago congress of mathematicians, held in connection with the World Fair of 1893, removed the last shadow of doubt from the minds of the most critical. In historical perspective, this mass conversion loses its suggestion of the miraculous.

Almost from the beginning of his career, all of Klein's unexcelled talents for organization and leadership, and all his wide knowledge of mathematics, were devoted to a crusade against what he considered the dangers of specialization. He would unite at least some of the specialties or perish trying. His magic was the theory of groups; and for about half a century it worked to a certain extent. But experience has shown that he might as well have attempted to wave back a rising tide on the Atlantic Ocean. Klein's (1872) most successful unification, that of the geometries of his day, as we have seen, proved inadequate for those of the twentieth century.

Even in the nineteenth century, but few of Klein's European contemporaries were willing to assimilate his singularly personal methods. Kleinian mathematics demanded too much knowledge of too many things for mastery in a reasonable time, and in addition it frequently presupposed a facility in spacial linguistics beyond the capacities of most mathematicians. With the turn of the century, a younger, more critical generation saw only confusion and an impressionistic lack of precision where the old master had sought to portray the many as but different aspects of a single whole. Klein himself had assisted ably in the development of the subjects he strove to unify; and his intuition was almost unique in his own generation. (Poincaré was a close rival.) Lacking these advantages, newcomers, finding Klein's works almost impossible to read, conserved time and effort by using more economical if less spectacular methods. With the advance of the abstract method in the early 1900's, the leader of the 1890's found himself lagging far behind the fashion of the times. Of course history may yet show that the lag was the other way, and that Klein was several generations ahead of his later contemporaries.

A strong advocate of mathematical 'schools' dominated by masterful leaders, Klein (1849–1925) at his death had outlived most of his own active disciples by a decade or more. Vigorous young individualists were busily seeking unity in more subtle ways, devising unifications that bore no resemblance to the attempted syntheses of the late nineteenth century. The only prospect of a Kleinian revival is a new school of impressionists as peculiarly gifted as Klein himself was, and with his conception of mathematics as an intuitive art rather than an exact science. Mathematics in the meantime, like some other activities of Western civilization, will continue its desperate struggle to avoid being smothered by its own riches.

Departure from intuition

The rather abrupt transition from intuitive to unintuitive thinking which distinguished much of the mathematics of the twentieth century from that of the nineteenth is strikingly evident in the newer approach to some of the geometry, particularly algebraic geometry, to which we have frequently referred. The work in question might be credited to geometry, to abstract algebra, to the theory of ideals in algebraic function fields and the theory of polynomial ideals as they have developed since about 1916, or partly to topology, according to preference. If to geometry, it demands a much deeper knowledge of the algebra of its epoch than sufficed in the analogous situation in the nineteenth century. It also exacts familiarity with the general arithmetic which evolved since about 1920 from Dedekind's theory of algebraic numbers, and with Hensel's theory of p-adic numbers as it has developed since about 1913, including its offshoot in valuation theory. All the necessary background material is readily accessible in standard treatises, but its mastery presupposes a marked aptitude for abstraction.

If this algebraic-arithmetic approach to algebraic geometry should ever become popular, the textbooks of the future on higher geometry will contain but few technical terms that geometers of the present and recent past would recognize as belonging to the vocabulary of their subject. The metaphors of geometry will be no less vivid than formerly, but they will be in a different language. Nor will they always mean what the older language struggled to express, for at least a little of that has been shown to be without much meaning, if any. Once more, as so often in our journey from the past, there is no implication in any of this that the last word has been spoken or even that what the newer language attempts to convey will be either wholly consistent or even mildly interesting to mathematicians half a generation hence. But so long as sharpened methods of proof continue to rectify actual oversights in earlier work and to secure ampler discoveries where even the most penetrating intuition confused what is provable with what is not, the painstaking analysis of the algebraic-arithmetic method will doubtless continue to survive. To some tastes it may be less brilliant than inspired divination; to others it has the pedestrian merit of being less frequently only fractionally correct.

The reduction of the singularities of an algebraic surface (or generally, of an algebraic variety) is one of the problems that the algebraic-arithmetic method has renovated. A geometric theory of the singularities of algebraic surfaces was begun in 1897 by the Italian geometer C. Segre (1863–1924), whose research and teaching strongly influenced the Italian school of geometers. B. Levi (1875–) of this school invalidated (1899) the attempted proof of G. Kobb (1892) that the complete neighborhood of a singular point of an algebraic surface is representable by a finite number of integral power series of two parameters. A sound proof was first obtained (1901) by the little-known American analyst C. W. M. Black, whose work, published in a journal but seldom consulted by mathematicians, was effectively overlooked.

The reduction problem for S_3 (space of three dimensions) is analogous to that already noted for algebraic curves. The capital theorem for curves states that an algebraic curve is always birationally transformable into a curve in S_3 free from singularities, and thence, by projection, into a plane curve whose only singularities are double points. Analogously, every algebraic surface is birationally transformable into a surface in S_5 with only ordinary singularities, that is, a nodal curve, on which there is a finite number of ordinary cuspidal points and of triple points for the curve, which are also triple and triplanar points for the surface. For further details we refer to O. Zariski (1899–, Poland, U.S.A.), *Algebraic surfaces* (1934). The most impressive attempt at a proof by more or less traditional methods of algebraic geometry, Severi's of 1914, was invalidated (1934) at an essential point by Zariski. The need for sharper precision and less slippery methods was apparent.

The algebraic-arithmetic method stems from two principal roots, among others: the arithmetical theory of algebraic functions of a single variable, initiated in 1880 by Dedekind and H. Weber; modern abstract algebra as developed since about 1920 by A. E. Noether, her pupils, and her followers. The Dedekind-Weber theory was recast (1907) by the German mathematicians K. Hensel (1861–) and G. Landsberg (1860–) whose work was carried on (1908) by H. W. E. Jung (Germany) to functions of two independent variables. Certain essential elements of the theory (in particular, divisors) were extended (1929) by B. L. v. d. Waerden (Holland, Germany) to functions of n independent variables. The arithmetical treatment of the last became possible only when the theory of ideals had been extensively developed in the Noether tradition. As always in this account, no attempt is made to enumerate all the tillers of this fertile field or to catalogue the fruits of their prolific labors.

A critical examination of the literature of algebraic surfaces convinced Zariski that in 1942 a general theory of birational correspondences did not exist. Having stated his conviction, Zariski continued:

> This may sound too reckless a statement or too harsh a criticism, especially if one thinks of the fundamental role which birational transformations are supposed to have in algebraic geometry. Nevertheless our conclusion is in exact agreement with the facts. . . . It is true that the geometers have a fairly good intuitive idea of what happens or what may happen to an algebraic variety when it undergoes a birational transformation; but the only thing they

know with any certainty is what happens in a thousand and one special cases. All these cases—and they include all Cremona transformations—are essentially reducible to one special but very important case in which the varieties under consideration are nonsingular (that is, free from singular points). One can give many reasons for regarding as inadequate any theory which has been developed exclusively for nonsingular varieties. . . . Were such a proof available, it would still be advisable to develop the theory of algebraic varieties, *as far as possible,* without restricting oneself to nonsingular projective models. This certainly would be the correct program of work from an arithmetic standpoint. . . . It turns out, as I have found out at some cost to myself, that we have to know a lot more about birational correspondences than we know at present [1942] before we can even attempt to carry out the resolution of the singularities of higher varieties. A general theory of birational correspondences is a necessary prerequisite for such an attempt.

The same year (1942) Zariski greatly simplified his arithmetical proof (1939) that an algebraic surface over a ground field of characteristic zero can be birationally transformed into one without singular points. In 1944 he published a detailed account of the reduction of the singularities of an algebraic surface.

The problem of the reduction of the singularities of an algebraic surface is as intricate as it is famous. Its successful solution may fittingly stand as a testimonial to the power of unintuitive reasoning in mathematics. Not that intuition is absent from such reasoning; it has merely descended to deeper levels than it penetrated in the past. Contrasted with the intuition exercised in the algebraic geometry of the 1930's–1940's, that which sufficed some of the geometers of the 1900's–1920's seems rather naive. In differential geometry, intuition has been less conspicuous—unless imagination be a form of intuition; some decades of rigorous analysis have influenced the development of that branch of geometry considerably.

CHAPTER 22

Through Physics to General Analysis and Abstractness

A scientific theory has made its principal contribution to the progress of mathematics when its central concepts have been formulated mathematically. Only if new methods are required to solve specific problems is the theory of further mathematical interest. Significant advances in the exact sciences have usually demanded improvements in calculation at least, and so have been indirectly responsible for progress in mathematics itself. Thus we have seen that eighteenth-century dynamics suggested the calculus of variations, in the direct line of descent to twentieth-century general analysis. An important episode in the same evolution was the controversy in the eighteenth century over the meaning of an arbitrary function. We shall consider this first.

Arbitrary functions

It was the problem of a vibrating string that first caused mathematicians to doubt whether they had apprehended what a function may be in mathematical analysis. The earliest to discuss the problem of the string was Taylor—of Taylor series fame—in 1713. To him is due the initial hint of a solution in terms of trigonometric functions. John Bernoulli (1727) translated the problem into a difference equation. From the physics of the situation, this could be only a crude approximation to the facts. Like the bold mathematical physicist he was, Daniel Bernoulli ignored all niceties involved in passing to the limit and replaced the difference equation by a partial differential equation of the second order. Incidentally, he (1753) emphasized the physical and mathematical significance of what has since been called the principle of superposition. We shall recur to this later; it was one of the longest steps forward ever taken in mathematical physics.

Under appropriate physical assumptions, the equation of motion for a string with fixed end-points, vibrating in a plane, is the one-dimensional wave equation, $\frac{\partial^2 y}{\partial x^2} = \frac{1}{c^2}\frac{\partial^2 y}{\partial t^2}$, where c is a constant. D'Alembert and Euler in 1747 obtained solutions of the form $y = f(x + ct) + g(x - ct)$ with f, g 'arbitrary' functions. Daniel Bernoulli's solution was by a sine series of the kind now named after Fourier. This apparently irreconcilable conflict between two mathematical solutions of the same 'natural' problem precipitated a controversy that was to last for twenty years.

Bernoulli contended that his solution must imply Euler's and d'Alembert's. Euler (1753) objected that if this were so, then it must be possible to expand an 'arbitrary' function in a sine series.[1] But this was paradoxical: not every function is odd and periodic, as seemed to be implied.

Lagrange (1759) *almost* resolved the paradox in his great memoir on the propagation and nature of sound. He regarded the material string as the limiting case of a finite set of equally spaced equal mass-particles on a weightless string. Increasing the number of particles indefinitely, he derived the partial differential equation of the material string as the limiting form of his system of difference equations for the finite case. As he had used trigonometric interpolation, he *almost* obtained the solution that would be given today by Fourier analysis. By postulating an 'arbitrary' initial displacement of the string, and an 'arbitrary' initial velocity in the finite case, Lagrange obtained the solution for the continuous, material string as a trigonometric series. One short step more, and he would have isolated the Fourier coefficients of the expansion. But his mathematical conscience forbade him to interchange the operators $\int_0^1$, $\sum_1^\infty$ without further investigation, and he missed anticipating Fourier analysis. Had Lagrange's objective been not merely to reconcile the solutions of Bernoulli and Euler, he might well have forestalled Fourier—provided his conscience had let him.

Lagrange's failure to make the capital discovery has been exhibited by competent judges[2] as a typical example of the temptation in historical studies to read the future into the past, and to discern ulterior motives where none existed. When, about fifty years later, Fourier took the decisive last step, Lagrange objected. As one of the jury consisting of himself, Legendre, and Laplace, he pointed out that Fourier's memoir on heat conduction, submitted to the Paris Academy, lacked proofs. If Fourier was competing for a prize in mathematics, he had misjudged the nature of the competition. If the prescribed conditions were lax enough to cover an unsubstantiated algorithm that gave empirically correct results, the jury may have been too exacting. In any event, the rejection of Fourier's memoir has been publicized as a deplorable example of the obtuseness of academicians and the petty jealousies of rival mathematicians. The history of Fourier series since 1807 has shown that the academicians in this particular instance knew what was what and acted accordingly.

Lagrange's trigonometric solution did not settle all of Euler's doubts concerning the analytic representation of 'arbitrary' functions, which seems rather strange in view of Euler's catholicity in analysis. Euler himself had encountered the Fourier coefficients in 1754 and had not objected to them, possibly because they appeared in connection with the heavenly bodies of celestial mechanics. Later (1777, 1793), he obtained these coefficients by the usual process. But it remained for Fourier, beginning in 1807 and ending with his treatise on the analytic theory of heat conduction in 1822, to get the representation of an 'arbitrary' function by a trigonometric series, if not generally accepted, at least seriously considered.

In an earlier chapter we noted the influence of Fourier's analysis on the evolution of rigor. The point of interest here is the emergence of the idea of a so-called arbitrary function as an early step toward modern abstract analysis. A function not initially imagined in the Eulerian tradition as the embodiment of an algorithm was shown nevertheless to have a definite mathematical representation. Both the concept of functionality and the potentialities of analysis were enormously extended.

The boundary-value problem associated with the vibrating string with fixed end-points was thus a source of much modern classical analysis. Other physical problems have been equally prolific of pure mathematics. As a typical and important example, we shall indicate next the main steps which led to the mathematical theory of elasticity, a department of the mechanics of continuous media. Here the mathematical contributions to be noticed are the generalization of rectangular and other coordinate systems of ordinary analytic geometry, and the new functions necessitated by the corresponding differential equations.

Contributions from elasticity

The empirical bases for the mathematical theory of elasticity date from the seventeenth century. Galileo's observations (published 1638) on the fracture of loaded beams, and Hooke's (1678) on the resilience of springs, suggested physical hypotheses capable of mathematical formulation. We note that Hooke's experiments were of economic origin, being undertaken in connection with marine chronometers. In 1694, and again in 1705, James Bernoulli gave a mathematical discussion of the flexure of an elastic lamina. This presented him with the transcendental curve known as the elastica, which has been studied as an application of elliptic functions. In the decade 1741–51, Daniel Bernoulli, investigating the vibrations of a bar clamped at one end, was led to a differential equation of the fourth order, a result prophetic of Navier's general equations to come seventy years later. In all of this early work, there were frequent appeals to plausible physical hypotheses and experiments to elicit the facts. The same cannot be said for some of Euler's purely mathematical investigations.

Always the daring algorist, Euler occasionally trusted his formulas too far, and was unperturbed when they prophesied material absurdities. Nevertheless, his effort marks the next long step forward in the early development of elasticity. In 1744 he studied the elastica by the calculus of variations, and stated its equation in a form involving an elliptic integral. In 1771 he investigated the mathematics of flexible strings and elastic rods. His theory of 1778 for the flexure of vertical columns led him through a partial differential equation to Bessel coefficients, and in 1779 he derived a fourth-order equation for the transverse vibrations of an elastic rod. Euler's work on columns has retained its utility.

Lagrange's less useful contributions were of greater interest mathematically than Euler's. In 1770–1 he reduced the problem of bent springs to the evaluation of elliptic integrals, where he left it, because the required integrations "elude all known methods." Extending Euler's work, Lagrange in 1770–3 discussed columns generated by the revolution of conics about their principal axes. Using the calculus of variations, he encountered a non-linear differential equation of the second order which he could not integrate.

Up to 1800, elasticians had invented hypotheses which enabled them to give partial solutions of numerous special problems; but there had been no progress toward the general equations for the equilibrium or vibrations of an elastic solid. The mathematical contribution of this early period was its prophetic suggestiveness rather than any tangible achievement, particularly for differential equations, boundary-value problems, elliptic functions, and the calculus of variations. Anything approaching a rigorous analysis of even the special problems attacked was beyond the mathematics of the age. Of numerous details from 1800 to 1820, we shall mention only one. S. D. Poisson (1781–1840, French) in 1818 solved G. A. A. Plana's (1781–1864, Italian) partial differential equation (1815) of the fourth order for the vibrations of an elastic lamina by infinite integrals involving two arbitrary functions. There was no discussion of convergence.

In thc decade 1820–30, the modern theory of elasticity was essentially completed in broad outline by M. H. Navier (1785–1836, French), Poisson, and Cauchy. (It will be noticed that all three were French.[3] Although Poisson's name is familiar to every student of mathematical physics the world over, he has received rather scant honor from his compatriots. The works of Cauchy and several other leading French mathematicians have been collected and published at the expense of the French Government. Poisson's are still dispersed in periodicals no longer readily accessible.) Navier, who in 1820–1 obtained the general equations of motion and equilibrium for an elastic solid, is usually regarded as the founder of the modern theory. By 1821, then, elasticity had made its principal contribution to the progress of mathematics.

Its subsequent career typifies that of other divisions of classical mathematical physics. General solutions in terms of integrals were given for the equations by Poisson in 1827–9, and by Cauchy in 1830. Mathematically interesting as these solutions may be, they seem to have attracted but few engineers handicapped by a sense of practicality. For further advances, generality had to be curtailed by the manufacture of approachable problems. Some were suggested by the demands of engineering; the majority appear to realistic mathematicians to have been fabricated by the elasticians for the express purpose of furnishing skilled analysts with solvable boundary-value problems sophisticated by a dash of academic reality.

If today the engineers on a reinforced concrete dam tarried for mathematical elasticians to supply them with a reliable stress analysis, the dam would not be built. Scale models and a photoelastic laboratory do easily in a few months what no mathematician could accomplish in a working lifetime. Yet this triumph of modern efficiency owed much to the theoretical labors of Navier, Poisson, Cauchy, and many others who developed mathematical elasticity and the closely allied analysis of wave motion. Cauchy in particular, by his elaboration of the elastic-solid theory of light in the 1840's, contributed to the physics from which the photoelastic technique evolved. And although the elastic-solid theory collapsed long since, like the luminiferous ether it once supported, it has left a substantial residue in widely used mathematical methods. These are applied to innumerable enterprises of modern civilization, ranging from the design of earthquake-proof buildings to that of long-range, heavy artillery for reducing the buildings to piles of rubbish.

The importance of coordinates

With one exception to be noted presently, progress in elasticity after 1830 was of greater interest to theoretical physicists than to mathematicians. Classic work on the entire subject was done by B. de Saint-Venant (1797–1886, French), the master coordinator and extender of the theory, especially in his investigation of torsion and flexure (1855–64) and in his detailed amplification of the revised mathematical treatment (1862) due to the German mathematician R. F. A. Clebsch (1833–1872). Franz Neumann (1798–1895, German) analyzed the elasticity of crystals, and G. R. Kirchhoff (1824–1887, German) improved the general theory in certain details. The exception of greater mathematical interest was the outstanding work of G. Lamé (1795–1870, French) who, in the estimation of Gauss, was the foremost French mathematician of his generation.

A graduate of the Ecole Polytechnique and a civil and railway engineer, Lamé in 1824 went to Russia with his colleague B. P. E. Clapeyron as a consulting engineer. Lamé's interest in elasticity was aroused by the problem of iron suspension bridges. From that he quickly proceeded to master the entire theory of elasticity and to make many improvements in its mathematical treatment. Only that contribution growing out of this work which Lamé himself considered his most important advance need be noted here: *Leçons sur les coordonnées curvilignes et leurs diverses applications* (to mechanics, heat, elasticity), published in 1859. This summarized and extended all of Lamé's previous work on his useful invention of curvilinear coordinates. As most introductory texts on mathematical physics contain some account of these coordinates and their application to physical problems, we shall not describe them, but give instead a free translation of Lamé's remarkable forecast of their scientific significance.

> Should anyone find it singular that we have been able to found a Course of Mathematics on the sole concept of a system of coordinates, he may be reminded that it is precisely these systems which characterize the phases and stages of science. Without the invention of rectangular coordinates, algebra might still be where Diophantus and his commentators left it, and we should lack both the infinitesimal calculus and analytic mechanics. Without the introduction of spherical coordinates, celestial mechanics would be absolutely impossible; and without elliptic coordinates, illustrious mathematicians would have been unable to solve several important problems of this theory. . . . Subsequently the reign of general curvilinear coordinates supervened, and these alone are capable of attacking the new problems [of mathematical physics] in all their generality. Yes, this definitive epoch will arrive, but tardily: those who first recognized these new implements will have ceased to exist and will be completely forgotten—unless some archaeological mathematician revives their names. Well, what of it, provided science has advanced?

Science has progressed, as Lamé hoped, and he himself is still remembered by others than antiquaries. His early work (1839) on the conduction of heat in ellipsoids led him to the functions appropriate for physical problems connected with general ellipsoids; and these have inspired much pure mathematics in differential equations and special functions, such as the (pseudo) doubly periodic functions of the second kind, in terms of which Hermite (1872) integrated the physically important differential equation named after Lamé.

We noted in an earlier chapter that spherical and cylindrical coordinates are appropriate for certain applications of Laplace's equation where there is symmetry about a point or about a straight line. We remarked also that on transforming Laplace's equation to these special coordinate systems, the variables in the transformed equations are separable, and that one of the resulting ordinary linear differential equations of the second order thus obtained defines the Bessel functions. Each of the similarly derived ordinary differential equations defines a species of functions adaptable to the solution of boundary-value problems associated with the original, unseparated, partial differential equation. Laplace's equation, for example, was thus responsible for the several kinds of harmonic functions of Legendre, Lagrange, Laplace, and others, which arose in Newtonian celestial mechanics, and which in the nineteenth century proved indispensable in the theory of electricity and magnetism and other departments of mathematical physics.

But none of these special coordinate systems with its proper functions was suitable for the type of symmetry associated with an ellipsoid having its principal axes all unequal. To attack the general problem, Lamé invented his coordinate systems of confocal quadrics, and was led to the ellipsoidal harmonics. As already indicated, these are intimately connected with the (pseudo) doubly periodic functions, and have given rise to much interesting pure analysis, whose applications range from the higher arithmetic to aerodynamics.

Lamé's work was partly responsible for two further items of considerable interest. An immediate generalization of Lamé's differential equation (1838) was shown by Klein[4] and M. Bôcher[5] (1867–1918, U.S.A.) in 1894 to include as special cases the ordinary linear differential equations of the second order defining the functions of Bessel, Legendre, Mathieu, and several others of classical mathematical physics. Thus a unified treatment of all was achieved.

The second item is of far greater significance. Since the abandonment of the ether except by a few who still find its picturesque contradictions useful, there is no uniquely distinguished coordinate system in cosmology. Gaussian coordinates provide the necessary degree of indifference, and these are akin to Lamé's curvilinears. Conversely, the tensor calculus provides the simplest algorithm for transforming the equations of Laplace, those of elasticity, and others in classical physics, to various systems of curvilinear coordinates. Lamé's insistence on the importance of coordinates has been justified in modern physics.[6]

In dwelling at some length on Lamé's contributions, we have not intended to exhibit either them or the theory of elasticity as the only physics of the nineteenth century responsible for memorable accessions to mathematics. Lamé's work is, however, typical of much: the impulse from science was directly responsible for extensive tracts of modern pure mathematics whose scientific affiliations subsequently became at best extremely tenuous.

Toward functional analysis

Another episode of Lamé's times illustrates even more forcibly the same trend from mathematical physics to pure mathematics: integral equations began to emerge. As this led to developments of capital importance in the evolution of both pure and applied analysis, also of geometry, we shall describe it in some detail. The ultimate source was again in the boundary-value problems of physics.

An algebraic analogy provides one approach to the history. Those values of x for which the function $f(x)$ vanishes, are called the zeros of $f(x)$. In an obvious sense, an irreducible polynomial is *characterized* by its zeros. J. C. F. Sturm (1803–1855, Swiss) in 1836, while analyzing the flow of heat in an inhomogeneous rod, was led to an ordinary differential equation of the second order, whose solutions defined a species of functions which were characterized by their zeros in a certain interval. A simple example of such a differential equation is that whose linearly independent solutions are sin nx, cos nx. In many boundary-value problems of physics, as noted in an earlier chapter, it is required to expand an 'arbitrary' function in a series of functions constructed from the solutions of certain ordinary differential equations, such as those appearing on the separation of variables in certain linear partial differential equations by a proper choice of coordinates. Fourier's trigonometric expansion of an 'arbitrary' function is an illustration. Sturm's discovery inaugurated a new epoch in this expansion problem of mathematical physics. Incidentally, a famous by-product of Sturm's researches was his theorem on the separation of the roots of an algebraic equation. With its hundred per cent efficiency, Sturm's theorem rendered the earlier attempt (1796) of Fourier, also unjustifiably associated in some textbooks with the name of F. D. Budan (French), partly obsolete.

A decade earlier (1823, 1826), Abel had made the first significant progress in a theory which was to join with Sturm's in a unification of certain scientifically important types of boundary-value problems. It is customary to attribute the origin of integral equations as a distinct department of analysis to Abel's solution and subsequent generalization of the classical problem of the tautochrone: a particle starting from rest at a point P in a vertical place, and moving under gravity in the plane, is to reach the lowest point O of a frictionless curve joining P, O in the time $f(h)$, where h is the height of P above O. The equation of the curve is required. If $f(h)$ is constant, the curve is the classical tautochrone—a cycloid. We need not reproduce the equations which Abel found and which he solved. His great merit was in recognizing that the more general of his equations, today called an integral equation of Volterra's first kind, presented an essentially new problem in analysis. An integral equation is one in which the dependent variable $u(\equiv u(x)$, x being the independent variable) occurs at least once under a definite integral sign.

Integral equations had appeared before Abel: Laplace (1782) encountered them in solving difference and differential equations; and Fourier, in the result now called his inversion formula, gave (1820, 1822) what is probably the first[7] solution of an integral equation. But no analyst before Abel recognized that integral equations presented a basically novel problem in analysis. We shall return presently to the connection between such equations and the boundary-value problems of physics.

A general linear integral equation is

$$f(x) = g(x)u(x) + \lambda \int_a^b K(x, t)u(t)dt,$$

in which: $K(x, t)$, called the kernel, is a known function of x, t; $f(x)$, $g(x)$ are known functions of x; a, b are known functions of x, the case of either or both constant being included; λ is either an absolute constant or a parameter. It is required to solve the equation for u as a function of the independent variable x. Four special cases, being those of widest utility, have been minutely investigated. Volterra's equation (1896) of the first kind has $g(x) \equiv 0$, $\lambda = 1$; his equation of the second kind has $g(x) \equiv 1$; and in both kinds, $a = 0$, $b = x$. Fredholm's (1900, 1903) of the first and second kinds differ from Volterra's only in that the limits a, b of the definite integral are both constant. If either or both limits become infinite, or if the kernel becomes infinite for one or more points in the given interval a, b, the equation is said to be singular. Abel's equation (1826) was singular in the latter respect. It will be noted that, in the linear equations described, the unknown u enters only to the first power. A generalization of this situation has also been discussed,

$$u(x) = f(x) + \lambda \int_a^b F[x, t, u(t)]dt,$$

where F is a known function. There are also extensions to several variables and the corresponding multiple integrals, also to systems of integral equations. The theory of a single linear equation occupies the central position in both the analysis and its applications.

The scientific significance of integral equations, noted by Liouville in the 1830's but first elaborated in 1904 by Hilbert, is that in many important instances one integral equation is equivalent analytically to a differential equation together with its boundary conditions. The solution of an integral equation gives that of a boundary-value problem in such instances. The advantage of reformulating certain physical problems in terms of integral rather than differential equations is that whereas an increase in the number of independent variables (say from two dimensions to three in Laplace's equation) greatly increases the mathematical difficulties in the differential formulation, there is no serious increase[8] in the corresponding integral reformulation. The integral equation precipitates once for all the essential difficulties[9] of the general physical problem instead of isolating, more or less fortuitously, those peculiar to special cases. The theory of integral equations, when applicable at all, also provides a uniform method for constructing the functions demanded for the expansions of mathematical physics by Sturm's method and its developments.

For reasons into which we need not enter here, the functions introduced by Sturm are called oscillating. Following Sturm, Liouville in 1837 was led to a linear integral equation, which he solved by the method (iteration) of successive substitutions. Liouville's problem was that of finding those solutions, if any, of a linear differential equation of the second order which assume preassigned values for given values of the independent variable. He was thus concerned with Sturm's oscillating functions, and it is customary to name the resulting theory after both men. Liouville's method of solving his integral equation, giving the unknown $u(x)$ as a power series in λ with functions of x as coefficients, was applied (1877) by C. Neumann to Dirichlet's problem in potential theory. The solutions thus obtained converge for $|\lambda| < c$, where c is a finite constant. Volterra (1896), Poincaré, and others in the 1890's also used this method.

The decisive advance was made in 1900 and amplified in 1903 by I. Fredholm (1866–1927, Swedish), who exhibited solutions for his equations as quotients of power series in λ, the denominator $D(\lambda)$ being independent of x. Both numerator and denominator converge for all finite λ; if $D(\lambda) = 0$, there is a solution only in exceptional cases, given by the method. It had been observed earlier by Volterra that the integral equation of Fredholm's second kind is obtainable as the limiting form of a system of linear algebraic equations. The peculiar significance of this detail will appear later. Fredholm carried this through and justified the solution by obviating the convergence proof, given by Hilbert in 1904, for this particular method. Two years later E. Schmidt proved Hilbert's results without appeal to a limiting process. Thus by 1906 the modern theory of integral equations was successfully launched, after eighty years of trial. Several distinguished analysts of the nineteenth century had contributed to the final success; those mentioned mark well-defined stages in the development of integral equations from puzzling curiosities to an understood and powerful discipline.

Shortly after the publication (1903) of Fredholm's fundamental paper, mathematicians of most civilized nations, including the U.S.A., swarmed into the new field of analysis thus unexpectedly opened up. The main highways of the theory were rapidly laid down, especially by E. Schmidt (German) in 1907, and from 1904 on by Hilbert, who in 1912 collected his contributions in a classic on the general theory of linear integral equations.[10] The outcome of less than a decade's intensive effort was a unified theory of the boundary-value problems of classical mathematical physics, and a practicable method for constructing the special functions associated with such problems.

To indicate briefly an important feature of this advance, we recall that it is the orthogonality of the sine and cosine which accounts for their applicability to the expansion of an 'arbitrary' function in a trigonometric series. Orthogonality is also the source of utility of the Legendre functions, and similarly for others of classical physics. The continuous functions $f_j(x)$, $j = 1, 2, \ldots$, are said to be orthogonal in the interval $[a, b]$ if

$$\int_a^b f_j(x)f_k(x)dx = 0, \qquad j \neq k;$$

if in addition, $\int_a^b [f_j(x)]^2dx = 1$, the functions are called normal. The first systematic investigation of orthogonal functions was by an almost forgotten analyst, R. Murphy (1806–1843, Irish), in 1833–5, shortly before Sturm and Liouville began their researches on differential equations. In the theory of integral equations, orthogonal functions[11] fit naturally into the systematic development of the general expansions connected with boundary-value problems.

Classical physics, hereditary phenomena, and linearity

Before considering the significance of the extraordinarily rapid development just sketched for the evolution of analysis as a whole, we shall glance back over what was accomplished in about a century.

The Sturm-Liouville theory of the 1830's was the first step toward a unified treatment of the numerous boundary-value problems and their solutions that had been accumulating in applied mathematics since the early eighteenth century. Integral equations entered significantly when analysis was sufficiently advanced to solve the difficult problems of existence (convergence of formal solutions), and became thoroughly effective only when Fredholm's spark, illuminating a chaos of details, showed a way through disorder to order. This signal advance might have come a decade earlier than it did; for Poincaré in 1894 had used the characteristic functions of the expansions in the Hilbert-Schmidt method. Thus it would seem that the algorist has his place no less than the analyst in modern mathematics. For although Fredholm was primarily an analyst, his long stride forward took off from an algorithm that Euler himself might have invented.

It has been sufficiently emphasized that the physical problems reduced to order by the theory of integral equations were in the *classical* tradition of mathematical physics. Fredholm himself (1905) applied his methods to "the solution of a fundamental problem of elasticity." But physics in 1905 was already in possession of the special theory of relativity, and had been familiar with Planck's form of the quantum theory since the turn of the century. Unless the reformulated mathematics of classical physics should prove competent to handle the new physics that was so shortly to follow, it would find itself as classical and as uninteresting to an advancing science as the luminiferous ether.

One of the most dramatic anticipations in the history of mathematical physics was the publication in 1924 of the *Methoden der mathematischen Physik* (Bd. 1) by R. Courant and Hilbert, in which the modern methods of linear transformations bilinear and quadratic forms, expansions of arbitrary functions, integral equations, and the calculus of variations were complected into a single implement of tremendous power for application to physics as it then was. The advance over previous classics of applicable mathematics marked by the *Methoden* is best appreciated by comparing this work with the earlier editions of Riemann-Weber's *Die partiellen differential-Gleichungen der mathematischen Physik*. But unless this systematized and more powerful analysis were more adaptable to an explosively expanding physics, it was out of date the day it was published.

With the advent of wave mechanics in the modern quantum theory only two years after the *Methoden* appeared, the more powerful analysis of modern mathematical physics came into its own. The book might have been written expressly for physicists whose major activity for more than a decade was to be quantum mechanics and its innumerable applications. It was not so written; neither author in 1924 had the slightest inkling of Schrödinger's equation of 1926 or of the physics which it inspired.

Incidentally, quantum mechanics redeemed integral equations from academic immortality by putting their offspring to productive scientific work. For although integral equations themselves appeared explicitly but seldom in the new physics, the characteristic functions of the Hilbert-Schmidt theory became a commonplace. So successful was the *Methoden* that the weird hybrids 'eigenvalue' and 'eigenfunction' were begotten to oust their all-English equivalents that had served efficiently for twenty years or more. But working physicists had been too busy to notice them. A fatalist might believe that this triumph of mathematical foresight was preordained; but there is a less mystical explanation which will occur to any skeptic on a moment's thought. A detail, but an important one, in the entire anticipation was the famous theorem (1885–6) of Lord Kelvin and Hadamard on the minimum of the absolute value of a determinant. Without this, the now classic development of Fredholm's theory would have been at least delayed, if not impossible.

Concurrently with integral equations, another, more complicated type of equation entered applied mathematics in the 1880's. The unknown in a given equation may appear at least once under an integral sign and at least once under a sign of derivation; such equations are called integro-differential[12] by Volterra, who inaugurated their theory in 1890.

In passing, it may be noted that Volterra's boldly creative work in functional analysis, from its beginning in 1887 to its maturity in the 1930's, ranks him with the most imaginative inventors in the history of analysis. He was constantly guided by scientific problems. As hinted in connection with Fredholm, analysis cannot live by rigor alone. Nor are there any historical grounds for anticipating that successive abstractions are to be the ultimate residue of analysis in mathematics. New inventions are constantly demanded to maintain vitality and to provide both rigor and abstraction with substance for their continued nourishment. Although Volterra's prolific contributions have not been accused of a lack of rigor, their peculiar importance for analysis after the age of Weierstrass is in their abundance of new ideas.

Integro-differential equations made possible a mathematical attack on such phenomena as magnetic and electric hysteresis, in which the state of a physical system depends not only upon its immediately preceding state, but also upon all its previous states. Volterra called such phenomena, conditioned by the entire previous history of the system concerned, hereditary. Familiar examples in elasticity will occur to anyone who has ever bent or twisted a wire.

Integral equations, integro-differential equations, and tracts of the calculus of variations are included in the yet more general theory of functionals with its related equations, originated by Volterra and extensively developed by him and other leading analysts of the twentieth century. This will be our concern presently. For the moment we note a most important limitation, so far ignored, of the classical analysis originating in mathematical physics. Much the greater part of applied analysis is subject to this limitation. Should the implied restriction be an oversimplification of physical science, there is at least the possibility that by the year 2000 the analysis which in 1945 was indispensable in science will be as antiquated and as impotent scientifically as the Babylonian mathematics of 2000 B.C. Science moves faster than it did four thousand years ago; and instead of achieving the appalling immortality of astrology and numerology, a false world picture is now forgotten in a decade.

Practically all of classical mathematical physics has evolved from the hypothesis of linearity. For a precise statement of what linearity means in this connection, we must refer to texts on physics; the following is a rough description by means of an example. The assumption of linearity is also called, in physical optics, the principle of superposition. As noted earlier, the principle goes back to Daniel Bernoulli.[13]

If the properties of an elastic medium are such that the stresses are linear in the vector displacements or in their space-derivatives, the disturbance at any point P of the medium is obtainable by superposing the disturbances reaching P from all centers of disturbance within a certain region. Thus the total effect of several displacements is obtainable by vector addition of the individual displacements. (This may hold only to a first approximation.) More concisely, the displacements are additive. In a similar sense, magnetic fields and Newtonian gravitational fields are additive or superposable. Einsteinian gravitational fields are not additive.

When the principle of superposition is valid for a physical situation which can be translated into a differential equation, the equation is linear. The mathematical equivalent of the principle appears in the elementary fact that if U, V are solutions of a linear differential equation, and a, b are constants, then $aU + bV$ is also a solution. This is at the root of the solution of boundary-value problems by expansions of arbitrary functions in series of suitably defined special functions.[14] In short, but for the principle of superposition, a vast tract of mathematical physics would be radically and at present unimaginably different from what it is. Should the principle fail to accord with the refinements of modern experience, the basic equations of mathematical physics cannot be linear, except possibly as a first and inadequate approximation; and all the ingenious analysis which has evolved from the hypothesis of linearity is at best a first approximation to the applicable mathematics of the future.

Against such a possibility it may be argued that the principle of superposition worked admirably from Daniel Bernoulli to Dirac, and therefore it must be in some sense[15] a correct description of certain physical phenomena. A similar argument validates the use of classical, statistical, and wave mechanics, each in its own domain of approximation to observable events. So far as linearity is concerned, the point of mathematical interest is not that it has led to remarkably accurate *quantitative* results, but whether it is *qualitatively* a sufficient description of living and growing science. The equations of general relativity are already non-linear; the fundamental equation (Schrödinger's) of the quantum theory is linear.

It has been suggested by at least one competent physicist that the basic equations of physics must be non-linear, and that mathematical physics will have to be done over again. Should this suggestion (Einstein's) prevail, the analysts of the next century will probably face harder problems than any their predecessors solved so successfully, and the outcome may well be a mathematics totally different from any now known.

The economic motive in the development of some departments of both pure and applied mathematics is by now rather a well-worn theme. Many of the conclusions seem likely to remain controversial for a long time. But there is one aspect of economic mathematics, germane to the present topic, which is clear.

If it is true that even the most idealistic of wars are clouded to a slight extent by materialistic motives, such as the lust for territorial or other loot, then to that extent military mathematics is motivated by economic impulses. And since time is often the essence of the contract between hostilities and gain or glory, it as often follows that mere expense is a negligible consideration in the mass production of mathematics for military purposes.

The problems posed by modern warfare are usually urgent and frequently hard—much harder, in fact, than the over-simplified idealizations of concrete situations imagined by mathematicians in times of peace and solved at their leisure. Again, the real problems of kill or be killed are but seldom as elegant mathematically as those of lethargic peace, demanding for their applicable solution a prodigious amount of numerical computation to obtain a close enough approximation to what may be the ungarbled facts. Few mathematicians in peacetime would find sufficient 'joy through work' in such repulsive drudgery to undertake it voluntarily. But spurred by the urgency of war, their only course is to proceed as fast as they can with the uncongenial necessity. Compulsory cooperation, the relaxation of financial restraints in such matters as the employment of regiments of robots (mechanical and human), and the construction of intricate new calculating machines to do the less skilled labor multiply the efficiency of the individual mathematician (and sometimes also his ego) by a factor which may be as high as ten—he accomplishes about ten times as much, when given the proper tools in the right kind of factory, as he would if working in polite penury and scholarly solitude. The desperate problems of survive or perish receive at least partial solutions. Concurrently, new mathematical vistas open up to distant regions which might never have been imagined, much less explored, had it not been for the compulsions of war. Mathematics and casualties mount simultaneously, and expense is not far behind.

In the second world war mathematical skill was at a much higher premium than in the first. And as many of the problems demanding urgent consideration dealt with fluids hampered by viscosities which could not be ignored, or with equally awkward facts in other manifestations of brute matter in the real world, the equations to be solved were seldom those of the classical tradition. In particular, non-linear equations received more attention in four or five years of war than they normally would have in half a century of peace. From all the masses of computation to meet immediate demands, at least the more obvious characteristics of a few types of equations may have emerged. If so, these types may guide the study of non-linear equations when there shall be an opportunity to think of other things than the next war, much as the problems of classical mechanics and Newtonian gravitation determined the pattern of a vast tract of the classical theory of linear differential equations, both ordinary and partial. Should this come about, the mathematical historian of the future may find in the second world war the same kind of good that realistic historians of technics have detected in the first. For had it not been for the demands of military aviation in the first world war, civil aviation might still be the hazardous sport of a few reckless amateurs that it was in 1913, and certainly it would have been totally inadequate for its duties and privileges in the second world war. So it may be, realistically, with the mathematical gains directly due to the second world war. But, as pointed out long ago by Charles Lamb in his biography of Ho-ti, it is not obligatory to rely exclusively on this particular realism for progress in aviation, mathematics, or anything else.

Generalized functionality

All of this continued activity in mathematical physics naturally left a substantial residue in pure mathematics. What at present appear to be the contributions most likely to endure for some decades or longer are functional analysis, general analysis, abstract spaces, and several specialized subdivisions of these inclusive theories. The major development in each took place in the twentieth century, although functional analysis might have been extensively cultivated in the 1890's had Volterra's pioneering work of 1887 been more widely appreciated.

Only the most summary indication of the nature of these fairly recent acquisitions can be given here. A bibliography of functionals alone to 1930 contains 540 titles, and there have been numerous contributions since. The literature of abstract spaces is even more extensive and is still (1945) expanding rapidly. General or abstract analysis also is flourishing. The partly scientific origin of all is clear, especially when their evolution is traced back through the calculus of variations and the connections between integral equations and boundary-value problems. Compared to the impulse from science, the influence of older problems in pure mathematics on the development of functionals, abstract spaces, and general analysis, now themselves divisions of pure mathematics, was all but negligible, with the outstanding exception of Cantor's theory in relation to abstract spaces. Before Cantorism, there appear to have been only two minor 'pure' sources of functional analysis worth noting, the second of which might also be assigned to applied mathematics.

Euler and others in the eighteenth century proposed geometrical and mechanical problems leading to equations in which the unknowns, as in the calculus of variations, are functions. For example, Euler's problem of finding the equation of the curve such that the square of the normal at any point exceeds the square of the corresponding ordinate by a constant C leads to an equation containing both an unknown function and its derivative. This is a functional ('functional' is here an adjective) equation, to be solved for the function. Again, the addition theorems for the circular functions suggest the problem of finding the 'most general' functions satisfying them. In this order of ideas, Weierstrass gave a precise definition of an algebraic addition theorem, and proved that elliptic functions and their degenerate cases are the only functions of one variable, subject to obvious restrictions of continuity, uniformity, etc., possessing such an addition theorem.

All through the first half of the nineteenth century, functional equations attracted analysts with a taste for ingenious devices. This work was characterized by a lack of precision and a failure to appreciate the very serious difficulties involved. It was as naively ambitious and as inconclusive as Euler's inordinately ambitious attack on diophantine analysis. There seems to have been no satisfactory investigation of a functional equation prior to Cauchy's (1832) study[16] of the extremely simple and important[17] example $f(x + y) = f(x) + f(y)$.

Another impulse in the same direction was Liouville's theory (1832) of fractional differentiation.[18] Many a beginner in the calculus, having grasped d^ny/dx^n for positive integers n, has asked himself what interpretation might be assigned to this symbol when n is not a positive integer. Liouville's ingenious solution came several decades too early, finding its proper place in analysis only in the twentieth century.

The second possible source in pure mathematics of modern functional analysis is the theory of probability. Laplace and many others in the eighteenth century translated questions on probability into difference equations, in which the unknown is a function and the range of the independent variable is the class of non-negative integers. But to ascribe the origin of functional analysis to such nebulous beginnings would be a brilliant example of *post hoc, ergo propter hoc*. The creators of the new analysis proceeded consciously on quite different lines.

Functional analysis originated with Volterra in 1887. In his earlier work, Volterra used the terminology "functions depending on other functions," and the special case of "functions of lines," for what were subsequently called functionals ('functional' is here a noun) by Hadamard, one of the foremost developers of the modern theory. The simplest instance of a functional is generated by a function $x(t)$ defined for all values of t in a given interval $[a, b]$: $F[x(t)]$ is called a functional of $x(t)$ when its value depends on all the values assumed by $x(t)$ in $[a, b]$. The special case $x(t) = t$ gives Dirichlet's definition of a function. Volterra's functionals are thus an infinitely wide generalization of the functions of classical analysis.

Instances of functionals had long been familiar in the calculus of variations when Volterra isolated the general concept and began its systematic elaboration. Even earlier, functionals had appeared in analysis, but had not been recognized as such, at the very beginning of the integral calculus. Thus $\int_a^b x(t)dt$, defined only in the domain of integrable functions $x(t)$, is an instance of the functional $F[x(t)]$ in the particular 'functional field' of integrable functions. Generally, $F[x(t)]$ is defined only when $x(t)$ is restricted to range over a prescribed class of functions. Ranges of particular interest are those of all analytic functions, or of the class of all continuous functions, each of which defines a 'space' of a denumerable infinity of dimensions.[19]

A powerful device of generalization for passing from classical algebra and analysis to the corresponding situations in functionals is that of replacing finite summations, in which the index or independent variable t ranges over the integers $1, \ldots, n$, by definite integrations wherein the range of t is an interval $[a, b]$ in which $f(t)$, $g(t)$ are continuous. Thus $\sum_{t=1}^{n} f(t)g(t)$ would be replaced by $\int_a^b f(t)g(t)dt$. An example of this technique was noted in connection with the passage from a system of linear algebraic equations to an integral equation; here we see the broader significance of that special case. The difficulty, of course, is to justify the limiting processes. Volterra has described the functional calculus as an instance of that transition from the finite to the infinite which characterized much of mathematics since the close of the nineteenth century.

The outcome of Volterra's inspiration was the functional calculus, developed by numerous analysts with unprecedented rapidity in the first third of the twentieth century. Possibly the most active schools were the Italian, French, German, and American. For an impartial estimate of their several contributions, we must refer to Volterra's exposition.[12] Among the leading contributors up to 1930, Volterra, S. Pincherle (1853–1936, Italian), Hadamard, Fréchet, Tonelli, P. Levy (French), and G. C. Evans (1887–, U.S.A.) may be mentioned.

The theory of functionals generalizes and extends much of classical analysis. Its applications include integral equations, parts of the calculus of variations, a generalization of analytic functions, integro-differential equations, generalizations of classic differentials and derivatives, functional-derivative equations, functional invariants; and, on the physical side, functional dynamics and the theory of hereditary phenomena.

A recent application is to mathematical economics; but it is still too early to predict whether expertness in the functional calculus will enable society as a whole to avoid depressions. It may be recalled in this connection how, shortly after the publication (1910–13) of *Principia mathematica*, optimistic mathematicians prophesied that symbolic logic would greatly assist the justices of the United States Supreme Court in their tortuous deliberations. Deadly serious, the prophets have not yet learned to laugh in the wilderness.

A less immodest proposal advocated the application of functional analysis to human bionomics. In this direction, Volterra (1927) had a somewhat disturbing success in the corresponding application to ravenous populations, say sharks of two different species, living together and maintaining themselves in stable economic equilibrium for indefinitely long periods by devouring one another. He found, as was humanly verified in the great depression in 1929, that the numbers of individuals in the two species fluctuate indefinitely. The feature of social interest in such applications is that the appropriate analysis is vastly more subtle and more complicated than any that the classical economists of the nineteenth century believed would suffice. To be a mathematical economist in the twentieth century demands a mastery of modern analysis. The younger generation declares that mathematical economics is one of the few divisions of applied mathematics where most of the classics might profitably be consigned to the flames.

General analysis, abstract spaces

The same passage from the finite to the infinite that characterized functional analysis appeared also in geometry with Hilbert's work on integral equations. The generalization of the algebraic theory of quadratic and bilinear forms (also of Hermitian forms) from three to any finite number n of variables was a favorite topic with many of the nineteenth-century algebraists. The resulting algebra of real forms found a ready interpretation in the geometry of hyperquadrics in n-space. In fact, familiar problems of the case $n = 3$, such as the determination of the principal axes of a quadric, suggested rather obvious things to do in the general finite case. Much of this work proved useful in analytic mechanics. What seems to be the first significant hint of an extension to space of an infinity of dimensions occurs in the researches (1877) of G. W. Hill (1838–1914, U.S.A.) on the lunar theory, in which appeared infinite systems of linear algebraic equations in an infinity of unknowns. The accompanying theory of infinite determinants was rigorized by Poincare in 1886, and further developed from 1896 on by H. von Koch (1870–1924, Scandinavian) and many others.

The generalization of quadratic and bilinear forms from three to any finite number of variables is an almost trivial project for a competent algebraist. With the extension to an infinity of variables, convergence must be considered[20] to validate the formal algebra, and the resulting analytic problem is trivial for nobody. Hilbert laid down the foundations of the theory for an infinity of variables in 1906. As in the finite algebraic generalization, the analytic extension to a denumerable infinity of variables is conveniently and suggestively described in spacial imagery. The geometry so defined is that of classical Hilbert space.[21] Its first applications were by Hilbert and his followers to integral equations; but it was early perceived that the analysis and geometry thus created were tending toward more inclusive theories.

For accounts of Hilbert space and its physical significance we refer to the works cited. From these the partly scientific origin of the theory is evident. For a sense in which classical Hilbert space is an ultimate in dimensional generalizations of geometry, we must also refer to a technical exposition,[22] where the applications to the quantum theory are developed, and it is shown by an example that in the sense mentioned "there is no essential distinction between spaces of a denumerable and of a non-denumerable infinitude of dimensions."

Our concern here is the outcome in pure mathematics for which this entire movement was partly responsible. A feature of great interest is the return in theories of general analysis, abstract spaces, and topology to the Euclidean methodology. There is nothing remarkable in this. As the final aim was inclusive abstraction, no other approach was possible.

Beginning in 1905–6 and continuing to the early 1920's, two widely divergent types of general analysis, due respectively to E. H. Moore and Fréchet, shared the new field. Moore does not seem to have relied much on spacial imagery; Fréchet, the founder of the theory of abstract spaces, appears to have been led to some of his most fruitful ideas in general analysis by skillfully abstracting the commonest notions of elementary geometry and those of the classical theory of functions of real variables as it had developed since Cantor.

If this is a just conclusion, it indicates that much remains to be done in other departments by a similar process of abstraction and generalization. But routine constructions and logical analyses of postulate systems are not enough if such generalizations are to yield more than trivial restatements or useful simplifications of known theories. The prime requisite is an instinct for what characteristics of a given situation are worth abstracting in the first place; and for this there are no rules. Examples of picking the 'right' thing were noted in connection with generalizations of absolute values. Those who have made happy choices in the art of abstraction agree that success comes only after many rejections.

What might be called the historical necessity for a general analysis of some sort in the early twentieth century was indicated by Moore in his lectures of 1906:

> Especially during the last decade the study of Integral Equations has brought to light numerous *analogies* between the n-fold algebra of real n-dimensional space and the theory of continuous functions varying over a finite interval of the real number system and the theory of certain types of functions of infinitely many variables. These analogies have their root in the classic analogy of a definite integral to an algebraic sum.
>
> We lay down a fundamental principle of generalization by abstraction:
>
> *The existence of analogies between central features of various theories implies the existence of a general theory which underlies the particular theories and unifies them with respect to those central features* . . .
>
> From the linear continuum with its infinite variety of functions and corresponding singularities G. Cantor developed his *theory of classes of points* (*Punktmengenlehre*) with the notions: limit-point, derived class, closed class, perfect class, etc., and his theory of *classes in general* (*allgemeine Mengenlehre*) with the notions: cardinal number, ordinal number, order-type, etc. These theories of Cantor are permeating Modern Mathematics. Thus there is a theory of functions on point-sets . . . while the arithmetic of cardinal numbers and the algebra and function theory of ordinal numbers are under development.
>
> Less technical generalizations or analogues of the continuous real variable occur throughout the various doctrines and applications of Analysis. A function of several variables is a function of a single multipartite variable; a distribution of potential or a field of force is a function of position on a curve or surface or region; the value of the definite integral of the Calculus of Variations is a function of the variable function entering the definite integral; a curvilinear integral is a function of the path of integration; a functional operation is a function of the argument function of function; etc., etc., . . .
>
> . . . there is [also] extension by direct generalization. Finite generalization, from the case $n = 1$ to the case $n = n$, occurs throughout Analysis. . . . The theory [Hilbert's] of functions of a denumerable infinity of variables is another step in this direction.
>
> We notice a more general theory dating from the year 1906. Recognizing the fundamental role played by the notion *limit-element* (number, point, function, curve, etc.) in various special doctrines, M. Fréchet has given,[23] with extensive applications, an abstract generalization of a considerable part of Cantor's theory of classes of points and the theory of continuous functions on classes of points. Fréchet considers a general class P of elements p with the notion *limit* defined for sequences of elements. The nature of the elements p is not specified [that is, the elements are completely general or abstract]; the notion *limit* is not explicitly defined; it is postulated as defined subject to specified conditions. For particular applications explicit definitions satisfying the conditions are given.

The earlier part of this quotation will recall several of those divisions of modern mathematics which were selected for description in previous chapters as typifying the broad development since 1800. Before continuing with general analysis, we note the reference to Cantor's theories, and recall that Moore was describing mathematics as it was in 1906, a third of a century ago as this is written.

The mere project of a general analysis has a certain sublimity about it that must appeal to any mathematician whose mind has emerged from the nineteenth century. But the mathematics of 1906 and that of 1945 are not exactly the same thing. According to taste or temperament, we may ignore the shift in basic interest as being of no importance for mathematics as practiced by its technicians; or we may agree with Galileo that "Still it moves" is as valid for mathematics as it is for our planet. Moore's motto was "Sufficient unto the day is the rigor thereof," a maxim with which the majority of professionals agree unreservedly. However, there is a difference of several days between 1906 and 1945; and to the critical school, the sublime projects of 1906 are somewhat reminiscent of Lucifer and Faust. Hilbert in the 1920's emphasized that, in his opinion, the foremost problems facing the oncoming generation of mathematicians were in the foundations of mathematics. Disregarding this for the present, we shall continue with what the great majority of professional mathematicians regard as likely to be of some enduring worth in the program of abstraction and generalization initiated by Fréchet and Moore, and developed since 1906 into theories of vast extent by scores of industrious workers.

Moore practically abandoned his first theory of 1906–15 in favor of a less postulational and more constructive approach in his second general analysis of 1915–22. One goal appears to have been an analysis that would yield the special theories cited in the above quotation as instances. The first theory concerned functions of a general variable on a general range: "this *general* is the *true general*, embracing every well-defined particular case of variable and range. . . . We define General Analysis more precisely as *the theory of systems of functions, functional operations, etc., involving at least one general variable on a general range.*"

The expansion theorems[24] in connection with integral equations, especially as developed by Hilbert in 1904–10, and the reformulation in 1909 of the theory of quadratic forms in an infinity of variables by E. Hellinger (German) demanded increasingly complicated modifications in the postulates of the first theory. Beginning in 1915, Moore recast his analysis to accommodate Hilbert's quadratic forms and the new type of integral introduced by Hellinger. This was accomplished at the expense of the Fredholm theory of integral equations; but there was a consequent gain in contacts with Lebesgue integration. By 1925, the improved theory had been broadened by a basic extension of the algebra of finite matrices to include the case where multiplication of elements is not necessarily commutative. The reconstructed theory retained the general range; the values of the functions were in a number system simply isomorphic with either real numbers, complex numbers, or quaternions with real coefficients.

Owing partly to the expressive but extremely condensed and novel notation in which Moore presented his analysis, his system made but little headway beyond his immediate circle. Publication also was delayed by successive modifications necessitated by a continually expanding analysis that, so far, has refused to surrender all its riches to any single generalization. It appears to be beyond human foresight to devise any mathematical discipline so powerful that it can dominate the future. This, however, is not the purpose of abstract theories. If in any epoch an abstract theory is sufficiently general to exhibit masses of disparate details as instances of underlying principles, it has fulfilled its function.

As pointed out by Moore in 1906, Fréchet's approach from its inception was more abstract than his own, in the wider generality of its basic elements. In its subsequent career Fréchet's theory,[25] consistently presented in a symbolism familiar to most mathematicians, also appears to have been less concerned with attempts to include new technicalities of special theories as they arose. It seems to have evolved from its own innate compulsions, implanted in it by a singularly fortunate choice of initial abstractions, such as those of 'distance' and 'neighborhood.'

The avowed objectives of Fréchet's theory in 1928 were the acquisition of new results and a unification of classic theories of functions and functional analysis. It just missed modern topology. The measure in which both of these objectives were attained can be estimated only by noting the scope of the subjects discussed in typical books devoted at least partly to abstract space and its numerous applications, and by scanning the literature since 1930.

It would leave a spurious impression of the origins of general analysis and abstract space if the scientific impulse alone were given the credit. The purest of pure mathematics, Cantor's theory of classes (*Mengenlehre*), shared equally and perhaps more directly in the evolution. The basic ideas of class-inclusion as in elementary mathematical logic (Boolean algebra) that appear frequently in abstract theories entered mainly through Cantor's point sets. The suggestive geometric terminology implied in calling a class of objects of unspecified nature, subject to certain postulates, a 'space,' as in the systems of Fréchet and Banach, appears to be due to F. Hausdorff[26] who, in his work on Mengenlehre (1914, 1917), referred to a 'metric space' in this connection.

This technique of 'geometrizing' classes proved extremely fruitful. To mention only one intensively cultivated field, S. Banach (1892–1941, Polish) in 1922 developed a postulational formulation of the system since called Banach space. This is concerned with a class of at least two completely arbitrary elements, the class being closed under addition and under multiplication by real numbers subject to certain postulates. Banach also postulated an 'absolute value' for which the triangular inequality holds. Among the numerous instances, that of the class of continuous functions had already been discussed in 1918 by F. Riesz (1880–, Hungarian). Other instances are complex numbers, vectors, quaternions, and Grassmann's forms. But as multiplication of elements in abstract space is not defined, the classical algebra of quaternions, for example, is unobtainable in the theory. Where abstract space does apply, the general theorems provide unified and simpler proofs in the special instances.

One concept, that of linear operations, which can be traced back through numerous sporadic appearances as far as Brisson in 1808, has proved particularly useful when abstracted. Let S, S_1 be two abstract spaces for which an associative addition and a zero element are defined; and let $y = U(x)$ be a function (operation, transformation) which makes an element y of S_1 correspond to an element x of S, for every x in S. If for every x_1, x_2 in S, $U(x_1 + x_2) = U(x_1) + U(x_2)$, the operation $U(x)$ is said to be additive. If also S, S_1 are metric (spaces in which a 'distance' relation is defined for each pair of elements), continuous operations $U(x)$ are definable. An operation which is both additive and continuous is called linear. Among the numerous applications of linear operations are many to modern theories of integration.

The fertility of Cantor's and Fréchet's initial conceptions seems inexhaustible. Thus attempts to resolve Cantor's seeming paradox that the points of a plane can be put in one-one correspondence with those of a straight line were partly responsible for the modern theory of dimensionality. The coordinate systems of analytic geometry had suggested an immediate generalization to space of n dimensions, for n any positive integer; and classical Hilbert space extended this to a denumerable infinity of dimensions. But Poincaré in 1912, emphasizing that the dimensionality of a space needed clarification, proposed a definition, later criticized adversely by L. E. J. Brouwer (1882–, Dutch). A satisfactory definition[27] was finally devised in 1921 by K. Menger (1902–, Austrian; U.S.A.). This detail alone led to an extensive theory.[27] Thus does modern mathematics breed upon itself with almost nightmarish fecundity. A 'space' may now have any dimensionality in the range of real numbers from -1 to $+\infty$.

As a last instance of what has issued from the modern postulational method, we cite the abstract topologic spaces which evolved from Hausdorff's work of 1914. A topologic space, we recall, is a class for which the neighborhood of any element in the class is defined—an abstraction of the familiar neighborhoods in the theories of functions of real and complex variables. Topologic spaces are thus the mathematical intersection of the theories that developed from the original inventions of Cantor and Fréchet.

In the development of abstract spaces we have a typical example of the internationalism of mathematics. The origins of the theory are mainly German and French; among its most active devotees are (or were, till September 1, 1939) numerous Poles, Russians, and Americans.

Three estimates

The preceding sketch is only the barest indication of some of the leading features of what in 1945 was one of the most active movements in all the long history of mathematics. Inadequate as the sketch is, it may nevertheless suffice to illustrate the profound differences in outlook between the mathematics of the nineteenth century and that of the twentieth. Few with the faintest stirrings of a historical instinct imagine that 1945 marks a finality in any direction. But it does seem from the record that the period of abstractness was an inevitable stage in the orderly development of all mathematics.

Some, who have grown rather satiated with Euclid's program, sigh for another Pythagoras or a modern Euler to inject just one strikingly novel invention for future generations of Euclideans to abstract and systematize. That new ideas will be welcomed if and when they arrive seems probable from the enthusiastic reception accorded integral equations in the early 1900's. A somewhat jaded analysis suddenly took a new interest in life after Volterra and Fredholm. For the future, there is always the hope or the fear that Einstein's non-linearity may be demanded in physics; and discreteness may stimulate a recovery in finite differences and combinatorial analysis. Should either of these possibilities materialize, it seems likely that but little of the abstract technique of 1945 will be applicable, as most of it evolved from qualitatively different origins.

Leaving the unpredictable future to write its own epitaph, we shall report here three estimates, by competent judges in 1945, of the abstract method in the first four decades (1899–1939) of its modern popularity. The earlier date marks Hilbert's revival and sharpening of the postulational technique in his work on the foundations of geometry. As might be expected in a matter affecting the problematical 'value' of the work of hundreds—abstractionists and non-abstractionists alike—opinion was divided. The line of cleavage followed, but not strictly, the personal interests of the judges. On two points, however, the verdict was practically unanimous.

It was agreed that the postulational or abstract technique had isolated and standardized many patterns of mathematical reasoning which recur frequently either in some one major division of mathematics, say geometry or analysis, or in more than one division, say in both geometry and analysis. Abstract reformulations of specific problems in a particular division disclose such patterns if present, and a single step from hypothesis to conclusion then replaces the many that would probably be necessary were the proof conducted in terms of the special data. The technique resembles industrial mass production, with a corresponding uniformity in the product. If several standardized abstract patterns are concealed beneath the data, abstract reformulation of the problem indicates the types of machinery to be applied individually as in the case of but one standardized pattern. Where a single pattern underlies several divisions, proof in any one suffices to indicate proofs in each of the others, and indeed renders separate proofs for all, individually, unnecessary. This is something quite different, because it is much more general, from the nineteenth-century unifications by the theory of groups. The abstract structure common to several problems need have no connection with groups, and many have none. On the other hand, if a connection through groups is present, then it is merely an instance of the possible identities in structure to be revealed uniformly by abstraction.

Any student who has plowed through a classical presentation of the theory of functions of a real variable before attacking some of the same material by the abstract method will grant that the strain on the memory is greatly reduced, and that insight is correspondingly increased. The chief dissenters (1945) to the implied doctrine were the die-hard mental disciplinarians who maintained that the proper way to impart mathematics is to make it as tedious and as repulsive as possible. But if the object is to get over the ground with a minimum of impedimenta, the abstract method has no competitor. It is true that numerous interesting details are discarded on the way. But so are they in mechanics when the subject is presented in a manner designed to reach the living parts of the physical sciences in the shortest possible time. If mathematics is to progress, it cannot encumber itself with all the antiquated baggage it has picked up on its travels through the past.

It was also generally conceded (1945) that the abstract attack had acquired many new results which might not have been won by classical methods, and which certainly were not so won. These very additions to mathematical knowledge provided the first bone of contention.

Abstractionists maintained that the quality of the new products was at least as high as that of the old. Anti-abstractionists hotly disputed this purely subjective judgment, contending with equal subjectivity that most of the new results were, essentially, trivialities. All the sap, these critics insisted, had been squeezed out of what they called 'real' mathematics, leaving only desiccated husks superficially like well-known elementary propositions in classical mathematics. In short, it was disputed that anything 'really new' had issued from the abstract method;

and abstractionists were freely challenged to produce at least one instance of their allegedly all-inclusive theorems significantly different from the trite classical results generalized by abstraction. History, not idle bystanders, will decide the issue.

The final difference of opinion concerned the relation of the modern abstract movement to all previous mathematics. Abstractionists saw the movement as the retarded climax of a halting evolution that had blundered through to 'real mathematics' after some six thousand years of not knowing where it was going. In spite of this blindness to its manifest destiny—in retrospect —mathematics had finally reached its unperceived goal. Moderates of this faction admitted that progress might possibly continue. But the abstract method of 1899–1945, they believed, had permanently fixed the direction of all future progress.

Anti-abstractionists stigmatized the entire movement as a sinister recurrence of the sterile period of criticism and commentary that marked the death of Greek mathematics. The inventive nineteenth century, these insisted, was the delayed echo of the Archimedean age; the Euclidean revival of the twentieth century was a reversion to the cramped mathematics which Archimedes had liberated, only for his historical successors to recapture and immobilize in the pedantries of formal logical analysis. Whichever contestant, if either, history is to uphold, it seems likely that the early twentieth century will be remembered as one of the major epochs in the development of mathematics.

While the abstractionists and their opponents were debating their private differences, a third faction, greatly inferior numerically to either of the others, contended that both were blind to the only thing of any vital significance that was happening to all mathematics. The abstract method, after all, was nothing basically new: the Greeks invented it, and modern practice excelled Euclid's only in its sharper precision. But the epistemological doubts concerning the conclusions of classical mathematical reasoning were new in kind. No previous age could match these. While believing that the outward aspect of mathematics would not be greatly changed by any doubts concerning its internal consistency, critics of the foundations prophesied a radically new conception of mathematics in the not-distant future.

According to the prophets, the last adherent of the Platonic ideal in mathematics will have joined the dinosaurs by the year 2000. Divested of its mythical raiment of eternalism, mathematics will then be recognized for what it has always been, a humanly constructed language devised by human beings for definite ends prescribed by themselves. The last temple of an absolute truth will then have vanished with the nothing it enshrined. Other temples may be built, but not by mathematicians, should the prophets be right. Mathematics as ideal truth expired somewhere on its journey from 1900 to 1945; and those who continue worshiping this ideal are attending its protracted obsequies in ignorance that their deity is dead.

Such a possible outcome of six thousand years of mathematical progress will depress some and exhilarate others. Once more history alone will decide. Less than a century ago, the possibility had probably not occurred to any mathematician. Some of the supporting evidence will be our next and last concern. Anyone who has followed thus far need not be cautioned that a suspended judgment is all that the evidence warrants.

CHAPTER 23

Uncertainties and Probabilities

From Pythagoras and Zeno to Hilbert and Brouwer, mathematicians have reveled in the flexibility of their reasoning, and some few have sought to understand the sources of its power. For centuries after the first great age of mathematics in ancient Greece, it was accepted without question that deductive reasoning, if properly applied, would never lead to inconsistencies. The doubtful elements in this simple creed are concealed in the words 'properly' and 'never,' of which the first begs the question, and the second disguises a potential infinite. With the intrusion of irrational numbers to disrupt the integral harmonies of the Pythagorean cosmos, a controversy that has raged off and on for well over two thousand years began: is the mathematical infinite a safe concept in mathematical reasoning, safe in the sense that contradictions will not result from the use of this infinite subject to certain prescribed conditions? (The 'infinites' of religion and philosophy are irrelevant for mathematics.)

We have seen what shapes this protean question assumed in the paradoxes of Zeno; in the proportion of Eudoxus; in the submathematical analysis of the Middle Ages; in the indivisibles of Cavalieri; in the calculus of Newton and Leibniz; in the *Mengenlehre* of Cantor; in the theories of the real number system proposed by Cantor, Dedekind, and Weierstrass; in the definition of cardinal numbers as classes by Frege and Russell; and finally, in Hilbert's problem of proving the consistency of geometry. We remarked also that these strictly technical questions of mathematics were responsible for a searching scrutiny of all deductive reasoning in the first four decades of the twentieth century; and it was observed that, without the aid of mathematical logic, some of the sharper analyses might not have been feasible. Accordingly, before reporting a few of the more suggestive conclusions, we shall note the principal stages by which mathematical logic developed into a major activity of recent mathematics. For reasons to be indicated in the proper place, we shall then do the like for the mathematical theory of probability. In this manner, we shall reach successive positions from which to describe the more popular of several creeds regarding the nature and meaning, if any, of mathematics, as of the year 1945. The experienced observer will not extrapolate to 1946 or beyond.

Prejudices and misconceptions

Before proceeding to details, we take a swift glance ahead. The matters to be described are possibly those of greatest interest to onlookers concerned with the significance of mathematics in the fluctuations of civilization, beyond a trite utility in science and technology. Inquiries into the foundations and meaning of mathematics have undoubtedly stimulated epistemology since 1900, and especially since 1910, when the first volume of *Principia mathematica* by A. N. Whitehead (1861–, English) and B. (A. W.) Russell (1872–, English) appeared. An earlier book by Russell, *The principles of mathematics*, 1903, probably did more than any previous work, with the possible exception of Kant's *Critique of pure reason*, 1781, to interest philosophers in mathematics. It also irritated a few and roused them from their philosophic slumbers with such forthright declarations as this (Russell's): "The Philosophy of Mathematics has been hitherto as controversial, obscure and unprogressive as the other branches of philosophy." If a man has something to say, there is no good reason why he should not say it so that those for whom it is intended cannot possibly misunderstand him, even at the risk of being less polite than is customary at an afternoon tea.

In the thirty-four years that elapsed between the first edition of the *Principles* and the second (1938, reprint except for new Introduction), the philosophy of mathematics had become as controversial, almost as obscure, but by no means as unprogressive, as the other branches of philosophy. It was in fact progressing with disconcerting rapidity in several directions at once, like an exploding shell.

Thirty-four years is a long time in recent mathematics. It is therefore rather curious to find that many philosophers and not a few mathematicians in 1938 seemed to be as hostile to mathematical logic and mathematical philosophy as they were in 1903. As late as 1939, a distinguished American liberal philosopher (unnamed, because he may wake before he dies) exulted that mathematical logic had contributed nothing new to either mathematics or epistemology. He therefore had spared himself the trouble of learning to read the necessary symbolism. The scholarly article in which this decision was handed down was cordially received by professed liberals as a competent judgment by an expert.

It is only for a philosopher to decide whether mathematical logic has made any contribution to philosophy. Those who wish to form an opinion on the other half of the above adverse decision are referred to the abstract-journals devoted to technical mathematics and to nothing else, or to *A bibliography of symbolic logic*[1] for the period 1666–1935, with four supplements. About 2,500 works are listed. All but a few are regarded by mathematicians as belonging to their subject. To cite but one detail, the entire field of Boolean algebra, with an extensive literature of its own, is of first-rate importance in the severely technical modern abstract algebra described in earlier chapters. In its modernized version, this algebra has significant applications to topology, as elaborated in 1936–7, by M. H. Stone (1903–, U.S.A.). It also appears to be basic in modern arithmetic, particularly in the theory of arithmetical structure. If none of this work is new, then all the specialists who write the abstracts of current research in mathematics are incompetent. This seems hardly probable.

Contrary to the scholarly verdict noted, many professional mathematicians since 1912 have acted as if they believed the epistemological questions discussed by the workers in the foundations of mathematics to be both relevant for technical mathematics and novel. In the 1920's, for example, two eminent mathematicians became so excited over each other's strange conceptions of mathematics that they were forced to resort to robust language to express their opinions. The war of words between Hilbert and Brouwer over the foundations will doubtless take its place in history with the greater classics of mathematical controversy. So disturbing were the tumult and the shouting that even the physicists dropped whatever they were doing, and listened apprehensively. Their reaction to the battle of the mathematical giants is another classic, illustrating how little the struggles of one set of human beings, warring to defend their most cherished possession, interest or affect another set. Meeting a colleague who might be supposed to know, Einstein asked him, "What is all this frog-and-mouse battle between the mathematicians about?"

Neutrals in the 1930's vacillated between nervousness and irritation. Technicians of the older generation persisted in their error, inherited from Poincaré, that mathematical logic is as sterile as a mule. Poincaré died in 1912. A few, vaguely disturbed by rumors even the deaf must have sensed, dismissed the debates over the foundations as so much metaphysics, and therefore irrelevant for mathematics. Exploding the popular delusion that a mathematical (or scientific) education is necessarily a valuable training in objectivity, some of the more influential conservatives deplored the publicizing of the so-called 'crisis' in mathematics, for the possible disturbing effect a knowledge of the facts in the matter might have on the minds of 'the young.' It was even insinuated by the bolder reactionaries that the entire critical movement was but one more assault of the communists, with their Marxian philosophy and Hegelian dialectic, on the established verities of their forefathers.

At first blush, this somewhat diseased theory might seem seductive. For it is undeniable that the critical movement in mathematics was rapidly approaching one of its climaxes in the 1920's, when much else that had appeared sound before 1914 was being revalued with a callous disregard for traditional sanctities. According to this theory, the 'revolution' in mathematical thought was but a minor skirmish between the old and the new in an incomparably vaster Revolution. But, attractive as this theory may be, it is negated by verifiable fact. The upheaval in mathematics began from forty to fifty years before the war of 1914–18 deflected political, religious, and economic thought into new and rougher channels.

Each of several definite incidents might be exhibited to mark a beginning of the modern age of uncertainty. Not to go too far back, and to instance only men who were primarily mathematicians, we recall that Dedekind hesitated at least two years before publishing his *Stetigkeit und irrationale Zahlen* (1872), the basic concepts of which had occupied him in 1858, because he was not convinced that his reasoning was sound. Again, in the preface to the third edition (1911) of *Was sind und was sollen die Zahlen?* (1888), Dedekind acknowledged the doubts that had arisen since 1893 regarding the tenability of his position in that classic; and he regretted that other work had prevented him from completing the very difficult research necessary to render the foundations of his theory unobjectionable. Craving indulgence for this remissness, Dedekind, eighty years old at the time, reaffirmed his belief in "the inner harmony of our logic." Dedekind, therefore, might be counted with the believers in the essential soundness of the reasoning in mathematical analysis, particularly in the use of the infinite as it occurs in limits and continuity. However, his own fundamental work in these directions was partly responsible for some of the more radical objections of the finitists, of whom Kronecker was the first. By 1880 at the latest, Kronecker had denied all meaning to such mathematics as occurs in Dedekind's theory of irrational numbers and continuity, without which classical analysis does not exist; and he had insisted that the logic accompanying such analysis is invalid.

Neither Kronecker nor Dedekind has yet been honored by being called a philosopher in the accepted sense. Nor can either be credited to communism, as one died in 1891 and the other in 1916. And, as was seen in another connection, it was Burali-Forti's paradox of 1897 that opened the attack on Cantor's theory of the infinite. It is interesting to recall that Cantor himself found this paradox in 1895, and communicated it to Hilbert in 1896. Cantor died in 1918; his last work on *Mengenlehre*, *Beiträge zur Begründung der transfiniten Mengenlehre*, appeared in 1895 and 1897. One of the three Latin mottoes under the title is Newton's famous "Hypotheses non fingo."

The unshakeable faith of Dedekind, the uncompromising skepticism of Kronecker, and the shadowy hypotheses which Cantor framed are three of the principal sources of doubts concerning the complete validity of traditional mathematical reasoning. All antedated the period of general uncertainty that began in 1914.

Once more, lest 'the young' be misled, we repeat that these doubts did not halt mathematical creation. Technicians working on the superstructure did not drop their tools and scurry down to the basement because some of the underpinning needed reinforcing. Continuing their own highly specialized labors, they left the necessary task to experts who understood what they were about. The building had not collapsed as late as 1945; and while those engaged in elaborating the superstructure but seldom concerned themselves with what the consolidators of the foundations were doing, they had at least come to tolerate their presence in the building. The misconceptions and recriminations of the 1900's gave way in the 1930's to a first crude approximation to harmony. It was almost as if the American Federation of Labor and the Committee for Industrial Organization had at last decided to bury the hatchet elsewhere than in either's skull, and get on with the job.

Mathematical logic from Leibniz (1666) *to Gödel* (1931)

A distinction is sometimes drawn between mathematical logic and symbolic logic. The latter is then "the formal structure of deductive reasoning investigated by the symbolic method."[1] Mathematical logic is deductive reasoning as it occurs in mathematics; it contains a great deal that is foreign to classical logic in the traditions of Aristotle and the Middle Ages, and it need not be symbolic. In what has already been said, we have ignored this distinction, and usually we shall continue to do so. The context will indicate which meaning is pertinent if the distinction is important. From the history of the symbolic method we shall select only the more outstanding of those episodes which have a direct bearing on creeds of the 1940's concerning the nature of mathematics, the validity of mathematical reasoning in general, and the consistency of arithmetic in particular, or which made possible logical analyses of the accompanying beliefs.

The history begins, with catastrophic abruptness, in 1666 with what Leibniz called his "schoolboy essay," *De arte combinatoria.* Although this contained nothing definite on symbolic logic, it is the historical source of both symbolic logic and mathematical logic. For Leibniz imagined the possibility of a 'universal characteristic,' or 'kind of universal mathematics,' in which a 'calculus of reasoning,' expressed in an efficient symbolism subject to appropriate rules of combination worked out once for all, would guide the reason. Errors, except of fact, could then be only slips in calculation. This is the situation in any mathematics. The reservation which Leibniz makes is most painfully evident in applied mathematics, where a slight error of fact in the scientific assumptions may[2] lead to worthless conclusions, although the 'calculus of reasoning' by which the conclusions are deduced is sound and correctly used.

Leibniz' misadventure with Huygens (1679) over the 'universal characteristic,' noted in an earlier chapter, was not without benefit for the ambitious dreamer. Evidently Leibniz got down to work; for in 1686 he recorded what he considered his first notable progress. From 1679 to the 1690's he devoted part of his spare (!) time to an attempt to create a symbolic logic. In current terminology, he stated the principal properties of logical addition, multiplication, negation, the null class, and class-inclusion; and he noted the similarity (abstract identity) of certain properties of inclusion for classes and implication for propositions, almost as stated today in texts on symbolic logic.

The next notable incident was J. H. Lambert's claim (1765) that he had shown certain of the rules of arithmetic and algebra to be interpretable in a wider domain of the kind imagined by Leibniz. Whatever the merits of Lambert's theory, it was abortive, as also was the 'logical algorithm' (1803) of G. F. Castillon.

Symbolic logic finally got itself born in Boole's epochal pamphlet of 82 pages, *The mathematical analysis of logic, being an essay toward a calculus of deductive reasoning*, 1847. This was followed in 1848 by a fifteen-page paper, *The calculus of logic;* and finally, in 1854, Boole gave a formal exposition of his system in his masterpiece, *An investigation into the laws of thought, on which are founded the mathematical theories of logic and probabilities.*

Russell has remarked (1901) that "Pure mathematics was discovered by Boole in a work which he called 'The Laws of Thought.' . . . His work was concerned with formal logic, and this is the same thing as mathematics." Although the identification of mathematics with formal logic was controverted between 1901 and 1945, it had not been questioned that Boole was the true founder of symbolic logic. He was also a mystic, in one usual technical sense of believing that knowledge comes by immediate intuitions of extrahuman existences.

The contents of the *Laws of thought* are gorgeously heterogeneous, ranging from "The fundamental principles of symbolical reasoning, and of the expansion[3] or development of expressions involving logical symbols," through an "Analysis of a portion of Dr. Samuel Clarke's 'Demonstration of the Being and Attributes of God,' and of a portion of the 'Ethica ordine geometrico demonstrata' of Spinoza," to a somewhat puzzling climax "On the nature of science, and the constitution of the intellect." In the closing paragraph, Boole observes that "Hence [not the mathematical 'hence'], perhaps, it is that we sometimes find juster conceptions of the unity, the vital connexion, and the subordination to a moral purpose, of the different parts of Truth, among those who acknowledge nothing higher than the changing aspect of collective humanity, than among those who profess an intellectual allegiance to the Father of Lights." Such was pure mathematics in the reign of Victoria the Good.

In creating his system, Boole was influenced by the British work, in which he participated, on the purely formal or abstract character of elementary algebra. It was therefore not desecrating the eternal verities to seek a logical interpretation of the algebraic 'laws' of addition and multiplication, with appropriate meanings of zero (the identity of addition) and unity (the identity of multiplication) to complete the interpretation in conformity with common algebra. Full and elementary accounts of the resulting system of symbolic logic being readily accessible in textbooks and treatises in nearly every civilized language, we shall not describe either Boole's original system or the improved (in some respects) Boolean algebra into which it evolved from 1854 to the 1940's.

Two details are of more than historical interest. Boole's interpretation of 'or,' the equivalent of logical addition, was less convenient than that in Boolean algebra; and he had no satisfactory inverse for logical addition. The latter defect was remedied in the so-called symmetric difference,[4] introduced in 1916 by P. J. Daniell (1889–, English). This made possible the incorporation of Boolean algebra into the modern algebra of rings, ideals, etc. To dispose of this aspect of the subject here, the technical modernization of Boolean algebra was largely done in the United States in 1924–35, by B. A. Bernstein (1881–), Stone, and J. von Neumann (1903–, Hungary, U.S.A.) among others. Of greater interest to professional mathematicians than to philosophers, Boolean algebra has been disparaged by some experts in the foundations as being too much like mathematics and too little like a calculus of mathematical logic to be of much use in an analysis of the fundamentals. Nevertheless, Boole's algebra gave Leibniz' dream its first substance, and it was a main inspiration for the more powerful mathematical logics which followed it.

Concurrently with Boole, De Morgan took another long step forward, in his treatise (1874) on *Formal logic: or, the calculus of inference, necessary and probable*. Proceeding beyond Boole, De Morgan initiated (1860) the logic of relations, in the fourth of a famous series of five papers (1846–60) on the syllogism. For its historical importance we recall the farcical feud between De Morgan and the Scottish anatomist, botanist, lawyer, and metaphysician, W. Hamilton[5] (1788–1856), over the quantification of the predicate. It was the sound and fury of this controversy that roused Boole's interest in formal logic to the point of creating his symbolic system. Losing his temper, and almost his mind, in his rage, the lawyer-metaphysician accused De Morgan of plagiarism. When the quality of Hamilton's contribution to logic is considered, the accusation was an insult.

After De Morgan, the next outstanding advance toward mathematical logic was made in the United States by C. S. Peirce. For reasons into which we need not enter, Peirce's work failed to make the immediate mark its penetrating quality should have made. Peirce's own explanation for his lack of adequate recognition is contained in the unacademic remark, attributed to him on good authority, "My damned brain has a kink in it that prevents me from thinking as other people think." A few items must suffice to indicate the quality of Peirce's work. He noted (1867) several dual pairs of propositions but apparently missed De Morgan's laws (1847). From these laws the principle of duality in Boolean algebra (to every proposition concerning logical addition and multiplication there is a correspondent concerning multiplication and addition), stated by Schröder in 1877, is an almost trivial deduction. In 1870 Peirce described a notation for the logic of relatives, "resulting from an amplification of the conceptions of Boole's calculus of logic." This was continued in 1880–3.

For their prophetic significance no less than their technical interest, two details may be specially mentioned. In a paper of 1885 with the subtitle, *A contribution to the philosophy of notation*, Peirce defined the 'truth-values' of a proposition: a true proposition has the constant value v, and false proposition the constant value f. This appears to have been the first hint of the so-called matrix-method (truth-tables) in mathematical logic, where the usual 'values' are 0 ('falsity'), 1 ('truth') in a two-valued logic, and $0, 1, \ldots, n - 1$ in an n-valued logic. Logics with a continuum of truth-values were also devised in the 1930's.

The other prophetic novelty was Peirce's reduction (1880) of the logical constants 'and,' 'or,' etc., of Boolean algebra to a single constant, in terms of which all are expressible. This reduction was rediscovered in 1913 by H. M. Sheffer (U.S.A.), and was said by Russell to afford a most notable simplification of certain parts of *Principia mathematica*. A like reduction may be effected in an n-valued logic. With the publication of Peirce's collected papers (vol. 3, 1933, vol. 4, 1934), the place that

Peirce might have occupied in the development of mathematical logic was realized. But as history would have it, others retraced his steps, unaware that he had gone before.

The 1870's saw hints of a new approach in the work of H. MacColl. Departing from the tradition of Boole, MacColl (1837–1909, England), all but hit upon the concept of a propositional function. Unaware of Boole's work when he began, MacColl invented his own symbolic logic as an aid to his investigations in mathematical probability. He called his system a calculus of equivalent statements, and applied it not only to probability, but to the change of the order of integration. The latter was a purely formal algorithm of no importance for analysis. In his later work of the 1890's, MacColl attempted to expel all absolute constants from logic and mathematics. His suggestion, not satisfactorily developed in his own system, that symbolized statements may have an indeterminate truth-value has reappeared in one of its possible variants in many-valued logics. In his insistence that propositions and implication are more useful in mathematical logic that classes and inclusion, MacColl was a lone wolf in his generation. His inability to push some of his novel ideas through to conclusions seems to have been partly due to a lack of mere technical facility in elementary algebraic manipulations and an allergy to usable, expressive symbolism.

In an estimate (1906) of MacColl's *Symbolic logic and its applications*, 1906, Russell observed that MacColl differed from other symbolic logicians in his preoccupation with verbal expressions rather than with their meanings. Had MacColl lived till the 1920's, he would have found himself in distinguished company in this very respect. Struggling to extricate themselves from the stickier quagmires of classical metaphysics, the formalists of the Hilbert school separated the symbolic game of mathematics from its 'meaning' or 'interpretation,' in an endeavor to reformulate the rules so that they would 'never' produce a contradiction. MacColl seems to have dimly recognized that there are at least two major problems in mathematical logic: to make the symbolism and the rules for manipulating it self-consistent; to interpret the symbolism and the operations by which one formula is transformed into another. The first is the formal problem of mathematics; the second is a problem in metamathematics. The distinction was not recognized in 1906 when MacColl finished his work. It became of the first importance in 1931, as will be seen, with the work of Gödel, which showed that in certain respects a system of mathematical logic has an interior and an exterior. Propositions can be exhibited within a system which are undemonstrable by the apparatus of the system, and it is necessary to go outside the system in search of a proof. This is only a crude analogy of the situation, which will be described in more detail later. For the present, it seems just to say that although MacColl may not have seen clearly where he was going, he had set foot on a promising road to what, in his day, was the future of mathematical logic. Categorical judgments in these controversial matters seem to suffer a greater risk of being reversed than those in any other department of mathematics.

The Boolean tradition ended, in a sense, in 1905 with the completion of the four-volume, 2,033-page treatise of F. W. K. E. Schröder (1841–1902, German), *Vorlesungen über die Algebra der Logik*, 1890, 1891, 1895, 1905. Symbolic logic as it existed at the turn of the century was here organized and expounded with that thoroughness which the professorial world had learned to anticipate from German scholarship in the nineteenth century. Had not the originality of Frege and Peano inaugurated a renaissance in mathematical logic, Schroder's massive masterpiece might have weathered the centuries as the tombstone of Leibniz' dream. But mathematical logic was too vigorous to submit to premature burial under anybody's two thousand pages of meticulous erudition. Once more it was demonstrated, as in the three weighty volumes that might have crushed Lie's theory, that the useful life of a contribution to the technical literature of mathematics sometimes varies inversely as the length of the contribution. Attempts in the twentieth century to be all-inclusive and definitive in any department whatever of mathematics seem somewhat futile, if not presumptuous. When, if ever, it becomes possible to print 2,000 or 10,000 pages on mathematical logic that will retain their vitality unimpaired for a generation, mathematics itself will be dead.

We have already observed in other connections the beginning of the recent period in mathematical logic: first, in the work of Frege (1879, 1884, 1893, 1903) on the symbolization and analysis of logical and mathematical concepts, and on the foundations of arithmetic; second, in Peano's postulational analyses, with a much ampler elaboration of logical symbolism than any known before his time, of arithmetic and geometry (1889), and in his recasting (with the help of Italian collaborators) of extensive tracts of mathematics in the new symbolism, in the *Formulaire de mathématiques* (1895–7–8–9, 1901–2–3, 1905–8); and third, in Hilbert's postulational treatment of the foundations of geometry. The great importance of these innovations for twentieth-century algebra, geometry, and analysis has already been noted; their importance for the development of mathematical logic after 1900 was equally great.

Peano's project in the *Formulaire* of translating technical mathematics into the symbols of mathematical logic which he had invented was partly responsible for the most comprehensive logical symbolization of mathematics yet (1945) attempted, the *Principia mathematica* (1910, 1912, 1913) of Whitehead and Russell. Here was executed in minute detail the program outlined by Russell in his *Principles* (1903), "to prove that all pure mathematics deals exclusively with concepts definable in a small number of fundamental logical concepts, and that all its propositions are deducible from a very small number of fundamental logical principles; to explain the fundamental concepts accepted in mathematics as indefinable" (postulated, and not further analyzed). This states the program of the Whitehead-Russell school for mathematics and its foundations, which R. Carnap[6] (Austria, U.S.A.) has since called logicalism. According to logicalism, mathematics is a branch of logic. While it is rash to the point of foolhardiness for anyone but the author(s) of a particular mathematical philosophy to make any assertion concerning either the aims or the achievements of that philosophy, it is a tenet of logicalism that logic is more basic than, and prior to, mathematics. This, if a correct statement of the logicalists' position, is of importance here; for Brouwer's more mathematical thesis is the opposite.

In Brouwer's intuitionism, mathematics is prior to logic. Again with the reservation noted above, it is a cardinal article of faith in the intuitionists' creed that human beings are born with an "original intuition" (*Urintuition*) of an unending sequence of objects (numbers?) generated by successive additions of one object at a time. That is, intuitionism replaces the original sin of the theologians by an original arithmetic of the mathematicians. The position of the intuitionists in this article of their faith may be unassailable; for just as no theologian has yet succeeded in demonstrating the existence of original sin to any human being not disposed to believe it, so no intuitionist has yet shown the existence of dormant numerical language habits in a newborn infant. But this may not be what is meant by the mystical concept 'original intuition'; and it may be that the 'inner meaning' is denied to any but a mathematical mystic. The primary revelations of the creed are veiled in the Dutch language. German and English expositions are available; but it is said by converts with expert knowledge in both the languages and the mathematics that only those who can think in Dutch can grasp the finer shades of meaning, without which intuitionism is incomprehensible. A correct understanding of this central mystery seems to be essential for a comprehension of the intuitionists' mathematical philosophy. The points at issue are not trivialities, as the few details to be considered in the sequel may indicate.

The intuitionist creed recalls Kant and his insistence on 'intuition' in mathematics. Brouwer at first imagined that Kant's philosophy had suggested his own but later emphatically denied any indebtedness and repudiated Kant and nearly all of his obsolete mathematical metaphysics. The imputation of mysticism, which indeed may be baseless, is also strongly resented by the intuitionists. Mathematics according to some intuitionists is identified (intuitively?) with "the exact part of our thought" and is held to be antecedent to both logic and philosophy. The source of mathematics is asserted to be "an intuition that presents mathematical concepts to us as immediately clear." It is denied that this intuition is in any sense mystical; it is merely "the ability to treat separately certain concepts and inferences which appear regularly in common thinking." It is interesting at this point to confront this denial and affirmation with a little of what a standard English dictionary (not a German or a Dutch dictionary, but an English dictionary) states under 'mysticism': "the doctrine or belief that direct knowledge of God, of spiritual truth, of ultimate reality, etc., is attainable through immediate intuition, insight, illumination, and in a way differing from ordinary sense perception or ratiocination." The 'objects' with which intuitionistic mathematics is concerned are said to be immediately apprehended in thought, possibly as the convinced mystic immediately apprehends the immanence of the deity or whatever it may be that he does apprehend. Reminiscent of Kant's discredited 'synthetic, a priori' geometry, though not quite the same, these objects of the intuitionists are independent of experience and have no existence independent of thought. The last may give the intuitionists their 'out.' But if so, it is a narrow escape. Vestiges of ancient and medieval dogmas reappear also in the intuitionists' insistence on the human capacity for imagining a sequence of "distinct, individual objects" obtainable by adjoining objects indefinitely, one at a time, to those already imagined. Starting with 'one' and the conceptual operation of 'adding one,' the intuitionist thus intuits the unending sequence of natural numbers by indefinite repetition of the operation. This is the 'original intuition' of intuitionism. Factually, it is anything but universal among primitives who, presumably, are human beings although not so far from the higher animals as the rest of human kind in the 1940's. Civilized peoples appear to have acquired this 'capacity' at the cost of considerable experience and some almost inspired inventiveness—unless, of course, this rather ad hoc 'capacity' is by hypothesis latent though unobservable. A thoroughgoing finitist (like Kronecker) might assert that the intuitionists' infinite sequence has only a meaningless verbal existence on paper, not an intuitive existence in the human mind. The strict finitist rejects the infinite as a pernicious futility inherited from outmoded philosophies and confused theologies; he can get as far as he likes without it.

To conclude this report on the suit of intuitionism versus mysticism, we may allow ourselves one of the very few anecdotes in this book. It echoes Gordan's outraged cry when he read Hilbert's finiteness proof in the algebra of quantics. A devout intuitionist closed his New Testament after reading *The Gospel according to Saint John* for the first time in his life with the ecstatic whoop, "This is not theology, it is mathematics!"

A basic section of the *Principia* develops the calculus of propositional functions. A statement containing a variable x, which is such that it becomes a proposition when x is given any fixed determinate meaning, is called a propositional function; and "a proposition is anything that is true or that is false." The last definition (Russell's) may be less innocent than it seems, as will appear later in connection with intuitionism.

One aim of the calculus of propositions is the solution of certain contradictions, such as Burali-Forti's, in mathematics. The proposed solution appeals to the so-called vicious-circle principle, that whatever involves *all* of a collection must not be one of the collection; or, conversely: "If, provided a certain collection had a total, it would have members only definable in terms of that total, then the said collection has no total." In addition to this principle, two further devices are used, a doctrine of types and a famous postulate, the axiom of reducibility. For safety, we describe the first in the authors' own words.

> We are thus ['after much argument about it and about'] led to the conclusion, both from the vicious circle principle and from direct inspection, that the functions to which a given object a can be an argument are incapable of being arguments to each other, and that they have no term in common with the functions to which they can be arguments. We are thus led to construct a hierarchy. Beginning with a and the other terms which can be arguments to the same functions to which a can be an argument, we come next to functions to which a is a possible argument, and then to functions to which such functions are possible arguments, and so on.

Proceeding from this, the authors of *Principia* (a 'description,' by the way, of a kind exhaustively analyzed in their work) separated propositional functions into types according to their allowable arguments. The axiom of reducibility is the assumption that each propositional function in any one of the types is equivalent to some propositional function in some of the lowest type. This is only a very crude description of a very subtle matter, which in the original requires five large pages of exposition. But for those who prefer an embodiment of the axiom in the language of classical logic, the authors state that their "axiom of reducibility is equivalent to the assumption that 'any combination or disjunction of predicates is equivalent to a single predicate,'" with the understanding that "the combination or disjunction is supposed to be given intensionally." This crude but verbatim statement of the axiom by its sponsors was promptly criticized as either inadequate or misleading, and what in 1910 (and in 1925) satisfied the authors of *Principia* failed to satisfy *Principia*'s critics. In more detail, but still not enough for some objectors, the authors summarized their chapter on the hierarchy of types and the axiom of reducibility in four sufficiently clear statements of their intentions. First it is necessary to recall the definition of 'matrix' as used in *Principia:* "Let us give the name *matrix* to any function, of however many variables, which does not involve any apparent variables." For the significance of 'apparent' we must refer to texts on mathematical logic. The x in such a statement as 'for all x, $f(x)$' is apparent, the range of x being over the whole set of values for which $f(x)$ is significant. A mathematical analogy is the variable of integration in a definite integral. The four statements: (1) "A function of the first order is one which involves no variables except individuals, whether as apparent variables or as arguments." (2) "A function of the $(n + 1)$th order is one which has at least one argument or apparent variable of order n, and contains no argument or apparent variable which is not either an individual or a first-order function or a second-order function or . . . or a function of order n." (3) "A predicative function is one which contains no apparent variables, *i.e.* is a matrix. It is possible, without loss of generality, to use no variables except matrices and individuals, so long as variable *propositions* are not required." (4) "Any function of one argument or of two is formally equivalent to a predicative function of the same argument or arguments." The cordial invitation to controversy extended by Russell's theory of types to philosophers, mathematical logicians, and others was eagerly accepted, and before the second edition (1925–7) of *Principia* appeared several modifications or substitutes for the pioneering theory were available. That none has escaped the ravages of less than a generation unscarred is indicative perhaps only of a greatly accelerated interest in mathematical logic since the first edition (1910–13) of *Principia* set a new and faster pace in the foundations of mathematics. A usable form of the theory of types as of the 1940's stated essentially that "each entity is regarded as belonging to precisely one of a hierarchy of 'types,' and any formula representing membership between entities of other than consecutive ascending types is rejected as meaningless together with all of its contexts." For further details Quine (1936, 1940) may be consulted.

By such heroic assumptions, the contradictions were solved to the partial satisfaction of some for a time. In more orthodox mathematics, Kronecker's God-created 1, 2, 3, . . . demanded the 'existence' of an infinity of propositional functions. Not being immediately forthcoming, the necessary functions were postulated in the so-called axiom of infinity which, however, left open the question of 'existence.' Logicalism, as far as the natural numbers were concerned, thus returned to the sub-mathematical analysis of the Middle Ages. And just as that analysis stirred up interminable controversy, so did the axioms of reducibility and infinity provoke all but endless disputation.

Other debatable points in the *Principia* proved equally stimulating, for example, the basic concept of 'proposition.' The difficulty here was the lack of a criterion for deciding whether certain statements are propositions or whether they are nonsense, or whether they are neither true nor false. Whatever is to remain of the revised *Principia* of 1925–7, the original of 1910–13 will retain its place in history as the work which began a new epoch in mathematical logic and the foundations of mathematics.

Our principal interest here in mathematical logic is its bearing on mathematics, particularly arithmetic. Controversies engendered by *Principia* at least cleared the air; and beliefs which seemed reasonable in 1913 were no longer held a quarter of a century later by those who could profit by the exercise of their reason. Russell, for example, was severely criticized by some philosophers for his propensity to renounce beliefs when he found them untenable. Ignoring criticism, he persisted in changing his mind when he believed the change conducive to mental health. In the second edition (1938) of the *Principles*, he recorded one such change which is of particular interest to mathematicians. Having recalled the influence of Pythagorean numerology on all subsequent philosophy and mathematics, Russell states that when he wrote the *Principles*, most of it in 1900, he "shared Frege's belief in the Platonic reality of numbers, which, in my imagination, peopled the timeless realm of Being. It was a comforting faith, which I later abandoned with regret." Possibly the distinguishing peculiarity of the human race is its capacity for getting comfort out of the most unpromising material. Russell's faith in logicalism, however, remained unshaken, although modified to accommodate competent criticism of details. Concurrently with the abandonment of Platonic numerology, "many apparent entities, such as classes, points, and instants, have been swept away." Classes were said to have been abolished in *Principia;* Whitehead is credited with the Spartan ruthlessness of abandoning instants of time and particles of matter to the mercy of the elements, and "substituting for them logical constructions composed of events." From one point of view, the mathematical physics of relativity with its intersections of world-lines as point-events, and quantum mechanics with its exclusive attention to observables, present a similar world picture, as unsubstantial as the dream of Pythagoras. If everything is no longer mere number, it is mathematics, and this, according to logicalism, no less than number itself, is logic. It is a comforting faith for those who can embrace it.

Every creed appears to breed heretics as freely as it spawns believers. Before *Principia mathematica* was off the press, Brouwer (1907, 1908, 1912) had challenged beliefs which nearly all of his predecessors in mathematics had accepted without question. The one outstanding exception to respectable orthodoxy had been Kronecker (1823–1891). In his merciless attack on Weierstrass, Kronecker declared that if he himself should not be granted the years and the strength to destroy mathematical analysis, his successors would finish what he had begun. Although neither Brouwer nor anyone else has yet (1945) destroyed mathematical analysis, it appeared by 1912 that a successor to Kronecker had at length arrived. Brouwer's doctoral dissertation of 1907 dealt with the foundations of mathematics. This was followed in 1908 by a blast of only six pages on "the unreliability of the principles of logic." A modest fifteen pages in 1912, on intuitionism and formalism,[7] consolidated Brouwer's creed—intuitionism—and simultaneously attacked its rival faiths. It will be remembered that David slew Goliath with a single well-aimed pebble from his trifling slingshot. Not that either logicalism or formalism expired under the impact of Brouwer's fifteen pages, for neither did. But both bestirred themselves.

Striking at the very heart of accepted beliefs in analysis, Brouwer denied the universal validity of the law of excluded middle (in the form that a proposition is either true or false) in deductive reasoning as it occurs in mathematics. In particular, it is forbidden, in intuitionism, to make a general assertion about an infinite class unless a method is prescribed for proving or disproving the assertion in a finite number of steps; and 'existence' without 'construction' is likewise banned. To illustrate by a simplified detail from one of Brouwer's examples: "the sequence of digits 123456789 occurs somewhere in the decimal representation of $\pi(=3.1415926\ldots)$" is not admitted to be either true or false, because no method is known (1945) for deciding. From this it is easy to describe in words a series which cannot be proved or disproved to be convergent. The statement that the series is convergent is not admitted as a proposition in the sense of logicalism; and there is left open the possibility that the statement is neither true nor false. Contrary to what Plato's Eternal Geometer appears to have believed, the logic of a possible 'mathematics' need not be hobbled by a two-valved 'truth'; 'mathematics' may be potentially richer than the system contemplated in logicalism. Another variant of traditional logic is Brouwer's theorem that "*A* implies the absurdity of the absurdity of *A*, but the absurdity of the absurdity of *A* does not imply *A*." At this point, some may feel inclined to abandon intuitionism as itself absurd. But if they do, they may miss a great deal of nourishing entertainment.

After Brouwer's assault, Weyl followed with a subversive critique, *Das Kontinuum, u.s.w.*, 1918, in which he reached the disturbing conclusion that "'the firm rock' on which the house of analysis is founded in the sense of formalism" appears on critical inspection to be illusory: "that house, in an essential part, is built on sand." Remembering a parable in the New Testament, we do not need to be warned how foolish it is to build a house on sand. Weyl abandoned the formalists' house and took up his residence with the intuitionists, as did several other distinguished mathematicians.

It cost them something to move. They could no longer believe, for instance, that the real numbers are more numerous than the rationals. What they retained of their classical mathematics was less even than the grudging Kronecker would have permitted them to keep. Had Kronecker lived till 1918, he too would have been induced to move out, dispossessed by the sheriff who arrived in response to the call for a disciple. In addition to all these severe deprivations, the intuitionists surrendered the good will of Hilbert, founder and uncompromising director of the formalist school.

The "frog-and-mouse battle" of the 1920's–30's between the formalists and the intuitionists has already been mentioned. We must now briefly answer Einstein's question, and say what the battle was about. Unlike the gaffer who could not remember the causes of the battle of Blenheim, but who did recall that somebody won a memorable victory, we are unable to state that anybody was victorious, athough we do know what the great fight was about. It was a battle to the death between Hilbert's formalism and Brouwer's intuitionism for the possession of mathematics. It does not seem to have occurred to either combatant that while he was engaged in trying to exterminate his enemy, some ragged camp follower might make off with the prize; or that it might not make the slightest difference to mathematics whether the battle for him was won, lost, or drawn. The entire fracas bore a singular resemblance to the wars of the Middle Ages over subtle questions of religious dogma that later and saner generations perceived to have been pseudo questions devoid of meaning.

Formalism denies logicalism and seeks to controvert the conclusions of intuitionism. It ignores the technical difficulties of logic as being irrelevant for mathematics, and seeks to reduce all mathematics to the manipulation of symbols, 'meaningless marks,' in accordance with definite rules. The rules are extremely simple, such as that permitting the substitution of one symbol for another in a deductive proof under certain prescribed conditions. Incidentally, the symbolism contains far fewer signs than Whitehead and Russell used in *Principia mathematica*. As has been said by many, mathematics is reduced in formalism to a game like chess. People play chess; they do not ask what a particularly game 'means.' Conceivably chess may some day be interpreted in terms of the weather, or politics, or psychiatry; it will then acquire a meaning. In applied mathematics there is no necessity, of course, to interpret the equations of mathematical physics, for example, in terms of observable phenomena in the physical universe; and formalists deliberately leave all such 'realizations' of their game aside, as being irrelevant for mathematics.

Nearly all mathematicians agree with the formalists up to a certain point; for nothing so exasperates a mathematician who knows his trade as to hear geometry called a physical science. Whatever goemetry as mathematicians understand it may be, and whatever scientific applications it may have, it is not a science in the sense that physical measurement is. Not all the gadgetry of all the scientific laboratories in the world would suffice for a proof that the sum of the angles of a plane Euclidean triangle is two right angles. If the formalists succeed in eradicating the ignorant misconception that "geometry is the simplest of the physical sciences," they will have justified their existence. Nor is there much point in berating the formalists, as some have done, because their program temporarily divorces mathematics from a possible practical utility. If the history of mathematics teaches our all but unteachable race anything, it is the futility of trying to take two steps at the same time in different directions. To advance at all, the formalists were forced to take one step at a time. The capital problem of formalism, that of proving the consistency of analysis and, more generally, of all the mathematics believed by a majority of competent experts to be essentially sound, is difficult enough. So also is the more philosophical problem of the 'meaning' of mathematics. It has yet to be shown that the second problem is not a pseudo problem, and, if it is not, to exhibit its exact relevance for the first problem. Whatever the outcome is to be, the formalists have proceeded on the assumption that 'divide and conquer' is as sound a strategy in mathematics as it is in war. Whether or not they have conquered so far will be seen when we consider the fate of consistency-proofs in arithmetic.

Having mentioned pseudo problems, we may note in passing one of the more disastrous consequences of Hilbertian formalism. There can be but little doubt that a principal source of the revived logical positivism of the twentieth century was the renewed interest in the Euclidean program that began with Hilbert's *Grundlagen* of 1899. Postulational technique in mathematics, and the consequent sharpened logical analysis of mathematical foundations, inspired one work in particular that suggested a wider attack on the problems of metaphysics: L. Wittgenstein's (Austria) *Logisch-philosophische Abhandlung*, 1921 (reprinted with an English translation as *Tractatus logico-philosophicus*, 1922). Wittgenstein's thesis is that mathematics is a vast tautology, that it repeats only '*a* is *a*' over and over again in endless convolutions. A complete pure mathematics would be a 'logical syntax' of all possible 'symbolic languages.' This, however, was no disaster; what issued from it, was.

Wittgenstein's views profoundly influenced the positivists of the so-called Vienna Circle, an extremely aggressive and equally progressive band recruited in the 1920's by M. Schlick (1882–1936), of whom R. Carnap (Vienna, Chicago) became the leader after Schlick's assassination. Carnap's program is much broader than any of those of other workers in the foundations of mathematics. It includes also logic, language, science, and, in a wry sense peculiar to itself, metaphysics. Broadly, the aim is an analysis of languages and their semantics. In 1934 Carnap summarized and further developed his conclusions, elaborating his syntax of "theorems about theorems" in his *Logische Syntax der Sprache* (English translation, 1937). The relevance of Carnap's logical syntax for metaphysics is described in untechnical language in a short popular work of 1935, *Le problème de la logique de la science. Science formelle et science du réel.* Here the adept in classical metaphysics will find much to interest him, among other things the thesis that the greater problems of metaphysics that have exercised many philosophers from Plato to the present are pseudo problems without meaning. We begin to sense disaster in the offing.

It is small wonder that the logical positivists were as soundly trounced as they no doubt deserved by irate philosophers. Our interest here in this wholesale spanking is that we should probably never have heard it, had it not been that the labors of a long succession of mathematicians from Pythagoras to Hilbert inspired a few free spirits to challenge the pontifications of traditional authority. And those who blithely imagine that debates over the meaning of mathematics are of no human significance may be reminded of the fate of Schlick. So disturbing to some reactionaries were the theses of the positivists, seeming as they did to make nonsense of more than one cherished belief, that an overzealous defender of established creeds put a bullet through Schlick. Although that shot was not heard round the world, it announced the opening of the most efficient campaign against reason that the world has seen since the days of Galileo. Mathematicians, along with other conscientious objectors to obscurantism, were effectively silenced. All this was only reasonable from the standpoint of the authoritarians; for where ignorance pays dividends, it is folly to enlighten. But it was somewhat of a shock to realize that the fear and hatred[8] of science which Galileo thought he had quenched had been smoldering for three hundred years.

On the logical-mathematical front, before more realistic arguments terminated academic debates, Hilbert and Brouwer fought valiantly for the honor of their respective creeds. There was something exhilarating to disinterested bystanders in this exchange of incivilities between the Prussian veteran, well past his prime, and the resolute Dutchman in his forties who refused to be shouted down by all the authority of a great tradition. Even granting that Hilbert's project of proving mathematics free of contradiction by his theory of proof might succeed, Brouwer roared that "nothing of mathematical value will be attained in this manner; a false theory which is not stopped by a contradiction is none the less false, just as a criminal policy unchecked by a reprimanding court is none the less criminal."

Strong language was answered in kind. "What Weyl and Brouwer are doing," Hilbert shouted, "is mainly following in the steps of Kronecker. They are trying to establish mathematics by pitching overboard everything that does not suit them and setting up an embargo." Alarmed by the mere magnitude of what the rebels were jettisoning, Hilbert in his anxiety gave intuitionism an unintentionally excellent advertisement that attracted many customers for its gruesome charms:

The effect is to dismember our science and run the risk of losing a large part of our most valuable possessions. Weyl and Brouwer condemn the general notions of irrational numbers, of functions—even of such functions as occur in the theory of numbers—Cantor's transfinite numbers, etc., the theorem [basic in analysis] that an infinite set of positive integers has a least, and even the law of excluded middle, as for example the assertion: Either there is only a finite number of primes or there are infinitely many. These are examples of forbidden theorems and modes of reasoning. I believe that impotent as Kronecker was to abolish irrational numbers (Weyl and Brouwer do permit us to retain a torso), no less impotent will their efforts prove today. No!

The final shout re-echoed to cheer the dismayed troops of formalism on to a final victory, "Brouwer's program is not a revolution, but merely the repetition of a futile *coup de main* with old methods, but which was then undertaken with greater verve, yet failed utterly. Today the State is thoroughly armed through the labors[9] of Frege, Dedekind, and Cantor. The efforts of Brouwer and Weyl are foredoomed to futility."

So all day long, and for many a day, the noise of battle rolled from Amsterdam to Göttingen and back again. Having recovered from their momentary shock of alarm at the unwonted uproar in the ranks of the pure mathematicians, the physicists stood back to avoid the missiles and shook with laughter. Meanwhile the opposing factions, their dumps of epithets squandered without scoring a single direct hit, set about consolidating their respective gains. The intuitionists showed that a surprisingly large tract of mathematics could be revised to fit their 'original intuition' without exceeding the drastic limitations imposed by the creed of finitism; and the formalists continued their labors in Hilbert's theory of proof, confident that they were about to establish the eternal freedom from contradiction of traditional mathematics. And simultaneously, mathematical logic began to go the way of all mathematics of the recent period. It started subdividing and multiplying upon itself like a horrible nightmare.

Instead of only the three creeds of logicalism, formalism, and intuitionism to bewilder introspective mathematicians with 'the will to believe' in something, no matter what, crossbreeds between the original three were competing in the 1940's with attractive mutations of the originals for universal acceptance. Much of the highest interest was developed; but the total effect was one of anarchy and ungovernable confusion. Unreconciled differences of opinion between equally competent experts hinted that the entire problem of the foundations of mathematics might be but another of the pseudo problems of metaphysics in a modern disguise. Of many undeveloped suggestions—or prophecies—for a possible way round the apparent stalemate, only the more mathematical are of interest to us here.

It has been suggested that the ultimate course of mathematics will resemble that followed by geometry after the invention of non-Euclidean geometries. There may be no necessity to have but one kind of mathematics, if the objective is scientific applications; and there are no reasons for supposing that a single, all-inclusive mathematics can embrace all the useful kinds without internal contradictions. Again, in the same direction, but going deeper, it has been very hesitantly suggested that complete self-consistency is more of a luxury than a necessity in a usable mathematics. Looking far ahead, some whose work entitles them to an opinion believe that the paradoxes which have infested mathematics since Zeno are unsolvable, and that the mathematics of continuity will be abandoned, to be succeeded by a type of reasoning in which unsolvable paradoxes cannot arise. What this is to be, provided it is possible, of course nobody can predict. It may be worth mentioning that one of the most distinguished mathematicians of the twentieth century practically abandoned analysis, in which he had done outstanding work, because he had convinced himself that the foundations of analysis are unsound beyond repair. (This is not the sort of opinion a man cares to publish today; he might change his mind tomorrow.) A last possibility is suggested by a similar situation in quantum mechanics. With the apparent necessity for some form of indeterminacy in the physical sciences, the classical dichotomy 'true, false' is modified by an admixture of probabilities. In analogy with this, mathematics may cease to insist on a two-valued logic, and develop in accordance with a logic of three or more truth-values.

A few items from the post-*Principia* period may be mentioned for their affiliations with the capital problem of arithmetic to which we shall pass presently. Among numerous modifications and rectifications of *Principia mathematica*, concerned principally with the theory of types and the axiom of reducibility, the application (1926) by F. P. Ramsey (1903–1930, English) of Wittgenstein's ideas gave some promise of freeing *Principia* from contradiction. A technical detail of considerable significance (the omission of scope indicators) in the *Principia* was shown to be inconsistent by L. Chwistek (Polish), who in 1925 proposed his theory of constructive types for salvaging that part of

Principia which remains if the axiom of reducibility is rejected (along with another assumption, which need not be stated here). Chwistek found it necessary to invent a new symbolism to express some of his ideas. In his efforts to escape the theory of types, Chwistek developed his own metamathematics (1929, 1932), substituting for a hierarchy of types one of languages, in which any language in the hierarchy becomes the subject matter of the next higher language. For reasons that will appear when we describe Gödel's work, these and other attempts to patch up *Principia* took on a new aspect in 1931.

Generalizing *Principia* in some respects, for example by a method for proving theorems concerning general relations among n terms, W. V. Quine (U.S.A.) developed his *System of logistic*, 1934, said by Whitehead to be a fundamental advance: "In the modern development of Logic, the traditional Aristotelian Logic takes its place in a simplification of the full problem presented by the subject. In this there is an analogy to arithmetic of primitive tribes compared to modern mathematics." The case $n = 2$ of Quine's system is that discussed in *Principia.*

It is interesting to record here Whitehead's tribute to H. M. Sheffer and "the great school of Polish mathematicians," for their "influence on Dr. Quine's thought." The tribute (dated October 8, 1934) closes with the comforting assurance that "There is continuity in the progress of ordered knowledge." Five years less one month later, "the great school of Polish mathematicians" was being bombed from the air in the progress of ordered ignorance, that is, in the general progress of European civilization. What had taken twenty years to gather was dispersed and in part obliterated in about twenty days. "The great school of Polish mathematicians" followed the Vienna Circle into death or exile.

The progress (in mathematical logic) just noted was in the general direction of logicalism. Intuitionism also advanced, particularly in A. Heyting's (Dutch) discussion (1930) of the formal rules of intuitionist logic. At the same time H. B. Curry (1900–, U.S.A.) began his extensive combinatorial logic, a new departure. Formal logics having a wide degree of freedom in the expression of their formulas as values of functions were constructed (1933) by Curry and by A. Church (1903–, U.S.A.). Illustrative of the rapidity with which symbolic logic advanced in the 1930's both of these logics were proved inconsistent in 1935 by S. C. Kleene (U.S.A.) and J. B. Rosser (U.S.A.). Another result (1942) of Rosser's attracted considerable attention from experts in mathematical logic on account of its unexpectedness. In attempting to avoid the Burali-Forti paradox (described here in an earlier chapter), Quine in his *Mathematical logic* (1940) effectively admitted the paradox in a disguised form, thereby invalidating parts of his system. As stated by Rosser the nature of the oversight was as follows. Four of the basic principles of the theory of ordinal numbers and well-ordered series are: (1) To every well-ordered series there corresponds a unique ordinal number. (2) The series of ordinal numbers is well-ordered. (3) If x is a term of a well-ordered series S, then the series consisting of all terms of S which precede x is also well-ordered, and has a smaller ordinal number than S. (4) Any ordinal number, α, is the ordinal number of the series of all ordinals which precede α. The Burali-Forti paradox asserts the incompatibility of (1)–(4). To avoid the paradox, two devices have been tried: a theory of types to invalidate (4); an effective invalidation of (1) in certain critical cases. Essentially a combination of these two was proposed by Quine. It failed to invalidate (4) and did not invalidate (1) at certain critical points. By a slight but significant omission from his system, Quine circumvented the paradox. The modified system was then compared with others, including Zermelo's. By the addition of two axioms, one of which postulates the existence of a most inclusive class, Zermelo's system becomes equivalent to Quine's up to the point where Quine imposes axioms of elementhood.

A fitting climax, logically but not historically, to the symbolic logic that evolved from the symbolism in *Laws of thought* (1854) into the system of mathematical logic in Whitehead and Russell's *Principia mathematica* (1910–13), suddenly materialized in J. Lukasiewicz's two-page paper of 1921 on three-valued logic. Practically simultaneously, and in complete independence of Lukasiewicz's work (published in Polish, as the author was a member of "the great school of Polish mathematicians"), E. L. Post (1897–, U.S.A.) passed at once from the two-valued logic of Aristotle to m-valued truth-systems, where m is any integer greater than one.

The mathematical significance of this advance needs no comment. The history, we recall, began with William of Occam's buried discovery of a three-valued logic. Thus the submathematical analysis of the fourteenth century was resurrected in the mathematical logic of the twentieth. It is a most remarkable fact that some of the subtlest reasoning of the Polish developers of many-valued logics, like Occam's, was carried on without symbols. To common mathematicians constrained to manipulate symbols in order to reason consistently, this feat of Occam's and the Polish logicians almost passes belief. Possibly the explanation in both cases is the same: several of the Polish logicians, like Occam, were of the Catholic faith, and no doubt had mastered the difficult technique of reasoning in words which usually accompanies a Jesuit or other Catholic training in logic as part of a liberal education. But it seems unlikely that purely verbal argument could transcend the finite m in Post's m-valued truth-systems to reach the infinite-valued logic, with a contiouous range of truth-values, devised in 1932 by H. Reichenbach (1891–, Germany, U.S.A.) as a basis for the mathematical theory of probability.

Another possible use for multivalued logics was noted (1933) by F. Zwicky (1898–, Swiss, U.S.A.), who observed that such logics should be applicable to the quantum theory. The reception of these logics by physicists was similar to that of all non-Euclidean geometries until Einstein discovered Riemann. A reconsideration of Kant's conception of time, complementary to the rejection of his theory of space, seemed in the 1940's to be demanded in physics. Pythagoras also appeared to be about to moderate the universality of his 'number' as the measure and meaning of all things, and likewise for Aristotle and his cherished 'identity.' In macroscopic phenomena there is no difficulty about counting and identifying the objects contemplated. But the quanta of light, it is asserted, are unidentifiable and uncountable, whereas electrons are unidentifiable but countable. Should these assertions be correct, the immemorial label of 'number' is no longer universally applicable. Space and time likewise shed their traditional universality when pursued to the atomic nucleus. In an attempt to put a workable logic into the metaphysics of the physics of the 1940's, Reichenbach published (1944) an account of his application of a three-valued logic to the quantum theory.

In concluding this sample of post-*Principia* mathematical logic, we remember the distinguished American philosopher who in 1939 informed his scholarly public that mathematical logic had contributed nothing new to either mathematics or epistemology. We pass on to that item which may prove to be of the greatest interest for all mathematics in the entire development since Leibniz. It is the outcome of attempts with the refinements of modern mathematical logic to prove the consistency of arithmetic. Hilbert in 1898–9, we recall, emphasized the basic importance of this problem for geometry. Its significance for analysis needs no emphasis.

What at the time seemed like a promising attempt was J. Herbrand's (French) begun in 1929 and essentially completed in 1931–2, as a concomitant of the author's technique for solving certain problems in metamathematics. This appears to have fallen by the wayside. G. Gentzen (German) followed in 1936 with a proof for the consistency of a part of arithmetic, which he simplified in 1938. In the latter work, Gentzen stated his belief that the lack of a constructive proof at one stage of the entire demonstration would shortly be supplied. The actual demonstration therefore did not fully meet the Kronecker-Brouwer demand for finite constructibility. Contrary to Hilbert's formalistic program of treating infinite classes by the same methods as those used for finite classes, Brouwer admits an infinite class only in the sense that, given any finite class, a larger finite class can be taken. Brouwer's restricted 'infinite' had not sufficed to bridge the gap in Gentzen's proof early in 1945. Its transfinite elements had been refined by Gentzen in 1943. Whether this proof, the most nearly complete up to its time, is to survive can be predicted only by one competent to finish or to destroy it. Not being in that fortunate position, we proceed to the end of what appears to be a blind alley in this direction.

Efforts to prove the consistency of arithmetic by mathematical logics of the genera developed in *Principia mathematica* and Hilbert's theory of proof (*Beweistheorie*) were believed by many in 1945 to be necessarily futile. The official manifesto of the formalist school, *Grundlagen der Mathematik*, by Hilbert and P. Bernays, appeared in two substantial volumes in 1934 and 1939. It was not clear in 1945 that the work of 1939 had righted the fiasco—the authors' word—in which the *Beweistheorie* seemed to have collapsed in 1934. As the authors observed, the final test of the theory will come when it is applied to prove the consistency of analysis. Gentzen's proof for arithmetic was accepted.

The apparent fiasco was occasioned by a disturbing theorem proved in 1931 by K. Gödel (Austria). According to Gödel's theorem, it is impossible in certain logical systems to prove, by the rules of the system, certain theorems belonging to the system which can be seen otherwise to be true. This is no mere existence theorem of the non-constructive type to which Kronecker objected: Gödel constructed a true theorem such that a formal proof of it leads to a contradiction. Undecidable statements exist: within the system certain assertions can be neither proved nor disproved. The logical systems for which Gödel's theorem holds include what is called the restricted calculus of propositional functions (quantification extends only to the arguments of the functions, not to the functions). Without technicalities it is impossible to be precise; but it may be said roughly that Gödel's theorem establishes the impossibility of proving the consistency of any system S containing either what is essentially the system of *Principia mathematica*, or arithmetic, by the methods of proof used in S.

Gödel's theorem has been called the most decisive result in modern mathematical logic. In proving it, Gödel invented a new device of symbolism: the usual symbols of the propositional calculus were replaced by the usual symbols for integers. A provable theorem, for example, is thus represented by a sequence of integers, and likewise for statements concerning the theorem. The outcome is a numerical algorithm. It might seem that this device when applied to theorems in arithmetic is a vicious circle. But obviously it is legitimate; any other recognizably distinct signs would serve the same purpose as the signs for the integers.

Gödel also proved a more general theorem than that mentioned, from which it follows that the calculus of propositional functions cannot be so symbolized as to furnish the kind of consistency-proofs contemplated in Hilbert's theory of proof. Another obstacle to the *Beweistheorie* of 1934 appeared in a theorem proved in 1936 by A. Church (1903–, U.S.A.), who demonstrated the impossibility of producing a general solution for a central problem (*Entscheidungsproblem*) of Hilbert's theory as originally formulated. The matter, too technical for description here, is discussed in the Hilbert-Bernays work of 1939, Supplement II, to which the reader is referred for a full account.

Another result cognate to the possibility of a consistency proof for arithmetic concerns the potential solvability of arithmetic problems. It seems reasonable to define an arithmetic theorem as a propositional function of integers constructible by the use of logical symbols from the three elementary relations $x = y$, $x = y + z$, $x = yz$. With this definition and proved propositions of Gödel and S. C. Kleene (1909–, U.S.A.), Skolem showed that every equation between general recursive functions is equivalent to an arithmetic proposition. The set of all arithmetic propositions is effectively enumerable. Skolem proved that no general method of solving all arithmetic problems is obtainable. Like Gödel's theorem, conclusions such as this were probably unthinkable to the mathematicians of the nineteenth century. Although it is no sense profound, as the preceding is, another supposition of the past may be easily set aside. Poincaré insisted that mathematical reasoning par excellence is reasoning by recurrence as in mathematical (or complete) induction. It might be imagined then that any arithmetic theorem is provable by mathematical induction. This is not so.

Skolem's theorem is in the general domain of completeness theorems and decision problems. A unified account of completeness theorems was given (1943) by Kleene, whose main nonequivalence theorem includes the theorems of Gödel and Church. These are to the effect that one-quantifier predicates exist for which there is no complete formal deductive theory, and that there also exist such predicates for which the decision problem is unsolvable.

From all this it would seem that the consistency of arithmetic had yet to be proved in 1945. The potential significance for all mathematics of such a proof, or of an acceptable statement of conditions under which it may be attainable, is obvious. The residue so far of the struggle to construct a proof is recognized as one of the most suggestive contributions that symbolic logic has made to mathematics in about two and a half centuries from Leibniz to Gödel.

Algebra of relations

Mathematics has been described (but not defined) as a study of relations. From the earliest systematic exposition of a mathematical theory, that of elementary plane geometry by Euclid in the fourth century B.C., to the twentieth century, relations have dominated extensive tracts of mathematical reasoning, and the simplest properties of a few relations were recognized abstractly from the earliest times. Among other relations admitted or used by Euclid, equality and congruence as in elementary geometry are instances of the general equivalence relation. Euclid incorporated some but not all of the equivalence properties of equality into his postulates. The corresponding properties of congruence were taken for granted in tacit assumptions wherever required in attempted proofs, as in the fourth proposition of the first book. Even as late as the first decade of the twentieth century, the postulates for equality were not always stated as a necessary part of the axiomatization of many disciplines. Classic papers on the axiomatics of algebra and geometry of the second decade of the twentieth century continued to ignore the necessity of defining equality by a set of postulates before using equality in proofs. Although the *Principia mathematica* of Whitehead and Russell undoubtedly was partly responsible for the sharpened precision of axiomatics after 1920, it seems to have been only in the early 1930's that the inclusion of a complete set of postulates for equality became customary. Until then the usual first postulate had been overlooked or included only by implication.

While an abstract theory of mathematical relations was thus hesitantly emerging, a theory of relations in the tradition of mathematical logic reached a certain maturity in a rudimentary but adequate algebra of such concepts as inclusion and relative product. This development possibly was of greater interest to logicians than to mathematicians. The techniques of modern abstract algebra seemed to suggest little of technical mathematical interest in the logical study of relations as such, whereas the algebra of classes had been fitted into the general scheme of modern algebra by the work of Stone and others. At last, in 1942, more than a beginning of an autonomous mathematical theory of relations appeared in Ore's general investigation of the properties of the several relations actually occurring in mathematical theories. The emphasis here was primarily algebraic, although some of the properties studied might be of interest to logicians.

The classical relations of mathematics, such as inclusion and betweenness, suggested to Ore a more general type, considered as correspondences $A \to R_A$ between sets, where the R_A corresponding to a set A may depend upon both the elements of A and a certain grouping or ordering of the elements. The notation aRA signifies that an element a in a set belongs to a subset R_A. Normal forms of relations are definable from combinations of such relations aRA. Automorphisms or endomorphisms of a relation R are those one-one or many-one correspondences α such that any relation aRA goes into another valid relation, say $a^\alpha RA^\alpha$. The problem of determining all automorphisms is connected with that of finding all correspondences commuting with a given correspondence; and this problem leads to a study of monomial groups and to problems in a general Galois theory, extending the Galois theory of equations.

On a more familiar level, the theory of binary relations, first systematically investigated by Peirce and by Schröder, becomes a theory of matrices (in the algebraic, not the *Principia*, sense) over a Boolean ring. In an alternative presentation, binary relations are pictured by graphs; symmetric relations correspond to symmetric matrices and to graphs whose edges are considered as undirected. The matric representation of relations combines the theory of relations with the well-developed theory of vector spaces over special rings. Sum, intersection, multiplication, and dual multiplication are definable for relations; whence there is a theory of rings and ideals of binary relations, and parts of the theory of linear algebras can be transposed to the theory of relations. But in other aspects the theory of relations differs from the classical theory of matrices. In this theory of binary relations, the automorphisms are of importance; the automorphisms correspond to the permutation matrices commuting with the relation matrix. The similarity here with somewhat analogous situations in projective geometry and the quantum theory is at least suggestive and, as for the debates of metaphysicians over the nature of 'space,' may hint at a concealed necessity of a two-valued logic.

Transitive relations R are characterized in Ore's theory, among other ways, as certain units or as certain sum relations; they are closely associated with the idempotents of the rings of relations. Any transitive relation is decomposable into an equivalence relation and a partial order relation. Complete solutions of several important algebraic problems are given for equivalence relations, so the interest passes to the partially ordered sets. As was seen in connection with lattices, and as is evident in much of twentieth-century abstract algebra, abstract geometry, and abstract analysis, partial ordering has become of basic importance in the mathematics of the recent period. Lattices or structures are among the most useful of partially ordered sets. Their appearance in the theory of relations complements their earlier recognition in classical Boolean algebra. It may be mentioned that in the domain of an n-valued truth system, a practicable algebraic algorithm is available for the discovery and exhibition of all relations in the system having prescribed characteristics, such as reflexivity, commutativity, transitivity, and the like.

Consistency proofs

Technical difficulties and obscurities in classical mathematical analysis were responsible for some of the more subtle developments of the mathematical logic of the twentieth century. The reactions of the resulting logic on analysis either clarified the obscurities or disclosed unsuspected flaws in accepted reasoning. Concurrently, the older programs of attempting to prove certain crucial hypotheses were replaced where feasible by proofs that sets of postulates containing at least one of the doubtful hypotheses are consistent. As a typical example, Cantor's generalized continuum hypothesis, fundamental for the real number system and all the centuries of analysis and metaphysics that have issued from it, passed in 1938 from the first phase to the second. Up to 1938 the first case of the hypothesis had been turned this way and that in numerous attempts to deduce from it either a contradiction or a valid conclusion. Should the latter succeed, and should the steps leading to the conclusion be reversible, the hypothesis would be proved. An exhaustive account of efforts up to 1934 in this direction was published by W. Sierpinski (Polish), himself one of the most prolific contributors.

At the Zurich Conference of 1938 on mathematical foundations and methodology, Sierpinski gave an elementary résumé of his own attempts and those of others to discover principles implying, or equivalent to, or implied by, the continuum hypothesis, the axiom of choice, and others of the same genus. Shortly thereafter a majority of the Polish mathematicians were concerned with matters more urgent than the speculations of two German mathematicians (Cantor, Zermelo) on the continuum. Banach, for example, after having been elevated by the Russian partitioners of Poland to the rectorship of the University of Warsaw, succumbed to the zeal of Cantor's and Zermelo's compatriots when the Russians precipitately withdrew from their share of the victim's homeland. But this is only in passing, as a mere footnote to European culture. It has little or no bearing on the advancement of mathematics, except possibly as a reminder that too many footnotes retard scholarship and are an evidence of pedantry rather than of learning. Pessimists of 1945 were confidently prophesying an avalanche of footnotes about the year 1960, halting the advance of mathematics indefinitely and obliterating whatever traces of logical and/or sane reasoning, mathematical or other, may then remain.

In its mildest form as originally proposed by Cantor, the continuum hypothesis asserts that every nondenumerable set has the power of the continuum. To recall the definitions: two sets are said to have the same power if there is a one-one correspondence between their respective elements; sets having the same power as the set of positive integers are called denumerable; sets having the same power as the set of all real numbers are said to have the power of the continuum. Attempts to invent or discover a set of real numbers which is not finite, nor denumerable, nor of the power of the continuum, have been unsuccessful; nor is it known whether one or other of these characteristics is an inherent property of a set. We recall that some of the controversial topics in Cantor's *Mengenlehre* stemmed from his meticulously phrased but dubiously comprehensive definition of *Menge* (set, class). An alternative statement of Cantor's continuum hypothesis is $2^{A_0} = A_1$, where for typographical convenience the Hebrew aleph is written A. Here A_0 is the power of a denumerable set, A_1 that of the continuum. All nondenumerable well-ordered sets all of whose segments are finite or denumerable have the power A_1. For the definitions of the alephs A_α, $\alpha > 1$, we must refer to any text on transfinite numbers, noting only that Borel and other skeptics early denied any conceivable meaning to these further alephs, while believers experienced no difficulty once they have thoroughly assimilated A_1. Including Cantor's hypothesis as stated above, the generalized continuum hypothesis asserts that $2^{A_\alpha} = A_{\alpha+1}$ for any α. Accompanying this is the strong form of the axiom of choice, permitting the simultaneous choice of an element from each nonempty set from the totality of sets, the choice to be a single relation. In 1938–9 (published, 1940), Gödel proved that if a certain set of axioms from *Mengenlehre* is consistent, then the set of axioms formed by adjoining to the original set both the generalized continuum hypothesis and the strong form of the axiom of choice, is also consistent. Some of the axioms of the original set have themselves been the occasion for controversy, for example the axiom of infinity. Nevertheless, experts in the foundations of analysis agreed that Gödel's proof marked a long step in advance in this much-disputed territory.

"*The solid ground of nature*"

"To the solid ground of nature trusts the Mind that builds for aye," Wordsworth declared, occasionally fitting deeds to the words; and again, "Come forth into the light of things; let nature be your teacher." Heeding Wordsworth's command, we shall ascend from the subterreanean regions of mathematics for a brief look about us, to see what became of the 'natural' half of the dream of Pythagoras.

Although profoundly altered after twenty-five centuries of incessant change, the dream is still recognizable. "The solid ground of nature" vanished from under our feet in the late 1920's in a haze of mathematical abstractions called the quantum theory. Whether it is ever to materialize again is not clear, even in "the light of things"; for 'things' have ebbed away in 'waves of probability,' and 'light' is a discontinuity of fleeting photons. Nevertheless, number still rules the universe of the exact sciences; and since the beginning of the twentieth century, number has begun to penetrate the less exact sciences. Wherever the statistical method is applied, mathematical probability is implicit; and this probability is measured by numbers.

If the application of mathematical probability to the physical sciences softened them into something less deterministically rigid than they were in the nineteenth century, a similar application stiffened the social sciences with a dash of determinism. An indivudual human being may still be as free as a famous document declares him to have been when he was created; but a hundred and thirty million individuals are no longer as free as they once imagined themselves. Mankind in the mass is more despotically governed by the laws of chance than it ever was by the decrees of any tyrant. If our shambling race is ever to get anything but suicidal destruction out of science, it may be a necessary first step that half a dozen human beings in every hundred thousand understand the mass-reactions of creatures who, as individuals, occasionally show that they can stand erect and walk like men. To grasp and analyze mass-reactions, whether of atoms or of human beings, a mastery of the modern statistical method is essential. The statistical method is social mathematics par excellence. A brief indication of the principal stages by which this humanized mathematics developed may therefore stand as the conclusion to our account of what mathematics had done—or may do—for the mass of mankind. On the technical side, the mathematical theory of probability seems to have contributed less to modern pure mathematics than any other major division of mathematics. It has been parasitic, deriving its vitality from the sciences to which it attached itself. No great method of analysis originated in the theory of probability, as several did in astronomy, mechanics, and mathematical physics.

Four contributions in the eighteenth and early nineteenth centuries overtop all the rest in importance for modern work. James Bernoulli's (1654–1705) posthumously published *Ars conjectandi* (1713) made a mathematical science of the elementary theory which had originated in a gamblers' dispute in the year of his birth. Bernoulli foresaw the social implications of probability, and is said to have had some notion of inverse probabilities. He also stated the theorem known by his name, that by sufficiently increasing the number of observations any preassigned degree of accuracy is obtainable. To anticipate, P. Tchebycheff (1821–1894), founder of the influential Russian school in mathematical probability, generalized Bernoulli's theorem in the law of large numbers.

Next, DeMoivre, in his *Doctrine of chances*, 1718, took several long steps forward, devising new methods for specific problems and inventing means for approximating to functions of large numbers. On the basis of his approximations (1733) to sums of terms of a binomial expansion, DeMoivre is now commonly credited with the all-important idea of the normal distribution curve.

Sociology and theology being barely distinguished from each other in his day, DeMoivre confined his more ambitious applications of probability to establishing the credibility of "A Great First Cause." Similar considerations led the Rev. T. Bayes (?–1761, English) to his theorem or formula—for inverse probability, published in 1763–4, relating to the probabilities of unknown causes inferred from observed events. Bayes' proof of his formula rested on an unsatisfactory postulate, and the entire subject of inverse probability remained more or less controversial until R. A. Fisher (1890–, English) put it on a sound basis in the early 1930's. Bayes, however, is assured of his immortality: he was the first to use mathematical probability inductively, "that is, for arguing from the particular to the general, or from the sample to the population."[9a]

The fourth outstanding advance of the period was Laplace's *Théorie analytique des probabilités*, 1812, the third edition of which appeared, with three supplements, in 1820. Before nearly seven hundred pages of not always lucid reasoning and intricate analysis, Laplace optimistically[10] remarks that "At bottom, the theory of probabilities is only common sense[11] reduced to calculation." This masterpiece, usually rated the greatest contribution by one man to probability, incorporates Laplace's previous work extending over about forty years. It is impossible to give an adequate idea of the richness of Laplace's work in short compass; but it may be said that it was distinguished by a free use of mathematical analysis. Laplace may not have been the first to conceive of a continuous distribution of probability, as T. Simpson (1710–1761, English) had introduced continuity into the theory of mathematical probability in 1756. But Laplace was the first to apply analysis consistently and extensively to what, from its very nature, is a department of combinatorial mathematics. The gain in power and flexibility was immeasurable.

Although the specific details of Laplace's synthesis of his own work and that of his predecessors in probability add up to an impressive total, their sum is scarcely comparable in over-all importance to this application of infinitesimal analysis to an essentially discrete department of mathematics. We have seen what analysis did for geometry in the seventeenth century and in the differential geometry of the eighteenth to the twentieth centuries; we saw also how analysis brought vast tracts of the theory of numbers under control. Laplace's advance in methodology similarly transformed mathematical probability. Always inventive of the effective algorithms appropriate for his problems, Laplace applied the method of generating functions to probabilities, and made free use of what are now called Laplace transforms. The last, as noted earlier, acquired a new importance in applied mathematics when it was observed that Heaviside's operational calculus is properly a topic in these transforms. With the excessive caution consequent on the work of Cauchy and Abel on convergence, the method of generating functions fell into disrepute. However, for many of Laplace's applications of the method, convergence is irrelevant, and parts of it are easily recast in terms of what Menger has called (1944) algebraic analysis.

On the historical side, Laplace in his masterpiece on probability overcomes—as usual—the temptation to give just credit to his contemporaries. It would be an interesting exercise for some historian of mathematics to separate what is indisputably Laplace's own from what is not. If Laplace did not originate the method of solving linear differential equations by definite integrals, he made such extensive use of it that it is usually credited to him, and likewise for the application of equations in partial differences, to which Lagrange has a claim.

The *Essai philosophique sur les probabilités* contains Laplace's reflections on the epistemology of probability. As this aspect of probability is still under dispute by metaphysicians and mathematicians, it seems unlikely that Laplace can have said the final word. One detail will suffice to show that it may be of practical importance to attempt to understand what 'probability' might mean in science and mathematics. If it is known only that an event has occurred n times and failed f times under given conditions, the probability that it will occur when these conditions next obtain is $(n + 1)/(n + f + 2)$. Both the reasoning by which this formula is derived and the formula itself have been rejected by some authorities and accepted by others equally competent. The profound cliché that we do not know whether the sun will rise tomorrow is disposed of by Laplace—as a deduction from the formula—by the comforting assurance that there is only one chance in 1826214 that the sun will not rise tomorrow. Lagrange, if asked, might have answered simply with his usual formula when he had no grounds for belief or disbelief, "I don't know." But the man who convinced himself that he had proved that the solar system is a mathematically determined perpetual motion machine (when excised from the rest of the universe of which it happens to be a part) had no doubts on the matter. If the critical examination of the postulates underlying probability no less than any other part of mathematics has done nothing else, it has made some mathematical astronomers of the twentieth century less Jehovahlike than their predecessors of the eighteenth and nineteenth centuries when legislating for the universe.

Laplace's interest in probability was subordinate to his passion for celestial mechanics. Problems in dynamical astronomy were thus indirectly responsible for some of the mathematics applied to economic affairs, from estimating by random sampling the marketability of a batch of telephone receivers manufactured by mass production, to forecasting by intelligence testing the fire-resistant qualities of a crop of machine-gun fodder harvested by conscription to make the world safe for democracy. Bayes' theorem is applicable in the first of these projects.

Mathematical astronomy was also responsible for another indispensable technique of the more exact social sciences, the calculation of probable errors. It also was the source of the method of least squares, in which it is required to find the distribution of the measurements of an observed datum when the measurements differ due to errors of observation. The method was used by Gauss in 1795, and reinvented in 1806 by Legendre. Its history is marred by a priority dispute, dignified on Gauss' side, bitter on Legendre's. In his astronomical work of 1809, Gauss postulated that when any number of 'equally reliable' measurements of an unknown 'magnitude' are given, their arithmetic mean is the most probable value.

Phrases like 'equally reliable,' 'equally probable,' generated much controversy over the 'meaning' of probability, and led in the twentieth century to attempts at a consistent theory. Early efforts to prove the normal distribution 'law' appear to be vitiated by a tacit assumption that a 'natural law' can be demonstrated mathematically without assuming anything. This may sound absurd; but it was only in the twentieth century that the Euclidean methodology began to be taken seriously by natural philosophers. Where it was recognized that postulates must precede proof, the postulates laid down as the 'natural' basis for a proof were sometimes no simpler than the 'law' which it was desired to prove. Modern practice tends more and more to accept the 'laws' themselves as facts of observation.

Dynamical astronomy was again responsible, at least historically, for one of the most extensive scientific applications of probability of the nineteenth and twentieth centuries. Maxwell in 1857 became interested in the kinetic theory of gases (it is said) as a direct consequence of his work on the stability of Saturn's rings. The kinetic theory may be traced back to the atomism of Democritus, Epicurus, and Lucretius; but as a mathematical discipline it originated with Daniel Bernoulli in 1738, and was first extensively developed by Maxwell, beginning in 1859. Others quickly followed Maxwell, particularly L. Boltzmann (1844–1906, Austria), beginning in the 1870's. The dates here are important: the earlier precedes the rigorization of analysis, and the later overlap Cantor's *Mengenlehre*. This hint should suffice to suggest that a fundamental part of the theory of statistical mechanics, which evolved from the kinetic theory of gases, could not possibly have been discussed adequately when it was first proposed. We refer to the ergodic theorem, or ergodic hypothesis, in any of its forms.[12] An historical sketch of the relevant matter would take us too far afield; our interest here is in the necessity for refined analysis in modern scientific applications of probability. The first satisfactory treatment of ergodic theorems[13] was delayed till 1932, when great progress was made by von Neumann and G. D. Birkhoff.

A rigorous discussion appeals to several of the theories noted in previous chapters, for example linear integral equations, nonlinear differential equations, harmonic analysis, Hilbert space, and modern theories of measure and integration. In somewhat the same direction, N. Wiener in 1937–8 attacked the problem of turbulence, mentioned in the Prospectus as basic for aerodynamics. From Saturn's rings to airplanes may seem quite a gap; but it is bridged by the modern theory of mathematical probability.

Until the 1930's, statistical mechanics was the most highly developed physical application of probability.[14] The more recent work appeared in 1945 to hint at some revision in the fundamental concepts of probability underlying classical statistical mechanics, either in its pre-quantum or modernized form. By a curious historical coincidence, Gibbs' classic *Elementary principles in statistical mechanics* was published in 1901, the same year as the older quantum theory, whose revised form, initiated in 1925, was to clear up much that had been obscure in the classical statistical mechanics. But no revisions are likely to affect Gibbs' reputation as the greatest mathematical physicist produced by the U.S.A. up to 1945, and one of the greatest of the nineteenth century. The deeper revisions of nineteenth-century applications of probability are of the same genus as those of the twentieth century in pure mathematics They affect our entire outlook on epistemology. But we must pass on; for it is still true, as Russell remarked[15] in 1929, that "probability is the most important concept in modern science, especially as nobody has the slightest notion what it means." The justice of this irony is evident on reading the discussions on free will and determinism touched off by Heisenberg's uncertainty principle in 1927. An apparently ineradicable indeterminacy in the physical sciences was seized upon to prove (1) that man has free will, (2) that he has not.

This brings us to that item in the development of probability which promises to be of greatest significance for mankind in the mass, the modern statistical method.[16] Whether or not individuals have free will, it seems to be established that crowds have not. Nevertheless, the simpler reactions of masses of human beings to external stimuli appear to be not entirely lawless. A typhoid epidemic will slay a roughly predictable percentage of the uninoculated and another roughly predictable percentage of the inoculated in the same environment. Again, philanthropists who insist on marrying their feeble-minded cousins to take care of them may expect a roughly predictable percentage of their (usually numerous) offspring to be feeble minded. And so on; there seems to be no escape from the mathematical laws of chance.

A history of the statistical method would fill a large book. We shall note only five of the main episodes partly responsible for modern developments. The first statistical breakdown of a national census was A. Quetelet's (1796–1874, Belgian), who in 1829 analyzed the first Belgian census, noting the influence on mortality of age, sex, season, occupation, and economic status. The bearing of such an analysis on life insurance is obvious. Of actuarial science in general it may be said that it has been so thoroughly explored that little remains to attract a professional mathematician. The run-of-the-mill actuarial mathematics requires no originality; and the valuable man is the one with imagination enough to devise new policies (both senses) that will make a profit for his company.

To Quetelet is due the valuable but somewhat damnable concept of "the average man"—"l'homme moyen." Quetelet's statistical studies convinced him that crime in a given population is to a certain extent mathematically predictable. Writing in the darkest age of pietistic moralizing, he had the courage to face down the sentimental obscurantists who called him a materialist—the ultimate in opprobrious epithets at the time—for suggesting that 'moral' and 'intellectual' qualities are measurable. Number had invaded the heavens and had replaced superstition by celestial mechanics. It was now about to attack the last stronghold of ignorance, the human mind. The desecration of this holy of holies was too much for the good people of Quetelet's day. A world war was required to teach homo sapiens that intelligence testing is not immoral. The U.S. Army tests in 1917 accomplished in a few months what Quetelet and his successors had been unable to do in seventy years. By 1919, intelligence testing in the United States was accepted without too violent opposition as a useful tool in education and criminology.

Among Quetelet's most ardent converts was Florence Nightingale, who, according to Pearson, believed that "to understand God's thoughts, we must study statistics, for these are the measure of his purpose." As it would be impossible to give a higher recommendation for the study of statistics, we pass on to the next notable manifestation of this divine mathematics, with the parting tribute to Quetelet that for thirty-five years (1837–71) he slaved at anthropometry—the measurement of man.

An obscure but highly intellectual contemporary of Quetelet's was the next to apply arithmetic to the inner secrets of living things. G. Mendel (1822–1884, Austria), abbot of the monastery at Brünn, in 1865–6 published the outcome of his beautifully scientific experiments on the hybridization of peas, *Versuch über Pflanzenhybriden*. For the first time in history heredity was linked to mathematics. Mendel's work was overlooked until the discovery of his buried paper in 1900. By then, others had found similar trails leading to the science of genetics; but Mendel is universally recognized as the founder of that vast science. It would be interesting to know what the abbot would have thought could he have foreseen that his arithmetic of heredity was to be used as an argument for compulsory birth control among the criminal and the feeble-minded.

An intellectual synthesis of both Quetelet and Mendel appeared in the peculiar genius of F. Galton[17] (1822–1911, English, a cousin of Charles Darwin), who made the next epochal contribution to the mathematics of man, in his investigation of correlations (1875–89). By 1877, Galton had arrived at the law of regression and the formula for the standard error of estimate. The experimental data from which he constructed his charts ranged from sweet peas through moths and hounds to human beings. It was in the period 1885–8, apparently, that Galton first worked out correlation coefficients. In 1889 he published a summary of his work on correlation in *Natural inheritance*. His earlier books, *Hereditary genius*, 1869, and *English men of science*, etc., 1874, still make good reading, especially for those who like to debate on environment versus heredity. Galton's work, on the whole, argues against Rousseau's conception of democracy, and possibly also against the somewhat diluted form of Rousseau's doctrine which inspired the Founding Fathers of the U.S.A.

Galton himself had an insatiable curiosity regarding everything under the sun. His interests embraced criminology and finger prints; numismatics and eugenics; and a few that can be described only in De Morgan's words as "the crockiest crotchets you ever saw." Galton's strength was twofold: he had inherited (or acquired?) a first-rate intelligence; his education had included mathematics, botany, and anatomy. This biologist-mathematician was the ideal candidate to invent a social mathematics, the modern statistical method. Galton's technical skill in mathematics was very limited. But his ideas were essentially mathematical, and they were new. When some elementary transformation bothered him, Galton called on his mathematical friends, of whom one was Cayley.

The decisive step took off from Galton's work, in W. F. R. Weldon's (1860–1906, English) studies of variation among individuals and local races. Weldon's 'populations' consisted of shrimps and shore crabs; his statistics was taught to him by the master himself, Galton. Measurements of the organs of shrimps (1890) convinced Weldon that certain biological characteristics are expressible in terms of correlation and variation. In 1892, he devised a method for computing correlation coefficients. Incidentally, the perfection of computing machines greatly accelerated the applications of statistical methods. Weldon was not a mathematician by training, but a biologist. In 1901 he and K. Pearson (1857–1936, English) founded *Biometrika*, a journal devoted to biometry, 'the science of the measurement of life.'

Continuing in Weldon's direction, Pearson gave the statistical method its strongest impulse since Quetelet. Pearson was a highly trained mathematician before he ever thought of biometry. The mathematics of statistics therefore caused him little trouble. His earlier work[18] was in mathematical physics, in such things as theories of ether-squirts, as he called them, to account for the existence of matter. But he quickly dropped these purely academic exercises when he found his true vocation in a new subject just opening up and suggestive enough to give even his restless imagination pause.

To Pearson more than to any other one man are due the mathematical techniques used by the majority of American educationists. Other methods are available, notably those developed with great skill and mathematical tact by the Scandinavian school; but in the main, Americans have preferred the procedures of the British school. The preference was confirmed by the work of R. A. Fisher, especially in applications to agricultural and horticultural experimentation. Although Fisher struck out in new directions of his own, he is usually regarded in the United States as Pearson's successor. Concerning the Scandinavian procedures, it seems just to say that the differences between them and the British are more a matter of techniques in calculation than any addition of significant statistical concepts. The basic ideas, the 'philosophy,' of the modern statistical method originated in the British school—not forgetting "the late Rev. Mr. Bayes."

Since the foregoing remarks on Fisher's work were written (1940), an uninformed criticism of his contribution to modern statistical methods appeared where it is likely to do most harm, namely, in a journal addressed primarily to educationists. Statistical analyses of test data forming so prominent a part of educational technique, it will be well here to quote a leading American expert on statistical theory regarding the criticism in question, so that others than the author of the mistaken judgment may not be misled. Replying (1943) to the misguided critic, H. Hotelling (1895–, U.S.A.) put the matter right:

> The statement that "except for the distribution of the coefficient of correlation for small samples, there is not a single thing that is fundamentally new in the Fisher system" does not do justice to the work of one of the greatest originators in the history of statistical methods. R. A. Fisher has been responsible for so many and such ingenious advances in the theory of statistics and in statistical methods that it is impossible in a brief note even to enumerate them. Among them are the systematic use and treatment of the method of maximum likelihood, the idea of sufficient statistics, the demonstration of the inefficiency of the method of moments which has been universally used and taught as the sole method of fitting frequency curves, the exact distributions, not only in small samples of any size, of the correlation coefficients, and also of the partial and multiple correlation coefficients, of the variance ratio, and of other statistics. Fisher was the principal leader in bringing about the use of the correct number of degrees of freedom in testing independence from contingency tables, and also in the introduction of valid and efficient experimental designs through the principle of randomization. He has made valuable contributions to computational techniques, including his useful method of finding an inverse matrix.

Incidentally, the mention of the inverse matrix is a reminder that matrix algebra, especially on its computational side (in which Hotelling himself has made notable advances among his numerous contributions to statistical theory and methods), has become a commonplace of modern statistics. Algebraists as a rule have been rather indifferent to the computational aspects of matrices. There is still much to be done; the recent interest of experts in modern algebra, for instance A. A. Albert (1905–,

U.S.A.), in this practical department of the theory of matrices promises that some of what is required will be forthcoming. The demands of atomic and molecular physics are equally compelling.

Some of Pearson's most important contributions are contained in the long series of memoirs beginning in 1894 and extending over about twenty years, with the rather curious title *Contributions to the mathematical theory of evolution.* Geneticists and other practical biologists who work with living animals and plants, from fruit flies and the greater families of the human feeble-minded to evening primroses and jimson weeds, look rather coldly on purely mathematical attempts to unravel all the tangled complexities of heredity. Nevertheless, even the most hardened experimentalist uses the statistical method to interpret his genetic findings. Pearson himself as he aged seems to have lost some of his youthful faith that mathematics can guide biology. Apparently it cannot, any more than it can render physical laboratories superfluous.

On the technical side, Pearson (1893) introduced the method of moments, and defined the normal curve and standard deviation. In 1900 he independently invented the extremely useful 'chi-squared' test for goodness of fit. In this he had been anticipated by Helmert in 1875. As pointed out by Fisher,[9a] "Pearson's paper contained a serious error, which vitiated most of the tests of goodness of fit up to 1921, yet the correction of the error" necessitates only minor numerical changes. This circumstance in itself suggests an interesting problem in probability.

Modern statistics makes heavy demands of knowledge and skill on its devotees in pure mathematics. Among the methods which have found applications in statistics are those of spherical trigonometry in space of n dimensions, topology, and abstract spaces. Two details of the modern pure theory will suffice. To A. Markoff (1856–1922, Russian) is due the concept of enchained probabilities. This concerns chains of successive probabilities, in which the value of the probability for a variable in a given place in the chain depends on the values obtained for the preceding variables. The Russian and French schools have investigated such probabilities most extensively. For an account of modern work, we refer to Fréchet's *Recherches théoriques modernes en calcul des probabilités*, 1936. Fréchet himself generalized the classical theory of probability as an application of his abstract spaces. Among general theorems, the following is one of the most curious: whenever the probability p of a determinate event is not changed by the knowledge of a finite number of results, p cannot be either 0 or 1. As a last specimen of the unexpected, Borel convinced himself that he had proved the astounding theorem that the human mind cannot imitate chance. Venturing no opinion on the reality of this diaphanous ghost of human free will, we merely note that a mastery of the modern statistical method demands a subtle mind and much mathematics.

The pure theory of the 1930's also appeared at last to be on the point of generating essentially new mathematics, for the first time in the history of probability. In his analysis of 'collectives,' R. von Mises (German) encountered novel problems in convergence. To circumvent endless metaphysical discussion on the meaning of 'equally likely' and 'random,' von Mises considered the actual sequences that might arise, say in throwing a die, such as 6, 6, 2, 1, 4, 3, ... Such a sequence is called a collective; and it is postulated that if, among the first n numbers of a collective, a specified number occurs m times, then m/n approaches a limit. 'Randomness' is replaced by the postulate that if from one collective another is formed by selection without knowledge of the value of a selected term, the limits described remain unchanged. It may therefore be possible to replace doubtful metaphysics of probability by sound mathematics. If so, we may some day begin to understand why our race, endowed with reason as it is supposed to be, occasionally behaves so unreasonably.

Retrospect

Here we reach the end of our journey. Looking back over the long and devious road, we see that mathematics in the six or seven thousand years of its known progress has contributed to civilization two things which promise to be of enduring worth: the method of deductive reasoning as it has developed in technical mathematics; the mathematical description of nature.

It was in mathematics that deductive reasoning first appeared, and it is from the same source that its successive extensions and refinements have issued. In its most powerful form, deductive reasoning is mathematics. The logical apparatus used in mathematics is incomparably more varied, more subtle, and more creative of new combinations than that associated with any other field of knowledge. And there has yet to be devised a method more efficacious than the mathematical for enabling human beings to reason about the results of scentific observations and experiments.

Compared with these two, all the rest is a matter of tactics. Thus, for example, the fact that analysis since the seventeenth century has proved more pliable than synthetic geometry in the mathematical description of nature is of historical interest, but it is not necessarily of any lasting significance. Geometry in the past held the same position relative to science that analysis holds today. A century hence, topology or some as yet uncreated mathematics may have revitalized geometry and restored it to scientific favor. But unless the mathematical method evolves into another, as distinct from what mathematics is now as was the empiricism which preceded mathematics from mathematics, it seems probable that the mathematical description of nature will retain its significance.

Alternatives are conceivable, indeed possible. Mystics, to whom the scientific habit of mind is only less repellent than the precision of mathematics with its hard, sharp clarity, prophesy a method more intuitive than those of science and mathematics. Adepts will perceive the universe as it 'is,' without an effort of sense or thought. Even stranger things have happened; and perhaps the strangest of all is the marvel that mathematics should be possible to a race akin to the apes.

Notes

(A) GENERAL REFERENCE WORKS; HISTORIES

1. *Encyklopädie der mathematischen Wissenschaften*, Leipzig, 1898–1935; revised, 1939–.

2. *Encyclopédie des sciences mathématiques pures et appliquées*, Paris, 1904–15 (incomplete).

3. *Bibliotheca mathematica*, Stockholm, Leipzig, 1884–1915.

4. *Bolletino di bibliografia e storia delle scienze matematiche*, Torino, 1898–1917; *Bolletino di matematica*, 1919–.

5. *Isis*, Bruges, 1913–.

6. *Quellen und Studien zur Geschichte der Mathematik*, Berlin, 1929–

7. M. Cantor, *Vorlesungen über Geschichte der Mathematik*, Leipzig, 1900–8, 4 vols. (to 1799 only).

8. D. E. Smith, *History of mathematics*, Boston, 1923–5, 2 vols. (to the calculus).

9. F. Cajori, *A history of mathematics*, ed. 2, New York, 1917.

10. W. W. R. Ball, *A short account of the history of mathematics*, ed. 5, London, 1912 (1927).

11. Others in bibliography in **8.**

12. G. Sarton, *Introduction to the history of science*, Washington, 1927–31, 3 vols. (to 1300; mathematics mentioned only incidentally).

13. J. Tropfke, *Geschichte der elementar-Mathematik*, u. s. w., ed. 2, 1–6 (1921–4).

14. R. C. Archibald, *Outline of the history of mathematics*, ed. 4, 1939 (Math. Assoc. of Amer.).

15. Vera Sanford, *A short history of mathematics*, Boston, 1930.

(B) NOTES FOR CHAPTERS

Chap. 2

1. O. Neugebauer, *Acta orientalia*, Copenhagen, 17, 1938, 169.

2. Neugebauer and others in **A 6**; R. C. Archibald, *Isis*, 71, 1936, 63.

3. *The Rhind mathematical papyrus*, A. B. Chace, R. C. Archibald, and others, Oberlin, Ohio, 1927–9.

4. S. Clarke and R. Engelbach, *Ancient Egyptian Masonry*, Oxford, 1930.

5. W. W. Struve in **A,** *Quellen*, 1.

6. Used by Antiphon (c. 430 B.C.), Bryson (5th cent. B.C.), Democritus (460?–370? B.C.).

7. O. Neugebauer, *Vorlesungen*, 1934, 203, seems to be more generous.

Chap. 3

1. General reference: **A** (Tropfke), T. L. Heath's editions of Euclid, Aristarchus, Archimedes, Apollonius, and his other writings, all cited in **A** (Smith); any edition of Plato's dialogues; the bibliography in **A 8.**

2. Nothing of first-rate importance issued from the Greeks' perfect numbers.

3. Claims by Hindu historians, B. Datta and A. N. Singh; *Hist. Hindu Math.*, Lahore, 1935–8. F. Cajori, *Scientific Monthly*, 9, 1919, 458. Place-system and zero attributed by some to abacus and other primitive apparatus for arithmetic.

4. V. F. Hopper, *Medieval number mysticism*, New York, 1939.

5. *Republic*, 546. *Timaeus*, 53–81, for Plato's ripest numerology; criticism by Aristotle, *Metaphysics*, A, 992a–, 1083b, 987b, 1084.

6. J. H. Jeans.

7. *Republic*, 527: " . . . this knowledge at which geometry aims is of the eternal, and not of the perishing and transient." Frequently quoted with approval by romanticists; dismissed as rubbish by mathematicians, who know that mathematics is made by human beings for human needs.

8. Controversial. Dinostratus, c. 330, is said (Tropfke, 4, 200) to have given the first extant proof by the indirect method. Hippocrates reputedly used the more general technique of equivalent reduction of propositions. Arguments ascribing the method to Plato are vague. It is used in discussing Zeno's paradoxes.

9. Mathematical end, 415 A.D., with death of Hypatia.

10. Arguments in Heath's *Aristarchus*.

11. Plato, *Theaetetus*, 147. On *mathematical* objections to certain *historical* conjectures, see G. H. Hardy and E. M. Wright, *An introduction to the theory of numbers*, Oxford, 1938, 42–3, 180–1; Tropfke, 2, 63–4.

Chap. 4

1. Gibbon's *Decline and fall of the Roman empire*.

2. 'Moslem' refers to a common religion (Mohammetanism), whatever the adherent's nationality.

3. More sympathetic estimates in Smith, **A 8,** Sarton, **A 12.**

4. From what is known of him, the Devil could never have been so stupid at figures as Gerbert, his alleged collaborator, was.

5. Doubtfully credited with a collection of slightly silly puzzle problems. Even "for his own times"—the trite historical cliché—he was lower than hundredth-rate mathematically.

7. **A 11, 12,** among many.

8. Bacon's rudimentary knowledge of mathematics, as represented in his published writings, scarcely guarantees his frequently quoted testimonial—"mathematics is the gate and key of the sciences," etc. Curiously, the most flattering testimonials for mathematics have been (and still are) the enthusiastic utterances of men who knew very little about the subject. To dispose here of another myth, Leonardo da Vinci's (1452–1591) published jottings on mathematics are trivial, even puerile, and show no mathematical talent whatever.

9. That the last phrase has no meaning, does not detract from whatever else it may have.

10. *Mysil katolica wobec logiki wspólczsnej* (catholic thought in modern logic), *Studia gnesiana*, 15, Posen, 1937.

Chap. 5

1. Readable account in Gibbon's *Decline and fall*.

2. Religious designation only; included Arabs, Persians, etc.

3. Chap. 2.

4. Almost abandoned.

5. L. E. Dickson, *Hist. theory of numbers*, Washington, 2, 1920, 347.

6. F. Cajori, **A 9,** 96.

7. Approvingly quoted from H. Hankel, *Geschichte der Mathematik*, u.s.w., Leipzig, 1874, by B. Datta, *Bull. Calcutta Math. Soc.*, 19, 1928,87 –; S. K. Ganguli, ibid., 151– for criticisms of European critics; approved by Sarton, **A 12.**

8. F. Cajori, *Hist. math. notations*, Chicago, 1928, 1, 84.

9. **A 9, 8, 12.**

Chap. 6

1. But not, unfortunately, in textbooks and elementary instruction, where masses of dead material and outmoded ways of thinking obscure the little worth knowing or remembering.

2. Contrary estimates readily available. As elsewhere, we here ignore textbooks and compilations which, however famous and useful in their day, made no substantial addition to mathematics. Likewise for their authors. Thus L. Pacioli (c. 1445–c. 1509, Tuscan), whose "great work [1494] summing up . . . the general mathematical knowledge of his time is a remarkable compilation with almost no originality"; **A 8.**

3. C. H. Gräffe's (1799–1873, Swiss) method (1837) preferred by some.

4. The libel laws prohibit publication of names.

5. In method, not in time. Euler's work in algebraic equations is in the same pre-Lagrangian tradition.

6. But not used. Modern algebra proceeds on lines totally different from those in any of Gauss' four proofs (of which two are inadequate).

Chap. 7

1. *Philosophiae naturalis principia mathematica.*

2. Such as Stirling's for $n!$.

3. Tropfke, **A 13,** 92–100, 104.

4. J. L. Coolidge, *Osiris*, 1, 1936, 231–50.

5. Facsimile reproduction and English translation by D. E. Smith and M. L. Latham, Chicago, 1925.

6. The so-called plane problem of Pappus inspired Descartes and several of his successors: PL_i, $PM_j (i = 1, \ldots, r; j = 1, \ldots, s)$ are the line segments drawn in a fixed direction from P to $r + s$ given straight lines; if

$$PL_1 \cdot PL_2 \ldots PL_r = c. PM_1 \cdot PM_2 \ldots PM_s,$$

where c is constant, to find the locus of P. Newton's "ad quatuor lineas" is an elegant solution for $r = s = 2$.

7. Appendix to first edition of *Opticks: Enumeratio linearum tertii ordinis.*

8. Fermat, *Oeuvres*, 1, 91–100; 3, 161.

9. The general French opinion; argument against, **4.**

10. *Oeuvres*, 1, 170–3; 2, 354, 457.

11. Heron of Alexandria knew the special case for reflection from a plane.

12. *Methodus ad disquirendum maximum et minimum*, *Oeuvres*, 1.

13. If x denotes a number, and 1 the unit 'length,' x^n denotes the number $x^n \times 1$.

14. Theory of vortices.

15. "The sort of story a little boy might make up,"—Madeleine Dmytryk.

16. *Leibnizens mathematische Schriften*, hsg. C. I. Gerhardt, Berlin, 1850, Abt. 1, Bd. 2, 20–35, espec. 21.

17. Among projected applications was a logical calculus of geometrical propositions called by Leibniz analysis situs. This has no connection with the modern meaning of analysis situs.

18. *Disquisitiones arithmeticae*, Lipsiae, 1801, 74–5, (§76): " . . . et cel. Waring fatetur demonstrationem eo difficiliorem videri, quod nulla *notatio* fingi possit, quae numerum primum exprimat.—At nostro iudicio huiusmodi veritates ex notionibus potius quam notationibus hauriri debebant." This is the one place in what he himself published that Gauss indulged in irony and sarcasm (also a pun), perhaps to show that he was more human than some of his admirers. Contrary opinion on notation, especially in regard to the calculus, by M. Cantor, **A 7.**

19. See **7,** where Newton explains his notation.

20. W. D. Ross, *Aristotle selections*, New York, 1927, espec. 88–99 (*Metaphysics*, 1066–7, 1068–9; *Physics*, 231–3, 239–40).
21. Correspondence in Fermat's *Oeuvres*.
22. *De ratiociniis in ludo alaeae*.
23. L. E. Dickson, *Hist. theory of numbers*, Washington, 1919, 1, 59.
24. For a spectacular instance, S. Skewes, *Jour. London Math. Soc.*, 8, 1933, 277.
25. H. J. S. Smith, *Coll. math. papers*, Oxford, 1894, 1, 42.
26. Cajori, in *Napier tercentenary vol.*, London, 1914.
27. *Discorsi e dimostrazioni matematice* etc., Leyden, 1638, E. Mach, *Die Mechanik in ihre Entwickelung*, 1883; (trans., T. J. McCormack) London, 1902; A. Einstein and L. Infeld, *Evolution of physics*, New York, 1938, Chap. 1.
28. A. S. Eddington, *The philosophy of physical science*, London, 1939.
29. Much scholarly and inconclusive disputation on both sides.
30. *Portsmouth papers*.
31. *Horologium oscillatorium*.

Chap. 8

1. Credited by some with exponents.
2. A proof of consistency is external to the formalism.
3. *Mémoir sur la théorie des équivalences algébriques*. (*Exercices d'analyse*, etc., 4, 94).
4. *Schriften* (ed., Gerhardt) II (3), 12; V (3), 218, 360.
4a. Euler's, 5 May 1777.
5. *Harmonia mensurarum*, Cambridge, 1722, 28.
6. Not to be confused with Horst Wessel.
7. Corrected by A. Ostrowski, Gauss' *Werke*, 10_2.
8. Texts on modern higher algebra written after 1929; concise summary by O. Ore, *L'algèbre abstraite*, Paris, 1936.
9. References and historical evaluation, Gauss, *Werke*, 10_2, 1923, 56–.
10. *Report, British Assoc. Adv. Sci.*, 3, 1834; *Symbolical algebra*, 1845.
11. The terms 'commutative,' 'distributive' were introduced in 1814 by F. J. Servois (Gergonne's *Annales*, 5, 1814–15, 93); 'associative' by Hamilton.
12. A **8**, 1, 460.
13. Dedekind, *Werke*, 3, 335.
14. *Allgemeine Function-Theorie*, 1882, 54.
15. *Compte rendu du* 2me (1900) *congrès internationale des mathématiciens*, Paris, 1902, 72.

Chap. 9

1. Foreseen by Gauss in 1831; *Werke* 2, 176: "Der Mathematik . . . "
2. Combinatorial multiplication.
3. Alternate numbers.
4. *Über der Zahlbegriff*.
5. *Die lineale Ausdehnungslehre, ein neuer Zweig der Mathematik*, 1844; *Die Ausdehnungslehre, vollständig und in strenger Form bearbeitet*, 1862.
6. *Werke*, 8, 357–62, espec. 360.
7. Möbius tried but failed to make Gauss publish his thoughts on analysis situs.
8. *Berichte . . . der Sächsischen Gesellschaft der Wissenschaften zu Leipzig*, 62, 1910, 189.
9. *Amer. Journ. Math.*, 4, 1881, 97–.
10. *Proc. Amer. Assoc. Adv. Sci.*, 35, 1886, 37; *Collected works* of J. W. G., New York, 1928, 2, 90. *Vector analysis*, New Haven, 1881, 1884.
11. *Life*, etc., *of P. G. Tait*, Cambridge, 1911, 164.
12. *An elementary treatise on quaternions*, ed. 3, 1890, vi.
13. *Delle derivazione covariante e contravariante*, Padova, 1888.

14. L. E. Dickson, *Algebras and their arithmetics*, Chicago, 1923, 200; L. E. D., *Trans. Amer. Math. Soc.*, 4, 1913, 13; E. V. Huntington, ibid., 6, 1905, 181.

15. *Oeuvres*, Christiania, 1881, 2, 224, 330.

16. *Oeuvres*, etc.; *Journ. des Math.*, 11, 1846, 395, 417–8.

17. First paper, 1853; *Werke*, 4, 1929, 1.

18. Beginning in 1858; *Werke*, 3, 439.

19. In the 1930's, 'structure' appeared with a different meaning in abstract algebra; 'structure' as explained here dates from 1912 at latest.

21. *Principia mathematica*, Cambridge, 1912, 2, *150; B. Russell, *The analysis of matter*, New York, 1927, Chap. 24.

Chap. 10

1. Kronecker's history (1875) is followed here; *Werke*, 1897, 2, 1.

2. Generalized in the theory of relative fields.

3. See H. J. S. Smith, *Report on the theory of numbers*, Oxford, 1894, 1, 93–. Kummer drew a bizarre analogy between the invisible activities of his implicit ideal numbers and those of the then-undiscovered element fluorine. Advances in arithmetic and chemistry quickly made nonsense of the analogy—retailed, with approval, as late as 1921 by a prominent mathematician.

4. The ideals of a given number field are classified with respect to a certain equivalence relation into a finite number of classes; a *usable* determination, in any given instance, of this 'class-number' is an outstanding desideratum.

5. G. Zolotareff's (Russian) of the 1870's–80's has been undeservedly neglected.

6. Historical-critical account in H. Hasse (German) *Bericht über . . . Zahlkörper*, 1,2, Leipzig, 1930. Modern generalizations originate in the class-field theory, begun by Hilbert in 1898; the landmarks are the theorems of P. Fürtwangler (German), 1902–28; T. Takagi (Japanese), 1922; E. Artin (Germany, U.S.A.), 1927–8; H. Hasse, 1924–30. The modernized Galois theory of fields is basic here.

7. Daughter of M. Noether, mathematician. According to E. Landau (German mathematician), "Emmy was the origin of coordinates in the N. family."

8. *Werke*, 1932, 1, 64.

9. *Acta eruditorum*, 2, 1683, 204.

10. Special case of G. B. Jerrard's (English) work 1832–5.

11. *Oeuvres*, 3. Similar progress by C. A. Vandermonde, *Mémoire sur la résolution des équations*, Paris, 1771; G. F. Malfatti (1731–1807, Italian), *De aequationibus* [of degree 6], Siena, 1771; *Tentativo per la risoluzione* [of equations of degree 5], ibid., 1772.

12. *Coll. Math. Papers*, 10, 402: "A group is defined by the laws of combination of its symbols." Ibid., 403, for what is sometimes called Dyck's theorem (*Math. Annalen*, 20, 1882, 30): every finite group is representable as a permutation group.

13. O. Hölder, 1889.

14. C. Jordan, *Traité des substitutions*, Paris, 1870, 42.

15. A modernized presentation of the Galois theory proceeds from the group of automorphisms of a normal field. (See A. A. Albert, *Modern higher algebra*, Chicago, 1937, Chap. 8).

16. Synoptic account, with historical notes, in F. Klein, *Lectures on the icosahedron* (trans., G. G. Morrice, London, 1913, of *Das Ikosaeder*, u.s.w., 1884).

17. Moore and Hölder almost simultaneously (1893) discovered the group of automorphisms of any finite group, basic in **15.**

18. Notably I. Schur (German).

Chap. 11

1. Algebras without a finite basis have been much less studied.

2. Principally in the class-field theory.

3. What God told Eve, as reported by Milton.

4. Exposition and bibliography to 1916 in L. E. Dickson, *Linear algebras*, Cambridge, 1916; for 1916–33, Dickson, *Algebras and their arithmetics*, Chicago, 1923; for

1923–34, M. Deuring, *Algebren*, Berlin, 1935. The like for the theory of ideals in W. Krull, *Idealtheorie*, Berlin, 1935.

5. *Amer. Jour. Math.*, 4, 1881, 229. Previously by Frobenius, *Journ. für Math.*, 84, 1878, 59.

6. Non-associative algebras have been less extensively investigated; a famous one is Cayley's of order 8, which found an application in the quantum theory.

7. This marks the beginning of the modern structural theory of linear associative algebra.

8. From **4,** 1916, 66.

9. *Werke*, 2, 169.

10. Son of the mathematician G. D. Birkhoff.

11. Ore's, with a few verbal changes to suit our terminology. See also S. MacLane, *Amer. Math. Monthly*, 46, 1939, 3.

12. The Jordan-Hölder decomposition theorem, described earlier, and its refinements, also Wedderburn's theorem on the structure of linear algebras, are instances.

Chap. 12

1. W. F. Osgood, *Functions of real variables*, Peking, 1936.

2. E. Landau, *Grundlagen der Analysis*, Leipzig, 1930.

3. Galileo Galilei, *Discorsi e dimostrazioni matematiche intorno à due nuove scienze*, Leida, 1638, espec. 32–3.

4. First English translation of **3**: Galilaeus Galilaeus, *Mathematical discourses and demonstrations*, etc., London, 1665, 25–7. This gives a much sharper rendition of Galileo's arguments on the infinite than any other translation.

5. Terms used interchangeably by several authors.

6. *Paradoxien des Unendlichen*, 1850 (posthumous ed., F. Přinhonsky).

7. *Phil. Werke* (ed., Gerhardt), 1, 338. Leibniz was right, but not for the reason he gave; see end of this chapter.

8. Unter einer "Menge" verstehen wir jede Zusammenfassung *M* von bestimmten wohlunterschiedenen Objecten *m* unsrer Anschauung oder unseres Denkens (welche die "Elemente" von *M* gennant werden) zu einen Ganzen.—G. Cantor, *Ges. Abhandlungen*, Berlin, 1932, 282. No two of a dozen mathematicians and scientists bilingual in English and German agreed on the meaning of this definition; two said it was meaningless. This definition is one source of trouble in the foundations of mathematics.

9. E. V. Huntington's.

10. E. Borel, *Théorie des fonctions*, Paris, 1914, 102–81. Similar doubts in 1940.

11. *Cours d'analyse de l'Ecole Polytechnique*, Paris (ed. 3, 1909, 1, 90).

12. K. Gödel (1939) worked in this field.

13. Kronecker has some claim to have been the founder of 'operationalism'—a popular (1945) philosophy in science.

14. Principal works: *Begriffschrift*, 1879; *Die Grundlagen der Arithmetik, eine logische-mathematische Untersuchung über den Begriff der Zahl*, 1884; *Grundgesetze der Arithmetik*, 1 (1893), 2 (1903).

15. It seldom is in mathematics.

Chap. 13

1. Also in C. Maclaurin's *Treatise on fluxions*, Edinburgh, 1742, the locus classicus for fluxions.

2. Ostwald's *Klassiker*, u.s.w., No. 211.

3. *Analyse des infiniment petits*, Paris, 1730.

4. Exasperated by some of his pupils from Canada, E. H. Moore called 'nothing' "the Canadian zero."

5. Exceptions noted in connection with asymptotic expansions.

6. The Diderot-D'Alembert *Encyclopédie, ou dictionnaire raisonné*, etc., Paris, 1754, Geneva, 1772; Arts. *Limite, différentiel.*

7. *Oeuvres*, 9, 15–20.

8. *Oeuvres*, 9, 141.

9. For a critical mathematician's defense of Lagrange's method when it was comparatively new, see A. De Morgan, *Penny Encyclopedia*, London, 1837; Art. *Differential Calculus*.

10. *Der Polynomische Lehrsatz* u.s.w., *von Tetens, Klügel, Krampf, Pfaff und Hindenburg*, Leipzig, 1796.

11. "Is. Uber," *L'Intermédiare des Mathématiciens*, 23, 1916, 164.

12. *Werke*, 3, 123.

13. Serious oversight noted in later chapter.

15. In one proof of the fundamental theorem of algebra, he *assumed* that if $f(x)$ is a polynomial in x, and if $f(a)$, $f(b)$ are of opposite signs, for a, b real and $a < b$, then $f(c) = 0$ for some c such that $a < c < b$.

16. Noted in later chapter.

17. *The analytical theory of heat* (trans., A. Freeman), Cambridge, 1878. The virtues ascribed (p. 7) by Fourier to mathematics are precisely those which this, his masterpiece, lacks.

18. ibid., 168.

19. ibid., 184.

20. The general French opinion, Darboux dissenting.

21. See **17**, 185. Euler also had *almost* found the theorem. This matter is resumed in a later chapter.

22. *Werke*, 1, 135.

23. Published posthumously, 1867.

24. H. J. S. Smith, (*Coll. papers*, 2, 313) in 1881 corrected and amplified some of Riemann's statements.

25. *Compte rendu du* 2me *congrès internationale des mathématiciens* . . . 1900; publ., Paris, 1902; 120–2.

Chap. 14

For references to the literature (exclusive of the analytic theory) to 1923, L. E. Dickson, *History of the theory of numbers*, Washington, 1919–20–23, 3 vols.

1. The disastrous fallacy in Art. 299 of the *Disquisitiones*, first detected by Dickson (1921), retarded or misled diophantine analysis for 120 years; see Dickson, *Modern elementary theory of numbers*, Chicago, 1939, Chap. 9.

2. Customarily misnamed the Pellian equation, after an obscure English mathematician, J. Pell (1610–1685).

3. Reproduced, 1913.

4. As pointed out first by Jacobi, *Journ. für Math.*, 30, 1845, 184. The Weierstrassian concept of primary factors was necessary.

5. Historical claims in behalf of the crystallographer Bravais, based on his work on crystal lattices, seem to be based on a misapprehension of what the geometry of numbers is about.

6. Used by Gauss, 1801, in the composition of binary quadratics.

7. Sufficient to assume the roots of $f(z)$ distinct and $c \neq 0$.

8. A brilliant suggestion by an eminent American historian of mathematics ascribes the invention of congruence to the 12-hour day and the 365-day year, facts with which Gauss undoubtedly was acquainted. The same authority detects finite groups in the Babylon of 2000 B.C., and even in prehistoric times. Not to be outdone, a German specialist ascribes a 'consciousness' of group theory to practically all peoples who ever made or used a pot, because of the repeat patterns decorating primitive pottery. Anyone personally acquainted with Indians may be pardoned a little skepticism.—Both conjectures were advanced as serious contributions to the history of mathematics. They may be.

9. *History*, 3, viii.

10. Posthumously published work of Gauss on the binary quadratic class-number indicates that he used analysis organically in 1834 and possibly earlier. Dirichlet has priority of publication and complete independence.

11. For example, the prime number theorem.

12. Proceeding from his work on functions defined by Taylor series, Paris thesis, 1892.

13. Outline history to 1909 incidentally in E. Landau's *Handbuch der Lehre von der Verteilung der Primzahlen*, Leipzig, 1909, 1, 2. Evaluations of Tchebycheff's contributions, ibid., 1, 11–18; and of Riemann's, 29–36. For more recent related work, Landau, *Vorlesungen über Zahlentheorie*, Leipzig, 1927, 2.

14. Six properties of $\zeta(s)$; Landau, *Primzahlen*, 31–3.

15. The crucial point is to find all positive integers n, or to show that none exist, for which

$$\left(\frac{3}{2}\right)^n - \left[\left(\frac{3}{2}\right)^n\right] = 1 - \left(\frac{1}{2}\right)^n \left\{\left[\left(\frac{3}{2}\right)^n\right] + 2\right\}$$

where $[N]$ denotes the greatest integer $\leqq$ the positive integer N.

Chap. 15

1. L. E. Dickson, in Miller, Blichfeldt and Dickson, *Finite groups*, New York, 1916.

2. Historical note in G. Salmon, *Geometry of three dimensions*, ed. 2, Dublin, 1865, 422.

3. As by the calculus of variations.

4. Stationary, as by **3.**

5. D. M. Y. Sommerville's, St. Andrews, 1911.

6. The *interpretation* of a geometry is not in that geometry. More on this point is implied in the concluding chapter.

Chap. 16

1. By count of the periodicals reported on in the abstract-journals.

2. An estimate, including trade journals, gave about 70,000 in 1930.

Chap. 17

1. The calculus of spinors, the nearest approach to novelty, demanded the invention of no new mathematics.

2. The much-quoted remark of d'Alembert, attributed by him to a nameless gentleman, that mechanics can be considered as a four-dimensional geometry, had no influence whatever on subsequent developments.

3. Almost extinct in living mathematics as contrasted with mathematical physics.

4. "The only object of theoretical physics is to calculate results that can be compared with experiment."—P. A. M. Dirac, *The principles of quantum mechanics*, Cambridge, 1930, 7.

5. If there is a potential function, the system is said to be conservative.

6. The brachistochrone; arc of a cycloid.

7. G. D. Birkhoff, *Dynamical systems*, New York, 1927.

8. Account, with historical references, in E. T. Whittaker, *Analytical dynamics*, etc., Cambridge, 1917. K. F. Sundman (Finland), *Acta Soc. Scientatis Fennicae*, 1906, 1909, proved the existence of an analytic solution, the single case of triple collision excluded; see also *Acta Mathematica* (Stockholm), 36, 1912, 105.

9. Mathematical comparison in **7,** 55–8.

10. Zenodorus (c. 150 B.C.), Archimedes.

11. Virgil's *Aeneid*, I, 369: " . . . taurino quantum possent circumdare tergo."

12. Stated and incorrectly solved by Galileo in his *Dialogues*, 1630.

13. Just apportionment of credit between James and John has been a source of much controversy.

14. Detailed account and references to original sources in G. A. Bliss, *Amer. Math. Monthly*, 43, 1936, 598.

15. Too technical for discussion here; see Bliss, *Calculus of variations*, Chicago, 1925, 131; O. Bolza, *Lectures on the calculus of variations*, New York, 1931 (reprint from 1904), 60–3.

16. First course not later than 1872.

Chap. 18

1. Accounts of the special functions mentioned here are given in texts on advanced calculus, or in introductory texts on mathematical physics.

2. Originating with Euler (1755) and Lagrange (1778), who obtained the general equations by different methods.

3. *Comm. Sci. Imp. Petrop.* (St. Petersburg) 6, publ. 1738.

4. *Leibnizens Ges. Werke*, 3d seq., Halle, 1855, 75.

5. *Novi Comm. Acad. Petrop.*, 10, publ. 1766, 243.

6. ∇^2 in three dimensions is the Laplacian operator $\partial^2/\partial x^2 + \partial^2/\partial y^2 + \partial^2/\partial z^2$, in two dimensions, $\nabla^2 = \partial^2/\partial x^2 + \partial^2/\partial y^2$.

7. κ denotes the diffusivity; t, time. Fourier's first paper was presented to the Paris Academy in 1807, and in improved form, 1811. In his *Théorie analytique de la chaleur*, 1822, Fourier practically ignored the objections raised by Laplace, Lagrange, and Legendre to earlier work.

8. Simplified requirement.

9. Either by modifying the restrictions imposed by scientific requirements, or by passing from 3 independent variables to n, as in immediate generalizations of Laplace's equation and the corresponding 'harmonic' functions. The latter type of generalization, in spite of its elegance, has been called pointless by modern analysts.

10. He also made notable contributions to the theory of permutation groups.

11. Detailed statement in any modern text on potential theory.

12. O. D. Kellogg, *Foundations of potential theory*, Berlin, 1929, (historical notes).

13. So dubbed by W. H. Young, *Proc. London Math. Soc.*, 24, 1925–6. 'Stokes' theorem' should be named after Kelvin (W. Thomson), who knew it by 1850; see G. G. Stokes, *Mathematical and physical papers*, Cambridge, 5, 320.

14. See Young, **13.**

15. This philosophy is not copyrighted. There is a partial generalization by abelian functions to time of any even number of dimensions. It is no sillier than the resurrected four-dimensional time-nonsense of the English engineer J. W. Dunne, or of the English novelist J. B. Priestley, and others, for the interpretation of dreams. Any trained mathematician could out-mystic the time mystics at their own crazy game, were it worth his trouble. At that it might be: educated people have paid from one to three dollars a head to listen to a lecture on this kind of 'higher mathematics.'

16. Son of the mathematician Count Riccati.

17. Most of Euler's work on elliptic integrals appeared in the publications of the St. Petersburg Academy. Publication was often delayed; so that the dates of printing and receipt of a completed manuscript sometimes differ by two or three years.

18. *London Roy. Soc. Phil. Trans.*, 1771, 1775.

19. De reductione . . . ac hyperbolae, *Novi Comm. Acad. Sci. Petrop.*, 1766. The relevant passage is: "Imprimis autem hic idoneus signandi modus desiderari videtur, cujus ope arcus elliptici aeque commode in calculo exprimi queant, ac jam logarithmi et arcus circulares ad insigne Analyseos per idonea signa in calculum sunt introducti."

20. Made as early as 1825.

21. In the letter to Bessel, 30 March, 1828, Gauss (*Werke*, 10_1, 247) seems to imply that Abel had anticipated only *a third* of his (G's) work. If this is what Gauss meant, it is not substantiated by anything thus far found in his papers. Abel went far beyond what is actually published (posthumously) of Gauss' work in elliptic functions.

22. *Werke*, 2, 29. Jacobi overlooked the necessary restriction to single-valuedness in his proof. A. Göpel's well-founded objection (*Journ. für Math.*, 35, 1847, 302) is dismissed by Jacobi with the incredulous footnote "!?."

23. It astonished Legendre.

24. Modernized account, to 1927, in R. Fueter, *Vorlesungen über die singulären Moduln*, u.s.w., Leipzig, 1924, 1927.

25. Their literature is a mass of conflicting notations.

26. Meromorphic functions.

27. Posthumously published; developed from his now almost abandoned theory of arithmetic-geometric means.

28. Thus there is the complete and popular theory of Weierstrass.

29. The wave surface in optics is an interesting example. The reasons underlying geometric applications will appear in connection with algebraic functions, discussed in a later chapter.

30. Jacobi gave the first treatment by elliptic functions. Since the earth is a top, fat books on the motion of tops and gyroscopes have been written, all liberally peppered with elliptic thetas and elliptic functions. Why anybody should master such works today is not clear.

Chap. 19

1. *Methodus fluxionum et serierum infinitarum*, publ. 1736, probably composed about 1671.

2. J. Francesco.

3. Leibniz' work (1692–4) on envelopes might have an earlier claim on geometric considerations.

4. First used, apparently, by John Bernoulli in 1697 to solve the general linear equation of the first order, $dy/dx + fy = g(f, g$ functions of x alone). Newton (1687) in his *Principia* used a geometric equivalent in discussing the motion of the moon.

5. An American author was severely lectured (1939) by an English critic for using the "Americanism" 'calculuses' instead of 'calculi.' The plural 'calculuses' originated in England. Lord Kelvin (died, 1907), among other scientists, staged several modernized grammarians' funerals for the schoolmasterish pedants afflicted with 'calculi.' 'Schoolmasterish' also is English

6. Often denoted now by an enormous capital 'S.'

7. For the formulas mentioned, see any text on interpolation or finite differences.

8. *Ars conjectandi*, Basel, 1713.

9. E. Goursat, *Leçons sur le problème de Pfaff*, Paris, 1922; J. M. Thomas, *Differential systems*, New York, 1937.

10. History, etc., in A. R. Forsyth, *Theory of differential equations*, Cambridge, 1890, vol. 1.

11. Published in abstract, 1835; more fully by l'Abbé Moigno, 1844, who reported Cauchy's lectures.

12. For partial differential equations, the method (1908) of W. Ritz, sharpening Lord Rayleigh's of 1870, 1899, has proved useful practically.

13. Through a generalization of Olivier's generalized circular and hyperbolic functions.

14. P. A. MacMahon, *Combinatorial analysis*, Cambridge, 1915, Vol. 1.

15. R. A. Fisher, *The design of experiments*, ed. 2, Edinburgh, 1937.

16. In later life a leader in French politics and holder of high offices.

17. Some of his original work is reproduced in his *Treatise on differential equations*, London, 1859, *Supplementary volume*, 1865.

18. N. W. McLachlan, *Complex variable and operational calculus with technical applications*, Cambridge, 1939 (with bibliography, 222 titles).

19. E. T. Whittaker, *Calcutta Math. Soc. Commemorative Vol.*, Calcutta, 1930, 216.

20. Exhibition in G. Doetsch, *Theorie und Anwendung der Laplace Transformation*, Berlin, 1937.

21. Existing work in differential equations sufficed to justify Heaviside's expansion theorem; see F. D. Murnaghan, *Bull. Amer. Math. Soc.*, 1927.

22. Criticism and history in A. R. Forsyth, *Atti del* IV (1908) *congresso internazionale dei matematici*, Rome, 1909, 86; *Theory of differential equations*, Cambridge, 1906, vol. 6.

23. J. Hadamard, *Lectures on Cauchy's problem*, New Haven, 1923.

24. References in F. R. Moulton, *Differential equations*, New York, 1930, 375.

25. O. Veblen.

Chap. 20

1. General definition of algebraic invariance from modern standpoint, H. Weyl, *Duke Math. Journ.*, 5, 1939, 493; also, Weyl, *The classical groups*, etc., Princeton, 1939.

2. Adapted from C. J. Keyser, *Hibbert Journ.*, 3, 1904–5, 313.

3. Square of determinant of the transformation; in Lagrange's example, the determinant is 1.

4. Modern usage favors 'invariant' for both species, as in 'theory of invariants.'

5. A quantic is an algebraic form as already defined. The theory as developed by Sylvester has a luxuriant vocabulary, the greater part of which is now a dead language.

6. Used in a technical sense different from that in algebra.

7. Galilean.

8. *Werke*, ed. 2, Leipzig, 1892, 402.

9. Specimen in Lie's *Theorie der Transformationsgruppen*, 3, xvii.

10. Sometimes assigned to the year 1870. Date of publication is 1872, and Klein himself (*Ges. math. abhand.*, Leipzig, 1921, 1, 411), makes no claim for an earlier origin, except that he imagined the incomplete project in 1871.

11. Such an invariant is an integral, extended over a certain region of space of one or more dimensions, whose value is unchanged by functional transformations of the coordinates of any point in the region.

12. O. Veblen and J. H. C. Whitehead, *Foundations of differential geometry*, Cambridge, 1932; O. Veblen, *Atti del congresso internat. dei matematici*, Bologna, 1928, 1, 81.

13. History to 1890 by F. Meyer, *Bericht*, u.s.w., in *Jahresbericht der deutschen Mathematiker-Vereinigung*, Berlin, 1, 1890–1.

14. **A 8,** 2, 433.

15. According to Seki's modest compatriots.

16. **A 8,** 2, 701.

17. For brevity here, the nature of the factor is included in the statement; it can be proved that the factor stated is necessary.

18. Cayley proved the finiteness for invariants of quantics up to degree 6, for covariants to degree 4, and found the complete systems in these cases.

19. Slight uncertainty in date due to publication later than explanation in lectures.

20. *Theorie der binären algebraischen Formen*, Leipzig, 1872; *Vorlesungen über Geometrie*, Leipzig, 1875, 1876.

21. Much of interest was found; for example, Sylvester's theory of canonical forms, and Hermite's law of reciprocity. But all now seem dead or, at liveliest, strangely antiquated.

22. As late as 1940, a few belated veterans and stragglers from the Grand Army of Finite Groups of the 1890's were still garnering special groups as if starvation were staring them in the face, which it was not.

23. H. Weyl, *Duke Math. Journ.*, 5, 1939, 500.

24. Possible priority dispute here; Lie has a strong claim.

25. Historically necessitated by early imprecise notions of a 'group' as now technically defined. Some of the earlier work needs restatement on this account.

26. *Theorie der Transformationsgruppen*, Leipzig, 1893, 3, Vorrede.

27. H. Weyl, *Göttingen Nachrichtung*, 1930, 293; 1931, 36.

28. For example, in a suitably restricted region, the functions $f_1, \ldots, f_n$ defining the transformation are analytic with non-vanishing Jacobian.

29. See **25.**

30. The r parameters must be 'essential'; not all the transformations are to be obtainable in terms of fewer than r parameters.

31. Inexact.

32. Rigorously.

33. This is only a rough description; see any modern text on transformation groups.

34. *Ges. Abhand.*, Leipzig, 1924, 5, 583. Lie's works give the best historical account of his theory, as he was always just, if occasionally a little less than generous, to his predecessors and contemporaries.

35. *Traité d'analyse*, ed. 2, vol. 3, Paris, 1908, Chap. 17.

36. Report to 1924 in *Proc. Internat. Math. Congress*, Toronto, 1924, 1, 473.

37. See **38.**

38. *Differential equations from the algebraic standpoint*, New York, 1932.

39. *Differential systems*, New York, 1937.

40. This is either a definition of a rigid body or a postulate of Euclidean space. Thus in a Euclidean plane, a triangle may be freely slid about without alteration in either its sides or its angles.

41. Weyl's work on the representation of continuous groups opened a new field.

42. For obvious reasons, Hilbert is not included in this statement; nor are the leaders in mathematical logic since 1909.

42a. O. Veblen, in Veblen and J. W. Young, *Projective geometry*, Boston, 1918, vol. 2, Chap. 3.

43. J. H. C. Whitehead, *Science Progress*, 34, 1939, 76.

44. Veblen's terminology.

45. Algebraic geometry also might have been included, had the theory of Cremona and birational transformations in higher space been sufficiently developed—'Steiner's transformation' is said to have been used before Steiner.

46. See **12,** V. and W., 38.

47. See **12,** O. Veblen, *The invariants of quadratic differential forms*, ed. 2, Cambridge, 1933; T. Y. Thomas, *Differential invariants of generalized spaces*, Cambridge, 1934. These contain abundant references.

48. A. S. Eddington, *The mathematical theory of relativity*, Cambridge, 1923, 70.

49. For example, early in October, 1939, the German universities, with the exception of Berlin, Munich, Jena, and Vienna were closed; 22 shut up shop.

50. See **12,** V. and W., 17.

51. See **12,** V. and W.

52. Following closely S. Lefschetz, *Topology*, New York, 1930.

53. *Werke*, 8, 271, 400; 10_2, Abh. 4, 46.

54. *Werke*, 5, 605; a generalization of Ampère's fundamental law. (No attempt is made here to enumerate all who have worked on knots.)

55. It contains incidentally a very curious generating function of a type but little studied.

56. He did not publish it.

57. Following closely J. W. Alexander, *Trans. Amer. Math. Soc.*, 28, 1926, 301.

58. As by G. D. Birkhoff, *Proc. Edinburgh Math Soc.*, 2, 1930, 83.

59. Cayley (1878) publicized the problem, already known (1840) to Möbius. Veblen (1912) reduced the problem to one in a finite geometry. There is an extensive and inconclusive literature on this 'Fermat's last theorem' of combinatorial topology.

60. So-called here as an equivalent of the German mengentheoretische.

61. Lefschetz, **52,** 391.

62. Going back to the theorems of Ampère, Green, Gauss, and Stokes (Kelvin) in electromagnetism.

63. P. Alexandroff and H. Hopf, *Topologie*, Berlin, 1935.

Chap. 21

1. From an uncompleted (1940) compendium by H. Bateman. Probably 1,400 is an upper limit.

2. 'Function' introduced by Leibniz, 1692, with meaning different from later usage. John Bernoulli, 1718, gave "a quantity composed in any manner from variables and constants"; Euler, 1748, repeated the foregoing with "any whatever analytic expression" instead of "quantity."

3. *Oeuvres*, 6, 348.

4. K. Knopp, *Theory and application of infinite series* (trans., R. C. Young), London, 1928. Historical notes.

5. *Analyse algébrique*, Paris, 1821.

6. Newton knew its importance in a special case.

7. *Leçons sur les séries divergentes*, Paris, 1901.

8. For other contributors, **A 8,** 2, 507–11.

9. Maclaurin is said to have given a 'geometrical equivalent'; but as the whole point here is rigor, it is difficult to see what the assertion could mean.

10. $\sum_{n=1}^{\infty} a_n e^{-\lambda_n s}$, the λ_n real, $\lambda_1 < \lambda_2 < \ldots < \lambda_n < \ldots$; s complex.

11. O. Perron, *Die Lehre von den Kettenbrüchen* Leipzig, 1913. Selected bibliography (to 1913), 511–17.

12. Applications of elementary parts of theory occur, as in the Gaussian theory of systems of lenses.

13. *Werke*, 3, 143.

14. The basic F. Riesz-Fischer theorem dates only from 1907; its powerful generalization, from 1923.

15. E. Saks, *Theory of the integral* (trans., R. C. Young), Warsaw, New York, 1937.

16. Numerous previous attempted proofs, perhaps satisfactory for their time, proved nothing.

17. Darboux (1875) showed that Riemann's definition produces an infinity of continuous functions having no derivatives.

18. As it is not yet (1945) settled by the principals who did exactly what, our account is purposely ambiguous.

19. Lebesgue's integration can be restated in terms of Riemann's, but the reduction is seldom practical.

20. *Journ. für Math.*, 74, 1872, 188; *Annales de l'Ecole Normale*, 1895, 51.

21. *Proc. London Math. Soc.*, 35, 1902, 387.

22. *Math. Annalen*, 59, 1904, 516.

23. *Bull. de la Société Math. de France*, 1905.

24. *Mémoire sur les intégrales définies prises entre des limites imaginaires.*

25. A theory of non-analytic functions also exists. Report to 1930 in E. R. Hedrick, *Bull. Calcutta Math. Soc.*, 20, 1933.

26. Many equivalent definitions. Analyticity is defined for a specified region of the z-plane.

27. Integration in the usual presentation is Riemann's (*Werke*, 239).

28. Definitions, etc., in G. A. Bliss, *Algebraic functions*, New York, 1933: A region or surface S is 'connected' if any two points on it can be joined by a continuous arc lying entirely on S. If S is connected, and if every closed contour C on S separates S into two connected parts, of which one has C as its complete boundary, S is simply connected. A cross cut joining two points on the rim (if there is one) of a simply connected surface S severs S into two simply connected surfaces.

29. Implicit in a memoir of 1814, on definite integrals, *Oeuvres*, 1, 319; explicitly in **24,** published, 1827.

30. A converse of Cauchy's theorem was proved by E. Morera, *Lombardi Rendiconti*, 22, 1889, 191.

31. But not preferable; it assumes the existence of the expansion noted presently. See any text for the integral definition.

32. Generalized by Poincaré to functions of two complex variables.

33. So named by Gauss, *Werke*, 5, 200.

34. *Werke*, 4, 189.

35. *Oeuvres*, 4, 189.

36. *Werke*, 4, 259.

37. Proof many times refined since 1901. Account to 1929 in O. D. Kellogg, *Foundations of potential theory*, Berlin, 1929.

38. Exact account in any modern text; only a description here.

39. Removable singularities are easily included.

40. The French put Cauchy first; the Germans and German-educated Americans, Weierstrass.

41. Cauchy had done practically the same.

42. Between any two there is another.

43. Term introduced by Briot and Bouquet (who declined to state their initials), *Théorie des fonctions elliptiques*, ed. 2, Paris, 1875, 15.

44. Paris, *Comptes rendus*, etc., 17, 1843, 938.

45. *Werke*, 5.

46. A function having this property is called monogenic; but there is no point here in multiplying technical terms.

47. The details of sign and units are immaterial here.

48. From unpublished historical notes by H. Bateman. The account by P. Stäckel, A **3**, 1900, 109, is incomplete.

49. *Oeuvres*, 1, 471.

50. *Werke*, 10_1, 365.

51. See Bliss, **28.**

52. H. F. Baker, *Abel's theorem*, etc., Cambridge, 1897, 209.

53. The theta functions are not strictly periodic. Simultaneous increase of the variables by integer multiples of the (pseudo-) periods restores the original function multiplied by a certain constant. Strictly $2p$-fold periodic functions are quotients of the thetas, as in the special case ($p = 1$) of elliptic functions.

54. He introduced theta functions of p variables in 1849; *Werke*, 1,111.

55. For history of algebraic geometry, *Bull. National* [U.S.A.] *Research Council*, Washington, 63, 1929; 96, 1934 (by a committee).

56. *Journ. für Math.*, 91, 1881, 301.

57. Bibliography (about 300 titles) to 1929 in L. R. Ford, *Automorphic functions*, New York, 1929.

58. *Vorlesungen über die Theorie der automorphen Funktionen*, (with R. Fricke), Leipzig, 1897, 1912.

59. Properly discontinuous.

60. Also claimed by some for F. Casorati (1835–1890, Italian).

61. *Darstellung und Begründung einiger neurer Ergebnisse der Funktionentheorie*, Berlin, 1929.

62. Brother of the physicist, N. Bohr.

63. A. S. Besicovitch, *Almost periodic functions*, Cambridge, 1932.

64. As a rather extreme instance, compare the discussion in Salmon's *Geometry of three dimensions*, any edition, with that of O. Zariski, *The reduction of the singularities of an algebraic surface*, *Annals of Math.*, 45, 1939, 639.

65. Klein's acceptance by his German colleagues was slow; some never altered their early conviction that, whatever it might be that Klein was doing, it was not mathematics. Others granted him standing as an able expositor of others' (Galois, Hermite, Clebsch, Lie, etc.) ideas in his peculiarly individual fashion.

Chap. 22

1. He overlooked the restricted interval in which the expansion is valid.

2. H. Burkhardt, *Entwicklungen nach oscillirenden Functionen*, u.s.w., (history), *Jahresbericht der deutschen Math.-Verein.*, Leipzig, 10, 1908, 33.

3. Sophie Germain (1776–1831, French) might also be mentioned, although the validity of her work was disputed.

4. *Ueber lineare Differentialgleichungen der zweiten Ordnung* (lithographed lectures), Göttingen, 1894.

5. *Ueber die Reihenentwicklungen der Potentialtheorie*, Leipzig, 1894, 193–4. (Historical sketch, 195.)

6. Contrary to the verdict in **2**, delivered in 1908 before 'modern' physics got well started.

7. Disputed by some who attribute Taylor's expansion with remainder to John Bernoulli. The expansion can be restated as an integral equation. Others exhibit this as a typical example of the kind of mathematical history which reads the future into the past.

8. According to some who should know. But there is the possible catch that the integral equation corresponding to the two-variable case may be difficult.

9. In the earlier development, experts waxed quite lyrical about this; faced with the unavoidable calculations involved, some lapsed into silence.

10. *Grundzüge einer allgemeine Theorie der linear Integralgleichungen*, Leipzig, 1912.

11. Also biorthogonal functions for boundary-value problems in which the differential equations are not self-adjoint.

12. Historical accounts, bibliographies, and expositions of this, and allied, topics in: V. Volterra, *Leçons sur les équations intégrales et les équations integro-différentielles*, Paris, 1913; *Theory of functionals and of integro-differential equations* (trans.), London, 1930; G. C. Evans, *Functionals and their applications*, etc., New York, 1918.

13. Huygens also has a claim; but Bernoulli seems to have been the first to state the principle explicitly.

14. It is sufficient; but whether it is also necessary seems to be an open question.

15. As an approximation up to 'small quantities' of the first order.

16. *Journ. de l'Ecole Polytechnique*, 13, 1832, 67.

17. In the theory of linear operators, noted later.

18. Proposed for consideration by Leibniz, 1695; Euler discussed it in 1729. Nothing adequate was possible at this primitive stage, and nothing came of the early attempts.

19. See **12,** Volterra.

20. If a metric is to be imposed.

21. Without qualification, 'Hilbert space' since 1927 means the space abstracted from classical Hilbert space by J. von Neumann. Theory in M. H. Stone, *Linear transformations in Hilbert space*, New York, 1932. The classical space is that of all real vectors $(x_1, \ldots, x_n, \ldots)$ such that $\sum_{n=1}^{\infty} x_n^2$ converges.

22. H. Weyl, *Theory of groups and quantum mechanics* (trans., H. P. Robertson) London, 1931, espec. 31–40.

23. *Sur quelques points du calcul fonctionel, Palmero Rendiconti*, 22, 1906, 1.

24. This description condensed from R. W. Barnard's synopsis in *General analysis* 1, 2, *Memoirs Amer. Phil. Soc.*, Philadelphia, 1935, 1939.

25. History to 1928, with critical evaluations, in M. Fréchet, *Les espaces abstraits et leur théorie considérée comme introduction à l'analyse générale*, Paris, 1928.

26. *Grundzüge der Mengenlehre*, Leipzig, 1914 (1917). But an earlier *explicit* use of 'space' in this sense may well have been overlooked in the voluminous literature.

27. For the historical and other relations of Menger's work to that of others, espec. P. Urysohn (Russian), see Menger, *Dimensiontheorie*, Leipzig, 1932.

Chap. 23

1. By A. Church, *Journ. Symbolic Logic*, 1, 1936; 3, 1938. All references to works on logic in the account here are listed in this bibliography.

2. By no means 'must.' It is an alluring pastime to disinter the theories of mathematical physics of the past, and try to imagine what kept them alive.

3. Boole's expansion in symbolic logic; an analogue of Maclaurin's theorem in the calculus for functions of several variables.

4. C. S. Peirce may have had it earlier. If so, it was not published where it would attract mathematical attention.

5. No relation to W. R. Hamilton, the mathematician.

6. *Foundations of logic and mathematics*, Chicago, 1939.

7. *Bull. Amer. Math. Soc.* (trans., A. Dresden), 20, 1913–14, 81.

8. There is plenty of it (1940) in the U.S.A.

9. If these labors represent the 'Siegfried line' of the formalists, their opponents are already overarmed.

9a. R. A. Fisher, *Statistical methods for research workers*, Edinburgh, 1938.

10. Or sardonically. Laplace was nobody's fool, not even Napoleon's. He came through the French Revolution and the restoration of the monarchy with his head on his neck, and with more money in his pockets than when all the fuss started.

11. 'Common sense' has changed meaning many times. Today (1945), 'common sense' is the one thing the so-called 'common man' never acquires. Witness the European masses: the kind of sense, or lack of it, which they exhibit is beyond 'calculation.' Laplace, then, must have been jesting.

12. See R. C. Tolman, *The principles of statistical mechanics*, Oxford, 1938.

13. Exposition and history in E. Hopf, *Ergodentheorie*, Berlin, 1937.

14. History of probability to 1899 in E. Czuber, *Die Entwicklung der Wahrscheinlichkeitstheorie* u.s.w., *Jahresbericht der deutschen Math.-Verein.*, Leipzig, 1899, vol. 7. This supersedes some of the older work of I. Todhunter.

15. In a lecture.

16. History, with attention to educational applications, in Helen M. Walker, *Studies in the history of statistical method*, Baltimore, 1929.

17. K. Pearson, *The life, letters and labours of Francis Galton*, Cambridge, 1914, 1924.

18. Pearson's enthusiasm for W. K. Clifford's intuitive dynamics and physics, also for Clifford's violent hostility to traditional beliefs, influenced at least his earlier thinking. Both Clifford and Pearson were creative mathematicians; neither fitted the milk-and-water, namby-pamby 'great man' ideal of the 'great mathematician' which seems to be the accepted norm in historical accounts of mathematicians; and at least one of them would have hooted at the idea that he was, or was to become, an object of reverence to generations of students.

Index

C

D

E

F

G

H

I

N

S